"十三五"国家重点出版物出版规划项目·航天先进技术研究与应用系列

THE VACUUM SUPER-UNIFIED THEORY

● Chen Shuqiao

内 容 简 介

There are eight chapters in this book. The intrinsic structure of basic particles formed by different forms of vacuum deformation is systematically described, and the super-unified field equation including gravity is obtained. Vacuum strain is the life gate of physics. It is incredible that the major puzzlement of physics today is a natural objective existence in the logic system of new physics, such as the invariance principle of light speed, mass, asymmetry of positive and negative matter, etc. Along the staircase of vacuum strain, we can climb the tower and have a bird's eye view of the whole physics. This book provides a new perspective for readers to appreciate the concise beauty of physics.

The book is equipped with a large number of illustrations, so readers can understand the new theory with the help of images. It is suitable for readers who are interested in modern physics and have basic knowledge of quantum physics and relativity. It can be used as a reference book for undergraduates or junior college graduates in physics, as well as a textbook for super-unified theory self-study.

图书在版编目(CIP)数据

真空超统一理论=The Vacuum Super-unified Theory:英文/陈蜀乔著. —哈尔滨:哈尔滨工业大学出版社,2020.4

"十三五"国家重点出版物出版规划项目. 航天先进技术研究与应用系列

ISBN 978-7-5603-8444-3

Ⅰ.①真… Ⅱ.①陈… Ⅲ.①统一场论-理论研究-英文 Ⅳ.①O412.2

中国版本图书馆 CIP 数据核字(2019)第 167279 号

策划编辑　王桂芝　刘　威

责任编辑　李长波　庞　雪　李青晏

出版发行　哈尔滨工业大学出版社

社　　址　哈尔滨市南岗区复华四道街10号　邮编150006

传　　真　0451-86414749

网　　址　http://hitpress.hit.edu.cn

印　　刷　黑龙江艺德印刷有限责任公司

开　　本　787mm×1092mm　1/16　印张 25.25　字数 774 千字

版　　次　2020年4月第1版　2020年4月第1次印刷

书　　号　ISBN 978-7-5603-8444-3

定　　价　88.00元

（如因印装质量问题影响阅读,我社负责调换）

前　言

本书介绍了基于真空场应变的超大统一场的新理论,四种力场的超统一机制来源于不同形式的真空应变。本书分类讨论了真空不同形式的应变及其相应的力场形式和相互作用;基于应变理论和规范场理论来构建理论体系,采用粒子内禀空间和背景观测空间来描述量子场,为新物理搭建了一个更广阔的平台。

1. 新理论对物理学基本常数的计算及解释

(1) 精细结构常数(1个):实验值 $\alpha \approx 1/137$。通过光子与电荷纤维化结构耦合可以直接算出 $\alpha \approx 1/138.6$。

(2) 强耦合常数(1个):通过夸克弦构成强作用通道,可以直接估算出强作用耦合强度的范围为 $1/3 \sim 3$。

(3) 弱相互作用强度(1个):根据内禀质量荷弱衰变分离过程,可计算出温伯格角度 θ_W,计算值 $\sin^2\theta_W = 0.213$,实验值 $\sin^2\theta_W = 0.2292 \pm 0.0013$。通过 θ_W 就可以确定弱相互作用强度。

(4) 引力相互作用强度(1个):引力为二阶应变 $[\mathscr{g}_{\mu\nu}] = \left(\pm\dfrac{\partial u_m}{\partial a_\mu}\right)\left(\pm\dfrac{\partial u_m}{\partial a_\nu}\right) \geq 0$(没有负引力),这种微小的应变会导致真空硬化,表现为时空弯曲。根据质量荷纤维结构球表面对背景时空的压迫,可以直接计算出引力场和弱相互作用强度的关系:$g_G = 5\pi[(g_W \cdot g_W)(g_W \cdot g_W)] \approx 0.4 \times 10^{-39}$。

(5) 6夸克与6轻子质量(12个):对于夸克和轻子的质量给出了统一的数学表达式:
$$m_i c = \frac{h_f}{\varphi_0 \cdot |Q_i^{-1}|\gamma_i},$$
并对夸克质量的反常特性给予解释。

(6) 夸克混合参数(4个):通过夸克弦通道结构解释了参数之间的关系,建立了夸克混合矩阵[①]的表达式:$U_{ij} = [(S_i - s_k)/S_i]^{1/2}[(S_j - s_k)/S_j]^{1/2}$。

(7) 中微子的振荡及中微子混合参数(4个):新理论认为中微子是自旋虚光子横向波动,这种波动体现为中微子振荡。根据中微子被探测的圆环面积及空间压缩对探测的影响,可以算出 $\sin^2 2\theta_{12} = S_1\gamma_e/(S_2\gamma_\mu) = 0.84$ 和 $\sin^2 2\theta_{23} = S_2\gamma_\mu/(S_3\gamma_\tau) = 0.95$,并根据隧穿效应判断 $\sin^2 2\theta_{13}$ 趋近于0,但大于0。

(8) 弱作用玻色子的质量(3个):目前根据实验进行粗略估算。

---

① 本书中矩阵、矢量、向量等均使用白体。

(9) 宇宙常数(1个):宇宙常数并不是一个常数,而是一个函数,表征的是暗能量。暗能量并非真正意义上的能量,而是由银河系时空弯曲与宇宙背景时空弯曲之间存在差异引起的,$\Lambda_{\mu\nu} = \rho_\Lambda (g^B_{\mu\nu} - g^M_{\mu\nu})$。

2. 新理论对现行物理学存在问题的解释

(1) 质量问题:真空的任何应变都会产生波动效应,应变会导致传播变差。费米子为涡旋波,其传播效应体现为惯性质量 m;光子为一维波动,其传播效应体现为能量 E,两者的关系为 $E = mc^2$。量子应变会对背景时空产生二阶应变效应,这种效应体现为引力质量。质量与 Higgs 粒子无关。

(2) 物质-反物质不对称的问题:带负电荷的基本粒子为正物质,带正电荷的基本粒子为反物质,正反物质数量是相等的。正反重子数不对称是由于负电荷为空穴,极其稳定;正电荷为游离态真空场物质,是不稳定的;游离态物质受到挤压之后产生质子,这导致了物质和反物质的不对称。

(3) 违反 CP 对称:为何弱相互作用可以违反 CP 对称,出现 CP 破坏,而强相互作用却不能违反?这是因为轻子衰变弱作用全程都在半向空间中,因此违背 CP 原理;而强子衰变是从半向空间中转变到对偶半向空间中,因此不违背 CP 对称。

(4) 时间:时间来自于量子场的应变波动性,是运动自由度,展现了量子运动存在的差异,没有几何性质。

(5) 暗物质:暗物质是真空裂纹,导致真空硬化,表现为引力。

(6) 暗能量:整个宇宙大的背景空间度规与地球观测的局部背景空间度规存在差异,这种差异构成暗能量。

(7) 量子波函数测量坍塌:量子场是应变波的叠加体。被测量子波与探测体某个原子发生相互作用,产生一个点扰动,对于观测者而言具有坍塌效应。

(8) 宇宙中 4 种力的合并:宇宙有 4 种基本自然力,即电磁力、强相互作用力、弱相互作用力和引力。在极高能量条件下,真空达到应变极限,完全硬化,4 种力归零,表现为 4 种力的合并。

(9) 裸奇点和虫洞:时间快慢由背景真空硬化度决定,时间没有几何性质,空间不可能无限弯曲,空间在达到应变极限后发生破裂,因此既不存在裸奇点,也不存在虫洞。

(10) 包含引力的超大统一场方程:根据真空应变守恒可以得到超大统一场方程

$$\sum_{n=i=0, l=0}^{N,1} \hat{\partial}_{n;i}^{1+\delta(i)\delta(l)} \Phi_\alpha(X) = 0。$$

上述公式中各物理量的含义详见本书对应的章节,在此不再赘述。本理论计算出来的结果与实验均存在 1% 的误差,但是能够计算出来,这本身就是把不可能变成了可能。

本书共有 8 章。第 1 章讨论了真空结构的性质,引入真空性质的基本假设,并介绍了最基本的应变理论。第 2 章介绍了光子结构和电磁场理论,由光子内禀空间的应变导出了麦克斯韦方程。第3章描述了电子内禀结构,引入自旋波的概念,解释轻子质量和自旋,计算电磁耦合常数。第4章描述了强相互作用,真空变形导致维度分裂,形成具有弦结构的强子。宇宙中正反物质的非对称性来源于正负电子内部中心结构的稳定性存在的差异。第5章基于真空应变的狭义相对论和广义相对论进行讨论,将时空弯曲理解为背景真空非均匀硬化,给出了真空破裂的引力场方程。第6章讨论了暗物质和暗能量。第7章讨论了规范不变性和弱作用场,提出了半空间概念来解释宇称不守恒。根据真空应变守恒给出超大统一场方程。第8章对新理论进行了回顾,以例子来帮助理解量子动力学的过程,给出了跑动耦合常数以解释真空极化,量子场的相互作用导致了反常的维度。为了读者阅读方便,给出该理论的基本结构图,如图1所示。

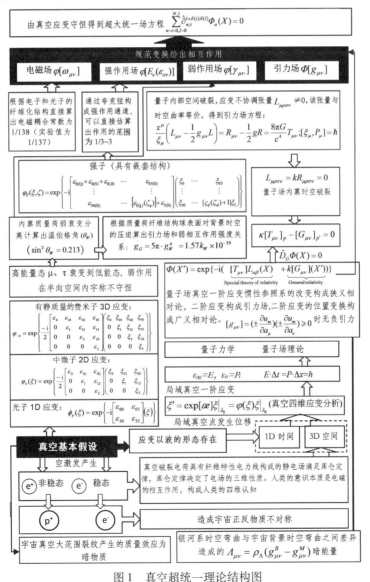

图1 真空超统一理论结构图

本书凝结了作者的最新研究成果,书中的新物理概念为物理学的研究人员提供了新的视角。

感谢两位恩师云南大学物理系赵树松教授和云南师范大学物理系李淮江教授对我的培养;感谢挚友汪一夫先生不断的鼓励。子欲养而亲不待,我不敢忘怀父母的教诲。因作者水平有限,书中疏漏之处在所难免,恳请读者批评指正。

陈蜀乔

2019 年 10 月

Preface

This book introduces a new theory of super-large unified field based on vacuum field strain. The super-large unification mechanism of the four force fields comes from different forms of vacuum strain. This book classifies and discusses strains in different forms of vacuum, corresponding forms of force field and interaction. Based on strain theory and gauge field theory, the theoretical system is constructed. The quantum field is described by particle intrinsic space and background observation space, which provides a broader platform for new physics.

1. Computation and interpretation of basic constants of physics by new theory

(1) Fine structure constant (one).

$\alpha \approx 1/138.6$ can be directly calculated by coupling photons with charge-induced fibrosis structure. And its experimental value $\alpha = 1/137$.

(2) Strong coupling constant (one).

The strong interaction range of $1/3 - 3$ can be directly estimated by the strong interaction channel formed by quark strings.

(3) Intensity of weak interaction (one).

According to the separation process of intrinsic mass charge-weakening decay, we can calculate the Weinberg angle θ_W, $\sin^2\theta_W = 0.213$, and the experimental value $\sin^2\theta_W = 0.2292 \pm 0.0013$. The strength of weak interaction can be determined by θ_W.

(4) Gravitational interaction strength (one).

Gravitation is second-order strain $[\mathscr{G}_{\mu\nu}] = \left(\pm\dfrac{\partial u_m}{\partial a_\mu}\right)\left(\pm\dfrac{\partial u_m}{\partial a_\nu}\right) \geqslant 0$ (without negative gravitation). The corresponding physical image is that the surface of the fiber structure results in staggered strain in the background space. This tiny strain will lead to vacuum hardening, which is manifested as space-time bending. The relationship between gravitational field and weak interaction strength can be directly calculated according to the pressure of the surface of the mass-loaded fiber structure on the background space-time $g_G = 5\pi[(g_W \cdot g_W)(g_W \cdot g_W)] \approx 0.4 \times 10^{-39}$.

(5) 6 quarks and 6 leptons mass (twelve).

A unified mathematical expression $m_i c = \dfrac{h_f}{\varphi_0 \cdot |Q_i^{-1}| \gamma_i}$ for the mass of quarks and leptons is given, and the abnormal characteristics of mass are explained.

(6) Quark mixing parameters (four).

The relationship between the parameters is explained by the structure of quark chord channels, and the expression $U_{ij} = [(S_i - s_k)/S_i]^{1/2}[(S_j - s_k)/S_j]^{1/2}$ of quark mixing matrix is established.

(7) Neutrino mass and oscillation and neutrino mixing parameters (four).

According to the new theory, neutrinos are spinning photon fluctuations and photons are transverse wave, which are reflected in neutrino oscillations. According to the detected area of the annulus and the influence of space compression on detection, we can calculate $\sin^2 2\theta_{12} = S_1\gamma_e/(S_2\gamma_\mu) = 0.84$ and $\sin^2 2\theta_{23} = S_2\gamma_\mu/(S_3\gamma_\tau) = 0.95$. Respectively, according to tunnel effect, $\sin^2 2\theta_{13}$ tends to zero, but it is bigger than 0.

(8) Mass of three large bosons (three).

At present, a rough estimate has been made based on experiments.

(9) Cosmic constant (one).

Cosmic constant is not a constant, but a function. It represents dark energy. Dark energy is not real energy. $\Lambda_{\mu\nu} = \rho_\Lambda(g^B_{\mu\nu} - g^M_{\mu\nu})$ is caused by the difference between the space-time curvature of the Milky Way and the space-time curvature of the cosmic background.

2. Explanation of existing problems in physics by new theory

(1) Quality problems.

Any strain in vacuum will produce wave effect, which will lead to poor propagation. This effect is expressed as inertial mass. Fermion is a vortex wave with static mass m and photon is a 1-dimensional wave. The relationship between them is $E = mc^2$. Strain will produce second-order strain effect on background space-time, which is reflected as gravitational mass. Mass has nothing to do with Higgs particles.

(2) Matter-antimatter asymmetry.

The basic particles with negative charge are positive matter, and the basic particles with positive charge are antimatter. The positive and negative matters are equal. The positive and negative baryon numbers are asymmetric, because the negative charges are holes, extremely stable and unchanged after being squeezed. Positive charge is a matter in free vacuum field, which is unstable. Positive charge is squeezed to produce protons, which leads to the asymmetry of matter and antimatter.

(3) CP violation.

Why can weak interaction violate CP symmetry and cause CP violation while strong interaction cannot? This is because the lepton decay weakening is in the half-directional space, so it violates the CP principle. And the hadron decay is transformed from the half-directional space to the dual half-directional space, so it does not violate the CP symmetry.

(4) Time.

Time comes from the strain fluctuation of the quantum field. It is a degree of freedom of motion and has no geometric properties.

(5) Dark matter.

Dark matter is a vacuum crack.

(6) Dark energy.

The difference between the large background space metric of the whole universe and the local background space metric of our earth observation constitutes dark energy.

(7) Measuring collapse of quantum wave function.

Quantum field is the superposition of strain wave. The detected quantum wave interacts with an atom of the probe body and produces a point perturbation, which has a collapse effect for the observer.

(8) The four forces of the universe merge into one.

The universe has four basic forces: electromagnetic force, strong nuclear force, weak interaction (also known as weak nuclear force) and gravity (different forms of vacuum strain). Under extremely high energy conditions, vacuum reaches strain limit, then vacuum hardens completely, and physics disappears. There is no difference among the four forces, which shows that the four forces merge into one.

(9) Bare singularities and wormholes.

The new theory holds that singularities do not exist, time has no geometric properties, space can not bend indefinitely, and space breaks up after reaching the strain limit, so there are neither bare singularities nor wormholes.

(10) Ultra-large unified field equation including gravity.

Ultra-large unified field equation $\sum_{n=i=0, l=0}^{N,1} \hat{\partial}_{n;i}^{1+\delta(i)\delta(l)} \Phi_a(X) = 0$ can be obtained according to the conservation of vacuum strain.

See the corresponding chapters of this book for the specific meaning of the above formula, which will not be repeated in the preface. There are 1% error between the calculated results and the experiments, but it is a great progress to be able to calculate the experimental parameters and make the impossible possible.

This book has eight chapters. In the first chapter, the properties of vacuum structure are discussed. By introducing the basic hypothesis of vacuum property, the basic strain theory is introduced. In the second chapter, the photon structure and electromagnetic field theory are introduced. Maxwell's equation is derived from the strain of photon intrinsic space. Chapter 3 describes the intrinsic structure of electrons. The concept of spin wave is introduced to explain lepton mass, spin and calculate electromagnetic coupling constants. Chapter 4 describes strong interaction. Vacuum deformation results in dimensional splitting to form hadrons with chord structure. The conclusion of the asymmetry of the positive matter and negative matter in the universe comes from the difference in the stability of the central structure of the positive and negative electrons. Chapter 5 discusses special relativity and general relativity based on vacuum strain. The space-time bending is understood as background vacuum hardening, and the gravitational field equation of vacuum rupture is given. Chapter 6 deals with dark matter and dark energy. Chapter 7 discusses gauge invariance and weak field of action, and puts forward the concept of semi-space to explain parity non-conservation. According to the conservation of vacuum strain, the super large unified field equation is given. Chapter 8 reviews the new theory and uses an example to understand the process of quantum electrodynamics, and gives the dimension of the runout coupling constant to explain the anomaly caused by the interaction between vacuum and quantum field. For the convenience of readers, the basic structure of the theory is given in Fig. 1.

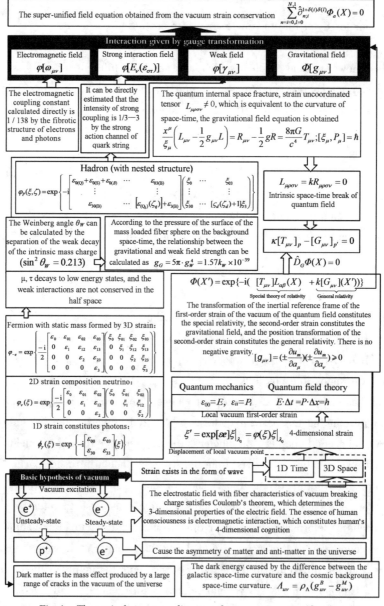

Fig. 1　Theoretical structure diagram of vacuum super-unification

Preface

This book gives the author's latest research results, then provides many clear illustrations, and attempts to illustrate the basic concepts of the new theory with images. The new physical concepts in this book provide a new perspective for the researchers of physics.

Thank my two teachers, Professor Zhao Shusong, Department of Physics of Yunnan University, and Professor Li Huaijiang, Department of Physics of Yunnan Normal University, for teaching me physics and life for decades. Also thank my close friend Mr. Wang Yifu for his spiritual encouragement. I miss my parents who love me deeply but have passed away. Due to my limited knowledge, deficiencies also exist in this book. Criticism and correction are hopefully welcomed.

Chen Shuqiao
Kunming University of Science and Technology, China
October, 2019

Contents

Chapter 1 Structure of the Vacuum 1

    1.1 Metric and spatial classification 5
    1.2 The basic assumptions of vacuum 8
    1.3 The basic unit of vacuum static strain analysis 16
    1.4 Space-time structure 28

Chapter 2 Structure of the Photons 39

    2.1 The image of photon 39
    2.2 Point translation strain is forming photonic in vacuum 42
    2.3 Intrinsic structure of photon 49
    2.4 The expression of photon strain in observational space 70
    2.5 Complete field functions of free photons 85

Chapter 3 The Structure of Lepton 99

    3.1 The strain $\varepsilon_{\mu\nu}$ of lepton 100
    3.2 Lepton mass 110
    3.3 The lepton spin strain $[\omega_{ij}]$ 117
    3.4 The intrinsic structure of lepton electric flux line 123
    3.5 The characteristics outside the intrinsic space of lepton electric flux line field ... 133
    3.6 Estimate of electromagnetic coupling constants 139
    3.7 The complete field function of lepton 146
    3.8 Several examples of generalized field equation of free particle 153
    3.9 Neutrino 163

Chapter 4 Hadron Structure 174

    4.1 The π meson structure 174
    4.2 Proton structure 184
    4.3 Quark 200
    4.4 Gluon and string 211

Chapter 5　Gravitational Field 221

5.1　The vacuum principle of special relativity 224
5.2　The spatial characteristics of gravitational field 229
5.3　The relationship between gravitational field and quantum field 236
5.4　The curved space-time description 247
5.5　Gravitational field equation 252
5.6　Gravitational waves 258

Chapter 6　Dark Matter and Dark Energy in the Universe 267

6.1　The universe come into being and extinction 270
6.2　Spherical stress waves that lead to vacuum fragmentation under the initial impact load of the explosion 281
6.3　The mechanism of the production of dark matter caused by vacuum rupture 287
6.4　The destruction scope of Big Bang to the vacuum 290
6.5　Energy distribution of dark energy stress wave in a vacuum 297
6.6　Dark energy of the universe 301
6.7　The species of dark energy in the universe 304

Chapter 7　Interaction and Super-unified Equation 313

7.1　Electromagnetic interaction 315
7.2　Weak interaction field 321
7.3　Higgs mechanism 332
7.4　Calculation of Weinberg angle θ_w 340
7.5　Weak interaction of quark 349
7.6　Super-unified field equation 359

Chapter 8　Supplementary Notes on Some Basic Issues 364

8.1　Basic idea of quantum field 364
8.2　Quantum fluctuations and space-time rupture 367
8.3　The difference between vacuum unified theory and string/ M theory 374
8.4　Vacuum interpretation of Dirac's "large number hypothesis" 385

References 388

Chapter 1   Structure of the Vacuum

Everything in universe comes from the vacuum (Fig. 1.0.1). The essence of physics should be simple and graceful. The vacuum should be simple and primitive. Field has an observable by deformation of vacuum. Space-time and mass are the effects of vacuum deformation.

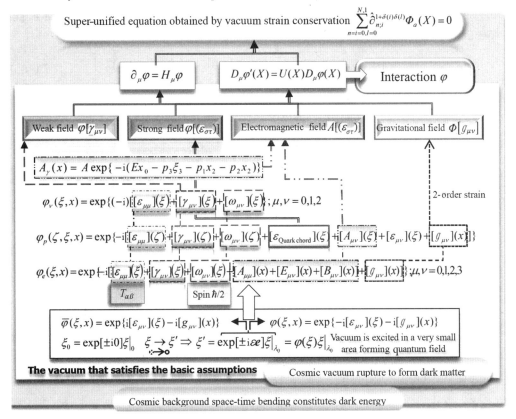

Fig. 1.0.1   Basic structure of vacuum super-unified theory

1. Green strain

$$E_{\mu\nu} = \frac{1}{2}\left[\frac{\partial u_\mu}{\partial \xi_\nu} + \frac{\partial u_\nu}{\partial \xi_\mu} + \frac{\partial u_m}{\partial \xi_\mu}\frac{\partial u_m}{\partial \xi_\nu}\right] = [\varepsilon_{\mu\nu}] + \frac{1}{2}[\mathscr{G}_{\mu\nu}] \quad (1.0.1)$$

$$\mathscr{G}_{\mu\nu} = \frac{\partial u_m}{\partial \xi_\mu}\frac{\partial u_m}{\partial \xi_\nu}, \quad g_{\mu\nu} = e_\mu e_\nu \quad (1.0.2)$$

Here, $\mathscr{G}_{\mu\nu}$ is two-order strain; $g_{\mu\nu}$ is metric.

2. The relationship between stress and strain

The existence of quantum field makes the two-order strain of background space-time produce gravitational effect. $[\varepsilon_{\mu\nu}]$ is the strain of the quantum field in expression (1.0.1).

$1/2\,\mathscr{g}_{\mu\nu}$ is the two-order strain of the background vacuum caused by $[\varepsilon_{\mu\nu}]$. The two-order strain of the vacuum causes the background space-time to bend. The metric of the curved space-time is $g_{\mu\nu}$. Set the internal flat vacuum coordinate frame is (ξ'^0, ξ'^1, ξ'^2, ξ'^3) and the flat space-time metric is $\eta_{\mu\nu}$, the internal bending vacuum coordinate frame is (ξ^0, ξ^1, ξ^2, ξ^3). The metric of internal bending space-time is $g_{\mu\nu}$. The existence of any quantum field has a weak influence on the background space-time. In order to describe quantum field, we need to establish a coordinate system in the quantum field, which is a curved space-time coordinate system. There are two parts — one part is the influence of the two-order strain of quantum field on the space-time of its own background, and the other part is the curvature of the original background.

$$\begin{cases} \Phi(\xi,X) = \exp\{-i([\varepsilon_{\mu\nu}](\xi) + \frac{1}{2}[\mathscr{g}_{\mu\nu}](X))\} \\ \xi_\mu = g_{\mu\nu}\xi'^\nu, \mathscr{g}_{\mu\nu} = kg_{\mu\nu}; X_\mu = g_{\mu\nu}X'^\nu \end{cases} \quad (1.0.3)$$

Background vacuum strain $[\mathscr{g}_{\mu\nu}]$ is expressed by the background space-time metric.

$$[\mathscr{g}_{\mu\nu}] = k[g_{\mu\nu}] \quad (1.0.4)$$

Here, k is the conversion coefficient.

The coordinates of the field function and wave equation in this book have a spatial structure. Here we make the following provisions, which stipulate that

$$[\varepsilon](X) = \begin{bmatrix} \varepsilon_{03} & \varepsilon_{01} & \varepsilon_{02} & \varepsilon_{03} \\ \varepsilon_{10} & \varepsilon_{11} & \varepsilon_{12} & \varepsilon_{13} \\ \varepsilon_{20} & \varepsilon_{21} & \varepsilon_{22} & \varepsilon_{23} \\ \varepsilon_{30} & \varepsilon_{31} & \varepsilon_{32} & \varepsilon_{33} \end{bmatrix} \begin{pmatrix} \xi_0 & \xi_0 & \xi_0 & \xi_0 \\ \xi_1 & \xi_1 & \xi_1 & \xi_1 \\ \xi_2 & \xi_2 & \xi_2 & \xi_2 \\ \xi_3 & \xi_3 & \xi_3 & \xi_3 \end{pmatrix}$$

$[\cdots](X)$ should not be confused with the function expression $F(X)$. In order to be distinguished from the function expression $F(X)$, the "$[\cdots](X)$" is separated by square brackets $[\cdots]$ and parentheses $(\cdots)$, and the coordinate quantity (X) is preceded by square brackets $[\cdots]$. In subsequent mathematical representations, the meanings are the same.

$$\Phi(\xi,X) = \exp\{-i[\varepsilon_{\mu\nu}](\xi)\} \cdot \exp\{-i\frac{1}{2}[\mathscr{g}_{\mu\nu}](X)\} = \underbrace{\Phi_q(\xi)}_{\text{Quantum field function}} \cdot \underbrace{\Phi_G(X)}_{\text{Gravitational field function}}$$

$$\Phi_q(\xi) = \exp\{\underbrace{-i[\varepsilon_{\mu\nu}](\xi_\alpha)}_{\text{Intrinsic strain}}\} = \exp\{-i[\varepsilon_{\mu\nu}]\underbrace{[g_{\alpha\beta}](\xi^\beta)}_{\text{Curved space-time coordinates}}\}$$

$$= \exp\{-i\underbrace{[\varepsilon_{\mu\nu}]}_{\text{Quantum field}} \cdot [\underbrace{\eta_{\alpha\beta} + h_{\alpha\beta}}_{\text{Gravitational space-time}}](\xi^\beta)\}$$

$$\Phi_q(\xi) = \exp\{-i[\varepsilon_{\mu\nu}][\eta_{\alpha\beta}](\xi^\beta)\} \quad \text{(Neglect space-time bending)} \quad (1.0.5a)$$

$$\Phi_G(X) = \exp\{-i\frac{1}{2}[\mathscr{g}_{\alpha\beta}](X)\} = \exp\{-i\frac{k}{2}[g_{\alpha\beta}](X)\} \quad (1.0.5b)$$

Here, $g_{\mu\nu} = \eta_{\mu\nu} + h_{\mu\nu}$. It can also be understood that the existence of the quantum field itself causes an extra bend in the space-time of its curved background. $\eta_{\alpha\beta}$ is the metric of the background space-time without the quantum field at P. When the P point exists in the quantum field, $h_{\mu\nu}$ is a minor change in the metric. The effect of gravitational field is negligible in the micro internal coordinate system $X^\mu(\xi)$. The above expression is the quantum field expression in curved space-time.

3. Wave equation

If the coordinate system used in the study is changed, the expression of the equation will be changed. The intrinsic field equation is

$$\Phi(\xi) = a\exp\{-i[\varepsilon_{\alpha\beta}]g_{\nu\mu}\xi^\mu\} \tag{1.0.6}$$

$$X^\nu = L^\nu_\mu X^\mu$$

Thus wave equation is

$$\Phi(X) = A\exp\{-i[\varepsilon_{\alpha\beta}]g_{\nu\mu}L^\nu_\mu X^\mu\} \tag{1.0.7}$$

The wave function expression of quantum field in 4-dimensional observation space-time is

$$\phi(x) = \int_{-\infty}^{\infty} \bar{\varepsilon}dV\exp\{-i[\bar{\varepsilon}_{\mu\nu}](X)\} \tag{1.0.8}$$

Fluctuation amplitude $\int_{-\infty}^{\infty} \bar{\varepsilon}dV = h$ satisfies the quantized condition.

4. The classification of quantum fields

(1) Quantum field structure of lepton.

$$\varphi_e(\xi) = h_f\exp\{-i[\underbrace{\varepsilon_{\mu\nu} + \Delta\varepsilon_{\mu\nu}}_{\text{Vacuum strain of electrons}}](\xi)\}$$

$$= h_f\exp\{-i[\underbrace{\varepsilon_{\mu\mu}}_{\text{4-momentum}} + \frac{1}{2}\underbrace{\gamma_{\mu\nu}}_{\text{Weakly acting strain}} + \frac{1}{2}\underbrace{\omega_{\mu\nu}}_{\text{Spin}}](\xi)\} \cdot$$

$$\exp\{-i[\underbrace{\Delta\varepsilon_{\mu\mu}}_{\substack{\text{4-momentum of electromagnetic field}\\ \text{derivative of scalar and vector potential}}} + \frac{1}{2}\underbrace{\Delta\gamma_{\mu\nu}}_{\substack{\text{Fiber shear}\\\text{strain}=0}} + \frac{1}{2}\underbrace{\Delta\omega_{\mu\nu}}_{\text{Electromagnetic field}}](\xi)\}$$

$$\tag{1.0.9}$$

The lepton formed the hadron after the occurrence of the fission of dimension.

(2) Quantum field structure of hadrons.

$$\psi_{\text{Hadron}}(X) = \exp\{-i[\underbrace{Q_{\mu\nu}}_{\text{Quark strain}}](X)\}$$

$$= h_f\exp\left[-i\left(\underbrace{\begin{bmatrix}\varepsilon^q_{00}+f(\varepsilon_{00}) & 0 & 0 & 0 \\ 0 & [q^B_{\alpha\beta}]+f(\varepsilon_{11}) & 0 & 0 \\ 0 & 0 & [q^G_{\mu\nu}]+f(\varepsilon_{22}) & 0 \\ 0 & 0 & 0 & [q^R_{\sigma\tau}]+f(\varepsilon_{33})\end{bmatrix}}_{\text{Three quark principal strain + Vacuum cell strain}} +\right.\right.$$

$$\left.\left.\underbrace{\gamma_{\mu\nu}/2}_{\substack{\text{Weakly acting}\\\text{strain}}} + \underbrace{\gamma^S_{\mu\nu}/2}_{\substack{\text{String shear}\\\text{strain}=0}} + \underbrace{\omega^H_{\mu\nu}/2}_{\text{Hadron Spin}}\right)(X)\right] \tag{1.0.10}$$

Here, $f(\varepsilon_{\mu\mu})$ is 1-dimensional strain wave function.

$$q(\xi) = \frac{h_f}{3}\exp\{-i[\underset{\text{4-momentum}}{\varepsilon_{\mu\mu}} + \frac{1}{2}\underset{\text{Weakly acting strain}}{\gamma_{\mu\nu}} + \frac{1}{2}\underset{\text{Spin}}{\omega_{\mu\nu}}][g_{\mu\nu}](\xi)\} \cdot$$

$$\exp\{-i[\underset{\substack{\text{4-momentum of electromagnetic field,}\\\text{derivative of scalar and vector potential}}}{\Delta\varepsilon_{\mu\mu}} + \frac{1}{2}\underset{\substack{\text{Fiber shear}\\\text{strain}=0}}{\Delta\gamma_{\mu\nu}} + \frac{1}{2}\underset{\text{Electromagnetic field}}{\Delta\omega_{\mu\nu}}][g_{\mu\nu}](\xi)\}$$

(1.0.11)

Here, $q(\xi)$ is quark instrinsic field function.

The above expressions include electromagnetic field, strong field and weak field.

5. Force field interaction

$$\Phi(X) = \exp\{i[E_1 X^1 + E_2 X^2 + \cdots + E_n X^n]\} \tag{1.0.12}$$

Here, $E_1 = \varepsilon_{\mu\mu}$, $E_2 = \gamma_{\mu\nu}/2$, $E_3 = \omega_{\mu\nu}/2$, $\cdots$. $X^\alpha = g^{\alpha\beta} X_\beta$ is the coordinate frame of the multi-degree of freedom space.

6. Gauge transformation

When a i-dimensional space X^i is bent, then $X^i_{;k} = X^i_{,k} + \Gamma^i_{jk} X^j$. Γ^n_{jk} is the difference between i-dimensional space and original space after X^i bending, this amount determines the strength of the interaction. We are used to operator $D_i = \partial_i + \Gamma^i$ to find eigenvalues.

$$D_i \Phi(X) = (E_i + \Gamma^i E_i)\exp\{-i[E_1 X^1 + E_2 X^2 + \cdots + E_n X^n]\}, \hat{\Gamma}^i \Phi = g^i \Phi$$

(1.0.13)

Here, g^i is the strength of the interaction of class i field. For the ordinary derivative of the equation, the eigenvalue is the source of i type force field which characterizes the strain of i-force field.

7. Gravitational field

Mass can be visually understood that there is a huge scalar particle in the universe, which is a cosmic scalar particle field function.

$$\Phi(X) = \exp\{-i(\kappa[\varepsilon_{\mu\nu}](X_T)|_p + R[g_{\mu\nu}](X_R)|_{p'})\} \tag{1.0.14}$$

Set $[\varepsilon_{\mu\nu}] = [T_{\mu\nu}]$, $R[g_{\mu\nu}] = G_{\mu\nu}$. The 4-momentum of matter mass is bending its own background space-time, then

$$G_{\mu\nu} = \kappa[T_{\mu\nu}] \tag{1.0.15}$$

8. Gravitational wave

When the visible matter is accelerated, the gravitational wave is emitted. It belongs to vacuum stress waves. The spherically symmetric mass source does not radiate gravitational waves in the observation space. The quantization condition of the gravitational field of point particles is $(p_1+p_2+p_3)\Delta V = h_G$. When the intrinsic space is extended to observation time and space, the gravitational field function in motion becomes

$$\Phi(x) = A\exp[\mathscr{g}_{\mu\mu}(x)] \tag{1.0.16}$$

9. Dark matter dark energy

The strain difference between the observer background vacuum and the cosmic background vacuum is dark energy. When the vacuum exceeds the deformation limit, the vacuum breaks

Chapter 1 Structure of the Vacuum

down and the crack is dark matter.

10. Super-unified field equation

$$\sum_{n=i=0,l=0}^{N,1} \hat{\partial}_{n;i}^{1+\delta(i)\delta(l)} \Phi_a(X) = 0 \tag{1.0.17}$$

Here, i is the spatial dimension of interaction; n is the generalized space dimension.

The above discussions are only summaries. The mathematical expression is very rough, and it's not a strict expression. Further discussions will be carried out in the subsequent chapters.

1.1 Metric and spatial classification

1.1.1 Natural units

The speed of light is a space-time measurement ruler in Vacuum Super-unified Theory. The movement of all substances compared with the photon motion. Photons travel slower, and so do clocks. If there is the curved path of photon propagation, the space is curved. And η is the total deformation of a quantum field, which is the basic amount of microscopic in quantum field theory. To avoid repeated formulas and equations, c and η frequently appear. We will introduct a new system of units— $\eta = c = 1$. This is called natural units. Newton's gravitational constant $G = 6.7 \times 10^{-39} \text{GeV}^{-2} = 1/m_{pl}^2$. Here, Planck mass is defined as $m_{pl} = 1.22 \times 10^{19} \text{GeV}$.

1.1.2 Metric

According to the covariant description in relativistic, space-time is symmetrical, and its coordinates are described by the vector in 4-dimensional space. The point x of Minkowski Space-time is described by inverter coordinates:

$$(x_\mu) = (x_0, x_1, x_2, x_3) = (ct, x_i) \tag{1.1.1}$$

Here, $c \equiv 1$. Metric tensor $g_{\mu\nu} = \eta_{\mu\nu}$ is defined as

$$\eta_{\mu\nu} = \eta^{\mu\nu} = \begin{bmatrix} 1 & 0 & 0 & 0 \\ 0 & -1 & 0 & 0 \\ 0 & 0 & -1 & 0 \\ 0 & 0 & 0 & -1 \end{bmatrix} \tag{1.1.2}$$

To introduce the following covariant coordinates:

$$x_\mu = g_{\mu\nu} x^\nu \tag{1.1.3}$$

$$x^\mu = g^{\mu\nu} x_\nu = g^\mu_\nu x^\nu \tag{1.1.4}$$

$$(x_\mu) = (x_0, x_1, x_2, x_3) = (x^0, -x^i) = (t, -x) \tag{1.1.5}$$

$$x^2 = x_\mu x^\mu = t^2 - x^2$$

$$g^{\mu\nu} g_{\nu\lambda} = g^\mu_\lambda = \delta^\mu_\lambda \tag{1.1.6}$$

Differential operators are expressed as

$$\partial_\mu \equiv \frac{\partial}{\partial x^\mu} = (\partial_0, \partial_i) = (\partial_t, \nabla_i) \qquad (1.1.7)$$

$$\partial^\mu \equiv \frac{\partial}{\partial x_\mu} = g^{\mu\nu}\partial_\nu = (\partial_0, -\partial_i) = (\partial_t, -\nabla_i) \qquad (1.1.8)$$

D'Alembert operator is expressed as

$$\Box = \partial^2 = \partial^\mu \partial_\mu = g_{\mu\nu}\partial^\mu \partial^\nu = \frac{\partial^2}{\partial t^2} - \nabla^2 \qquad (1.1.9)$$

The scalar product of 4-momentum vector and coordinates are expressed as

$$p \cdot x = p^\mu x_\mu = g_{\mu\nu} p^\mu x^\nu = Et - p \cdot x \qquad (1.1.10)$$

The 4-momentum operator is expressed as

$$p^\mu = i\partial^\mu = (i\partial^0, i\partial^i) = \left(i\frac{\partial}{\partial t}, -i\nabla_i\right) = (p^0, p^i) = (p^0, p) \qquad (1.1.11)$$

$$p_\mu = g_{\mu\nu} p^\nu = (p^0, -p) \qquad (1.1.12)$$

$$p^2 = p^\mu p_\mu = E^2 - p^2 = m^2 \qquad (1.1.13)$$

Here, ∇_i is often abbreviated as the ∇ in the following discussions.

1.1.3 A brief introduction

1. The kinds of the force field

(1) Gravitational field and Electromagnetic field.

Gravity and electromagnetic force are long-range force. Gravity between objects is described by Newton's Law of universal gravitation, $F_G = GM_1M_2/r^2$. The relationship between space-time and gravity is described by Albert Einstein's General Relativity. Clocks slow down in gravitational fields, but not in electromagnetic fields. There are positive and negative charges in the electromagnetic field. Gravity is more important in the range of the universe. The electromagnetic field theory is described by Maxwell's equations and quantum electrodynamics.

(2) Strong interaction field.

Atoms are constituted by the extra nuclear electron and the central part of the nucleus, which is composited by protons and neutrons. The strong force pulls them together (Fig. 1.1.1). Protons and neutrons composition by smaller particles are called quarks. They are composed of three quarks, and strong force pulls the three quarks. Therefore, strong force is the basic interaction between quarks. The intensity of strong force is 100 ~ 1,000 times more than electromagnetic force. The range of strong force is very short. The force range is within 10^{-13} cm. If it is more than this range, strong force weakens rapidly. There are great differences with long-range force which are electromagnetic force and gravity. Describing the strong interaction, field

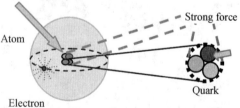

Fig. 1.1.1   Strong interaction field

Chapter 1 Structure of the Vacuum

theory is called quantum chromo dynamics.

(3) Weak field.

It exists in the internal of fundamental particles. The weak force acts on all of quarks and leptons (Fig. 1.1.2). The main impact of the weak force is to change the particle, the particle mass will change. For example, weak force causes τ lepton decay into μ, and it is lighter than τ lepton. Under the action of the weak effect, μ lepton changes into electrons. The electromagnetic force is much stronger than weak force, and it is only one thousandth of electromagnetic field. Weak force range is very short. Currently, it is less than 10^{-16} cm. Quantum flavor dynamics is built on the basis of quantum electrodynamics, which describes the weak interaction between fundamental particle theories.

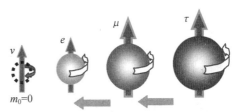

Particle is changed in weak interactions

Fig. 1.1.2 Weak interaction fields

The natures of each field are very different (Table 1.1.1), and each field has a corresponding field theory.

Table 1.1.1 Fundamental interaction

Items	Relative strength (Use of strong interaction as the standard)	Nature (range size)	The range (meters)	Intermediate boson (transfer interaction)
Strong interaction	1	$1/r^7$	10^{-15}	Gluon
Electromagnetic interaction	1/137	$1/r^2$	∞	Photon
Weak interaction	10^{-5}	$1/r^5 - 1/r^7$	10^{-18}	W and Z boson($W^\pm, Z^0$)
Gravitational interaction	10^{-39}	$1/r^2$	∞	Graviton

2. Case study of unified field theories

Physicists have made unremitting efforts to explore the essence of the material world and to achieve the unification of the four interaction fields.

(1) Maxwell realized the unification of electric field and magnetic field in 1861.

(2) Before his death (1926-1955), Einstein had devoted himself to the unification of gravitational and electromagnetic fields, but he failed.

(3) American physicists Grashaw, Weinberg and Pakistani physicist Abdus Salam established a unified theory of weak field, which unified electromagnetic field and weak field, in the late 1960s. So, they won the Nobel Prize in Physics in 1979.

(4) Physicists tried to unify the three fields of strong and weak electricity from the late

1970s to the early 1980s. This theory was called the Great Unification Theory, which was unsuccessful. At present, the most influential unified theory is the superstring theory. This theory is still being explored.

1.2 The basic assumptions of vacuum

1.2.1 Dirac vacuum

Dirac put forward relativistic electron equation in 1928. To overcome the difficulty of negative energy states, Dirac proposed the hypothesis of electronic sea. Dirac believes that in our world, all negative states are filled with electrons to form an electronic sea, and Pauli Exclusion Principle prevents electrons transition from ground state to lower negative state. The electronic sea is the background of our world, and it is also called vacuum. We excite out the electrons from the electron of the sea, and the electrons momentum is p. Its energy is $E = \sqrt{c^2p^2 + m^2c^4}$. When Electronic is excited to positive energy state (Fig. 1.2.1). We observed $a - e$, and $a+e$ left in electronic sea, so that is a hole. The holes form $a+e$. Its energy is $-E = +\sqrt{c^2p^2 + m^2c^4}$. Momentum is $-p$ and spin is $-S_p$. An ordinary positive energy electron can release energy transition to the negative energy state to fill a hole in the electron sea. This is the electron annihilation.

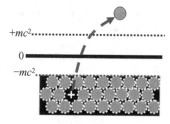

Fig. 1.2.1 Dirac vacuum

1.2.2 The picture of vacuum quantum field theory

The vacuum is not "empty" in quantum field theory. The whole space is filled with all kinds of virtual particles in vacuum state, however, there are no real particles. When vacuum ground state is excited, there is both positive and negative particle production. Quantum field theory describes a picture of the unity of physics of field with the particle — the whole space is full of a variety of overlapping fields at the same time, and each field corresponds to a particle.

Field on lowest energy state is called the ground state. When the field is on ground state, the field state changes that no energy is released, then it cannot output any signal and show any direct physical effect, so the observer cannot observe particles. When the ground state field is excited, it transitions to a higher energy state, which is called excited states. Field excites states corresponding to different number of particles and their state of motion is different. Creation and annihilation of particles corresponds to the excitation and de-excitation of quantum field. Particle and virtual particle states can be represented by lines (Fig. 1.2.2). Horizontal lines are used to represent the ground state of the field (Fig. 1.2.2(a)). The space filled with the ground state of virtual particles is called vacuum. The states of protons and electrons are shown in Fig. 1.2.2(b). Vacuum is not "empty", but a ground state field is full of various virtual particles, which has no observability.

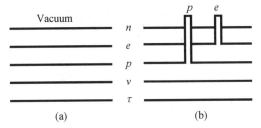

Fig. 1.2.2 Vacuum quantum field

In vacuum, there are positive and negative pairs of virtual particles on the micro-space-time scale. If the outside world does not input energy, the virtual particles will quickly annihilate. Various virtual particles are constantly produced. They are annihilated and transformed into each other in vacuum, which is called vacuum fluctuation. According to quantum field theory, the interaction between particles comes from their corresponding fields, and the interaction between particles is transformed into the interaction between fields.

Different theories have different understandings of vacuum.

1.2.3 Difficulties in experiments and methods

The conclusion that vacuum is not empty has become the consensus of modern physics. How to study vacuum by experiment is a difficult problem. Because all the fundamental particles come from vacuum, we can understand the nature of vacuum indirectly, after understanding the internal structure of elementary particles. Standard model theory constitutes that the most basic particles of our material world is the electrons and quarks.

Now, when we want to probe the internal structure of fundamental particles, make the two high-energy fundamental particles collide each other by high-energy particle accelerator. After the particles are destroyed, we can analyze the debris. However, the two high-energy particle collisions are not fragments but fundamental particles (Fig. 1.2.3). For example, a positron and an anti-electron collision will annihilate and become a pair of photons. Physical particles disappear in the vacuum in high-energy particle accelerator. The energy is very high and a large number of elementary particles will be produced. This shows that the vacuum is not empty. The particles can be produced from the vacuum, and also disappear in the vacuum.

Obviously, this experimental method cannot detect the internal structure of electrons and quarks. But apart from that, we don't have better experimental tools. We have established the standard model theory of quantum field which is based on the theoretical analysis of experimental results. Standard model can explain all the existing experimental datas of the fundamental particles. It correctly represents the overall nature of fundamental particles, so we can almost

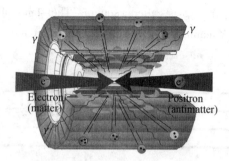

Fig. 1.2.3   Electronic collision occurred in the accelerator

say it is the most successful theory of physics. We did not involve any internal structure of electrons and quarks by existing ways and means to understand the world. We can only know the overall properties of electrons and quarks. Electrons and quarks have charge and mass, which satisfy Pauli Exclusion Principle. We don't know why electrons and quarks have these properties, because we don't know anything about the intrinsic structure of electrons and quarks. Standard model theory is the point particle theory, and it is not involved of the intrinsic structure of particles. This book will attempt to explore the intrinsic structure of particles and unify these four fields.

1.2.4   The basic assumptions of vacuum

The primary task of the vacuum unified theory is to understand the vacuum. To understand the vacuum, we face many difficulties. For experiments, flat vacuum properties are non-observable. This is similar to that a congenital blind wants to correctly understand the concept of color. The specific difficulties are as follows.

(1) Difficult 1.

Vacuum is not empty. There is substance. This substance is one of the most primitive material, but it has no observability and cannot be defined the measure. In other words, the flat vacuum cannot define space-time, space and time can only be defined on the deformation of vacuum.

(2) Difficult 2.

Vacuum is non-existent dynamic characteristics, such as mass, inertia, etc. Dynamics comes from the experimental observation of macroscopic objects, and it's not from vacuum. Therefore, it is obviously unreasonable to explain more basic properties by macroscopic physical properties, just as it explains atomic properties by molecular properties, but it can only explain molecular properties by the atom properties. There are some basic commonalities between the two, but they do not satisfy the necessary and sufficient conditions.

(3) Difficult 3.

Vacuum should not have any nature of fundamental particles, such as charge, spin and other properties. Vacuum with the particle and kinetic properties makes theory fall into the circulation of logic which explains characteristic of a particle by another particle. In essence, we

Chapter 1 Structure of the Vacuum

do not advance to understand of the vacuum.

(4) Difficult 4.

Existence of vacuum material inevitably introduces an absolute frame of reference, but if an absolute frame of reference exists, it will return to the era of Newton's classical mechanics.

There are four basic assumptions of vacuum(Fig. 1.2.4).

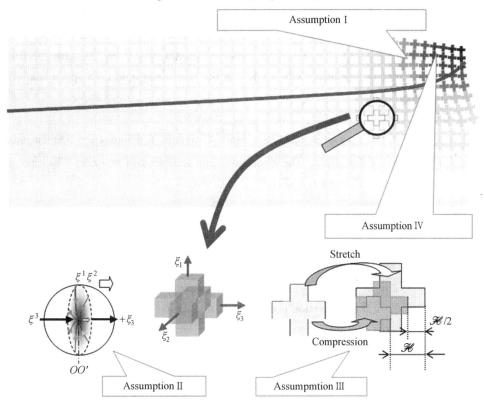

Fig. 1.2.4 The basic assumptions of vacuum

1. Assumption I

The vacuum is constituted by a countable infinite number of dense, uniform and continuous materials, which size is h_f. In other words, h_f is the "adhesive" composition of vacuum, and it is known as the basic unit of vacuum.

Based on the above difficulties, can we understand the vacuum? At present, we know the vacuum can be observed when it's disturbed. Particle physics experiments confirm that particles can be disappearing in vacuum or can be produced from the vacuum. Further, if the vacuum property had nothing to do with the existing physics, it would be difficult to imagine where today's physics comes from, and physics could not be created. We can carefully infer that the vacuum should exist more fundamental properties which are similar with macroscopic physical properties. We take these fundamental properties as the basic assumptions of vacuum unified theory. Based on this, a new theory is deduced. The new theory is consistent with the basic concepts and laws of the existing physics, and uses the new theory to reveal the internal structure of quantum fields, which will promote our understanding of vacuum.

How to understand the vacuum by the existing knowledge? What causes the vacuum produce ever-changing? Our basic considerations are the vacuum can be deformed, and it should have the basic property of vacuum. Physics is out of the question without the property. We know that the invariance of space-time comes from the conservation of energy and momentum according to Nother's theorem. This indicates that the vacuum has the uniformity, continuity and isotropy. Unlike the mathematics, if there is segmentation to the end, the minimum of matter that the original vacuum property can be preserved always exists. So, we give the first hypothesis. Experiments of Coulomb's Law show that the electromagnetic field space has strict 3-dimensional properties.

2. Assumption Ⅱ

Show the dimension properties after basic unit of vacuum h_f deformation. Dimensional direction is determined by the strain. The dimensions are independent mutually. Multiple strain superposition produces 3-dimensional effects.

(1) Primitive vector.

Setting a basic unit o_i of vacuum, which deforms u_i and gives an ideal scale x_i, then strain $du_i/dx_i = e_i$ constitutes a primitive vector of 1-dimensional space of the basic unit of vacuum. The existence of base vectors in vacuum enables vacuum to construct spatial dimensions. The strain of the basic unit of vacuum results in the fluctuation of vacuum, and waves propagate at the speed of light c, so the conversion coefficient between time and space is c. There is $ct = x$. There are differences among the fluctuations of different quantum fields. The essence of time is the degree of freedom of vacuum motion, which has no geometric properties. Time and space are integrated and cannot exist independently. Dimensions are divided into intrinsic dimension and measurement dimension.

(2) Intrinsic dimension.

The total strain of photons has 2-dimensional space-time characteristics and can be marked as $e_\gamma = \begin{bmatrix} e_0 & 0 \\ 0 & e_n \end{bmatrix}$. Neutrinos have 3-dimensional space-time structure and electrons have 4-dimensional structure.

(3) Measurement dimension.

Φ dimension comes from constraints. The dimension of field Φ can be determined by spherically symmetrical placement of the same field for constraints. If field Φ is constrained and needs $n/2$ same fields, then the spatial dimension of Φ is $N = n/2$.

Electrostatic field of 6 electrons in spherical symmetry can restrain the central electrons (Fig. 1.2.5). Therefore, the spatial dimension of electrons is 3-dimensional. From an

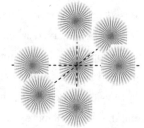

Fig. 1.2.5 Observational dimensions of elections

Chapter 1 Structure of the Vacuum

experimental point of view, electrons strictly follow Coulomb Theorem, which indicates that the electric field space is 3-dimensional. Our space-time dimension originates from the observation dimension of electric charge, because the interaction force between substances and molecules is essentially electromagnetic, and the consciousness of our brain is also electromagnetic interaction. For human observation and consciousness, our space-time is 4-dimensional space-time. Since then, human beings have always worn "three-dimensional space colored glasses", and we will always look at the world from the perspective of three-dimensional space.

(4) Set representation of vacuums.

Firstly, the strained vacuum is defined as a field, and the undisturbed vacuum field is defined as a flat vacuum without field. The vacuum is represented by symbol $V=(0)$, which is open set with no boundary.

When we introduce observation, just like flashlight irradiation (Fig. 1.2.6), only irradiation can observe objects, but in physics it means introducing disturbance. By defining a set, we mean that we observe the "flashlight irradiation" to a certain area, do not see any objects, what we see is "nothing". In other words, the corresponding physical facts in the area are no-observability, such as a vacuum set is empty set, written as $X = \{\varnothing\}$.

Fig. 1.2.6 Observation of flashlight irradiation

In fact, inspecting an area means that each point of the area being inspected is endowed with an extremely small strain $\lim_{|\varepsilon_0| \mapsto 0} \varphi_n(\varepsilon_0) = \varnothing_n$ by the observer, forming an empty set element $\varnothing_n \in X$, which is physically defined as

$$X = \{[\varnothing_i],[\varnothing_j],\cdots,[\varnothing_n],\cdots,[\varnothing_N]\}$$

Here, $i \neq j \neq \cdots \neq n \neq \cdots \neq N$. The arrangement is out of order. From an intuitive point of view, X is the area illuminated by the light. The set of the areas which are not illuminated (the area without reflection) is defined as $V_0 = (0)$. No measure can be defined on the set. It is the most primitive set. Because there is no topological structure, it is meaningless in physics.

(5) Strain mapping constitutive field.

First of all, we should consider that a strain set, which becomes $E = \{\varepsilon_1, \varepsilon_2, \cdots, \varepsilon_i, \cdots, \varepsilon_N\}$. When a strain is applied to a vacuum, it will form a field $F = \varphi[\varepsilon]$. This is a multi-point mapping. A point in the vacuum can withstand multiple strains:

$$\varphi : \varepsilon \mapsto x, x = \varphi(\varepsilon)$$

A set is called field, $F = \{\varphi_1(\varepsilon), \varphi_2(\varepsilon), \cdots, \varphi_n(\varepsilon), \cdots, \varphi_N(\varepsilon)\}, \varepsilon \neq 0, |\varepsilon| > |\varepsilon_0|$, when $|\varepsilon| > |\varepsilon_0|$ indicates that strain is not negligible. The vacuum field is represented by symbols V, $V = \{V_0 \cap X \cap F\}$, which refers to all sets of strained and non-strained vacuum. For prospective discussions, the discussion is limited to $\{X \cap F\}$. Vacuum filed is abbreviated as vacuum.

(6) Set of coordinate spaces.

When these points are arranged along ε, they form an ordered arrangement
$$X = \{[\varnothing_1],[\varnothing_2],[\varnothing_3],\cdots,[\varnothing_n],\cdots,[\varnothing_N]\}$$
Such elements form ordered points which form a line. Such a line forms a set of real numbers R. The length of x_N is defined as
$$|x_N| = \{[\varnothing_1] \cup [\varnothing_2] \cup [\varnothing_3] \cup \cdots \cup [\varnothing_n] \cup \cdots \cup [\varnothing_N]\}$$
The positive and negative definitions of elements $x_n = [\varnothing]_n$, $R = \{x \mid x \text{ are real numbers}\}$, x is defined as follows (positive strain (i.e. tension) is defined as positive, and negative strain (i.e. compression) is defined as negative):
$$X = \{x_i \in R \mid x_i = [\varnothing]_i = \lim_{\varepsilon_0 < 0, \varepsilon_0 \to 0} \varphi_i(\varepsilon_0) < 0; x_i = [\varnothing]_i = \lim_{\varepsilon_0 > 0, \varepsilon_0 \to 0} \varphi_i(\varepsilon_0) > 0\}$$
Apply a strain ε_0' to the same point $[\varnothing]_i$, if the strain is not collinear ($\varepsilon_0' \perp \varepsilon_0$), you can get a set Y. The Cartesian product of X and Y is
$$X \times Y : = \{(x,y) \mid x \in X, y \in Y\}$$
That is $R^2 : = R^1 \times R^1$, similar to $R^3 : = R^1 \times R^1 \times R^1$ and $R^n : = R^1 \times R^1 \times \cdots \times R^1$.

(7) Conclusion.

A coordinate system can be established in vacuum. The coordinate space itself is the strain vacuum. Since the vacuum strain at a certain point can be arbitrarily superimposed, the vacuum in the micro-world is infinite-dimensional, and we can establish an infinite-dimensional coordinate system in the vacuum. The space of the microcosmic world is a multi-dimensional space. Once a complete quantum field is formed, the interaction between quantum fields will occur. There is Pauli Incompatibility Principle in macroscopical view. The macroscopic material interaction is essentially the electromagnetic interaction between molecules. It only needs three spatial dimensions to constrain a macroscopic object. This is the characteristic of the electromagnetic field between molecules. Therefore, human beings have established the concept of 3-dimensional space in the macro world according to their own observations. Our consciousness will naturally have a 3-dimensional coordinate frame. The simplest method is to establish a 3-dimensional rectangular coordinate system with a point as the origin in macro analysis. The blocks of matter consisting of molecules have 3-dimensional volumes, and these materials can strain in three directions, so there is $R^3 : = R^1 \times R^1 \times R^1$. "Strain" exists in the form of "wave". The characteristic of "wave" is the propagation, therefore the macroscopic material must have the freedom of motion. This degree of freedom of motion is expressed as T, which is time dimension without geometric property. Therefore, the macro-material formed by vacuum strain leads to the generation of 4-dimensional space-time, $R^4 : = X \times Y \times Z \times T$.

The smallest particles in the material world are quarks and electrons. Electric field has 3-dimensional spatial characteristics, which makes the material world composed of charged particles. It has 3-dimensional characteristics. Therefore, we can give the third basic hypothesis of the vacuum field.

Chapter 1 Structure of the Vacuum

3. Assumption III

The basic unit of vacuum h_f can have a small deformation, and the existence limit of tension and compression, the limit deformation along a certain direction is $\mathcal{H}/2$. Elastic deformation occurs within the limit range of tension and compression, if the deformation limit is exceeded, the vacuum will break.

When a substance is deformed (Fig. 1.2.5), the deformation becomes relatively more difficult again, so we give the fourth hypothesis.

4. Assumption IV

Vacuum deformation will reduce their ability of propagation. If the propagation ability of vacuum reduces, the spreading speed of the photon propagation will slow down. Non-uniformity of the speed of light will cause light refraction, as light entering the water will slow down (refraction), and the light will bend.

With these four basic assumptions, we have the basic nature of vacuum. The new super-unified theory based on vacuum strain has found a starting point, and we have come out of the most difficult step. The establishment of basic hypothesis is the premise of successful theory. After that, we need to study the properties of vacuum basic unit. Constituting the basic unit of vacuum. h_f is the smallest unit of the universe, and it is a very tiny amount $h_f = 2h$. Here, h is the Planck constant.

1.2.5 Dimensional nature of the vacuum

Everything comes from the vacuum. The most fundamental characteristic of vacuum deformation is the field. It has three kinds of measure — time, space and mass. The dimension of field deformation is

$$[\text{Time}]^{-1}[\text{Length}]^2[\text{Mass}] = [J][s] = [p][x]$$

This dimension is the dimensions of the Planck constant $\hbar$.

1.2.6 Particles of the basic unit of vacuum

Vacuum constitute by the basic units of vacuum. Considering a basic vacuum unit as a particle in the vacuum ground state, there is no observability in non-excited state. The vacuum ground state wave function of the particle is defined as φ.

$$\varphi = \begin{bmatrix} \varphi_1 \\ \varphi_2 \\ \vdots \\ \varphi_n \end{bmatrix}, \quad \varphi^* = \begin{bmatrix} \varphi_1^* \\ \varphi_2^* \\ \vdots \\ \varphi_n^* \end{bmatrix} ; \quad \varphi_i = \exp[-i\varepsilon_\mu \xi^\nu], \quad \varphi_i^* = \exp[i\varepsilon_\mu \xi^\nu] \quad (1.2.1)$$

Vacuum strain $\varepsilon_\mu = 0$ meets

$$\varphi_0 = \varphi_0^* = \varphi_0 \varphi_0^* = 1 \quad (1.2.2)$$

The scalar particles with zero mass is known as Goldstone particles (see vacuum weak symmetry breaking model). From the point of view of particles, vacuum can be seen as bose condensed matter which is composed by positive and negative particles of the neutral with zero rest mass.

1.3 The basic unit of vacuum static strain analysis

1.3.1 Deformation

Deformation includes volume change and shape distortion. Here, we don't consider the reason why vacuum deforms. If the whole vacuum has no defects, it should satisfy the continuity hypothesis (Fig. 1.3.1). It requires every point in the region D respectively corresponding to the points in the region D_1. Specifically, if P is any point within D_1, after the object was deformation, it is through a shift and change to the point P_1 in D_1, and using $(\xi_0, \xi_1, \xi_2, \xi_3)$ to represent the coordinates of point P and point P_1. ξ_0, ξ_1, ξ_2 and ξ_3 must be single-valued as continuous functions of ξ_0, ξ_1, ξ_2 and ξ_3. Now the coordinates at point P_1 and point P are corresponding to the four-point subtraction. It is available that the displacement components of point P is $PP_1 = U$. There are four components. t, u, v and w are displacement components.

$$\begin{cases} t = \xi_0'(\xi_0,\xi_1,\xi_2,\xi_3) - \xi_0 = u_0(\xi_0,\xi_1,\xi_2,\xi_3) = u_0 \\ u = \xi_1'(\xi_0,\xi_1,\xi_2,\xi_3) - \xi_1 = u_1(\xi_0,\xi_1,\xi_2,\xi_3) = u_1 \\ v = \xi_1'(\xi_0,\xi_1,\xi_2,\xi_3) - \xi_2 = u_2(\xi_0,\xi_1,\xi_2,\xi_3) = u_2 \\ w = \xi_1'(\xi_0,\xi_1,\xi_2,\xi_3) - \xi_3 = u_3(\xi_0,\xi_1,\xi_2,\xi_3) = u_3 \end{cases}$$

(1.3.1)

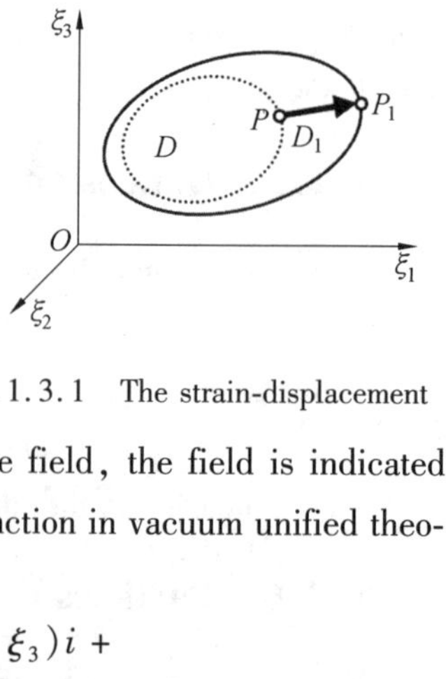

Here, t, u, v, w is the vacuum deformation, which can be expressed as u_μ, $\mu = 0, 1, 2, 3$. u_μ constitute the deformation space. Deformation space constitute by an observable effect of the field, the field is indicated by u. $u = u(u_0, u_1, u_2, u_3)$ is defined as the field displacement function in vacuum unified theory. Coordinate bases are as follows:

Fig. 1.3.1 The strain-displacement

$$u = u_0(\xi_0, \xi_1, \xi_2, \xi_3)t + u_1(\xi_0, \xi_1, \xi_2, \xi_3)i + u_2(\xi_0, \xi_1, \xi_2, \xi_3)j + u_3(\xi_0, \xi_1, \xi_2, \xi_3)k$$

$$e_0 = \partial u/\partial \xi_0, \quad e_1 = \partial u/\partial \xi_1, \quad e_2 = \partial u/\partial \xi_2, \quad e_3 = \partial u/\partial \xi_3$$

1.3.2 Vacuum strain

4-dimensional vacuum (time dimension cannot be expressed) can be thought as a tiny cube. Obviously, if we know each differential surface deformation to hexahedral, then the deformation of the entire field will be known. We will use the positive strain (symmetric relative elongation) and the shear strain to represent edge elongation and edge angle changes.

Let's look at the cube $ABDC$ of vacuum projection on the surface ξ_1, ξ_3 (Fig. 1.3.2). Before deformation, the cube point A of coordinates are ξ_1, ξ_2, ξ_3. When the cube is deforming, the projection is point A to point A′, point B to point B′, point C to point C′, point D to point D′, and the rectangle ABDC moved to the position A′B′D′C′, and the A displacement is u and w. They are the function of the coordinates.

$$u = u_1(\xi_0, \xi_1, \xi_2, \xi_3), w = u_3(\xi_0, \xi_1, \xi_2, \xi_3)$$

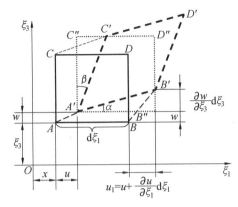

Fig. 1.3.2 The strain and the displacement relate schematic drawing

The ξ_1 coordinate of point A is different from B, so the displacement of the point B along ξ_1 axis is different. According to Taylor series expansion to omit the high-level, its expression should be

$$u_1(\xi_0, \xi_1 + d\xi_1, \xi_2, \xi_3) = u + \frac{\partial u}{\partial \xi_1} d\xi_1 \qquad (1.3.2)$$

$$u = u(\xi_0, \xi_1, \xi_2, \xi_3) \rightarrow u + du = f[(\xi_0 + d\xi_0), (\xi_1 + d\xi_1), (\xi_2 + d\xi_2), (\xi_3 + d\xi_3)]$$

After using Taylor series expansion, we can get

$$u + du = u(\xi_0, \xi_1, \xi_2, \xi_3) + \frac{\partial u}{\partial \xi_0} d\xi_0 + \frac{\partial u}{\partial \xi_1} d\xi_1 + \frac{\partial u}{\partial \xi_2} d\xi_2 + \frac{\partial u}{\partial \xi_3} d\xi_3 + \text{Higher order item}$$

If the side length $AB = d\xi_1$, then the projection of elongation of the whole on the ξ_1 axis is $u_1 - u = \frac{\partial u}{\partial \xi_1} d\xi_1$. If ε_1 is the relative elongation along the ξ_1 axis, there is

$$\varepsilon_1 = \frac{u_1 - u}{d\xi_1} = \frac{\partial u}{\partial \xi_1} \qquad (1.3.3a)$$

Using the same method can get the relative elongation of side length which parallel to the ξ_2 and ξ_3 axis:

$$\varepsilon_2 = \frac{\partial v}{\partial \xi_2}, \varepsilon_3 = \frac{\partial w}{\partial \xi_3} \qquad (1.3.3b)$$

Here, we look at the angular change by the rectangular strain. Take the right angle BAC or $B''A'C''$, the $A'B'$ rotates an angle α, and the $A'B'$ rotates an angle β. The edge of $\xi_1\xi_3$ plane has a rotation. The angular strain is ε_{31}, and the value of angle is

$$\gamma_{31} = \alpha + \beta \qquad (1.3.4)$$

As the deformation is small, these solutions description can use the sum of tangent or displacement function. If point A in ξ_3 axis direction of the displacement function is $w = f_3(\xi_0, \xi_1, \xi_2, \xi_3)$, point B in the ξ_3 axis displacement is $w_1 = f_3(\xi_0, \xi_1 + d\xi_1, \xi_2, \xi_3) = w + \frac{\partial w}{\partial \xi_1} d\xi_1$.

If point A in ξ_3 axis direction of the displacement function is $w = f_3(\xi_0, \xi_1, \xi_2, \xi_3)$, point B in the ξ_3 axis displacement is $w_1 = f_3(\xi_0, \xi_1 + d\xi_1, \xi_2, \xi_3) = w + \frac{\partial w}{\partial \xi_1} d\xi_1$. Because the transition from point B to point A, the coordinates ξ_1 has changed, and corresponding displacement will also change. The displacement along the ξ_3 axis difference between point B and point A is $B''B' = W'_1 - W = \frac{\partial w}{\partial \xi_1} d\xi_1$. In the right triangle $AB''B'$, get

$$\alpha \approx \tan \alpha = \frac{B''B'}{A'B''} = \frac{\frac{\partial w}{\partial \xi_1} d\xi_1}{d\xi_1 + \frac{\partial u}{\partial \xi_1} d\xi_1} = \frac{\frac{\partial w}{\partial \xi_1}}{1 + \frac{\partial u}{\partial \xi_1}}$$

In the decomposition, $\frac{\partial u}{\partial \xi_1}$ compared with 1 is the small amount. It can be omitted, which was $\alpha = \frac{\partial w}{\partial \xi_1}$. Using the same method, $\beta = \frac{\partial u}{\partial \xi_3}$. Finally, by the (1.3.4), $\gamma_{31} = \alpha + \beta = \frac{\partial w}{\partial \xi_1} + \frac{\partial u}{\partial \xi_3}$. With the same method, we can get the shear strain in $\xi_0 O \xi_1$, $\xi_0 O \xi_2$ and $\xi_2 O \xi_3$ plane.

$$\gamma_{01} = \frac{\partial u}{\partial \xi_0} + \frac{\partial t}{\partial \xi_1}, \gamma_{12} = \frac{\partial u}{\partial \xi_2} + \frac{\partial v}{\partial \xi_1}, \gamma_{23} = \frac{\partial v}{\partial \xi_3} + \frac{\partial w}{\partial \xi_2} \quad (1.3.5)$$

From the above analysis, we get

$$\varepsilon_{\mu\nu} = \varepsilon_{\nu\mu} = \frac{1}{2}\gamma_{\mu\nu} = \frac{1}{2}\left(\frac{\partial u_\mu}{\partial \xi_\nu} + \frac{\partial u_\nu}{\partial \xi_\mu}\right) \quad (1.3.6a)$$

The strain tensor is the symmetric tensor.

$$\varepsilon_{\mu\nu} = \begin{bmatrix} \varepsilon_{03} & \varepsilon_{01} & \varepsilon_{02} & \varepsilon_{03} \\ \varepsilon_{10} & \varepsilon_{11} & \varepsilon_{12} & \varepsilon_{13} \\ \varepsilon_{20} & \varepsilon_{21} & \varepsilon_{22} & \varepsilon_{23} \\ \varepsilon_{30} & \varepsilon_{31} & \varepsilon_{32} & \varepsilon_{33} \end{bmatrix} \quad (1.3.6b)$$

When we study three mutually perpendicular planes, these planes have no shear strain, and this plane is the main plane, which plane normal direction as the main direction. The direction corresponding to the principal strain is called the principal strain.

For the strain ε_x, if the u increases with the x, ε_x is positive, which is equivalent to that the field of the basic unit dx is stretched. If the function u decreases with the increase of x, ε_x is negative, then it is the equivalent compression of vacuum dx. This rule is the same as the elastic theory of strain.

The right angle xOy is considerably smaller with the positive shear strain γ_{xy}. The reduction of the hexahedron angle corresponds to the positive shear strain. The increase of the angle corresponds to the negative shear strain. We can derive the geometric equation of the cylindrical coordinate strain (Fig. 1.3.3).

Chapter 1 Structure of the Vacuum

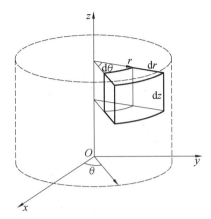

Fig. 1.3.3 The cylindrical coordinate system

$$\begin{cases} \varepsilon_r = \dfrac{\partial u}{\partial r}, \gamma_{r\theta} = \dfrac{\partial v}{\partial r} + \dfrac{1}{r} \cdot \dfrac{\partial u}{\partial \theta} - \dfrac{v}{r} \\ \varepsilon_\theta = \dfrac{1}{r} \cdot \dfrac{\partial v}{\partial \theta} + \dfrac{u}{r}, \gamma_{\theta z} = \dfrac{1}{r} \cdot \dfrac{\partial w}{\partial \theta} + \dfrac{\partial v}{\partial z} \\ \varepsilon_z = \dfrac{\partial w}{\partial z}, \gamma_{zr} = \dfrac{\partial w}{\partial r} + \dfrac{\partial u}{\partial z} \end{cases} \qquad (1.3.7)$$

Geometric equations of spherical coordinate system are

$$\begin{cases} \varepsilon_r = \dfrac{\partial u_r}{\partial r}, \varepsilon_\theta = \dfrac{1}{r}\dfrac{\partial u_\theta}{\partial \theta} + \dfrac{u_r}{r}, \varepsilon_\varphi = \dfrac{1}{r\sin\theta}\dfrac{\partial u_\varphi}{\partial \varphi} + \dfrac{u_r}{r} + \dfrac{\cot\theta}{r}u_\theta \\ \gamma_{r\theta} = \dfrac{1}{r}\dfrac{\partial u_r}{\partial \theta} + \dfrac{\partial u_\theta}{\partial r} - \dfrac{u_\theta}{r} \\ \gamma_{\theta\varphi} = \dfrac{1}{r\sin\theta}\dfrac{\partial u_\theta}{\partial \varphi} + \dfrac{1}{r}\dfrac{\partial u_\varphi}{\partial \theta} - \dfrac{\cot\theta}{r}u_\varphi \\ \gamma_{\varphi r} = \dfrac{\partial u_\varphi}{\partial r} + \dfrac{1}{r\sin\theta}\dfrac{\partial u_r}{\partial \varphi} - \dfrac{u_\varphi}{r} \end{cases} \qquad (1.3.8)$$

The geometric equations of the cylindrical coordinate system are $x_i = (r,\theta,z)$ and $u_i = (u,v,w)$. u,v,w represent the component of a point displacement in the direction of the meridian (n direction) and the circumferential (θ direction) and the axial (z direction).

The spherical coordinate system $x_i = (r,\vartheta,\varphi)$, and $u_i = (u_i, u_\theta, u_\varphi)$. Assuming that the radius of the plane object is r, the same displacement u will take place on the circular segment (Fig. 1.3.4). After deformation, the length of the arc is $(r + u)\mathrm{d}\theta$, and the original length is $r\mathrm{d}\theta$, so the relative elongation is

$$\varepsilon_\theta = \frac{(r+u)\mathrm{d}\theta - r\mathrm{d}\theta}{r\mathrm{d}\theta} = \frac{u}{r}$$

It is known from the upper form that u/r in formula (1.3.8) indicates the circumferential strain component caused by the radial displacement.

In the case of axisymmetric, $v = 0$, formula (1.3.7) can be simplified to

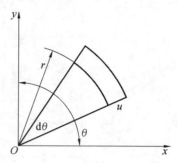

Fig. 1.3.4 A circular arc with the same radial displacement

$$\varepsilon_r = \frac{\partial u}{\partial r}, \varepsilon_\vartheta = \frac{u}{r} \tag{1.3.9a}$$

In the spherically symmetric problem, the geometric equation of the deformation is

$$\varepsilon_r = \frac{\partial u_r}{\partial r}, \varepsilon_\vartheta = \varepsilon_r = \frac{u_r}{r} \tag{1.3.9b}$$

1.3.3 Rigid body rotation

Because $u = f(\xi_0, \xi_1, \xi_2, \xi_3)$ is the very small quantity, its increase is also the very small quantity, therefore the $d\xi_0, d\xi_1, d\xi_2, d\xi_3$ higher order item may neglect, then

$$\begin{aligned}
du &= \frac{\partial u}{\partial \xi_1} d\xi_1 + \frac{\partial u}{\partial \xi_2} d\xi_2 + \frac{\partial u}{\partial \xi_3} d\xi_3 + \frac{\partial u}{\partial \xi_0} d\xi_0 \\
&= \frac{\partial u}{\partial \xi_1} d\xi_1 + \frac{1}{2}\left(\frac{\partial u}{\partial \xi_2} + \frac{\partial v}{\partial \xi_1}\right) d\xi_2 + \frac{1}{2}\left(\frac{\partial u}{\partial \xi_3} + \frac{\partial w}{\partial \xi_2}\right) d\xi_3 + \frac{1}{2}\left(\frac{\partial u}{\partial \xi_0} + \frac{\partial t}{\partial \xi_3}\right) d\xi_0 + \\
&\quad \frac{1}{2}\left(\frac{\partial u}{\partial \xi_2} - \frac{\partial v}{\partial \xi_1}\right) d\xi_2 + \frac{1}{2}\left(\frac{\partial u}{\partial \xi_3} - \frac{\partial w}{\partial \xi_2}\right) d\xi_3 + \frac{1}{2}\left(\frac{\partial u}{\partial \xi_0} - \frac{\partial t}{\partial \xi_3}\right) d\xi_0
\end{aligned} \tag{1.3.10a}$$

In the above equation, $\varepsilon_{12} = \varepsilon_{21} = \frac{1}{2}\left(\frac{\partial u}{\partial \xi_2} + \frac{\partial v}{\partial \xi_1}\right) = \frac{1}{2}\gamma_{12}$ is the shearing strain, its geometry image is that if v increases along with ξ_1, $\partial v / \partial \xi_1$ value is positive.

Obviously, the angle $\xi_1 O \xi_2$ reduces quite positive shearing strain ε_{12}, and the hexahedron included angle reduces corresponds to the positive shearing strain. The included angle increases corresponding to negative shearing strain.

$$\omega_{\mu\nu} = \frac{1}{2}\left(\frac{\partial u_\mu}{\partial \xi_\nu} - \frac{\partial u_\nu}{\partial \xi_\mu}\right) \tag{1.3.10b}$$

$\omega_{\mu\nu}$ is the component form of rotation tensor. For rotational strain (Fig. 1.3.5), consider $\omega_{21} = \frac{1}{2}\left(\frac{\partial u}{\partial \xi_2} - \frac{\partial v}{\partial \xi_1}\right)$, for the small strain tensor, $\frac{\partial v}{\partial \xi_1} \approx \alpha, \frac{\partial u}{\partial \xi_2} \approx \beta$.

If it's non-rigid body rotation, $\alpha \neq \beta$ (Fig. 1.3.6).

Chapter 1 Structure of the Vacuum

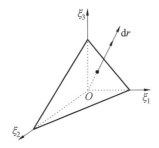

Fig. 1.3.5 The principal strain and strain increment

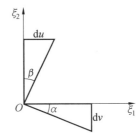

Fig. 1.3.6 The schematic diagram of the rotation tensor

1.3.4 Principal strain

There is no shear strain in the mutually perpendicular plane, which constitutes the main plane and the plane normal direction is the main direction. In equation (1.3.10), the rotation of a rigid body does not cause strain, so $\alpha = \beta$. It can be neglected in calculating strain

$$du_\mu = \varepsilon_{\mu\nu} d\xi_\nu \quad (\mu, \nu = 0, 1, 2, 3) \tag{1.3.11}$$

In principal strain space $\varepsilon_0, \varepsilon_1, \varepsilon_2, \varepsilon_3$, a strained condition is expressed in Fig. 1.3.5. If r increased a quantity dr, and its direction maintains invariable, by now r and dr is proportional in $O\xi_0, O\xi_1, O\xi_2$ and $O\xi_3$ direction of the projection, and the strain expression should satisfy the following relations:

$$\varepsilon = \frac{dr}{r} = \frac{dt}{d\xi_0} = \frac{du}{d\xi_1} = \frac{dv}{d\xi_2} = \frac{dw}{d\xi_3} \tag{1.3.12}$$

So we can get

$$dt = \varepsilon d\xi_0, du = \varepsilon d\xi_1, dv = \varepsilon d\xi_2, dw = \varepsilon d\xi_3 \tag{1.3.13}$$

According to the above, get

$$\begin{cases} (\varepsilon_0 - \varepsilon) d\xi_0 + \varepsilon_{01} d\xi_1 + \varepsilon_{02} d\xi_2 + \varepsilon_{03} d\xi_3 = 0 \\ \varepsilon_{10} d\xi_0 + (\varepsilon_1 - \varepsilon) d\xi_1 + \varepsilon_{12} d\xi_2 + \varepsilon_{13} d\xi_3 = 0 \\ \varepsilon_{20} d\xi_0 + \varepsilon_{21} d\xi_1 + (\varepsilon_2 - \varepsilon) d\xi_2 + \varepsilon_{23} d\xi_3 = 0 \\ \varepsilon_{30} d\xi_0 + \varepsilon_{31} dx + \varepsilon_{32} d\xi_2 + (\varepsilon_3 - \varepsilon) d\xi_3 = 0 \end{cases} \tag{1.3.14}$$

If the above equation determinant of coefficient is zero, this equation will have non-zero solution, then

$$\begin{vmatrix} \varepsilon_0 - \varepsilon & \varepsilon_{01} & \varepsilon_{02} & \varepsilon_{03} \\ \varepsilon_{10} & \varepsilon_1 - \varepsilon & \varepsilon_{12} & \varepsilon_{13} \\ \varepsilon_{20} & \varepsilon_{21} & \varepsilon_2 - \varepsilon & \varepsilon_{23} \\ \varepsilon_{30} & \varepsilon_{31} & \varepsilon_{32} & \varepsilon_3 - \varepsilon \end{vmatrix} = 0 \tag{1.3.15}$$

After this determinant expansion, we obtain

$$(\varepsilon_0 - \varepsilon)(\varepsilon - \varepsilon_1)(\varepsilon - \varepsilon_2)(\varepsilon - \varepsilon_3) = 0$$

Along the principal direction, take out the hexahedral which the length of side are $d\xi_0, d\xi_1$,

$d\xi_2, d\xi_3$. After the distortion, its relative change of volume (leaving out the higher order small amount) is

$$\frac{dV' - dV}{dV} = \frac{(1+\varepsilon_0)d\xi_0(1+\varepsilon_1)d\xi_1(1+\varepsilon_2)d\xi_2(1+\varepsilon_3)d\xi_3 - d\xi_0 d\xi_1 d\xi_2 d\xi_3}{d\xi_0 d\xi_1 d\xi_2 d\xi_3}$$

$$\approx \varepsilon_0 + \varepsilon_1 + \varepsilon_2 + \varepsilon_3 \qquad (1.3.16)$$

Therefore, first strain invariantis expressed after the unit cube distorts the change in volume. It is also called the volumetric strain.

1.3.5 Displacement solution obtained by the strain

Considering two points p and p' (Fig. 1.3.7), their coordinates are ξ_k and ξ'_k. Where p is an interior point with unidirectional displacement and u_i is its displacement, p' is an arbitrary interior point. The displacement $\bar{u}_i$ can be obtained by

$$\bar{u}_i = u_i + \int_p^{p'} du_i = u_i + \int_p^{p'} \frac{\partial u_i}{\partial \xi^k} d\xi^k \qquad (1.3.17)$$

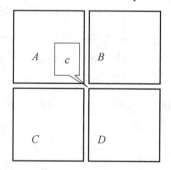

Fig. 1.3.7 Displacement solution obtained by the strain

1.3.6 Strain equations

Vacuum rupture is shown in Fig. 1.3.8. In adjacent hexahedrals A and B (Fig. 1.3.9), adjacent edges ab and $a_1 b_1$ should be the same when they are extended or shortened, because they are common edges. Hexahedral $ABCD$ intersect at point c (Fig. 1.3.9). Before the deformation, the total angle of point c is 360°. After deformation, the angle will change—some increase, while others reduce, but the sum of angles should be equal to 360°. Thus, strains should be related to each other in a certain relationship. This is the strain relation between continuous objects, which satisfies the strain compatibility equation.

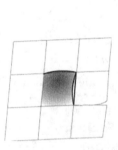

Fig. 1.3.8 Vacuum rupture   Fig. 1.3.9 Strain coordinated schematic drawing

Consider an object which coordinates of the point is (t, x, y, z), its displacement are t, u, v and w, and strain are $\varepsilon_\mu, \gamma_{\mu\nu}$. The known strain is expressed by displacement u_μ. If displacement u_μ is eliminated in the expression, then all 10 strain tensors ε_{tt}, ε_{xx}, ε_{yy}, ε_{zz}, γ_{tx}, γ_{ty}, γ_{tz}, γ_{xy}, γ_{xz}, γ_{yz} always represent the same displacement. In other words, the equation description's path is a continuous uninterrupted track, which can always remain continuous, non-cracking, and non-overlapping. We can get the relationship between strains. Strain ε_x take two order partial derivative of y, and ε_y take the two order partial derivative of x.

$$\frac{\partial^2 \varepsilon_x}{\partial y^2} = \frac{\partial^3 u}{\partial x \partial y^2}, \quad \frac{\partial^2 \varepsilon_y}{\partial x^2} = \frac{\partial^3 v}{\partial y \partial x^2}$$

For the above equations, both sides add together. It may result in

$$\frac{\partial^2 \varepsilon_x}{\partial y^2} + \frac{\partial^2 \varepsilon_y}{\partial x^2} = \frac{\partial^3 u}{\partial x \partial y^2} + \frac{\partial^3 v}{\partial y \partial x^2} = \frac{\partial^2}{\partial x \partial y}\left(\frac{\partial u}{\partial y} + \frac{\partial v}{\partial x}\right)$$

The right expression of the above equation in the parenthesis is shearing strain γ_{xy}, therefore

$$\frac{\partial^2 \varepsilon_x}{\partial y^2} + \frac{\partial^2 \varepsilon_y}{\partial x^2} = \frac{\partial^2 \gamma_{xy}}{\partial x \partial y}$$

Similarly, it may obtain other 6 equations with the same method.

$$\begin{cases} \dfrac{\partial^2 \varepsilon_t}{\partial x^2} + \dfrac{\partial^2 \varepsilon_x}{\partial t^2} = \dfrac{\partial^2 \gamma_{tx}}{\partial t \partial x} \\ \dfrac{\partial^2 \varepsilon_t}{\partial y^2} + \dfrac{\partial^2 \varepsilon_y}{\partial t^2} = \dfrac{\partial^2 \gamma_{ty}}{\partial t \partial y} \\ \dfrac{\partial^2 \varepsilon_t}{\partial z^2} + \dfrac{\partial^2 \varepsilon_z}{\partial t^2} = \dfrac{\partial^2 \gamma_{tz}}{\partial t \partial z} \end{cases}, \quad \begin{cases} \dfrac{\partial^2 \varepsilon_x}{\partial y^2} + \dfrac{\partial^2 \varepsilon_y}{\partial x^2} = \dfrac{\partial^2 \gamma_{xy}}{\partial x \partial y} \\ \dfrac{\partial^2 \varepsilon_y}{\partial z^2} + \dfrac{\partial^2 \varepsilon_z}{\partial y^2} = \dfrac{\partial^2 \gamma_{yz}}{\partial y \partial z} \\ \dfrac{\partial^2 \varepsilon_z}{\partial x^2} + \dfrac{\partial^2 \varepsilon_x}{\partial z^2} = \dfrac{\partial^2 \gamma_{zx}}{\partial y \partial z} \end{cases} \quad (1.3.18)$$

If it takes the shearing strain, the expressions are

$$\gamma_{xy} = \frac{\partial u}{\partial y} + \frac{\partial v}{\partial x}, \quad \gamma_{yz} = \frac{\partial v}{\partial z} + \frac{\partial w}{\partial y}, \quad \gamma_{zx} = \frac{\partial u}{\partial z} + \frac{\partial w}{\partial x}, \quad \text{i.e.} \quad \gamma_{\mu\nu} = \frac{\partial u_\mu}{\partial x_\nu} + \frac{\partial u_\nu}{\partial x_\mu}$$

In the above equation, derivative of z is for the first equation, and derivative of x is for the second equation, then derivative of y is for the third equation.

$$\frac{\partial \gamma_{xy}}{\partial z} = \frac{\partial^2 u}{\partial y \partial z} + \frac{\partial^2 v}{\partial x \partial z}, \quad \frac{\partial \gamma_{yz}}{\partial x} = \frac{\partial^2 v}{\partial z \partial x} + \frac{\partial^2 w}{\partial x \partial y}, \quad \frac{\partial \gamma_{zx}}{\partial y} = \frac{\partial^2 u}{\partial y \partial z} + \frac{\partial^2 w}{\partial x \partial y}$$

The first equation adds the third equation, and subtracts the second equation, then

$$\frac{\partial \gamma_{xy}}{\partial z} + \frac{\partial \gamma_{zx}}{\partial y} - \frac{\partial \gamma_{yz}}{\partial x} = 2 \frac{\partial^2 u}{\partial y \partial z}$$

With x derivation, it meets

$$\frac{\partial}{\partial x}\left(\frac{\partial \gamma_{xy}}{\partial z} + \frac{\partial \gamma_{zx}}{\partial y} - \frac{\partial \gamma_{yz}}{\partial x}\right) = 2 \frac{\partial^3 u}{\partial x \partial y \partial z} = 2 \frac{\partial^2 \varepsilon_x}{\partial y \partial z}$$

With the same method, it may result in

$$\frac{\partial}{\partial x_\nu}\left(\frac{\partial \gamma_{\mu\nu}}{\partial x_\lambda} + \frac{\partial \gamma_{\nu\lambda}}{\partial x_\mu} - \frac{\partial \gamma_{\lambda\mu}}{\partial x_\nu}\right) = 2 \frac{\partial^2 \varepsilon_\mu}{\partial x_\nu \partial x_\lambda} \quad (1.3.19)$$

Equation (1.3.19) is usually called the Strain coordinated equations. Strain compatibility equation gives the boundary conditions of vacuum strain.

1.3.7 The analysis of vacuum flow

This section introduces the strain rate tensor and the physical meaning of each component of the moving fluid.

1. The fluid motion: translation, rotation and deformation

Fluid movement is complex. We can analyze the fluid of micro-body and get some regularity (Fig. 1.3.10). Take a point of the flow field at the time of t, then $M_0(r) = M_0(t,x,y,z)$. For any point in the neighborhood, $M(r + \delta r) = M(t + \delta t, x + \delta x, y + \delta y, z + \delta z)$. Set v_0 as the speed of point M_0, and δv is the relative speed of point M and point M_0.

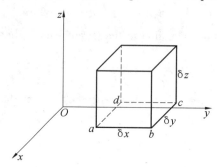

Fig. 1.3.10 Fluid micro body

$$v(M) = v_0 + \frac{\partial v}{\partial t}\delta t + \frac{\partial v}{\partial x}\delta x + \frac{\partial v}{\partial y}\delta y + \frac{\partial v}{\partial z}\delta z = v_0 + \delta v \quad (1.3.20)$$

$$\delta v = \frac{\partial v}{\partial t}\delta t + \frac{\partial v}{\partial x}\delta x + \frac{\partial v}{\partial y}\delta y + \frac{\partial v}{\partial z}\delta z \quad (1.3.21)$$

t', u, v, w is regarded as the velocity along t, x, y, z direction.

$$\begin{bmatrix} \delta t' \\ \delta u \\ \delta v \\ \delta w \end{bmatrix} = \begin{bmatrix} \frac{\partial t'}{\partial t} & \frac{\partial t}{\partial x} & \frac{\partial t}{\partial y} & \frac{\partial t}{\partial z} \\ \frac{\partial u}{\partial t} & \frac{\partial u}{\partial x} & \frac{\partial u}{\partial y} & \frac{\partial u}{\partial z} \\ \frac{\partial v}{\partial t} & \frac{\partial v}{\partial x} & \frac{\partial v}{\partial y} & \frac{\partial v}{\partial z} \\ \frac{\partial w}{\partial t} & \frac{\partial w}{\partial x} & \frac{\partial w}{\partial y} & \frac{\partial w}{\partial z} \end{bmatrix} \begin{bmatrix} \delta t \\ \delta x \\ \delta y \\ \delta z \end{bmatrix}$$

$$t', u, v, w = u_0, u_1, u_2, u_3 \quad (1.3.22)$$

According to the matrix algorithm, it meets

$$\begin{bmatrix} \frac{\partial t'}{\partial t} & \frac{\partial t}{\partial x} & \frac{\partial t}{\partial y} & \frac{\partial t}{\partial z} \\ \frac{\partial u}{\partial t} & \frac{\partial u}{\partial x} & \frac{\partial u}{\partial y} & \frac{\partial u}{\partial z} \\ \frac{\partial v}{\partial t} & \frac{\partial v}{\partial x} & \frac{\partial v}{\partial y} & \frac{\partial v}{\partial z} \\ \frac{\partial w}{\partial t} & \frac{\partial w}{\partial x} & \frac{\partial w}{\partial y} & \frac{\partial w}{\partial z} \end{bmatrix} = \begin{bmatrix} \frac{\partial t'}{\partial t} & 0 & 0 & 0 \\ 0 & \frac{\partial u}{\partial x} & 0 & 0 \\ 0 & 0 & \frac{\partial v}{\partial y} & 0 \\ 0 & 0 & 0 & \frac{\partial w}{\partial z} \end{bmatrix} + $$

Chapter 1 Structure of the Vacuum

$$\begin{bmatrix} 0 & \frac{1}{2}\left(\frac{\partial t'}{\partial x}+\frac{\partial u}{\partial t}\right) & \frac{1}{2}\left(\frac{\partial t'}{\partial y}+\frac{\partial v}{\partial t}\right) & \frac{1}{2}\left(\frac{\partial t'}{\partial z}+\frac{\partial w}{\partial t}\right) \\ \frac{1}{2}\left(\frac{\partial u}{\partial t}+\frac{\partial t'}{\partial x}\right) & 0 & \frac{1}{2}\left(\frac{\partial u}{\partial y}+\frac{\partial v}{\partial x}\right) & \frac{1}{2}\left(\frac{\partial u}{\partial z}+\frac{\partial w}{\partial x}\right) \\ \frac{1}{2}\left(\frac{\partial v}{\partial t}+\frac{\partial t'}{\partial y}\right) & \frac{1}{2}\left(\frac{\partial v}{\partial x}+\frac{\partial u}{\partial y}\right) & 0 & \frac{1}{2}\left(\frac{\partial v}{\partial z}+\frac{\partial w}{\partial y}\right) \\ \frac{1}{2}\left(\frac{\partial w}{\partial t}+\frac{\partial t'}{\partial z}\right) & \frac{1}{2}\left(\frac{\partial w}{\partial x}+\frac{\partial u}{\partial z}\right) & \frac{1}{2}\left(\frac{\partial w}{\partial y}+\frac{\partial v}{\partial z}\right) & 0 \end{bmatrix} +$$

$$\begin{bmatrix} 0 & \frac{1}{2}\left(\frac{\partial t}{\partial x}-\frac{\partial u}{\partial t}\right) & \frac{1}{2}\left(\frac{\partial t'}{\partial y}-\frac{\partial v}{\partial t}\right) & \frac{1}{2}\left(\frac{\partial t'}{\partial z}-\frac{\partial w}{\partial t}\right) \\ \frac{1}{2}\left(\frac{\partial u}{\partial t}-\frac{\partial t'}{\partial x}\right) & 0 & \frac{1}{2}\left(\frac{\partial u}{\partial y}-\frac{\partial v}{\partial x}\right) & \frac{1}{2}\left(\frac{\partial u}{\partial z}-\frac{\partial w}{\partial x}\right) \\ \frac{1}{2}\left(\frac{\partial v}{\partial t}-\frac{\partial t'}{\partial y}\right) & \frac{1}{2}\left(\frac{\partial v}{\partial x}-\frac{\partial u}{\partial y}\right) & 0 & \frac{1}{2}\left(\frac{\partial v}{\partial z}-\frac{\partial w}{\partial y}\right) \\ \frac{1}{2}\left(\frac{\partial w}{\partial t}-\frac{\partial t'}{\partial z}\right) & \frac{1}{2}\left(\frac{\partial w}{\partial x}-\frac{\partial u}{\partial z}\right) & \frac{1}{2}\left(\frac{\partial w}{\partial y}-\frac{\partial v}{\partial z}\right) & 0 \end{bmatrix}$$

$$= [\varepsilon_{\mu\mu}] + [\varepsilon_{\mu\nu}] + [\omega_{\mu\nu}] \qquad (1.3.23a)$$

Here, $[\varepsilon_{\mu\nu}] = \frac{1}{2}[\gamma_{\mu\nu}]$, and

$$\begin{bmatrix} 0 & \omega_{01} & \omega_{02} & \omega_{03} \\ \omega_{10} & 0 & \omega_{12} & \omega_{13} \\ \omega_{20} & \omega_{21} & 0 & \omega_{23} \\ \omega_{30} & \omega_{31} & \omega_{32} & 0 \end{bmatrix} = \begin{bmatrix} 0 & -\Omega_{x0} & -\Omega_{y0} & -\Omega_{z0} \\ \Omega_{x0} & 0 & -\Omega_z & \Omega_y \\ \Omega_{y0} & \Omega_z & 0 & -\Omega_x \\ \Omega_{z0} & -\Omega_y & \Omega_x & 0 \end{bmatrix} \qquad (1.3.23b)$$

Here, the spatial component has a clear physical meaning. Three of them represent the relative elongation (velocity) of the straight line segment and three represent the angular velocity of the fluid itself. Divergence of velocity $\nabla \cdot V = \frac{\partial u}{\partial x} + \frac{\partial v}{\partial y} + \frac{\partial w}{\partial z}$ represents the volume of the relative expansion of the fluid (Fig. 1.3.11). The equation (1.3.15) can also be written as

$$\delta V = E \cdot \delta r + \Omega \times \delta r$$
$$\Omega = \Omega_x i + \Omega_y j + \Omega_z k$$

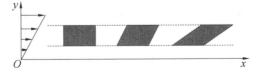

Fig. 1.3.11 Plannar flow field motion

It is the angular velocity vector of the rotation of fluid (Fig. 1.3.12).

$$V(M) = V_0 + \delta V = V_0(M_0) + E \cdot \delta r + \Omega \times \delta r$$

$V_0(M_0)$ is the translational velocity same with point M_0, $\Omega \times \delta r$ is the speed of rotation around the point M_0 caused by point M, $E \cdot \delta r$ is the velocity caused by fluid deformation at the point

M. This is the velocity decomposition theorem of Helmholtz.

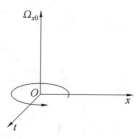

Fig. 1.3.12　The rotation of space-time

2. Analyze the motion of micro fluid

For ease of analysis, we study some special motion of fluid.

During time t, choose hexahedron micelles, which is shown in Fig. 1.3.10. Study the one side $abcd$. If the speed of point a is u, then $\dfrac{\partial u}{\partial x} = \varepsilon_{xx}$ represents the relative elongation rate of the line segment δx (the speed of relative to elongation). Similarly, $\varepsilon_{yy} = \dfrac{\partial v}{\partial y}$ and $\varepsilon_{zz} = \dfrac{\partial w}{\partial z}$ represent the relative elongation rate of the line segment in the direction of y, z.

For example, plane flow field $u_x = ky, u_y = 0$ (k is a constant greater than 0). We analyse the motion characteristics of flow field.

Streamline equation: $y = c$ (The streamline is a straight line parallel to the x axis)

Line deformation:

$$\varepsilon_{xx} = \frac{\partial u}{\partial x} = 0, \quad \varepsilon_{yy} = \frac{\partial v}{\partial y} = 0 \quad (\text{No line deformation})$$

Angular distortion:

$$\gamma_{xy} = \frac{1}{2}\left(\frac{\partial v}{\partial x} + \frac{\partial u}{\partial y}\right) = \frac{k}{2} \quad (\text{Angular distortion exists})$$

Angular velocity of rotation:

$$\Omega_z = \frac{1}{2}\left(\frac{\partial v}{\partial x} - \frac{\partial u}{\partial y}\right) = -\frac{k}{2} \quad (\text{Clockwise direction is negative})$$

It is worth noting that rotational motion or flow irrotational motion depends only on each fluid element itself is rotated. It has nothing to do with whole movement of fluid and the path of fluid element.

Field motion is considered as a vacuum mass flow. From the above analysis, we can see that the mass flow velocities u, v, w correspond strictly to the vacuum shape variables u, v, w in mathematical form. In the following discussion, we will know that the deformation of the vacuum has the same physical essence with the velocity of the vacuum flow.

The 4-dimensional velocity is $u^\mu \equiv \dfrac{\mathrm{d}x^\mu}{\mathrm{d}\tau}$. In relativity, proper time τ along a timelike world line is defined as the time as measured by a clock following that line. Written in component

form:
$$u^\mu = (\gamma c, \gamma v_x, \gamma v_y, \gamma v_z)$$

Here
$$\gamma = 1/\sqrt{1-v^2} \quad (1.3.24)$$

To analyze the vacuum mass flow, the reference frame used in our study is the follow-up reference frame. The time is proper time τ.

For the 4-dimensional space, mass flow of vacuum can be directly obtained by the 4-dimensional speed $u^\mu \equiv dx^\mu/d\tau$.

$$\varepsilon_{\mu\nu} = \frac{1}{2}\left[\frac{\partial}{\partial x_\nu}\left(\frac{dx^\mu}{d\tau}\right) + \frac{\partial}{\partial x_\mu}\left(\frac{dx^\nu}{d\tau}\right)\right], \quad \Omega_{\mu\nu} = \frac{1}{2}\left[\frac{\partial}{\partial x_\nu}\left(\frac{dx^\mu}{d\tau}\right) - \frac{\partial}{\partial x_\mu}\left(\frac{dx^\nu}{d\tau}\right)\right] \quad (1.3.25)$$

In 3-dimensional space, it meets

$$\Omega_k = \frac{1}{2}\left[\frac{\partial}{\partial x_j}\left(\frac{dx^i}{d\tau}\right) - \frac{\partial}{\partial x_i}\left(\frac{dx^j}{d\tau}\right)\right] \quad (i \neq j \neq k = 1,2,3)$$

1.3.8 Vacuum deformation

For flat vacuum (opposed to observer) without any observable, there are 4-dimensional flat space-time coordinates of frame $\xi(\xi^0, \xi^1, \xi^2, \xi^3)$ in the flat vacuum. After vacuum deformation, the bending 4-dimensional space-time coordinate system is $\xi'(\xi'^0, \xi'^1, \xi'^2, \xi'^3)$. Considering the resulting deformation ε at point λ_0 of vacuum $\xi(\lambda)$, the extrusion point λ_0 along ξ direction makes the vacuum deformed, and the deformed amount is ε.

$$\xi' = \xi(\lambda_0 + \varepsilon) = \xi(\lambda_0) + \varepsilon\left(\frac{d\xi}{d\lambda}\right)_{\lambda_0} + \frac{1}{2!}\varepsilon^2\left(\frac{d^2\xi}{d\lambda^2}\right)_{\lambda_0} + \cdots$$

$$= \left(1 + \varepsilon\frac{d}{d\lambda} + \frac{1}{2!}\varepsilon^2\frac{d^2}{d\lambda^2} + \cdots\right)\xi\bigg|_{\lambda_0} = \left(1 + \varepsilon e + \frac{1}{2!}\varepsilon^2 e^2 + \cdots\right)\xi\bigg|_{\lambda_0}$$

$$1 + \varepsilon e + \frac{1}{2!}\varepsilon^2 e^2 + \cdots = \exp[\varepsilon e]$$

$$\xi \to \xi' = \exp[\varepsilon e]\,\xi\big|_{\lambda_0} = \varphi(\xi)\,\xi\big|_{\lambda_0} \quad (1.3.26)$$

A point λ_0 strain of vacuum $\xi(\lambda)$ is $\dfrac{\xi(\lambda_0 + \varepsilon)}{\xi(\lambda_0)} = \exp[\varepsilon e]$, and it constitutes the point of field function, $\varphi_0(\xi) = \exp[\varepsilon e]\big|_{\lambda_0}$. The point λ_0 is in vacuum deformation production quantum field $\varphi(\xi)$, and this is one of the simplest translations caused by deformation. It is no longer a geometric sense of the space translation. Here, ε is a space small variable, and it's a geometric amount. Considering that the parameters change from λ_0 to λ_n, the expansion of points $\exp[\varepsilon e]\big|_{\lambda_0} \to \exp[\varepsilon e]\big|_{\lambda_n}$ in space becomes a continuous field function.

$$\varphi(\xi) = \exp[\varepsilon \xi] \quad (1.3.27)$$

In order to study the intrinsic structure of fundamental particles, we introduce the intrinsic field of fundamental particles. A single quantum field $\varphi(\xi)$ is treated as a point of neighborhood, the whole neighborhood is considered a point, and the neighborhood is the intrinsic field of quantum field $u^\mu(\xi)$. Based on this, we can build a structure image of fundamental particles which does not violate the Relativity, and discussed the reasons why fundamental particles have mass, charge, spin and other freedom by the quantum nature of the intrinsic field.

1.3.9 Green strain expressed by displacement

We know the linear element length square change amount is $dS^2 - dS_0^2 = 2E_{ij}da_i da_j$, $dS^2 = \frac{\partial x_m}{\partial a_i}\frac{\partial x_m}{\partial a_j}da_i da_j$. The interval is not related to the selection of the reference frame. The definition of Green's strain is

$$E_{ij} = \frac{1}{2}\left(\frac{\partial x_m}{\partial a_i}\frac{\partial x_m}{\partial a_j} - \delta_{ij}\right), \quad x_m = a_m + u_m(a_i)$$

Here, u_m is a deformation displacement, so there is $\frac{\partial x_m}{\partial a_i} = \delta_{mi} + \frac{\partial u_m}{\partial a_i}$, and it takes advantage of the permutation properties of δ_{ij}.

$$E_{ij} = \frac{1}{2}\left[\left(\delta_{mi} + \frac{\partial u_m}{\partial a_i}\right)\left(\delta_{mj} + \frac{\partial u_m}{\partial a_j}\right) - \delta_{ij}\right] = \frac{1}{2}\left[\delta_{ij} + \delta_{mi}\frac{\partial u_m}{\partial a_j} + \frac{\partial u_m}{\partial a_i}\delta_{mj} + \frac{\partial u_m}{\partial a_i}\frac{\partial u_m}{\partial a_j} - \delta_{ij}\right]$$

$$E_{ij} = \frac{1}{2}\left[\frac{\partial u_i}{\partial a_j} + \frac{\partial u_j}{\partial a_i} + \frac{\partial u_m}{\partial a_i}\frac{\partial u_m}{\partial a_j}\right] \quad (1.3.28)$$

It is the Green strain expressed by the displacement component. In the Cartesian coordinate system, the normal form after the expansion is

$$E_{11} = \frac{\partial u_1}{\partial a_1} + \frac{1}{2}\left[\left(\frac{\partial u_1}{\partial a_1}\right)^2 + \left(\frac{\partial u_2}{\partial a_1}\right)^2 + \left(\frac{\partial u_3}{\partial a_1}\right)^2\right], \quad E_{22} = \frac{\partial u_2}{\partial a_2} + \frac{1}{2}\left[\left(\frac{\partial u_1}{\partial a_2}\right)^2 + \left(\frac{\partial u_2}{\partial a_2}\right)^2 + \left(\frac{\partial u_3}{\partial a_2}\right)^2\right],$$

$$E_{23} = \frac{1}{2}\left[\frac{\partial u_2}{\partial a_3} + \frac{\partial u_3}{\partial a_2} + \frac{\partial u_1}{\partial a_2}\frac{\partial u_1}{\partial a_3} + \frac{\partial u_2}{\partial a_2}\frac{\partial u_2}{\partial a_3} + \frac{\partial u_3}{\partial a_2}\frac{\partial u_3}{\partial a_3}\right]$$

Green strain can be applied to finite deformation cases. For large shape variables, when we need to consider the two order small quantities, we need to use the Green strain to express it. Green strain is the basis of describing the gravitational field.

1.4　Space-time Structure

Space-time is the geometric effect of vacuum strain. Space-time has nested structure. For example, the intrinsic space-time of quarks is embedded in the inner space-time of neutrons. The inner space-time of neutrons constitutes the background space-time of quarks, and the intrinsic space-time of neutrons is embedded in the observation space-time. The observation space-time constitutes the background space-time of neutrons. For each strain of different shape, the different geometric properties of space-time are no longer the 4-dimensional space-time which we are familiar with, such as half-directional phase space, axisymmetric space and so on. Therefore, we need to solve the problem of space nesting and spatial geometric properties to realize the complete space expression. It should be noted that no matter how nesting space is still a 4-dimensional space-time. For example, there are other small concave surfaces on a large concave surface.

For spatial nesting, we introduce scale structure factor э and use γ matrix to describe spatial geometric properties.

1.4.1 Quantum field intrinsic space-time nesting into background space-time

1. Nested expressions for different range of strain matrices

Now let's consider a simple example. There is a tiny tubule, the length is x, and the axis of the small round tube is x. The line is 1-dimensional and the tube is 2-dimensional. The dimension of a thin tube is curled up into a straight line when it is viewed from a distance. If there is strain in a tiny tube, it can be described by a strain matrix. The coordinate frame of the tiny tube is set to (x, y). The detailed description is as follows

$$[E](X,Y) = \begin{bmatrix} 1 + \begin{bmatrix} \varepsilon_x & 0 \\ 0 & \varepsilon_y \end{bmatrix} э \begin{bmatrix} x\varepsilon(X) & 0 \\ 0 & y\varepsilon(Y) \end{bmatrix} & 0 \\ 0 & 1 \end{bmatrix} \begin{bmatrix} X & 0 \\ 0 & Y \end{bmatrix} \quad (1.4.1)$$

Here, $\begin{bmatrix} x\varepsilon(X) \\ y\varepsilon(X) \end{bmatrix}$ is coordinate scale factor function. The simplest way to describe the function is useing a piecewise function. э is scale structure factor.

$$\varepsilon(X) = \begin{cases} 0 \\ 1 \end{cases} \quad (X_0 < X < X_0 + \Delta X), \quad \varepsilon(Y) = \begin{cases} 0 \\ 1 \end{cases} \quad (-\Delta Y < Y < \Delta Y) \quad (1.4.2)$$

If there is an extremely small 2-dimensional point strain field at X_0 on the X axis, we can introduce $\delta(X - X_0)$ functions

$$[E](X,Y) = \begin{bmatrix} 1 + \begin{bmatrix} \varepsilon_x & 0 \\ 0 & \varepsilon_y \end{bmatrix} \cdot э \begin{bmatrix} x\int_{-\infty}^{+\infty} \delta(X - X_0)\mathrm{d}X \\ y\int_{-\infty}^{+\infty} \delta(X - X_0)\mathrm{d}X \end{bmatrix} & 0 \\ 0 & 1 \end{bmatrix} \begin{bmatrix} X & 0 \\ 0 & Y \end{bmatrix} \quad (1.4.3)$$

We can extend this concept to the description of protons.

2. Quantum intrinsic space-time

For the coordinate system based on the center of a quantum field $(\xi_0, \xi_1, \xi_2, \xi_3)$, that's 4-dimensional space-time. It is used to describe the internal structure of the quantum field. This coordinate space is representation of the intrinsic space-time structure of quantum field. That's the intrinsic space-time structure of elementary particles (Fig. 1.4.1).

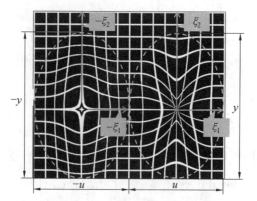

Fig. 1.4.1 The intrinsic space-time structure

$$[\xi] = \begin{bmatrix} \xi_0 & \xi_0 & \xi_0 & \xi_0 \\ \xi_1 & \xi_1 & \xi_1 & \xi_1 \\ \xi_2 & \xi_2 & \xi_2 & \xi_2 \\ \xi_3 & \xi_3 & \xi_3 & \xi_3 \end{bmatrix}, \quad \xi_\mu = g_{\mu\nu}\xi^\nu \qquad (1.4.4)$$

This is the 4-dimensional spatial coordinates of the quantum field. The metric $g_{\mu\nu}$ describes the bending degree of intrinsic spatial space.

3. Observation Background Space

A quantum field is regarded as a point. Observation background space (Fig. 1.4.2) is used to describe the motion characteristics of these points. The coordinate is the representation for the nature of particle motion in the 4-dimensional space-time. The amplitude of the wave function in observation space is the probability amplitude. Relativistic space-time is the observation space-time. The intrinsic space of quantum field is a small distortion space-time embedded in ordinary smooth space-time. When we get into the quantum field, we can choose different coordinate frames according to the structure of the quantum field to be described. For example, for Descartes coordinate system or spherical coordinate system and so on, the probability distribution of point particles and the motion of point particles are in observation background space. If the same kind of coordinate frame is used, intrinsic spatial coordinates can be extended to the observation space.

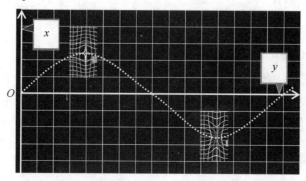

Fig. 1.4.2 The observation background space-time of quantum

Chapter 1 Structure of the Vacuum

$$\left\{\begin{array}{l}[x] = \begin{bmatrix} x_0 & x_0 & x_0 & x_0 \\ x_1 & x_1 & x_1 & x_1 \\ x_2 & x_2 & x_2 & x_2 \\ x_3 & x_3 & x_3 & x_3 \end{bmatrix}, x_\mu = g_{\mu\nu}x^\nu \\ [x_\mu] = \begin{bmatrix} x_0 & 0 & 0 & 0 \\ 0 & x_1 & 0 & 0 \\ 0 & 0 & x_2 & 0 \\ 0 & 0 & 0 & x_3 \end{bmatrix} \end{array}\right. \quad (1.4.5)$$

4. Mixing coordinate

The most commonly used method for describing the field function is the mixing coordinate, i.e. the coordinate systems of the observation space and the inner space appear simultaneously in a field function to give the strain of the inner space and the observation space of the quantum field in this book. The advantage of this expression is that it does not need to consider the scale factor in the nested structure.

$$\varphi(\xi, X) = \exp\{\varepsilon_{\mu\nu}(\xi)[\xi] + \varepsilon_{\mu\nu}(x)[x]\} \quad (1.4.6)$$

5. Intrinsic space-time nesting into background space-time

$[\xi]$ is embedded in the x_1-dimension of background space-time, which can be expressed as

$$[x_\mu(\xi)] = [g_{\mu\nu}]\begin{bmatrix} x_0 & 0 & 0 & 0 \\ 0 & (\partial\mathscr{G}_{\alpha\beta}[\xi]+1)x_1 & 0 & 0 \\ 0 & 0 & x_2 & 0 \\ 0 & 0 & 0 & x_3 \end{bmatrix} \quad (1.4.7)$$

Here, ∂ is scale factor, $|\partial| \ll |e_x|$; e_x is background space coordinate basis; $\mathscr{G}_{\alpha\beta}$ is intrinsic spatial metric; $[\xi]$ is intrinsic space-time coordinate system.

6. The change of the motion state of the intrinsic space extends to the observation space

Eigenspace is a follow-up, which is stationary relative to quantum field. After extending to the observation space, the particle is in the motion state in the observation space, and its field function is as follows

$$\varphi(\xi) = \exp\{\varepsilon_{\mu\nu}(\xi)[\xi]\} \Rightarrow \varphi(x) = \exp\{\varepsilon_{\mu\nu}(\xi)[L_{\alpha\beta}][x]\} \quad (1.4.8)$$

Here, $[L_{\alpha\beta}]$ is Lorentz coordinate transformation matrix.

7. Relationship between coordinate space and strain tensor

Here we stipulate that the X_i coordinate space has the same spatial structure as the strain matrix, such as

$$\begin{bmatrix} \varepsilon_{03} & \varepsilon_{01} & \varepsilon_{02} & \varepsilon_{03} \\ \varepsilon_{10} & \varepsilon_{11} & 0 & \varepsilon_{13} \\ \varepsilon_{20} & 0 & \varepsilon_{22} & 0 \\ \varepsilon_{30} & \varepsilon_{31} & 0 & \varepsilon_{33} \end{bmatrix} \xrightarrow{\text{The corresponding coordinate space structure of this strain matrix is}} (X_i) = \begin{pmatrix} \xi_0 & \xi_0 & \xi_0 & \xi_0 \\ \xi_1 & \xi_1 & 0 & \xi_1 \\ \xi_2 & 0 & \xi_2 & 0 \\ \xi_3 & \xi_3 & 0 & \xi_3 \end{pmatrix}$$

(1.4.9)

(X_i) is no longer a traditional matrix, but a generalized coordinate with spatial information.

1.4.2 The properties of space

1. The concept of half-space

The compression vacuum is negative space, while stretching vacuum is positive space. Positive and negative space-time corresponds to stretching and compression vacuum. Tension and compression always exist in the form of duality, which is manifested by space-time symmetry. The space-time symmetry comes from the symmetry of vacuum strain. Of course, space-time is not always symmetrical. If a local vacuum exists only in tension or compression, the whole local vacuum consists of positive or negative space, and the local space forms a unique asymmetric space. In the stretch space Ω, local vacuum strain

$$\varepsilon_i \geqslant 0 \quad (i=1, 2, 3), \text{ Quantum field: } \exp(-i\varepsilon_i x) \quad (1.4.10)$$

The compressed background space corresponding to negative half-space, local vacuum strain

$$\varepsilon_i \leqslant 0 \quad (i=1, 2, 3), \text{ Quantum field: } \exp(i\varepsilon_i x) \quad (1.4.11)$$

Charged fermions intrinsic space is half-space that within a very small region. The overall characteristics of the half-space are represented by particle interaction parity.

The most common is ordinary Cartesian coordinates, also known as omni-directional space, in which any vector rotates to 180° in the opposite direction (Fig. 1.4.3(b)). When a pair of leptons is produced, anti-lepton is compressed field without stretching field, while the positive lepton is stretching field without compression field. An omni-directional physical space divides into two independent half to space, which respectively are the positive half space and anti-half space (Fig. 1.4.4). This split state of space is very strange, the dimension maintains its original characteristics, but the space angle is only half of the original undivided one. In this half space, the vector rotation of 360° is equivalent to that of 180° in Cartesian coordinate space. The single positive or single negative space is defined as half-space. Lepton intrinsic space is half the space.

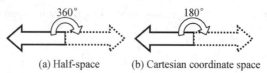

(a) Half-space   (b) Cartesian coordinate space

Fig. 1.4.3 Half-space and Cartesian coordinate space schematic

Chapter 1 Structure of the Vacuum

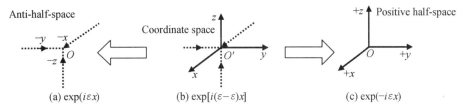

(a) $\exp(i\varepsilon x)$ (b) $\exp[i(\varepsilon-\varepsilon)x]$ (c) $\exp(-i\varepsilon x)$

Fig. 1.4.4 The formation half-space

$\Phi(x,y,z)$ is in coordinate space. The spatial inversion satisfies

$$\hat{P}\Phi(x,y,z) = \Phi(-x,-y,-z) \quad (1.4.12)$$

Here, $\hat{P}$ is the space inversion operator. The half-space is split by the coordinate space, then there should be an operation operator. The operator makes the coordinate space into the half-space.

2. Half space with parity nonconservation

The intrinsic space of charged fermions is half-space. Provision Φ_L is the positive half-space.

$$\Phi_L = \hat{P}_L \Phi \quad (1.4.13)$$

Here, $\hat{P}_L$ is the positive half-space operator, which acts on ordinary space to separate the positive half-space Φ_L. Similarly, the marked anti-half space is Φ_R.

$$\Phi_R = \hat{P}_R \Phi \quad (1.4.14)$$

Here, $\hat{P}_R$ is anti-space operator. Add to the positive and negative half-space which is back to normal space.

$$\Phi_R + \Phi_L = \Phi \quad (1.4.15)$$

Equation (1.4.13) and equation (1.4.14) can obtain

$$\hat{P}_L = \frac{1}{2}(1+\gamma_5), \hat{P}_R = \frac{1}{2}(1-\gamma_5) \quad (1.4.16)$$

This is the half-space operator.

$$P_L P_R = \frac{1+\gamma_5}{2}\frac{1-\gamma_5}{2} = \frac{1-\gamma_5^2}{2} = 0, P_L P_L = \frac{1+\gamma_5}{2}\frac{1+\gamma_5}{2} = \frac{1+2\gamma_5+\gamma_5^2}{4} = P_L$$

$$P_R P_R = \frac{1-\gamma_5}{2}\frac{1-\gamma_5}{2} = \frac{1-2\gamma_5+\gamma_5^2}{4} = P_R, P_L + P_R = \frac{1+\gamma_5}{2} + \frac{1-\gamma_5}{2} = 1$$

3. γ matrix representation of geometric properties of quantum space

Each vacuum strain produces a unique spatial geometry. The geometric properties of space are represented by gamma matrix (Table 1.4.1).

Table 1.4.1 Spatial properties

γ Matrix	Spatial character	Strain and physical image
γ_0	Time	$\varepsilon_{00} \longrightarrow$
$\gamma_1, \gamma_2, \gamma_3$	3D vector space	
γ_4	Spherical symmetry 3-dimensional space	
γ_5	Axisymmetric space (1-dimensional string strain)	
$\gamma_6 = \gamma_5 \gamma_\mu$	Axial vector space: under the mirror symmetry transformation, the vector of the normal component unchanges while the tangent component changes (parity is a positive vector)	
$\gamma_7 = 1-\gamma_5$ $\gamma_7' = 1+\gamma_5$	Semi-space	
I	Scalar space: a space that can be completely represented by its size (seen as a point in the observation space)	

The spatial characteristics are expressed by γ_μ matrix. The specific forms of the eigenvalue operator are as follows.

Eigenvalue operator:

$$\hat{\partial}_\mu = -i \frac{\partial}{\partial X_\mu} \tag{1.4.17}$$

Eigenvalue operator with interaction:

Chapter 1   Structure of the Vacuum

$$\hat{D}_\mu = -i\left(\frac{\partial}{\partial X_\mu} + \hat{\Gamma}_\mu\right) \qquad (1.4.18)$$

The standard representation (block form) of γ^μ is

$$\gamma^0 = \begin{bmatrix} I & 0 \\ 0 & -I \end{bmatrix}, \quad \gamma^i = \begin{bmatrix} 0 & \sigma^i \\ -\sigma^i & 0 \end{bmatrix} \qquad (1.4.19)$$

Where $(I_{\alpha\beta} = \delta_{\alpha\beta})$ is 4×4 matrix, which is usually omitted. The matrixs γ_μ are

$$\gamma_j = \begin{bmatrix} 0 & -i\sigma_j \\ i\sigma_j & 0 \end{bmatrix} \quad (j=1,2,3), \quad \gamma_4 = \begin{bmatrix} I_2 & 0 \\ 0 & -I_2 \end{bmatrix}, \quad \gamma_5 = \begin{bmatrix} 0 & -I_2 \\ -I_2 & 0 \end{bmatrix}$$

Here, $I_2 = \begin{bmatrix} 1 & 0 \\ 0 & 1 \end{bmatrix}$; γ_1, γ_2, γ_3 correspond to the coordinate basis of 3-dimensional space of particle degree of freedom space. The time dimension should also be considered, and $\gamma_0 = \begin{bmatrix} I_2 & 0 \\ 0 & I_2 \end{bmatrix}$ should be defined as the coordinate base of time dimension. γ_4 is as the coordinate base of degree of mass freedom, which can be projected to the time dimension of 4-dimensional space-time. γ_5 is as the strong interaction of the coordinate base of degree freedom. σ is called Pauli matrix.

$$\sigma_x = \sigma_1 = \sigma^1 = \begin{bmatrix} 0 & 1 \\ 1 & 0 \end{bmatrix}, \quad \sigma_y = \sigma_2 = \sigma^2 = \begin{bmatrix} 0 & -i \\ i & 0 \end{bmatrix}, \quad \sigma_z = \sigma_3 = \sigma^3 = \begin{bmatrix} 1 & 0 \\ 0 & -1 \end{bmatrix}$$
$$(1.4.20)$$

For example, $\begin{bmatrix} 0 & 0 & 0 & 0 \\ 0 & x_1 & 0 & 0 \\ 0 & 0 & x_2 & 0 \\ 0 & 0 & 0 & x_3 \end{bmatrix}$ is vector space, and we can express it as $\widetilde{\gamma}_\mu \begin{bmatrix} 0 & 0 & 0 & 0 \\ 0 & x_1 & 0 & 0 \\ 0 & 0 & x_2 & 0 \\ 0 & 0 & 0 & x_3 \end{bmatrix}$.

The γ matrix does not operate with the coordinate matrix $[x]$, it is only used to mark the nature of space.

4. Interacting space of vertex

For the $\bar{\psi}O\psi$, the interaction vertex of $\bar{\psi}$ and ψ is O.

(1) $O = I$, $\mathcal{R}\bar{\psi}(x)\psi(x)) = \bar{\psi}(x)\psi(x)$.

(2) $O = \gamma^\nu$, $\mathcal{R}\bar{\psi}(x)\gamma^\nu\psi(x)) = \bar{\psi}(x)\gamma_\nu\psi(x)$.

(3) $O = \sigma^{\mu\nu}$, $\mathcal{R}\bar{\psi}(x)\sigma^{\mu\nu}\psi(x)) = \bar{\psi}(x)\sigma_{\mu\nu}\psi(x)$.

(4) $O = \gamma^5$, $\mathcal{R}\bar{\psi}(x)\gamma^5\psi(x)) = -\bar{\psi}(x)\gamma^5\psi(x)$.

(5) $O = \gamma^5\gamma^\nu$, $\mathcal{R}\bar{\psi}(x)\gamma^5\gamma^\nu\psi(x)) = -\bar{\psi}(x)\gamma_\nu\psi(x)$.

(6) $O = \gamma^0\sigma^{\mu\nu}$, $\mathcal{R}\bar{\psi}(x)\gamma^0\sigma^{\mu\nu}\psi(x)) = -\bar{\psi}(x)\gamma^0\sigma_{\mu\nu}\psi(x)$.

Two interacting particles are in the common space Ω, and the interacting vertex γ_μ has symmetry. The physical facts are that the strain is equal to anti-strain, which causes the trace of γ_μ is zero (Fig. 1.4.5).

For antisymmetric tensor $\sigma^{\rho\sigma} = \frac{i}{2}[\gamma^\rho, \gamma^\sigma]$, the first three categories respectively are real scalar, vector and tensor, and last three are pseudo.

Two interacting particles are in the normal 4-dimensional space-time, and the interacting vertex is γ_μ. Interacting vertex in the half-space is $\gamma_\mu P_L$, and interacting vertex in the anti-space is $\gamma_\mu P_R$.

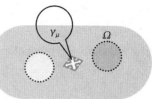

Fig. 1.4.5 The duality particles and the action of vertex
(In region Ω, the interaction vertex is strain field which is constant equal to the anti-strain field, and the mack of γ_P is zero)

5. Summary

For example, $[\xi]$ is embedded in the x_1-dimension of background space-time. The background space is half-space, and constitute generalized space X_i can be expressed as

$$(X_i) = (\check{\gamma}_5 + 1)[g_{\mu\nu}]\begin{bmatrix} x_0 & 0 & 0 & 0 \\ 0 & (\mathscr{D}_{\alpha\beta}[\xi] + 1)x_1 & 0 & 0 \\ 0 & 0 & x_2 & 0 \\ 0 & 0 & 0 & x_3 \end{bmatrix} \quad (1.4.21)$$

Here, $\ni$ is scale factor, $|\ni| \ll |e_x|$; e_x is background space coordinate basis; $\mathscr{G}_{\alpha\beta}$ is intrinsic spatial metric; $[\xi]$ is intrinsic space-time coordinate system; $[g_{\mu\nu}]$ is observation space-time metric.

1.4.3 Exchange symmetric physical images of identical particles

Consider a system consisting of two identical particles, whose states are described by wave functions $\psi(q_1, q_2)$. q_1, q_2 represent all coordinates of two particles respectively. When two particles exchange $\psi(q_1, q_2) \to P_{12}\psi(q_1, q_2) = \psi(q_2, q_1)$, if there is any difference, it is only the exchange of roles between the original first particle and the second particle. Since the properties of the two particles are the same, they are indistinguishable. So, we can only think that $\psi(q_1, q_2)$ and $\psi(q_2, q_1)$ describe the same quantum state. In this way, the form of wave function will be strongly restricted. More generally, we consider a multi-particle system consisting of N identical particles whose states are described by wave function $\psi(q_1, q_2, \cdots, q_N)$. P_{ij} denotes the operator of the exchange between the i-particle and the j-particle, i.e.

$$P_{ij}\psi(q_1, \cdots, q_i, \cdots, q_j, \cdots, q_N) \equiv \psi(q_1, \cdots, q_j, \cdots, q_i, \cdots, q_N) \quad (1.4.22)$$

The quantum states described by $P_{ij}\psi$ and ψ are exactly the same, so the maximum difference is a constant factor λ, $\lambda = \pm 1$. So P_{ij} has two eigenvalues. In this way, the wave functions of all particles must satisfy the following relations:

$$P_{ij}\psi = +\psi \quad (1.4.23a)$$
$$P_{ij}\psi = -\psi \quad (i \neq j = 1, 2, \cdots, N) \quad (1.4.23b)$$

Satisfaction formula (1.4.23a) is called symmetric wave function, and satisfaction formula

(1.4.23b) is called anti-symmetric wave function.

Experiments show that the exchange symmetry of wave functions of all particle systems is definitely related to the spin of particles.

Particle spin ($s = 0, \hbar, 2\hbar, \cdots$) is an integral multiple of $\hbar$, the wave function is always symmetrical for the exchange of two particles. The wave function is always symmetrical for the exchange of two particles. They obey the Bose-Einstein statistical method in statistical physics and are called Bosons.

Particle spin ($s = \hbar/2, 3\hbar/2, 5\hbar/2, \cdots$) is half odd times $\hbar$, the wave function is always asymmetric for the exchange of two particles. They obey the Fermi-Dirac statistical method in statistical physics and are called Fermions.

When particles are exchanged, it is equivalent to rotating 360° in particle space. The intrinsic space of boson is omni-directional space, and the particle space of boson is omni-directional space. So, when two particles exchange, such as black and white, it is equivalent to black particle rotation 180° and white particle rotation 180° (Fig. 1.4.6), so the cumulative rotation is

$$180° + 180° = 360°$$

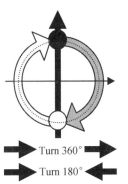

Thus, an arrow made of black and white particles rotates 360° in the omni-directional particle space, satisfying $P_{ij}\psi = +\psi$. If two fermions are exchanged, the total rotation is 360°, but the actual situation is that the intrinsic space of fermions is half-directional. The particle space composed of fermions is a half-directional space, so the cumulative rotation is only half of the omni-directional space, i.e. 180°. In this way, an arrow composed of black and white fermions rotates only 180° in particle space, and the arrow reverses phase, satisfying $P_{ij}\psi = -\psi$.

Fig. 1.4.6 Particle exchange is equivalent to particle space rotation

The difference of Boson and fermion exchange symmetry is due to the difference between omni-directional space and semi-directional space in vacuum field, which directly explains why boson and fermion are so different.

1.4.4 The state space of basic particles

Each field function containing both positive and negative elementary particles can be written as $\Phi(\xi) = \exp\{\pm i[\varepsilon_{\mu\nu}](\xi)\}$. For physics experiments, we can only observe particles in a state ψ_i. In the observation space $\{x\}$ $\Phi(\xi) \Rightarrow \Phi(x)$. The field function can be written in the form of splitting $\Phi(x) = \exp\{\mp i[\varepsilon_{\mu\mu}](x) \mp i[\omega_{\mu\nu}](x) \mp i[\gamma_{\mu\nu}](x)\}$. Some intrinsic properties can be shown in observation space, such as energy momentum, spin state and charge state. For example, in the simplest case

$$\psi_1(x) = \begin{bmatrix} 1 \\ 0 \\ 0 \\ 0 \end{bmatrix} \exp\{-i[\varepsilon_{\mu\mu}](x) - i[\omega_{\mu\nu}](x)\}$$

$$\psi_2(x) = \begin{bmatrix} 0 \\ 1 \\ 0 \\ 0 \end{bmatrix} \exp\{-i[\varepsilon_{\mu\mu}](x) + i[\omega_{\mu\nu}](x)\}$$

$$\psi_3(x) = \begin{bmatrix} 0 \\ 0 \\ 1 \\ 0 \end{bmatrix} \exp\{+i[\varepsilon_{\mu\mu}](x) - i[\omega_{\mu\nu}](x)\}$$

and

$$\psi_4(x) = \begin{bmatrix} 0 \\ 0 \\ 0 \\ 1 \end{bmatrix} \exp\{+i[\varepsilon_{\mu\mu}](x) - i[\omega_{\mu\nu}](x)\}$$

$$\psi = \begin{bmatrix} \psi_1 \\ \psi_2 \\ \psi_3 \\ \psi_4 \end{bmatrix}$$

It is a 4-dimensional vector space. The coordinate bases are γ_0, γ_1, γ_2, γ_3. The strain of n-type quantum field indicates that the vector space is n-dimensional. The newly added coordinate systems are γ_4, γ_5, $\cdots$, γ_N.

The above discusses the properties of the elementary particle state space, which are described by γ_i matrix. Elementary particle interaction space is called vertex space. The properties of the state vector space are described by γ_i matrix.

Note: In order to simplify the marking, in $\begin{pmatrix} x_1 & x_{12} \\ x_{21} & x_2 \end{pmatrix}$, set $x_{12} = x_1$, $x_{21} = x_2$. The case of higher-order matrix is similar.

The fermion field function can also be written as follows:

$$\varphi_F = \exp\{-i[\varepsilon](\xi)\} \equiv \exp\left\{-i\frac{1}{2}\begin{bmatrix} \varepsilon_0 & \cdots & \varepsilon_{0n} \\ \vdots & \ddots & \vdots \\ 0 & \cdots & \varepsilon_n \end{bmatrix} \widetilde{\gamma} \begin{pmatrix} \xi_0 & \cdots & \xi_{0n} \\ \vdots & \ddots & \vdots \\ 0 & \cdots & \xi_n \end{pmatrix}\right\};$$

$$\overline{\varphi}_F = \exp\{i[\varepsilon](\xi)\} \equiv \exp\left\{i\frac{1}{2}\begin{bmatrix} \varepsilon_0 & \cdots & 0 \\ \vdots & \ddots & \vdots \\ \varepsilon_{n0} & \cdots & \varepsilon_n \end{bmatrix} \widetilde{\gamma} \begin{pmatrix} \xi_0 & \cdots & 0 \\ \vdots & \ddots & \vdots \\ \xi_{n0} & \cdots & \xi_n \end{pmatrix}\right\}$$

The physical meaning of such expression is clearer. However, in this paper, in order to facilitate the reading habits of readers, the written form still follows the original format.

Chapter 2　Structure of the Photons

Why do photons have no rest mass and without charge? What is the energy and momentum of photon? What is the nature of the wave-particle duality? What are the probability wave characteristics and electromagnetic wave characteristics of photon? What is the spin of the photon? What is virtual photon? Current theories are powerless to deal with these problems, so a new breakthrough is needed.

Photons are the simplest vacuum local strain and strain contains all the information of photons. The strain tensor of the photon is composed of two parts. The first part is the internal strain, and the second part is the electric flux line strain tensor caused by fiber shift caused by internal strain.

$$[\varepsilon]_{\text{Photon}} = \underbrace{\left[\frac{\partial u_\mu}{\partial \xi_\nu}\right]}_{\substack{\text{Microcosmic space-time} \\ \text{rupture region, internal strain}}} + \underbrace{\left[\frac{\partial A_\mu}{\partial x_\nu}\right]}_{\substack{\text{The fiber field in the observational} \\ \text{space, electric flux line strain}}}$$

$$= \underbrace{\varepsilon_{\mu\mu}(\xi)}_{\text{4-momentum}} + \frac{1}{2}\underbrace{\gamma_{\mu\nu}(\xi)}_{\text{Weakly acting strain}} + \frac{1}{2}\underbrace{\omega_{\mu\nu}(\xi)}_{\text{Spin}} + \underbrace{\Delta\varepsilon_{\mu\mu}(x)}_{\substack{\text{Electromagnetic field of} \\ \text{scalar and vector potential}}} + \frac{1}{2}\underbrace{\Omega_{\mu\nu}(x)}_{\text{Electromagnetic field}}$$

The task in this chapter is to analyze the strain of photons one by one.

2.1　The image of photon

2.1.1　Electromagnetic wave image of photon

1. Maxwell image of electromagnetic wave

In 1863, Maxwell published the famous theory of electromagnetic field, focusing on the changing electric field to produce magnetic field and the changing magnetic field to produce electric field. Changing electric field → changing magnetic field → changing electric field..., constituting electromagnetic waves (Fig. 2.1.1). Consider the electromagnetic radiation from an oscillating dipole (electric moment rapidly changing electric dipole is called the oscillating dipole). The simplest oscillating dipole is electric moment to do the cosine vibration and dipole electric moment p can be expressed as

$$p = p_0 \cos \omega t \quad (2.1.1)$$

Here, p_0 is the amplitude; ω is the circular frequency. Oscillating dipole radiation is an electromagnetic wave, but leaving the dipole far away can be expressed by plane wave equation.

$$E = E_0 \cos 2\pi\left(\frac{t}{T} - \frac{x}{\lambda}\right) \quad (2.1.2)$$

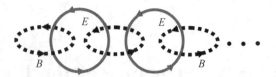

Fig. 2.1.1   Schematic diagram of electromagnetic field propagation

$$H = H_0 \cos 2\pi \left(\frac{t}{T} - \frac{x}{\lambda} \right) \qquad (2.1.3)$$

Electric and magnetic fields are perpendicular to each other $E \perp H$ (Fig. 2.1.2). Here, x is the propagation distance; λ and T are respectively wavelength and time period, meeting $\lambda = vT$. In classical electromagnetic theory, electromagnetic waves can propagate in vacuum without any medium, and the speed in vacuum is $c = \dfrac{1}{\sqrt{\varepsilon_0 \mu_0}} = 3 \times 10^8$ m/s. It is the speed of light. The optical is electromagnetic wave.

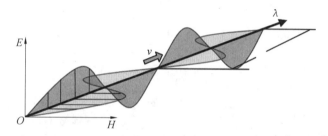

Fig. 2.1.2   Plane electromagnetic wave

2. Characteristics of the electromagnetic wave

(1) Electromagnetic waves transverse with polarization. The propagation direction v is perpendicular to E and H, which constitutes the right-handed spiral system.

(2) The value of E and H is proportional to $\sqrt{\varepsilon} E = \sqrt{\mu} H$.

(3) The velocity of electromagnetic wave propagation is equal to the speed of light in vacuum.

(4) The energy of electromagnetic waves is $E = h\nu$.

Electromagnetic wave propagation process is energy transmission process, and it is namely the process of radiant energy.

Energy density of electric field is $w_e = \dfrac{1}{2}\varepsilon E^2$.

Energy density of magnetic field is $w_m = \dfrac{1}{2}\mu H^2$.

Electromagnetic energy density is $W = W_e + W_m$.

Electromagnetic wave range is very wide. The essence is the same, and the differences are just frequency (wavelength) and characteristics. It can be divided into radio waves, infraredray, visible light, ultraviolet, X-ray, γ-ray and so on. Especially for visible light, its wavelengths range is between 400—760 nm. This electromagnetic wave can make us have the feeling of light (Fig. 2.1.3), so it is called light.

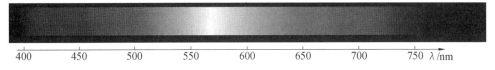

Fig. 2.1.3 Wavelengths of visible light

2.1.2 Particle characteristics of photon

Planck (1858—1947, German) proposed in 1900 that the radiation energy absorption or emission of electromagnetic wave exists in the form of quantum. Quantum energy is $\varepsilon = h\nu$. Here, ν is the frequency; h is Planck constant.

Bohr studied the spectrum. In 1913, he pointed out that the atomic structure has quantum states. The two quantum states m and n of energy are E_m and E_n, and the energy difference and spectral frequency have the following relationship:

$$\nu_{m \to n} = \frac{E_m - E_n}{h} \tag{2.1.4}$$

In 1905, Einstein departed from the Planck radiation quantum theory, and kept on in-depth study on the radiation. For the radiation absorption or emission, the quantum states exist, and energy of photons (ε) is

$$\varepsilon = h\nu = \hbar\omega \tag{2.1.5}$$

For the corresponding momentum p, the value is $p = h/\lambda$. Here, λ is the wavelength, and the wave vector k is expressed as

$$p = \hbar k, \quad |k| = \frac{2\pi}{\lambda}, \quad \lambda\nu = c \tag{2.1.6}$$

Here, ω and k are 4-dimensional wave vectors; c is the speed of light. This explains the photoelectric effect. Photons are electromagnetic waves, but they also have quantum properties. Establishing photon images is a difficult task. The difficulty lies in the unclear relationship between light waves and light particles.

2.1.3 The speed of light constitutes the space-time ruler of observation

Now we briefly review the main points of the space-time concept. For the motion of all substances compared with photon motion, time and space, a conversion factor c exists. $r \equiv ct$ constitute the light cone. In the real world, the speed of photons is the fastest, so using light as a scale has innate advantages in space-like space. Photon slows down, and the clock slows down, then photon propagation paths curve, and the space is also curved. Photon is the best measurement standard of space-time, and the path of light is the geodesic line of space-time. Space and time could not be independent without the other. Space and time are two expression ways of vacuum nature. For the observation of space-time, $(ct, x) = x^\mu = (x^0, x^1, x^2, x^3)$, and

$x(x^1, x^2, x^3)$ is vacuum of space. $x^0 = ct$ is the quantity through vacuum in the propagation time t of a single quantum field.

2.2 Point translation strain is forming photonic in vacuum

2.2.1 Point simple translation along ξ^1 in vacuum

1. 4-dimensional geometric properties of vacuum

Considering the vacuum principal strain tensor is $\varepsilon_{\mu\mu}$. From the whole point of view, space-time will not break up. Considering 4-dimensional characteristics of vacuum, $\varepsilon_1^2 + \varepsilon_2^2 + \varepsilon_3^2 = \varepsilon_0^2$, $\varepsilon_\mu = \partial u_\mu / \partial \xi_\mu$. That is light cone condition, and strain tensor ε_μ is a relative deformation along μ. We can obtain

$$\varepsilon_{00}^2 = \sum_{i=1}^{N} \varepsilon_{ii}^2 \tag{2.2.1a}$$

This is the basic law of the vacuum super-unified field theory, which is expressed as energy conservation. It can be expressed as

$$\Box u_\mu = 0 \tag{2.2.1b}$$

Its physical meaning is the conservation of 4-momentum.

2. The law of conservation of vacuum strain

Considering the spatial and temporalmetric structure, the volume deformation of space-time is

$$\Delta V_{\text{Time}} = \text{tr}\begin{bmatrix} \varepsilon_0 & 0 & 0 & 0 \\ 0 & 0 & 0 & 0 \\ 0 & 0 & 0 & 0 \\ 0 & 0 & 0 & 0 \end{bmatrix}, \quad \Delta V_{\text{Space}} = \text{tr}\begin{bmatrix} 0 & 0 & 0 & 0 \\ 0 & \varepsilon_1 & 0 & 0 \\ 0 & 0 & \varepsilon_2 & 0 \\ 0 & 0 & 0 & \varepsilon_3 \end{bmatrix} \tag{2.2.2a}$$

From the whole point of view, the vacuum strain tensor is 0. It is known by

$$\Delta V_{\text{Time}} \equiv \Delta V_{\text{Space}} \tag{2.2.2b}$$

This is the basic law of the vacuum super-unified field theory, which is expressed as energy conservation. It can be expressed as

$$u_{\mu,\mu} = 0 \tag{2.2.2c}$$

3. The definition of $\mathcal{H}$

Set the volume change of a basic unit which along 1-dimension is ΔV_i (It can be understood as a slender tube), then

$$\mathcal{H} = K_{\text{GP}} \sum_{i=1}^{N} |\Delta V_i| \tag{2.2.3}$$

Here, K_{GP} is the conversion coefficient between the geometric and the physical quantities.

Considering the vacuum principal strain tensor ε_μ, single fiber deformation has axisymmetric properties. The field basic unit can be regarded as a small cube $V_{\text{basic unit}} = dV$, space principal strain $\dfrac{dV' - dV}{dV} \approx \varepsilon_1 + \varepsilon_2 + \varepsilon_3$. We can obtain $\Delta V_i \approx (\varepsilon_1 + \varepsilon_2 + \varepsilon_3) V_i$. Here, V_i is the basic unit volume of vacuum i in the first; V_i' is the basic unit volume after the deformation. These basic units' volume with no deformation are equivalent, and that is $V_i = V_j = V_0 = h_f$. We set the basic unit volume is h_f, so

$$\Delta V_i = h_f \sum_{i=1}^{3} \varepsilon_i \tag{2.2.4}$$

Because of the small volume element V_i' along the 1-dimensional linear array, we can simplify the problem of volume to the problem of length. The shape of ΔV_i is a long tube, and the length of the tube is Δl_i. Setting the area of cross section of the basic unit is s. The radial cross-sectional area of the non-deformation basic unit V_i is s, which is a constant, then volume change is $|\Delta V_i| = |\Delta l_i| \cdot s$. Here we set

$$h_f / s \equiv 1 \tag{2.2.5}$$

Then the change of the volume was characterized by volume length changes Δl_i. For example, for a thin and long flexible tube, its volume changes in the unit length, and it is similar to a line from a distance. The volume change of the small body element can be incorporated in the length variation. In the following discussion, we will use the length change to represent the volume change of the 1-dimensional linear array of small body element.

$$\mathscr{H} = K_{\text{GP}} \sum_{i=1}^{N} |\Delta l_i| \tag{2.2.6}$$

The dimension of the Planck constant on the left is $[\text{Time}]^{-1}[\text{Length}]^{2}[\text{Mass}]$. The dimension of the right is $[\text{Length}]$, and $L = ct$, so we can get the dimension of the conversion coefficient

$$[K_{\text{GP}}] = [\text{Time}]^{-1}[\text{Length}] \ [\text{Mass}] \tag{2.2.7}$$

In order to facilitate the discussion of the problem in the future, we stipulate

$$K_{\text{GP}} \equiv 1 \tag{2.2.8}$$

and this coefficient will not appear in the future unless it is specially needed. We directly consider the Planck constant as the geometric quantity describing the vacuum deformation.

$$\mathscr{H} = \sum_{i=1}^{N} |\Delta l_i| \tag{2.2.9}$$

4. Analysis of figures

The envisaged is the ideal basic unit along the straight line consisting ξ^1-dimensional flat vacuum (Fig. 2.2.1). It is particularly important to note that it is impossible to draw multi-dimensional images, so we must use 2-dimensional images as sketches. Every little basic unit of vacuum is marked up by $a, b, c, d, o, f, g, i, j$. Now let us consider the case of point o. Vacuum basic unit o moves forward, and it will compress the vacuum in front, with stretching the vacuum behind. The closer it comes to the point o', the greater deformation will happen in the vacuum. Stretching and compression limit exist in the vacuum. We define the maximum value of point o' deviation from the point o as stretch or compression limit value. The volume change of basic unit of all vacuum along ξ^1-dimension is $\mathcal{H}$.

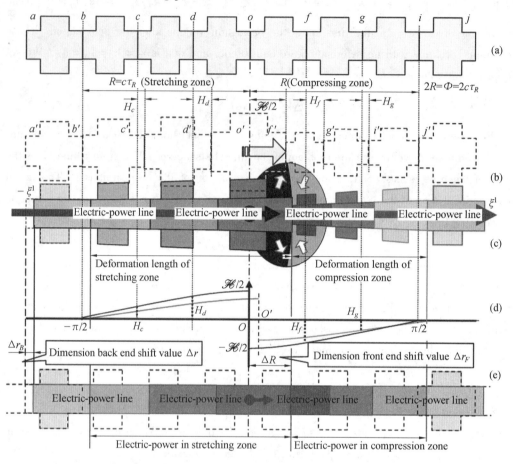

Fig. 2.2.1 1-dimensional photon intrinsic structure

Point o' does not reach the stretching and limit, and does not form longitudinal waves. In this case, o' should restore its natural state, and the vibration energy should be returned to the vacuum to apply deformation into local vacuum vibration. For experimental observation, it is the vacuum fluctuation or the vacuum foam. Because waves cannot be formed, and the deformation at o' point does not exist, so we only consider the formation of waves, which reach the limits of tension and compression. When the compression limit is reached, the point o will force at point f, and point f will compress point $g, \ldots$, then oi segment is shorter, and the shortening amount is $\mathcal{H}/2$.

According to the previous assumption, the tension and compression limits are inherent characteristics of vacuum, which are constants. The compression limit and the tension limit are also reached. The quantity of stretching is $\mathcal{H}/2$.

$$o'b' - ob = \mathcal{H}/2 \qquad (2.2.10)$$

The segment is from b' to i', which forms an independent longitudinal wave. Outside of segment $b'i'$, the vacuum remains, which is not affected by the moving of point o'. P-wave is composed of tension-compression region, and propagates in vacuum are different from classical solitary wave. Waves can't stay.

2.2.2 The physical meaning of deformation limit $\mathscr{H}$ in vacuum

Let's consider the 1-dimensional static structure of photons (Fig. 2.2.1). In center of intrinsic space of the photon, it is the origin coordinate frame ξ_μ. Photon intrinsic displacement is $u(\xi_\mu)$, $\mu = 0, 1, 2, 3$. The center o to i segment is defined as a compression region oi, and it is denoted as $R_{\text{Compression}}$. The ob segment is defined as stretching region. R_{Stretch}. $R_{\text{Compression}} \equiv c\tau_R \equiv R_{\text{Stretch}} = R$ is the radius of photons in intrinsic field ξ_1. Obviously only for a common intrinsic time, the geometric length is measured by time×speed. The speed of light is slow in the stretch region and compression region. The measuring lengths of two are exactly the same by the light ruler. The length of sum total of stretching and compression are defined as wavelength Φ of ξ_1 (longitudinal wave wavelength), $\Phi = 2R$.

The maximum of stretching in vacuum is $-\mathscr{H}/2$ and the maximum of compression is $-\mathscr{H}/2$. If the accumulative deformation in the field ξ_1 is less than $\mathscr{H}/2$, it is unable to form a longitudinal wave. Of course, the maximum value should not be exceeded, because the maximum value will form a longitudinal wave propagating at the speed of light. The maximum of total deformation is the sum of the maximum absolute values of tension and compression. $\mathscr{H}$ is called the 4D total deformation value of photon ξ_1-dimensional vacuum. $\mathscr{H}$ is defined as the maximum value of 1-dimensional vacuum deformation and it is a constant, which is the basic property of vacuum.

Single photon intrinsic space is the deformation area and the deformation region is spherical.

1. Intrinsic time

The intrinsic time τ is the time that it takes for the quantum field and grows out of nothing, or the time it takes to create a quantum field, which indicates the average hardening degree of the internal space of the quantum field.

When the intrinsic space in quantum field is formed, deformation wave at the speed of light diffused toward the center (or shrink to the center). The intrinsic space radius R needs a time τ_R, which is intrinsic time. There is a strict relationship between time and space. Intrinsic time τ_Φ can be expressed as

$$\Phi = 2c\tau_\Phi \qquad (2.2.11)$$

Here, Φ is the diameter of sphere intrinsic space.

$\mathscr{H}$ is the maximum deformation value in 1-dimensional vacuum. Then $\mathscr{H}_\xi$ is the maximum value of total deformation in space and $\mathscr{H}_\tau$ is the maximum value of total deformation time, which essence is the total deformation of the vacuum.

$$\mathscr{H} \equiv \mathscr{H}_\xi = \mathscr{H}_\tau \qquad (2.2.12)$$

2. The limit deformation

It can be seen from the definition that the vacuum in the compression minimum zone is only a basic unit of vacuum, while that in the tension minimum zone is only a basic unit of vacuum, and the maximum of compression is $-\mathcal{H}/2$, then the maximum of stretch is $\mathcal{H}/2$.

2.2.3 The energy of 2-dimensional space-time

Here, we discuss 2-dimensional space-time, and set the 2-dimensional space-time displacement as $u(\xi) = u(\xi_0, \xi_1)$. Here, $c \equiv 1$. The energy function is defined as

$$E(\tau) = \frac{\partial u(\xi)}{\partial \xi_0} \tag{2.2.13}$$

We can obtain the average of change rate in range $0 - \Phi_0$.

$$E = \bar{E}(\xi) = \frac{1}{\Phi_0}\int_0^{\phi_0} \frac{\partial u(\xi)}{\partial \xi_0} d\tau = \frac{\mathcal{H}}{\Phi_0} \tag{2.2.14}$$

This value is the photon ξ_1-dimensional energy value E, which meets

$$\mathcal{H} = E\Phi_0 \tag{2.2.15}$$

2.2.4 Momentum of 2-dimensional space-time

Define the photon $u(\xi^\mu)$ momentum function as

$$p(\xi) = \frac{\partial u(\xi^\mu)}{\partial \xi^1} \tag{2.2.16}$$

If you take the average of the rate of change in range $0-\Phi$, then

$$p = \bar{p}(\xi) = \frac{1}{\Phi_1}\int_0^{\phi_1} \frac{\partial u(\xi)}{\partial \xi_1} d\xi_1 = \frac{\mathcal{H}}{\Phi_1} \tag{2.2.17a}$$

This is known as the photon momentum, $\mathcal{H} = p\Phi_1$. According to above formulas, get

$$\mathcal{H} = p\Phi_1 = E\Phi_0 \tag{2.2.17b}$$

Here, p is the average value of $p(\xi)$ in the range $0-\Phi_0$. $\mathcal{H}$ is the total deformation maximum value H along the direction of ξ_1. Momentum p can be said as the average value of ξ_1 space deformation rate. Energy E is the average value of intrinsic time deformation rate. If intrinsic range is smaller than photon, ξ_1 momentum and energy is greater. In summary, the 4-momentum function of the photon is

$$p_\mu(\xi^\nu) = \frac{\partial u(\xi^\nu)}{\partial \xi^\mu} = T_{\nu\mu} \tag{2.2.18}$$

Photonic 4-momentum can be uniformly written as

$$p_\mu = \frac{1}{\Phi_\mu}\int_0^{\phi_\mu} \frac{\partial u(\xi^\nu)}{\partial \xi^\mu} d\xi^\mu \tag{2.2.19}$$

This is the component of intrinsic field deformation. p is the average value of $p(\xi)$, so find out $p(\xi)$ is the key to understand the intrinsic structure of the photon.

2.2.5 Establishment of intrinsic displacement function $u(\tau, \xi)$ of 2-dimension

In local coordinate system ξ, a point moves in 1-dimensional generation $u(\xi^i)$, so $\xi^i(\lambda + k) =$

Chapter 2 Structure of the Photons

$\exp[k\xi]\xi^i = \phi(\xi^i)\xi^i$. The intrinsic field function $\phi(\tau,\xi)$ of photon is

$$u(\tau,\xi) = e^{k\xi} \quad (2.2.20a)$$

Because the quantum field has its own dual field, it is necessary to expand the complex space.

$$\varphi(\tau,\xi) = e^{-ik\xi} \quad (2.2.20b)$$

$$u(\tau,\xi) = \exp(-i)(k_0 + k_i\xi_i) = \exp(-i)k_\mu \eta_{\mu\nu}\xi^\nu = \exp[-i(k_0\xi^0 - k_i\xi^i)] \quad (2.2.21)$$

Considering the amplitude u, introduce the operator:

$$\hat{E}(\tau) = -\mathscr{H}\frac{1}{i}c\frac{d}{d\xi^0}, \hat{p}(\xi) = \frac{1}{i}\mathscr{H}\frac{d}{d\xi^i} \quad (2.2.22)$$

Energy eigenvalue equation is

$$\hat{E}(\tau)\phi(\xi) = \mathscr{H}k_0\exp(-i)k_\mu\xi^\mu, \quad k_\mu\xi^\mu = k_0\xi^0 - k_i\xi^i \quad (2.2.23a)$$

Momentum eigenvalue equation is

$$\hat{p}(\xi)\phi(\xi) = k\exp(-i)k_\mu\xi^\mu \quad (2.2.23b)$$

To satisfy equations (2.2.14) and (2.2.17a), it meets

$$k = \frac{\pi}{\Phi}, k_0 = \frac{\pi}{\tau_\Phi} \quad (2.2.24)$$

Here, Φ is the intrinsic space diameter ($\Phi = c\tau_\Phi$). By equation (2.2.21), k_0 can be obtained, and the relationship between P and E is

$$P = \frac{\mathscr{H}}{\pi}k = \bar{\varepsilon}_{11}, E = \frac{\mathscr{H}}{\pi}k_0 = \bar{\varepsilon}_{00} \quad (2.2.25)$$

Above equations correspond to $P_\mu = (h/2\pi)k_\mu = \hbar k_\mu$ in quantum mechanics.

$u(\xi^\mu)$ is considered as a 2-dimensional structure particle. It can be understood as the basic unit of vacuum ϕ_0 moving along ξ^μ, and the amount of movement is k_μ. This will constitute the displacement function

$$u(\xi) = U_\gamma u_0 = \mathscr{H}\exp(-ik_\mu\xi^\mu)\exp(i0) \quad (2.2.26)$$

Here, $\mu = 0,1$. When the basic unit of vacuum is considered as the point, it can be intuitively understood that generated photon is a point moving in the vacuum. Defining the field function, its description basic unit of vacuum is

$$u_0 = \exp(i0) = 1 \quad (2.2.27)$$

It is no observation.

In order to describe the space deformation, use $\cos\beta + i\sin\beta = e^{i\beta}$. The above equation is written as follows, $u(\xi) = \mathscr{H}\exp(-ik_\mu\xi^\mu)$. It can be decomposed and kept real. Get $u(\xi) = \mathscr{H}\cos(-k_\mu\xi^\mu)$. Now we look at specific expressions of intrinsic displacement function $u(\tau,\xi)$. Restricted by the principle of uncertainty, we cannot enter the photon interior, nor can we find a probe smaller than the photon to detect. We can only try to construct photon intrinsic functions that satisfy the experimental observation. Because the vacuum is the elastic field, similarities should exist with the macroscopic world of elastic material, and similarities exist with the wave in elastic medium (Fig. 2.2.1). There is easy deformation for vacuum when it is in the free state, and close to proximity of point o'. Vacuum is hard, and it is difficult for deformation. At point o', vacuum reaches deformation limit, $u(\xi)_{max} = \mathcal{H}$. There is no deformation. Here, use cosine function to construct and express the displacement function u.

In the compression interval:
$$u(-\xi^\mu) = \frac{\mathcal{H}}{2}\cos k_\mu \xi^\mu$$

In the stretch interval:
$$u(+\xi^\mu) = -\frac{\mathcal{H}}{2}\cos k_\mu \xi^\mu \qquad (2.2.28)$$

The total deformation value $\mathcal{H}$ of ξ^1 dimension space-time of vacuum is
$$u(\xi^\mu) = \mathcal{H}\left[\int_{-\Phi/2}^{0}(-k_\mu \sin k_\mu \xi^\mu)\,\mathrm{d}\xi^\mu + \int_{0}^{\Phi/2}(-k_\mu \sin k_\mu \xi^\mu)\,\mathrm{d}\xi^\mu\right] = \mathcal{H} \qquad (2.2.29)$$

It is noteworthy that the time dimension and space dimension of the deformation are separately considered. The space-time displacement function is
$$u(\xi^\mu) = \mathcal{H}\cos k_\mu \xi^\mu \qquad (2.2.30)$$

The 2-dimensional quantum field displacement function is $u(\xi) = \mathcal{H}\cos[i(k_0\xi^0 - k_1\xi^1)]$.

2.2.6 2-dimensional quantum intrinsic field function and its wave function

Space-time strain is directly defined as the field function in ξ coordinate system, $\varphi(\xi^\mu) = \exp\{i[\varepsilon_{\mu\nu}][g_{\mu\nu}](\xi^\mu)\}$.

$$\varphi(\xi^0,\xi^1) = \exp\left(-i\begin{bmatrix}\varepsilon_0(\xi) & 0 \\ 0 & \varepsilon_1(\xi)\end{bmatrix}\begin{bmatrix}1 & 0 \\ 0 & -1\end{bmatrix}\begin{bmatrix}\xi^0 & 0 \\ 0 & \xi^1\end{bmatrix}\right), \int_{-R}^{+R}\varepsilon(\xi)\,\mathrm{d}\xi = \mathcal{H}$$
$$(2.2.31)$$

The displacement function is
$$u(\xi) = \int_{-R}^{+R}[\varepsilon_{ij}(\xi)]\,\mathrm{d}V\mathrm{Re}[\varphi(\xi)] \qquad (2.2.32)$$

Quantum wave function is
$$\Psi(\xi) = \underbrace{\int_{-R}^{+R}[\varepsilon_{ij}(\xi)]\,\mathrm{d}V}_{\text{Wave amplitude}}\varphi(\xi) \qquad (2.2.33)$$

The displacement function of 4-dimensional quantum field in intrinsic space-time is
$$u(\xi) = \int_{-R}^{+R}[\varepsilon_{ij}(\xi)]\,\mathrm{d}V\exp\{[\varepsilon_{\mu\nu}][g_{\mu\nu}](\xi^\mu)\}, \mathrm{d}V = \mathrm{d}\xi^1\mathrm{d}\xi^2\mathrm{d}\xi^3 \qquad (2.2.34)$$

The standard expression of 4-dimensional quantum field in observational space-time is
$$\varphi(x) = \exp i[E_{\alpha\beta}(\xi)L_\nu^\mu x^\nu] \qquad (2.2.35)$$

This function includes all the information:
$$E_{\alpha\beta}(\xi) = \varepsilon_{\alpha\beta}(\xi) + \Delta\varepsilon_{\alpha\beta}(\xi) + \mathscr{g}_{\alpha\beta}(\xi)$$

Here, $\Delta\varepsilon_{\alpha\beta}(\xi)$ is the strain of background spatiotemporal; $\mathscr{g}_{\alpha\beta}(\xi)$ is the strain caused by its self-strain on the surrounding space-time; L_ν^μ is Lorentz transformation matrix. The reference system changes from static state to wave system at the speed of v.

2.3 Intrinsic structure of photon

Here, the intrinsic structure of particles in new theory can be seen as neighborhood of a point (Fig. 2.3.1). To establish cartesian coordinate frame at point O in intrinsic space, that is a static reference frame. Around this point, the arrangement of basic unit of vacuum is changed when point O moves to O', along ξ^1 direction stretching or compression reach to limit, i.e. $OO' = \mathscr{H}/2$. It begins propagating along direction ξ^1, and forming photon. The intrinsic coordinate frame on photon center can simplify the problem.

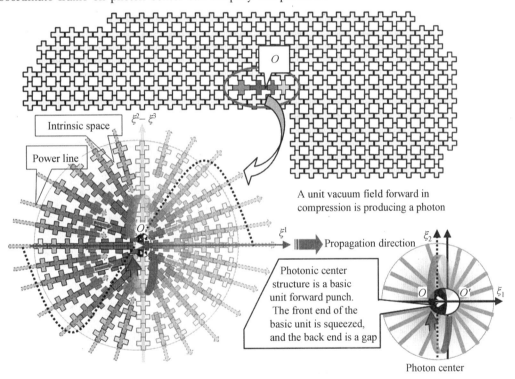

Fig. 2.3.1 The fibre structure of photon

Vacuum in flat or natural state has no deformation. If there is a strong disturbance somewhere in the vacuum, if it makes a basic unit to move forward. And it will compress the front vacuum, stretch the back vacuum, and form longitudinal wave. The stretch quantity equals to compress quantity. R indicates the deformation regional of sphere radius, and the outside is without deformation. $2R$ is the diameter of the photon, and it is also the photon longitudinal wavelength λ.

The movement of point O will result in extremely bent near the center of the field. According to the basic assumption, we know that the limit of stretching and compression exist in the vacuum. The extreme bending leads to the field deformation limit. When the bending curvature of vacuum is excessive, the tearing appears, which produces the fiber structure in vacuum, and the fiber structure is defined as the electric flux line. Electric flux line constitutes the electric field.

The longitudinal wave of overall effect of fiber structure is called photons, which is caused by the basic unit of vacuum moving. The deformation of vacuum constitutes displacement function $u(\xi)$, and the intrinsic space $\xi^\mu (\mu=0,1,2,3)$ is 4-dimensional space-time. When $u(\xi)=0$, the corresponding vacuum deformation is zero, such as the vacuum ground state field.

In order to have a visual image of the particle intrinsic space, we imagine a large piece of sponge (Fig. 2.3.2), which has a small ball. The small ball makes sponge occur deformation, and then a small concave surface appears. There is no deformation outside the concave. The deformation regions are called internal space or intrinsic space. Similarly, the whole region of vacuum deformation portion is composed of a single quantum field. Deformation area radius R is quantum field intrinsic space radius R. Here we just roughly introduce the photonic new image. Then analysis the image specifically.

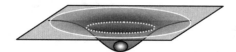

Fig. 2.3.2 Schematic intrinsic 2-dimensional space

The strain tensor of the photon $\varepsilon(X)$ is composed of two parts. The first part is the internal strain $\varepsilon(\xi)$, and the second part is strain $\varepsilon(x)$ in the observed space, including four momentum and electric flux lines strain. The electric flux line strain tensor derives from the fiber structure shifts, which is caused by internal strain.

$$\varepsilon(X) = \varepsilon(\xi) + \varepsilon(x) \qquad (2.3.1)$$

$$[\varepsilon]_{\text{Photon}} = \underbrace{\left[\frac{\partial u_\mu}{\partial \xi_\nu}\right]}_{\substack{\text{Microcosmic space-time} \\ \text{rupture region}}} + \underbrace{\left[\frac{\partial A_\mu}{\partial x_\nu}\right]}_{\substack{\text{The fiber field in the} \\ \text{observational space}}}$$

$$= \underbrace{\varepsilon_{\mu\mu}(\xi)}_{\text{4-momentum}} + \frac{1}{2}\underbrace{\gamma_{\mu\nu}(\xi)}_{\text{Weakly acting strain}} + \frac{1}{2}\underbrace{\omega_{\mu\nu}(\xi)}_{\text{Spin}} + \underbrace{\Delta\varepsilon_{\mu\mu}(x)}_{\substack{\text{Electromagnetic field of} \\ \text{scalar and vector potential}}} + \frac{1}{2}\underbrace{\Omega_{\mu\nu}(x)}_{\text{Electromagnetic field}}$$

$$(2.3.2)$$

Photon field function is

$$\varphi_{\text{Photon}}(\xi,x) = \exp(-\mathrm{i})\{[\varepsilon_{\mu\mu} + \frac{1}{2}\gamma_{\mu\nu} + \frac{1}{2}\omega_{\mu\nu}](\xi) + [\Delta\varepsilon_{\mu\mu} + \frac{1}{2}\Omega_{\mu\nu}](x)\}$$

$$(2.3.3)$$

The above field function contains all the information of photons.

2.3.1 The photonic internal field strain

Firstly, we discuss the intrinsic strain

Chapter 2 Structure of the Photons

$$\varepsilon_{\text{Internal}}(\xi) = \begin{bmatrix} \varepsilon_{00} & 0 & 0 & 0 \\ 0 & \varepsilon_{11} & 0 & 0 \\ 0 & 0 & \varepsilon_{22} & 0 \\ 0 & 0 & 0 & \varepsilon_{33} \end{bmatrix} + \frac{1}{2}\begin{bmatrix} 0 & \gamma_{01} & \gamma_{02} & \gamma_{03} \\ \gamma_{10} & 0 & \gamma_{12} & \gamma_{13} \\ \gamma_{20} & \gamma_{21} & 0 & \gamma_{23} \\ \gamma_{30} & \gamma_{31} & \gamma_{32} & 0 \end{bmatrix} + \frac{1}{2}\begin{bmatrix} 0 & \omega_{01} & \omega_{02} & \omega_{03} \\ \omega_{10} & 0 & \omega_{12} & \omega_{13} \\ \omega_{20} & \omega_{21} & 0 & \omega_{23} \\ \omega_{30} & \omega_{31} & \omega_{32} & 0 \end{bmatrix}$$

(1) The 4-momentum p of photon (2) Photonic power lines shear strain constitute the weak field (3) Photonic rotational strain constitute photon-spin

$$= \underset{\text{4-momentum}}{\varepsilon_{\mu\mu}} + \frac{1}{2}\underset{\text{Weakly acting strain}}{\gamma_{\mu\nu}} + \frac{1}{2}\underset{\text{Spin}}{\omega_{\mu\nu}} \qquad (2.3.4)$$

We can establish spherical coordinates in photonic intrinsic space (Fig. 2.3.3) to describe the electric flux line of photon by displacement function $u(\theta,\varphi,r)$. Single electric flux line is the most basic unit of electromagnetic field. The relationship between spherical coordinates and rectangular coordinates are

$$\begin{cases} (\xi^1)^2 + (\xi^2)^2 + (\xi^3)^2 = R^2 \\ \xi^1 = R\sin\theta\cos\phi \\ \xi^2 = R\sin\theta\sin\phi \\ \xi^3 = R\cos\theta \end{cases} \qquad (2.3.5)$$

$$[\varepsilon](\tau,r,\theta,\phi) = \left\{ -\frac{1}{2}\begin{bmatrix} \varepsilon_\tau & \gamma_{\tau r} & \gamma_{\tau\theta} & \gamma_{\tau\phi} \\ \gamma_{r\tau} & \varepsilon_r & \gamma_{r\theta} & \gamma_{r\phi} \\ \gamma_{\theta\tau} & \gamma_{\theta r} & \varepsilon_\theta & \gamma_{\theta\phi} \\ \gamma_{\phi\tau} & \gamma_{\phi r} & \gamma_{\phi\theta} & \varepsilon_\phi \end{bmatrix} + \frac{1}{2}\begin{bmatrix} \varepsilon_\tau & \gamma_{\tau r} & \gamma_{\tau\theta} & \gamma_{\tau\phi} \\ \gamma_{r\tau} & \varepsilon_r & \gamma_{r\theta} & \gamma_{r\phi} \\ \gamma_{\theta\tau} & \gamma_{\theta r} & \varepsilon_\theta & \gamma_{\theta\phi} \\ \gamma_{\phi\tau} & \gamma_{\phi r} & \gamma_{\phi\theta} & \varepsilon_\phi \end{bmatrix} \right\}\begin{bmatrix} c\tau & r & \theta & \phi \\ c\tau & r & \theta & \phi \\ c\tau & r & \theta & \phi \\ c\tau & r & \theta & \phi \end{bmatrix}$$

Front-end hemisphere, compression zone 4-momentum, weak field, spin, charge   Posterior hemisphere, stretch zone, 4-momentum, weak field, spin, charge

(2.3.6)

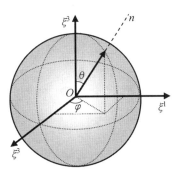

Fig. 2.3.3 Spherical coordinates

We consider that the total deformation of all the electric flux line for the photon ball. Set the total electric flux line of photon, which equals to e, and the cross section of the electric flux line is $e/4\pi r^2$. According to conservation of space-time strain, $\Delta V_{\text{Time}} \equiv \Delta V_{\text{Space}}$. The electron and the lepton are very similar in structure. Propagating along ξ_3 direction, consider strain distribution along the propagation direction. So take $\cos\varphi$ on the basis of the lepton, the total deformation of photon field (only consider the absolute value) is

$$u_r = \frac{e\mathcal{H}}{8\pi r^2}\cos\frac{\pi r}{\Phi} \quad (0 \leqslant r \leqslant R) \qquad (2.3.7)$$

Taking account of spin:

$$u_\phi = \frac{\mathscr{H}}{2}\cos\frac{\varphi}{2}\cos\varphi \quad (-\frac{\pi}{2} \le \varphi \le \frac{\pi}{2})$$

$$u_\theta = \frac{\mathscr{H}}{2}\sin\theta \quad (0 \le \theta \le 2\pi)$$

$$\varepsilon_r = \frac{\partial u_r}{\partial r} = -\mathscr{H}\frac{\pi}{\Phi}\sin\frac{\pi r}{\Phi}$$

$$\varepsilon_\varphi = \frac{1}{r}\frac{\partial u_\varphi}{\partial \varphi} + \frac{u_r}{r} = \frac{\mathscr{H}}{r}(\cos\frac{\pi r}{\Phi} - \frac{1}{2}\sin\frac{\varphi}{2}\cos\varphi - \sin\varphi\cos\frac{\varphi}{2}) \quad (2.3.8)$$

$$\varepsilon_\theta = \frac{1}{r\sin\varphi}\frac{\partial u_\theta}{\partial \theta} + \frac{u_r}{r} + \frac{\cot\varphi}{r}u_\varphi = \frac{\mathscr{H}}{r}\left(\cos\frac{\pi r}{\Phi} + \cot\varphi\cos\frac{\varphi}{2}\cos\varphi + \frac{\cos\theta}{\sin\varphi}\right)$$

$$\begin{cases}\gamma_{r\varphi} = \frac{1}{r}\frac{\partial u_r}{\partial \varphi} + \frac{\partial u_\varphi}{\partial r} - \frac{u_\varphi}{r} = -\frac{\mathscr{H}}{r}\cos\frac{\varphi}{2}\cos\varphi \\ \gamma_{\varphi\theta} = \frac{1}{r\sin\varphi}\frac{\partial u_\varphi}{\partial \theta} + \frac{1}{r}\frac{\partial u_\theta}{\partial \varphi} - \frac{\cot\varphi}{r}u_\theta = -\frac{\mathscr{H}}{r}\cot\varphi\sin\theta \\ \gamma_{\theta r} = \frac{\partial u_\theta}{\partial r} + \frac{1}{r\sin\varphi}\frac{\partial u_r}{\partial \theta} - \frac{u_\theta}{r} = -\frac{\mathscr{H}}{r}\sin\theta\end{cases} \quad (2.3.9)$$

Planck constant theory has a very clear physical meaning in vacuum, which is unified as the total deformation of single photon.

$$\Omega_{\text{Sphere}} = \int_0^{2\pi} d\varphi \int_0^\pi d\theta \int_0^R F(r,\theta,\varphi)r^2\sin\theta dr \quad (2.3.10)$$

Considering integral convenience, the range of φ is: $-\frac{\pi}{2} \le \varphi \le \frac{\pi}{2}, 0 \le \theta \le 2\pi$. The tensile strain and the compression strain are not distinguished. Only consider the distribution of strain

$$p_r = \frac{\partial u_r}{\partial r} = -\frac{e\mathscr{H}\pi}{8\pi r^2\phi}\sin\phi\cos\phi\cos\frac{\theta}{2}\sin\theta\sin\left(\frac{\pi r}{\phi}\right) = F(r,\theta,\phi) \quad (2.3.11)$$

$$h = -\int_{-\pi/2}^{\pi/2}\sin\varphi\cos\varphi d\varphi\int_0^{2\pi}\cos\frac{\theta}{2}\sin\theta d\theta\int_0^R\frac{e\pi}{4\pi r^2\phi}\left(\frac{\mathscr{H}}{2}\right)^3\sin\left(\frac{\pi r}{\phi}\right)r^2 dr$$

$$= \int_0^{\pi/2}\frac{1}{2}\sin 2\varphi d2\varphi \cdot \frac{8}{3} \cdot \frac{e}{4\pi}\left(\frac{\mathscr{H}}{2}\right)^3\int_0^R\left[-\sin\left(\frac{\pi r}{\phi}\right)\right]d\left(\frac{\pi r}{\phi}\right) = \frac{2}{3}e\mathscr{H}$$

$$h = e\left(\frac{2\mathscr{H}}{3}\right) \quad (2.3.12)$$

Tensile field and compressed field deformation is equal and opposite in direction. The total deformation of photon $h = h_p/2$ is half of the total deformation of the lepton.

2.3.2 Displacement function of electric flux line in photon single internal area

Considering the characteristics of quantum field vacuum fibrosis, we can merge ε_θ and ε_ϕ into a single fiber.

Chapter 2 Structure of the Photons

$$\frac{\mathscr{H}}{R} = \frac{\mathscr{H}_{\text{Radial}}}{R} + \frac{\mathscr{H}_{\text{Ring}}}{R} = |\bar{\varepsilon}_r| + |\bar{\varepsilon}_\theta + \bar{\varepsilon}_\phi| \qquad (2.3.13)$$

The circumferential strain ε_θ and ε_ϕ of a single fiber are incorporated into a single fiber volume strain in the intrinsic space. Considering the space structure, it can be obtained as

$$\begin{bmatrix} \varepsilon_\tau & 0 & 0 & 0 \\ 0 & \varepsilon_r & 0 & 0 \\ 0 & 0 & \varepsilon_\theta & 0 \\ 0 & 0 & 0 & \varepsilon_\phi \end{bmatrix} \begin{bmatrix} c\tau & 0 & 0 & 0 \\ 0 & r & 0 & 0 \\ 0 & 0 & \theta & 0 \\ 0 & 0 & 0 & \phi \end{bmatrix} \Rightarrow \begin{bmatrix} \varepsilon_\tau & 0 & 0 & 0 \\ 0 & \varepsilon_r & 0 & 0 \\ 0 & 0 & 0 & 0 \\ 0 & 0 & 0 & 0 \end{bmatrix} \begin{bmatrix} c\tau & 0 & 0 & 0 \\ 0 & r & 0 & 0 \\ 0 & 0 & 0 & 0 \\ 0 & 0 & 0 & 0 \end{bmatrix}$$

$$(2.3.14)$$

The fiber structure of photon is composed by two segment fibers which are the internal area electric flux line and electric flux line. The internal area electric flux line is the circumferential strain lines within the photonic intrinsic space. It's the front end of electric flux line. The shift of 1-dimensional fiber generates the electric flux line, and it's an infinite ray. Let's start by discussing the internal area electric flux line.

Photonic center structure is a basic unit forward punch. The front end of the basic unit is squeezed, and the back end is a gap or can be seen as a tiny hole (Fig. 2.3.1), which is equivalent superposition state of positive and negative charge. In this way, there is a hole in the front of the stretching area of the momentum, which is in the front of the stretching area, and compression zone front end is compressed. The length of the internal area electric flux line has been introduced, which is the change of the volume of the momentum fiber. The radial volumetric deformation is larger than that of the ring. The photon intrinsic radius is smaller, and the circumferential volume deformation is greater. The actual photon single fiber deformation is shown in Fig. 2.3.1.

Without considering the deformation of the shape, only 1-dimensional variable is used to characterize the single fiber volume of the different forms. Due to the fiber deformation limit in intrinsic space, we call the simplified fiber which length is R (The intrinsic space radius), as the internal area electric flux line.

When the point O along ξ^1 axis move $\mathscr{H}/2$, ξ^1-axis can be regarded as a 1-dimensional vacuum, now O' works as the center of sphere, $\xi_o^1 - \xi_{o'}^1 = \mathscr{H}/2$, considering in stretching district $0 - R$. The displacement function of the single internal area electric flux line is

$$u_{r(\text{single})}(\xi^1) = \frac{\mathscr{H}}{2}\cos\frac{\pi\xi^1}{\phi} \quad (-R \leqslant \xi' \leqslant 0)$$

$$u_{r(\text{single})}(\xi^1) = -\frac{\mathscr{H}}{2}\cos\frac{\pi\xi^1}{\phi} \quad (R \geqslant \xi' \geqslant 0) \qquad (2.3.15)$$

Deformation is in the intrinsic space, so it is necessary to introduce θ function, and the above equation can be written as

$$u_{r(\text{single})}(\xi^1) = \theta(0 - \xi^1)\frac{\mathscr{H}}{2}\cos\frac{\pi\xi^1}{\phi} - \theta(\xi^1 - 0)\frac{\mathscr{H}}{2}\cos\frac{\pi\xi^1}{\phi} \quad (\xi^1 \leqslant |R|)$$

$$(2.3.16)$$

Here, $\theta(\xi) = \begin{cases} 1 & (\xi > 0) \\ 0 & (\xi < 0) \end{cases}$. Without deformation, $\mathcal{H}$ is zero on $\xi^1\xi^2$ surface. Considering in this surface, when $\varphi = 0$, the deformation along ξ^1 is $\mathcal{H}/2$, and the distribution of $\mathcal{H}$ shall satisfy

$$u_{r(\text{single})}(r,\varphi) = \theta(\varphi - \pi)\frac{\mathcal{H}}{2}\cos\varphi\cos\frac{\pi r}{\phi} - \theta(\pi - \varphi)\frac{\mathcal{H}}{2}\cos\varphi\cos\frac{\pi r}{\phi} \quad (r \leq |R|) \tag{2.3.17}$$

Considering θ direction, $\theta = 0$, $\mathcal{H} = 0$. In the sphere coordinates (deformation of space):

$$u_{r(\text{single})}(r,\varphi) = \theta(\varphi - \pi)\theta(\vartheta - \pi)\frac{\mathcal{H}}{2}\sin\vartheta\cos\varphi\cos\frac{\pi r}{\phi}$$
$$- \theta(\pi - \varphi)\theta(\pi - \vartheta)\frac{\mathcal{H}}{2}\sin\vartheta\cos\varphi\cos\frac{\pi r}{\phi} \quad (r \leq |R|; \vartheta,\varphi \leq 2\pi) \tag{2.3.18}$$

This is formulation of a single internal area electric flux line of photon in the deformation zone. R is intrinsic spatial radius, and $r_0 = c\tau$, $\xi = (-\xi_0, \xi_1, \xi_2, \xi_3)$, and the deformation of space-time is

$$u_{r(\text{single})}(r,\tau,\theta,\varphi) = \theta(\varphi - \pi)\theta(\vartheta - \pi)\frac{\mathcal{H}}{2}\sin\theta\cos\varphi\cos\left(\frac{\pi r}{\phi} - \frac{\pi\tau}{\phi_0}\right)$$
$$- \theta(\pi - \varphi)\theta(\pi - \vartheta)\frac{\mathcal{H}}{2}\sin\theta\cos\phi\cos\left(\frac{\pi r}{\phi} - \frac{\pi\tau}{\phi_0}\right) \quad (r \leq |R|; \vartheta,\varphi \leq 2\pi)$$
$$\tag{2.3.19}$$

Here, $\phi_0 = 2R/c$. Considering the space-time deformation, we can use Euler formula, which can be written as

$$u_{r(\text{single})}(\tau,\varphi,\theta) = [\theta(\varphi - \pi)\theta(\vartheta - \pi) - \theta(\pi - \varphi)\theta(\pi - \vartheta)]\frac{\mathcal{H}}{2}\sin\theta\cos\varphi\exp[i(kr - k_0\tau)]$$
$$(r \leq |R|; \quad \vartheta,\varphi \leq 2\pi) \tag{2.3.20}$$

Here, $k = \pi/\phi$, $k_0 = \pi/\phi_\tau$. This is the inner field displacement function. Space-time deformation constitutes intrinsic field function.

2.3.3 Electric flux line

1. The structure of the electric flux line

The field near compression zone O', the basic units are arranged and composed of a fibrosis field compressed on the $\xi^2\xi^3$ surface, but this deformation is less than the deformation which is along the direction ξ^1, so the vacuum along the direction ξ^1 to move forward a tiny amount a^1 (Fig. 2.3.4(b)). It does not change the 3-dimensional characteristics of basic unit of electric flux line. It's closely linked to the basic unit of vacuum, which will move forward a small amount a, so that it continues to form an electric flux line in the direction ξ^1. Compression electric flux line is formed in the compression zone. The stretching electric flux line is formed in the stretching area (For contrast electric lines, the mechanism is the same).

Here, we also note that the deformation of electric flux line with independence in intrinsic

Chapter 2　Structure of the Photons

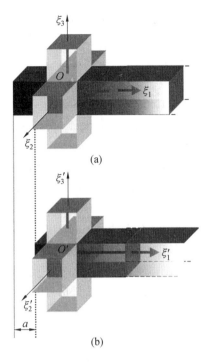

Fig. 2.3.4　Formation of the electric flux line

space consists by ξ^1, ξ^2, ξ^3 3-dimensional field. Along 1-dimensional vacuum ξ^1, the offset of a small amount a does not change the nature of ξ^2 and ξ^3. They are independent of each other, i. e. vacuum has independence of dimensions. The amount of stretching fiber field (negative electric field electric flux line) and the amount of the compression fiber field (positive electric field electric flux line) are strictly equal. The electric flux line is the product of the vacuum strain in the quantum field (Fig. 2.3.5).

$$u_\mu(\xi) = u_{\mu,\text{intrinsic}}(\xi) + a_{\mu,\text{Electric flux line}}(x^\mu) \qquad (2.3.21)$$

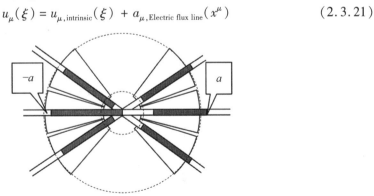

Fig. 2.3.5　Diagram of electric flux line displacement

The electric flux line can be understood the effect produced by flux line dimension shifting a^μ. It is a tiny quantity. The amount of shifting is the additional effect produced by the strain of the quantum fiber. For any single photon, a^μ is a constant, but from the point of view of the whole space area, the density of a^μ distribution in space is different. The smaller the radius is, the bigger the density is. So from the point of view, the whole space a^μ is a variable $a^\mu(r)$. The electromagnetic field of photons is a physical effect generated by the movement of quantum strain dimensions.

For electric flux line of photon, the characteristics are as follows.

(1) The basic unit of vacuum compression then will move forward a small amount $\mathscr{H}_{\text{Radial}}$, and its closely linked to the basic unit of vacuum is also followed to move a small amount $\mathscr{H}_{\text{Radial}}$. So it continues to form a ray of electric flux line. Since three dimensions are independent, the minimal dimension shuttle does not change the nature of the others dimensions, and there is no effect in vacuum, thus the photon existence in the surrounding space has no effect.

(2) From the macroscopic point of view, because the vacuum is elastic, the small movement of the basic unit of vacuum will gradually attenuate, and the length of electron flux line is less than the radius of particle gravity.

(3) When the fiber field meets the anti-fiber field, there is a restored flat trend in vacuum and trying to make the two together which is known as the electric field.

(4) In the inner region of the electron, the initial end of the electric flux line reaches the deformation limit, while the electric flux line of photon does not reach the deformation limit except for the propagation axis. $\mathscr{H}_{\text{Radial}}$ is proportional to the electric flux line radial strain rate. But the total radial strain of photon is a constant ($\mathscr{H}_{\text{Radial}}$ = constant).

2. The electric flux line of internal area

The radial shifting of the strain in the photon intrinsic space constitutes the photon electric flux line (Fig. 2.3.6). The offset of single electric flux line is a_r, and it's a tiny invariance. The total offset of electric flux line is

$$A_\gamma = ea_r \qquad (2.3.22)$$

It is an invariant, which is reinterpreted as the vacuum strain of photons.

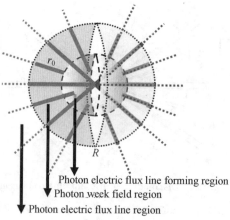

Photon electric flux line forming region
Photon week field region
Photon electric flux line region

Fig. 2.3.6 The partition of photonic force lines

(1) Displacement function in the internal formation region of electric flux line of photonic

Chapter 2 Structure of the Photons

$r_0 \leqslant r \leqslant R$.

The total number of single photon fiber is e in closed spherical ($4\pi r^2$), and the two hemispheres occasions as one by using step function $\theta(x) = \begin{cases} 1 & (x > 0) \\ 0 & (x < 0) \end{cases}$. In the internal area, the number of electric flux lines is directly proportional to the area, $n = e \dfrac{4\pi r^2}{4\pi r_0^2}$. The field intensity of the electron flux line in the internal formation region of photon is

$$\frac{A_\gamma(\xi,\theta,\varphi)}{\Delta S} = [\theta(\varphi - \pi)\theta(\vartheta - \pi) - \theta(\pi - \varphi)\theta(\pi - \theta)]\frac{er^2}{r_0^2}a_r \times \sin\theta\cos\varphi$$

$$(r \leqslant |R|; \vartheta, \varphi \leqslant 2\pi) \qquad (2.3.23)$$

Here, r_0 is radius of formation region. Photon internal space is ball structure $\xi^1 i + \xi^2 j + \xi^3 k = r$.

(2) Electric flux line in photon internal space $r_0 \leqslant r \leqslant R$.

The field intensity of electric flux line in photon internal space is

$$\frac{A_\gamma(\xi,\theta,\varphi)}{\Delta S} = [\theta(\varphi - \pi)\theta(\vartheta - \pi) - \theta(\pi - \varphi)\theta(\pi - \theta)]\frac{er}{4\pi r^3}a_r\sin\theta\cos\varphi$$

$$(r \leqslant |R|; \vartheta, \varphi \leqslant 2\pi) \qquad (2.3.24)$$

(3) Photon electric flux line area $R \leqslant r \leqslant \infty$.

Photon electric flux line is the fiber which has 1-dimensional offset a_r. The electric field is a long-range force. The displacement function of the electric flux line is represented by $A_\gamma(\xi, \theta, \phi, \tau)$, $r > R$. Outside the photon interior space, the momentum of electric flux line is zero. Since the electric flux line is the dimension offset, itself has no deformation and has no effect on observation space-time. The total number of photon electric flux lines is e. Photonic electric field intensity outside photon interior space is

$$\frac{\Delta u_\gamma(\xi,\theta,\varphi)}{\Delta S} = [\theta(\varphi - \pi)\theta(\vartheta - \pi) - \theta(\pi - \varphi)\theta(\pi - \theta)]\frac{er}{4\pi r^3}a_r\sin\theta\cos\varphi$$

$$(r > |R|; \vartheta, \varphi \leqslant 2\pi) \qquad (2.3.25)$$

$$A_\gamma = \frac{1}{2}(+A_{\gamma/2}) + \frac{1}{2}(-A_{-\gamma/2}) \equiv 0 \qquad (2.3.26)$$

Photon fiber field total strain is conservation.

With fiber characteristic, thus photons of ball structure is simplified to a 1-dimensional structure (Fig. 2.3.7). When we consider the fiber structure, photons electric flux line constitutes by e lines, the average value is $\dfrac{\mathscr{H}}{2} = \sum_{i=1}^{e} \dfrac{\mathscr{H}_i(\theta,\varphi)}{2e}$.

(1) $\varepsilon_{00} \equiv E$ is the total energy of photon $E = h\nu$, $E\Phi_0 = h$. $\varepsilon_{ii} \equiv P_i$ is 3-momentum of photon, and $p = h\lambda$, $p\Phi = h$. $A_\mu(\xi) = \dfrac{e\mathscr{H}}{4}\exp\left(\dfrac{\pi}{r_\mu}\xi^\mu\right)$. Considering integral results, the total energy of the single photon can be obtained.

$$E_{\text{EH}} = \frac{\int_0^{r_0} A_0 d\xi^0}{\phi_0} = \frac{h}{\phi_0} = \hbar\omega \qquad (2.3.27)$$

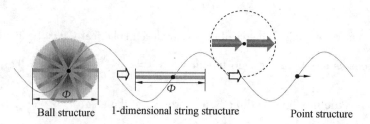

Fig. 2.3.7 Simplified photons

(Photon sphere structure simplified to a 1-dimensional string structure)

For $\sum_{i=1}^{e}[R_i \cdot p_i(\theta,\varphi)]$ and an average p, it can also be found that a photon R has nothing to do with θ and ϕ. P_{Photon} and E_{Photon} are the total momentum and total energy of a photon, and h is total deformation, further simplifying

$$h = P_{\text{Photon}} \cdot \Phi = E_{\text{Photon}} \cdot \Phi_0 \qquad (2.3.28)$$

This is the expression of the simplified photon images. Interestingly, the total number of electric flux lines is independent of the size of photosphere, and the total electric flux line is a constant. Formula (2.3.28) is called the definition of Photon Planck constant. $h = \lambda \cdot p = T \cdot E$ and $T = 1/\nu$ form vibration cycle.

(2) In the existing theory, it is stipulated that the direction of photon propagation is x^3. ε_{11} and ε_{22} are transverse momentum of photon p_1 and p_2. This makes photon has shear wave characteristics, photonic transverse fluctuation and probability wave.

(3) $\varepsilon_{33} = p_3$ is the longitudinal momentum of the photon, which constitutes the photon longitudinal wave.

Photon space-time displacement constitutes a longitudinal wave of single photon in the vacuum medium whose vibration direction ξ^3 is the direction of propagation. This longitudinal wave is not directly observable. Photon spherical radius R shows the deformation of Photon. Photon longitudinal displacement wave function is

$$u_r(\xi,\varphi,\theta) = \frac{e}{4\pi r^2}\frac{\mathscr{H}}{2}\sin\theta\cos\varphi\exp[i(k_i\xi^i)] \quad (r \leq |R|; \theta,\varphi \leq 2\pi) \qquad (2.3.29)$$

Photon longitudinal wave deformation amplitude is $\frac{e}{4\pi r^2}\frac{\mathscr{H}}{2}\sin\theta\cos\phi$ at r, is 1-dimensional simplify structure of the electric flux line in the photonic interior space is enclosed in the photonic spherical shell, the total deformation of photon constitute the amplitude h. Here, $k = k_1 + k_2 + k_3$. Photonic intrinsic field function in intrinsic coordinate system is $u_r(\xi) = e\left(\frac{2\mathscr{H}}{3}\right)\exp[i(k_0\xi^0 - k\xi)]$. Without any disturbance in the ideal case, photon propagation is along ξ_3-direction, and the photonic intrinsic longitudinal wave function is

$$u_{\text{Longitudinal}}(\xi) = h\exp\left[i\left(\frac{\pi}{\Phi_0}\tau - \frac{\pi}{\Phi}\xi_3\right)\right] \qquad (2.3.30)$$

Chapter 2 Structure of the Photons

It limits in intrinsic space, there is no-observability, corresponding to longitudinal photon or virtual photon in our present quantum field theory. In this image, the photon momentum has a very clear physical meaning which is the average deformation within longitudinal wave wavelength.

Photon is larger, average deformation is smaller and momentum is lower. Photon is smaller, and the momentum is greater, $h = p \cdot \Phi$. Here, Φ is the diameter of intrinsic photon sphere.

①ε_{0i} is flux of photon energy through the surface ξ^i.

②For free photon ε_{0i} and ε_{i0}, due to the shear strain is limited in intrinsic space, observation in outer space x^i is zero.

③$\varepsilon_{ij}(i \neq j)$ is shear strain of photon momentum which causes the photonic ring is breaking to form the structure of fiber, constituting a weak field B.

④ε'_{12} and ε'_{21} are the spin of photon which respectively are the positive and reverse rotation with constituting the magnetic field.

⑤ε'_{3j} and ε'_{i3} are longitudinal rotation. Due to photon along the propagation direction reach ultimate strain, it is impossible to somersault along the propagation direction which has no physical effects.

The shear strain of photons is

$$(\varepsilon)(\tau,r,\theta,\varphi) = \left\{ -\frac{1}{2}\begin{bmatrix} \varepsilon_\tau & \gamma_{\tau r} & \gamma_{\tau\theta} & \gamma_{\tau\varphi} \\ \gamma_{r\tau} & \varepsilon_r & \gamma_{r\theta} & \gamma_{r\varphi} \\ \gamma_{\theta\tau} & \gamma_{\theta r} & \varepsilon_\theta & \gamma_{\theta\varphi} \\ \gamma_{\varphi\tau} & \gamma_{\varphi r} & \gamma_{\varphi\theta} & \varepsilon_\varphi \end{bmatrix}_{\substack{\text{Front-end hemisphere, compression zone}\\ \text{4-momentum, weak field, spin, charge}}} + \frac{1}{2}\begin{bmatrix} \varepsilon_\tau & \gamma_{\tau r} & \gamma_{\tau\theta} & \gamma_{\tau\varphi} \\ \gamma_{r\tau} & \varepsilon_r & \gamma_{r\theta} & \gamma_{r\varphi} \\ \gamma_{\theta\tau} & \gamma_{\theta r} & \varepsilon_\theta & \gamma_{\theta\varphi} \\ \gamma_{\varphi\tau} & \gamma_{\varphi r} & \gamma_{\varphi\theta} & \varepsilon_\varphi \end{bmatrix}_{\substack{\text{Posterior hemisphere, stretch zone}\\ \text{4-momentum, weak field, spin, charge}}} \right\} \begin{bmatrix} c\tau \\ r \\ \theta \\ \varphi \end{bmatrix}$$

(2.3.31)

Go to the Cartesian coordinate system. The bare mass of photon is

$$\gamma_{\mu\nu}(\xi) = \underbrace{\begin{bmatrix} 0 & \gamma_{01} & \gamma_{02} & \gamma_{03} \\ \gamma_{10} & 0 & \gamma_{12} & \gamma_{13} \\ \gamma_{20} & \gamma_{21} & 0 & \gamma_{23} \\ \gamma_{30} & \gamma_{31} & \gamma_{32} & 0 \end{bmatrix}}_{\text{Photonic power lines shear strain constitute the weak field}}$$

$$= \left\{ \underbrace{\begin{bmatrix} 0 & \gamma_{01} & \gamma_{02} & \gamma_{03} \\ 0 & 0 & 0 & 0 \\ 0 & 0 & 0 & 0 \\ 0 & 0 & 0 & 0 \end{bmatrix}}_{\substack{\text{The energy flux of photon}\\ \text{through the surface } \xi^i}} + \underbrace{\begin{bmatrix} 0 & 0 & 0 & 0 \\ \gamma_{10} & 0 & 0 & 0 \\ \gamma_{20} & 0 & 0 & 0 \\ \gamma_{30} & 0 & 0 & 0 \end{bmatrix}}_{\substack{\text{The momentum flux}\\ \text{through the surface } \xi^i}} + \underbrace{\begin{bmatrix} 0 & 0 & 0 & 0 \\ 0 & 0 & \gamma_{21} & \gamma_{31} \\ 0 & \gamma_{12} & 0 & \gamma_{32} \\ 0 & \gamma_{13} & \gamma_{23} & 0 \end{bmatrix}}_{\text{Cyclic shear strain of photon}} \right\}$$

Photon "bare" dynamic quality item, showing the particle properties of photons

(2.3.32)

The shear strain region is the overlapping region of the fiber field (Fig. 2.3.8). Shear strain and internal rotation strain belong to circumferential strain. The bare mass of photons is formed by the circumferential strain and the longitudinal strain in photon interior space.

2.3.4 Intrinsic four momentum of photon

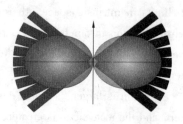

Fig. 2.3.8 The distribution of photon shear strain in space

Because photon has fiber structure, strain ε_3 is in the direction of propagation. The photon is propagating along the x_3 axis. The strain ε_3 is the intrinsic property of the photon and is confined to the interior space. The strains of the whole photon along the direction perpendicular to the propagation direction are $\Delta\varepsilon_1$ and $\Delta\varepsilon_2$, which are manifested as shear wave in the observation space.

$$[\varepsilon_{\mu\mu}](\xi,x) = \underbrace{\begin{bmatrix} \varepsilon_{00} & 0 & 0 & 0 \\ 0 & 0 & 0 & 0 \\ 0 & 0 & 0 & 0 \\ 0 & 0 & 0 & \varepsilon_{33} \end{bmatrix} \begin{bmatrix} \xi_0 & 0 & 0 & 0 \\ 0 & 0 & 0 & 0 \\ 0 & 0 & 0 & 0 \\ 0 & 0 & 0 & \xi_3 \end{bmatrix}}_{\substack{(2)\,\text{Longitudinal 4-momentum }p_3,E \\ \text{of photon electric filed}}} + \underbrace{\begin{bmatrix} \Delta\varepsilon_{00} & 0 & 0 & 0 \\ 0 & \Delta\varepsilon_{11} & 0 & 0 \\ 0 & 0 & \Delta\varepsilon_{22} & 0 \\ 0 & 0 & 0 & 0 \end{bmatrix} \begin{bmatrix} x_0 & 0 & 0 & 0 \\ 0 & x_1 & 0 & 0 \\ 0 & 0 & x_2 & 0 \\ 0 & 0 & 0 & 0 \end{bmatrix}}_{\substack{(3)\,\text{Transverse momentum }p_1p_2 \\ \text{of photon electric filed}}}$$

The 4-momentum of pure electric field of photon

(2.3.33)

Photon strain field function is

$$\phi_\gamma(\xi,x) = \exp\left[-i\left(\begin{bmatrix} \varepsilon_{00} & 0 & 0 & 0 \\ 0 & 0 & 0 & 0 \\ 0 & 0 & 0 & 0 \\ 0 & 0 & 0 & \varepsilon_{33} \end{bmatrix} \begin{bmatrix} \xi_0 & 0 & 0 & 0 \\ 0 & 0 & 0 & 0 \\ 0 & 0 & 0 & 0 \\ 0 & 0 & 0 & \xi_3 \end{bmatrix} + \begin{bmatrix} \Delta\varepsilon_{00} & 0 & 0 & 0 \\ 0 & \Delta\varepsilon_{11} & 0 & 0 \\ 0 & 0 & \Delta\varepsilon_{22} & 0 \\ 0 & 0 & 0 & 0 \end{bmatrix} \begin{bmatrix} x_0 & 0 & 0 & 0 \\ 0 & x_1 & 0 & 0 \\ 0 & 0 & x_2 & 0 \\ 0 & 0 & 0 & 0 \end{bmatrix}\right)\right]$$

(2.3.34)

Here, $E_0 = \varepsilon_0, p_i = \varepsilon_i$. Time dimension ξ_0 is equivalent to x_0, and time has no geometric properties, it can't be curled like space.

4-momentum field function of photon is

$$\phi_\gamma(\xi,x) = \exp\left[-i\left(\begin{bmatrix} E_0 + \Delta E & 0 & 0 & 0 \\ 0 & 0 & 0 & 0 \\ 0 & 0 & 0 & 0 \\ 0 & 0 & 0 & 0 \end{bmatrix} \breve{\gamma}_0 \underbrace{\begin{bmatrix} x_0 & 0 & 0 & 0 \\ 0 & 0 & 0 & 0 \\ 0 & 0 & 0 & 0 \\ 0 & 0 & 0 & 0 \end{bmatrix}}_{=X^0} + \right.\right.$$

$$\left. \begin{array}{l} \begin{bmatrix} 0 & 0 & 0 & 0 \\ 0 & 0 & 0 & 0 \\ 0 & 0 & 0 & 0 \\ 0 & 0 & 0 & p_3 \end{bmatrix} \overset{\smile}{\gamma_3} \underbrace{\begin{bmatrix} 0 & 0 & 0 & 0 \\ 0 & 0 & 0 & 0 \\ 0 & 0 & 0 & 0 \\ 0 & 0 & 0 & \xi_3 \end{bmatrix}}_{=-X^3} + \\ \begin{bmatrix} 0 & 0 & 0 & 0 \\ 0 & \Delta p_1 & 0 & 0 \\ 0 & 0 & 0 & 0 \\ 0 & 0 & 0 & 0 \end{bmatrix} \overset{\smile}{\gamma_1} \underbrace{\begin{bmatrix} 0 & 0 & 0 & 0 \\ 0 & x_1 & 0 & 0 \\ 0 & 0 & 0 & 0 \\ 0 & 0 & 0 & 0 \end{bmatrix}}_{=-X^1} + \\ \begin{bmatrix} 0 & 0 & 0 & 0 \\ 0 & 0 & 0 & 0 \\ 0 & 0 & \Delta p_2 & 0 \\ 0 & 0 & 0 & 0 \end{bmatrix} \overset{\smile}{\gamma_2} \underbrace{\begin{bmatrix} 0 & 0 & 0 & 0 \\ 0 & 0 & 0 & 0 \\ 0 & 0 & x_2 & 0 \\ 0 & 0 & 0 & 0 \end{bmatrix}}_{=-X^2} \end{array} \right) \quad (2.3.35)$$

$\overset{\smile}{\gamma}$ is used to describe the properties of observable space. We stipulate that B matrix is only used to display the characteristics of space without any calculation. It's a well-known γ-matrix after operator derivation.

The above representation is a hybrid space representation, which is closest to the familiar one. The disadvantage is the spatial structure is not intuitive. Using generalized coordinates can be written as $\phi_\gamma(\xi,x) = \exp[-\mathrm{i}(EX^0 - p_1X^1 - p_2X^2 - p_3X^3)]$. Considering intrinsic momentum is not observable in observation space, 4-momentum field function of photon can be further abbreviated as

$$\phi_\gamma(x) = \exp[-\mathrm{i}(E\overset{\smile}{\gamma}_0 x^0 - p_1\overset{\smile}{\gamma}_2 x^1 - p_2\overset{\smile}{\gamma}_2 x^2)] \quad (2.3.36)$$

Using the spatial nesting structure, the photon field function can be written as

$$\phi_\gamma(\xi,x) = \exp\left\{-\mathrm{i}\left[\begin{bmatrix} \Delta\varepsilon_0 & 0 & 0 & 0 \\ 0 & \Delta\varepsilon_1 & 0 & 0 \\ 0 & 0 & \Delta\varepsilon_1 & 0 \\ 0 & 0 & 0 & \varepsilon_3 \end{bmatrix} \overset{\smile}{\gamma}_\mu \left(\begin{bmatrix} x_0 & 0 & 0 & 0 \\ 0 & x_1 & 0 & 0 \\ 0 & 0 & x_2 & 0 \\ 0 & 0 & 0 & (\ni\begin{bmatrix} \xi_0 & 0 & 0 & 0 \\ 0 & 0 & 0 & 0 \\ 0 & 0 & 0 & 0 \\ 0 & 0 & 0 & \xi_3 \end{bmatrix} + 1)x_3 \end{bmatrix}\right)\right]\right\} \quad (2.3.37)$$

The advantage of spatial nesting is that the spatial structure is very intuitive, but determining the scale factor is a difficult task.

2.3.5 Intrinsic spin of photon

The physical meaning of rotation stain of rigid body is obtained by analyzing the change of position of infinite adjacent two points in elastic body. Let P be infinitely close to O, and the region around P and O rotate rigidly at a small angle. Set rotation angular velocity vector is ω, the distance vector between O and P is r

$$\omega = \omega_1 i + \omega_2 j + \omega_3 k, \quad \rho = \xi_1 i + \xi_2 j + \xi_3 k \tag{2.3.38}$$

Laplasse operator vector:

$$\nabla = \frac{\partial}{\partial \xi_1} i + \frac{\partial}{\partial \xi_2} j + \frac{\partial}{\partial \xi_3} k$$

The displacement vector of point P is U: $U = ui + vj + wk$. Because the displacement vector can be expressed as $U = \omega \times \rho$ (Fig. 2.3.9), then

$$\nabla \times U = \nabla \times (\omega \times \rho) = (\nabla \cdot \rho)\omega - (\omega \cdot \nabla)\rho$$

$$= 3\omega - \left(\omega_1 \frac{\partial}{\partial \xi_1} + \omega_2 \frac{\partial}{\partial \xi_2} + \omega_3 \frac{\partial}{\partial \xi_3} \right)(\xi_1 i + \xi_2 j + \xi_3 k)$$

$$= 3\omega - (\omega_1 i + \omega_2 j + \omega_3 k) = 2\omega$$

i. e.

$$\omega = \frac{1}{2} \nabla \times S = \frac{1}{2} \begin{vmatrix} i & j & k \\ \frac{\partial}{\partial \xi_1} & \frac{\partial}{\partial \xi_2} & \frac{\partial}{\partial \xi_3} \\ u_1 & u_2 & u_3 \end{vmatrix} \tag{2.3.39}$$

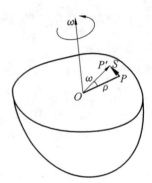

Fig. 2.3.9   Rigid rotation

Here, $\omega_{12} = \omega_3 = \frac{1}{2}\left(\frac{\partial u_1}{\partial \xi_2} - \frac{\partial u_2}{\partial \xi_1} \right)$; $\omega_{23} = \omega_1 = \frac{1}{2}\left(\frac{\partial u_2}{\partial \xi_3} - \frac{\partial u_3}{\partial \xi_2} \right)$; $\omega_{13} = \omega_2 = \frac{1}{2}\left(\frac{\partial u_1}{\partial \xi_3} - \frac{\partial u_3}{\partial \xi_1} \right)$.

$\omega_1, \omega_2, \omega_3$ are rotational components, which are functions of coordinates, representing the rigid rotation of differential elements in elastic bodies.

2.3.6   Intrinsic spin wave function of photon

$$\phi_S = \exp\left\{ i \begin{bmatrix} 0 & \omega_{01} & \omega_{02} & \omega_{03} \\ \omega_{10} & 0 & \omega_{12} & \omega_{13} \\ \omega_{20} & \omega_{21} & 0 & \omega_{23} \\ \omega_{30} & \omega_{31} & \omega_{32} & 0 \end{bmatrix} \begin{bmatrix} 0 & \xi_0 & \xi_0 & \xi_0 \\ \xi_1 & 0 & \xi_1 & \xi_1 \\ \xi_2 & \xi_2 & 0 & \xi_2 \\ \xi_3 & \xi_3 & \xi_3 & 0 \end{bmatrix} \right\}$$

(3) Photonic rotational strain
constitute photon-spin

$$= \exp\left[i \begin{bmatrix} 0 & \omega_{01} & \omega_{02} & \omega_{03} \\ 0 & 0 & 0 & 0 \\ 0 & 0 & 0 & 0 \\ 0 & 0 & 0 & 0 \end{bmatrix} \begin{bmatrix} 0 & 0 & 0 & 0 \\ \xi_1 & 0 & 0 & 0 \\ \xi_2 & 0 & 0 & 0 \\ \xi_3 & 0 & 0 & 0 \end{bmatrix} + \right.$$

The spin energy ω flux through the
surface of photon internal space

Chapter 2 Structure of the Photons

$$\begin{bmatrix} 0 & 0 & 0 & 0 \\ \omega_{10} & 0 & 0 & 0 \\ \omega_{20} & 0 & 0 & 0 \\ \omega_{30} & 0 & 0 & 0 \end{bmatrix} \begin{bmatrix} 0 & \xi_0 & \xi_0 & \xi_0 \\ 0 & 0 & 0 & 0 \\ 0 & 0 & 0 & 0 \\ 0 & 0 & 0 & 0 \end{bmatrix} +$$

The spin angular-momentum ω flux through the surface of photon internal space

$$\begin{bmatrix} 0 & 0 & 0 & 0 \\ 0 & 0 & \omega_{12} & 0 \\ 0 & \omega_{21} & 0 & 0 \\ 0 & 0 & 0 & 0 \end{bmatrix} \begin{bmatrix} 0 & 0 & 0 & 0 \\ 0 & 0 & \xi_1 & 0 \\ 0 & \xi_2 & 0 & 0 \\ 0 & 0 & 0 & 0 \end{bmatrix} +$$

The spin angular-momentum ω of photon internal space ξ_3

$$\begin{bmatrix} 0 & 0 & 0 & 0 \\ 0 & 0 & 0 & \omega_{13} \\ 0 & 0 & 0 & 0 \\ 0 & \omega_{31} & 0 & 0 \end{bmatrix} \begin{bmatrix} 0 & 0 & 0 & 0 \\ 0 & 0 & 0 & \xi_1 \\ 0 & 0 & 0 & 0 \\ 0 & \xi_3 & 0 & 0 \end{bmatrix} +$$

The spin angular-momentum ω of photon internal space ξ_2

$$\left. \begin{bmatrix} 0 & 0 & 0 & 0 \\ 0 & 0 & 0 & 0 \\ 0 & 0 & 0 & \omega_{23} \\ 0 & 0 & \omega_{32} & 0 \end{bmatrix} \begin{bmatrix} 0 & 0 & 0 & 0 \\ 0 & 0 & 0 & 0 \\ 0 & 0 & 0 & \xi_3 \\ 0 & 0 & \xi_2 & 0 \end{bmatrix} \right) \quad (2.3.40)$$

Spin angular momentum ω of photon internal space ξ_1

Intrinsic spin wave function of photon is

$$\phi_S = \exp(i[(\omega_{[0i]} + \omega_{[i0]})X_{S0} + [\omega_{[12]}]X_{S3} + [\omega_{[13]}]X_{S2} + [\omega_{[23]}]X_{S1}])$$

(2.3.41a)

$\omega_{[0i]} + \omega_{[i0]}$ characterizes the propagation properties of the spin waves. There are 3 pairs of antisymmetric eigenvalues. The square brackets of subscript denote antisymmetric.

For $\omega_{[ij]}$, the spin angular momentum axis is $\pm\xi_k$, and there are 3 pairs of antisymmetric eigenvalues. The square brackets of the subscript are antisymmetric.

Spin operator can be used to extract effective information of wave function:

$$\hat{S}_\mu = -iI_\mu \frac{\partial}{\partial X_{S\mu}}$$

$$\hat{S}_\mu \phi_S = -iI_\mu \frac{\partial}{\partial X_{S\mu}} \exp[i([\omega_{[0i]}]X_{S0} + [\omega_{[12]}]X_{S3} + [\omega_{[13]}]X_{S2} + [\omega_{[23]}]X_{S1})]$$

(2.3.41b)

Spin eigen value is $\hbar$. According to the present theory, there is

$$I_3 = \begin{bmatrix} 0 & 0 & 0 & 0 \\ 0 & 0 & 1 & 0 \\ 0 & -1 & 0 & 0 \\ 0 & 0 & 0 & 0 \end{bmatrix}, I_2 = \begin{bmatrix} 0 & 0 & 0 & 0 \\ 0 & 0 & 0 & -1 \\ 0 & 0 & 0 & 0 \\ 0 & 1 & 0 & 0 \end{bmatrix}, I_1 = \begin{bmatrix} 0 & 0 & 0 & 0 \\ 0 & 0 & 0 & 0 \\ 0 & 0 & 0 & 1 \\ 0 & 0 & -1 & 0 \end{bmatrix} \quad (2.3.42)$$

$$[\hat{S}_j, \hat{S}_k] = i\varepsilon_{jkl}\hat{S}_l \quad (2.3.43)$$

$$\hat{S}_3^2 = \begin{bmatrix} 0 & 0 & 0 & 0 \\ 0 & 1 & 0 & 0 \\ 0 & 0 & 1 & 0 \\ 0 & 0 & 0 & 0 \end{bmatrix}, \hat{S}_2^2 = \begin{bmatrix} 0 & 0 & 0 & 0 \\ 0 & 1 & 0 & 0 \\ 0 & 0 & 0 & 0 \\ 0 & 0 & 0 & 1 \end{bmatrix}, \hat{S}_1^2 = \begin{bmatrix} 0 & 0 & 0 & 0 \\ 0 & 0 & 0 & 0 \\ 0 & 0 & 1 & 0 \\ 0 & 0 & 0 & 1 \end{bmatrix}. \quad (2.3.44)$$

$$\hat{S}^2 = \hat{S}_1^2 + \hat{S}_2^2 + \hat{S}_3^2 = 2 = 1 \times (1+1)$$

Thus, the spin of the photon is $1\hbar$.

Spin is the intrinsic property of the photon, and the spin angular momentum in the observed space is $1\hbar$. The spin energy flow and the spin momentum flow show the existence of the spin wave in the observational space, which embodies the motion characteristic. In other words, the vacuum strain exists in the form of wave. Spin wave equation is

$$(\hat{S}_1^2 + \hat{S}_2^2 + \hat{S}_3^2)\phi_\gamma(X) = 2\phi_\gamma(X) \quad (2.3.45)$$

It is important to note that the spin wave is similar to the electron spin wave and does not cause the photon itself to rotate.

2.3.7 The photonic fiber field strain

The fiber structure of photons is divided into two parts:

①The first part is the effect of photon strain in intrinsic space, which is $\varepsilon_{\mu\nu}(\xi), 0 \leq \xi^\mu \leq r_0$.

②The second part is outside the intrinsic region, which is the strain effect of photon fiber field in the macroscopic region, namely, $\Delta\varepsilon_{\mu\nu}(x), x^\mu \geq r_0$. This is the observation space, and the intrinsic space is regarded as a point in the observable space.

(1) The photon electric flux line strain $[\varepsilon]_{\text{Electric flux line}}$.

The strain of photon electric flux line is

$$\frac{\partial A(\xi^0)}{\partial \xi_0} = A_{0,0}, \frac{\partial A(\xi)}{\partial \xi_i} = A_{i,i}, A(\xi) = ea(\xi) \quad (2.3.46)$$

Here, $a(\xi)$ is single electric flux line offset; $A(\xi)$ is total electric flux line offset; $A_{\mu,\mu}(\xi)$ is electromagnetic 4-potential in the internal space of photon.

$$A_{\mu,\mu} = (A_{0,0}, A_{1,1}, A_{2,2}, A_{3,3}) = (\underbrace{A_{0,0}}_{\text{Scalar potential }\phi}, \underbrace{A_{i,i}}_{\text{Vector potential}}) \quad (2.3.47)$$

The total strain of the photon electric flux line

$$h_{\text{Electric flux line}} = \int_0^R \int_0^\theta \int_0^\varphi A_{i,i}(r,\theta,\varphi) \, dr d\theta d\varphi = \text{Constant} \quad (2.3.48)$$

The distribution of photonelectric flux lines is not uniform, which is dense inside and sparse outside. So in macro space, the electric flux lines density per unit volume is different. In the internal space of the photon (Fig. 2.3.10), the total deformation of electric flux lines deformation is h_γ, which is understood that the total deformation of electromagnetic field in the whole observation space is h_γ. The interior space of a photon is simplified as a point in the observation space and is often set as the origin of the interior space coordinates.

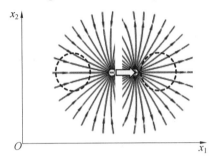

Fig. 2.3.10 Schematic diagram of photon electric flux line

Photonic potential vector and electron potential vector are different. The photon vector potential does not constitute a "source", that is $\nabla A(\xi) = 0$. If compressing is as "outflow", stretching is as "inflow", and the tension is as "outflow", then inflows equal to outflows. This is a passive field in the outer observation space, which obviously satisfies:

$$\nabla \cdot A(x) = 0 \tag{2.3.49}$$

This is the Coulomb gauge condition.

The strain is conserved intrinsic 4-dimensional space-time that is the tensile strain equal to the compression strain, expressed as $\mathrm{Tr}[\varepsilon_{\mu\mu}] = 0$. Extending to the outer observation space, the strain is conserved, i.e.

$$\partial_\mu A_\mu = 0 \tag{2.3.50}$$

This is the Lorentz gauge condition.

$$[\varepsilon]_{\text{Electric flux line}} = \begin{bmatrix} \dfrac{\partial A_0}{\partial x_0} & \dfrac{\partial A_0}{\partial x_1} & \dfrac{\partial A_0}{\partial x_2} & \dfrac{\partial A_0}{\partial x_3} \\ \dfrac{\partial A_1}{\partial x_0} & \dfrac{\partial A_1}{\partial x_1} & \dfrac{\partial A_1}{\partial x_2} & \dfrac{\partial A_1}{\partial x_3} \\ \dfrac{\partial A_2}{\partial x_0} & \dfrac{\partial A_2}{\partial x_1} & \dfrac{\partial A_2}{\partial x_2} & \dfrac{\partial A_2}{\partial x_3} \\ \dfrac{\partial A_3}{\partial x_0} & \dfrac{\partial A_3}{\partial x_1} & \dfrac{\partial A_3}{\partial x_2} & \dfrac{\partial A_3}{\partial x_3} \end{bmatrix}$$

$$= \underbrace{\begin{bmatrix} A_{0,0} & 0 & 0 & 0 \\ 0 & A_{1,1} & 0 & 0 \\ 0 & 0 & A_{2,2} & 0 \\ 0 & 0 & 0 & A_{3,3} \end{bmatrix}}_{\substack{(1)\text{ The 4-momentum } p \text{ of photonic electric flux line} \\ \text{is the 4 vector of electromagnetic field}}} + \frac{1}{2} \underbrace{\begin{bmatrix} 0 & \gamma_{01} & \gamma_{02} & \gamma_{03} \\ \gamma_{10} & 0 & \gamma_{12} & \gamma_{13} \\ \gamma_{20} & \gamma_{21} & 0 & \gamma_{23} \\ \gamma_{30} & \gamma_{31} & \gamma_{32} & 0 \end{bmatrix}}_{\substack{(2)\text{ Photonic shear strain of} \\ \text{electric flux line is zero}}} +$$

$$\frac{1}{2}\begin{bmatrix} 0 & \omega_{01} & \omega_{02} & \omega_{03} \\ \omega_{10} & 0 & \omega_{12} & \omega_{13} \\ \omega_{20} & \omega_{21} & 0 & \omega_{23} \\ \omega_{30} & \omega_{31} & \omega_{32} & 0 \end{bmatrix} \quad (2.3.51)$$

(3) Photonic electric flux line rotational strain constitute electric magnetic field

There are three items.

① First item ($A_{\mu,\mu} = \varepsilon_{\mu\mu}$) is the intrinsic 4-momentum of the photon electromagnetic field. This is the potential vector and scalar potential of the electromagnetic field.

② $[\gamma_{\mu\nu}]$, $\mu \neq \nu$: $\gamma_{\mu\nu} = \dfrac{\partial A_{\mu}}{\partial x^{\nu}} + \dfrac{\partial A_{\nu}}{\partial x^{\mu}} = 0$.

We are currently studying electricflux line outside and inside the region. Since shear strain only exists in the intrinsic region, there is no overlap of electric flux line outside the intrinsic region, and each electric flux line is independent. Therefore, the shear strain of electric flux line is zero outside the intrinsic region.

③ $[\Omega_{\mu\nu}]$ is electromagnetic tensor of single photon.

Electric flux line rotational strain: $\Omega_{\mu\nu} = \dfrac{\partial A_{\mu}}{\partial x^{\nu}} - \dfrac{\partial A_{\nu}}{\partial x^{\mu}} = F_{\mu\nu}$ is rotation tensor (antisymmetric two rank tensor). $F_{\mu\nu} = -F_{\nu\mu} = \partial_{\mu}A_{\nu} - \partial_{\nu}A_{\mu}$ is called electromagnetic tensor.

$$F_{uv} = \begin{bmatrix} 0 & E_1 & E_2 & E_3 \\ -E_1 & 0 & -B_3 & B_2 \\ -E_2 & B_3 & 0 & -B_1 \\ -E_3 & -B_2 & B_1 & 0 \end{bmatrix} \quad (2.3.52)$$

Here, $\omega_{0i} = E_i = F_{0\mu}$; $\omega_{ij} = B_i = F_{ij}$.

The "rotation" strain of electric flux line between time and space generates photon electric field. The rotation strain of electric flux line between two dimensions space generates magnetic field.

$$F_{0\mu} = -F_{\mu 0} = \begin{bmatrix} 0 & E_1 & E_2 & E_3 \\ -E_1 & 0 & 0 & 0 \\ -E_2 & 0 & 0 & 0 \\ -E_3 & 0 & 0 & 0 \end{bmatrix}, \quad F_{ij} = \begin{bmatrix} 0 & 0 & 0 & 0 \\ 0 & 0 & -B_3 & B_2 \\ 0 & B_3 & 0 & -B_1 \\ 0 & -B_2 & B_1 & 0 \end{bmatrix} \quad (2.3.53)$$

$$[\Omega_{\mu\nu}](x) = \begin{bmatrix} 0 & \Omega_{01} & \Omega_{02} & \Omega_{03} \\ \Omega_{10} & 0 & 0 & 0 \\ \Omega_{20} & 0 & 0 & 0 \\ \Omega_{30} & 0 & 0 & 0 \end{bmatrix} \begin{bmatrix} 0 & x_0 & x_0 & x_0 \\ x_1 & 0 & 0 & 0 \\ x_2 & 0 & 0 & 0 \\ x_3 & 0 & 0 & 0 \end{bmatrix} +$$

(1) The time and space rotation of power lines which is shown as the dimensional shift

The item is the electric field of photon

Fiber move, having no effect of mass

Chapter 2 Structure of the Photons

$$\begin{bmatrix} 0 & 0 & 0 & 0 \\ 0 & 0 & \Omega_{12} & 0 \\ 0 & \Omega_{21} & 0 & 0 \\ 0 & 0 & 0 & 0 \end{bmatrix} \begin{bmatrix} 0 & 0 & 0 & 0 \\ 0 & 0 & \xi_1 & 0 \\ 0 & \xi_2 & 0 & 0 \\ 0 & 0 & 0 & 0 \end{bmatrix} +$$
(2) Photon spin around ξ^3 axis
The item is the magnetic field of photon

$$\begin{bmatrix} 0 & 0 & 0 & 0 \\ 0 & 0 & 0 & \Omega_{13} \\ 0 & 0 & 0 & 0 \\ 0 & \Omega_{31} & 0 & 0 \end{bmatrix} \begin{bmatrix} 0 & 0 & 0 & 0 \\ 0 & 0 & 0 & \xi_1 \\ 0 & 0 & 0 & 0 \\ 0 & \xi_3 & 0 & 0 \end{bmatrix} +$$
(3) Photonic spin around ξ^2 axis is
shown as the magnetic field of photon

$$\begin{bmatrix} 0 & 0 & 0 & 0 \\ 0 & 0 & 0 & 0 \\ 0 & 0 & 0 & \Omega_{32} \\ 0 & 0 & \Omega_{23} & 0 \end{bmatrix} \begin{bmatrix} 0 & 0 & 0 & 0 \\ 0 & 0 & 0 & 0 \\ 0 & 0 & 0 & \xi_2 \\ 0 & 0 & \xi_3 & 0 \end{bmatrix} \qquad (2.3.54)$$
(4) Photonic spin around ξ^1 axis is
shown as the magnetic field of photon

(2) Electric charge of photon.

Considering a closed sphere, the photonic fiber strain of back hemisphere is $[\varepsilon]_{Back}$, and the photonic fiber strain of front hemisphere is $[\varepsilon]_{Front}$, and then $[\varepsilon]_{Back} + [\varepsilon]_{Front} \equiv 0$. Electric field in positive hemisphere is $E_{-\gamma/2}$. The total electric field is as follows.

$$E_\gamma = E_{+\gamma/2} + E_{-\gamma/2} = 0 \qquad (2.3.55)$$

The electric flux line points outward in the front half compression space (Fig. 2.3.1), and the same as electric field of e^+ is positive electric field. The electric flux line in the behind part points to the center in stretch space, and the electric field is the same as the e^- negative electric field. The front half part and the behind part of the electric field are equal with opposite direction, meet $\nabla(E_{+\gamma/2} + E_{-\gamma/2}) = 0$. The photon electric flux lines of inward and outward offset, photon charge is zero. In other words, photon is without charge. The Maxwell equation is expressed as

$$\nabla \cdot E_\gamma = 0 \qquad (2.3.56)$$

Photon is a field with no source.

(3) Photon magnetic field item.

Photon spin born out of the electron spin. Electron spin is caused by intrinsic vacuum shear strain $\varepsilon = B = \nabla \times u(\xi)$. Electrons vibrate to emit photons, and electrons have shear strain, which makes photons also have shear strain, showing photon spin.

The magnetic field lines in the front half compression space are the same as e^+ magnetic field. The magnetic field lines in the behind part stretch space are the same as e^- magnetic field lines. The front half part and the behind part of magnetic field lines are equal with opposite direction, which meets $\nabla \cdot B_{+\gamma/2} + \nabla \cdot B_{-\gamma/2} = 0$. In addition, according to the closed characteristic of the magnetic line, we know that the magnetic monopole does not exist, $\nabla \cdot B_\gamma = 0$. The magnetic field is in the superposition, because the center position is continuous polarization, and magnetic field $\varepsilon_{ij}(\xi)$ is also a periodic alternating field.

(4) The relationship between electric field and magnetic field.

Photon has a fibrous structure, and the relationship between electric field and magnetic field of a photon is described by following Maxwell equations:

$$\nabla \times E + \frac{1}{c}\frac{\partial B}{\partial t} = 0, \nabla \times B - \frac{1}{c}\frac{\partial E}{\partial t} = 0 \qquad (2.3.57)$$

Meeting $\partial_\mu A^\mu(x) = 0$, total strain conservation is Lorentz gauge condition in photonic intrinsic space. Hence, $\nabla \cdot A(x) = 0$ and $\partial_\mu A^\mu(x) = 0$ are the intrinsic properties of photons.

$$F^{0i} = -E, \quad F^{ij} = -\varepsilon^{ijk} B^k \qquad (2.3.58)$$

Here, $\varepsilon^{ijk} \equiv \varepsilon_{ijk}$ is the all antisymmetry symbol $\varepsilon^{123} = \varepsilon_{123} = 1$.

$$\partial_\mu F^{\mu\nu} = 0 \qquad (2.3.59)$$

(5) The amplitude of electromagnetic field function of photon.

The photon intrinsic electromagnetic tensors constitute by the tensile zone $F_{\mu\nu}$ and compression zone $F_{\nu\mu}$. The intrinsic degrees of freedom of photon electromagnetic are unified as

$$\phi_\gamma(F) = \exp\left[(\frac{1}{2}\underbrace{F_{\mu\nu}}_{\text{Front hemisphere}} - \frac{1}{2}\underbrace{F_{\mu\nu}}_{\text{Back hemisphere}})(x^\mu)\right] = \exp[0] = 1 \qquad (2.3.60)$$

The electromagnetic field is no observable effect in the observation space. The electromagnetic amplitude has no contribution. The contribution of freedom degree in magnetic field of photon is zero.

(6) The image of photon transfer electromagnetic interaction.

The exchanged photons between positive and negative charges attract each other, and the exchanged photons between the same charges repel each other (Fig. 2.3.11).

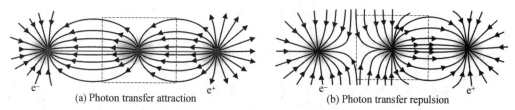

(a) Photon transfer attraction (b) Photon transfer repulsion

Fig. 2.3.11 The image of photon transferring electromagnetic interaction

(Realize the interaction by the exchange of photons, coupling constant $\alpha_e = 1/137$)

(7) Magnetic charge of photon.

$$\nabla \cdot B_\gamma = 0 \qquad (2.3.61)$$

There is no magnetic charge for photon.

2.3.8 Quantum entanglement and its measurement

The freedom degree of quantum states is usually incomplete. Quantum entanglement occurs when multiple quantum states interact in local space.

The Pauli Incompatibility Principle can be reinterpreted as that the other electron can be

Chapter 2 Structure of the Photons

constrained by three space quantities of an electron and four freedom degrees of the spin space in an electron. This is the same region for the 3-dimensional space of an electron, but the electron spin is the opposite. In this sense, the outer electrons of the nucleus have four dimensions, for the electrons of the same energy level.

These two electrons are entangled states. If there is a way to separate the two electrons without interfering with them, no matter how far they travel, it is certain that the pair of electrons spin is opposite. This is because the wave itself should maintain its propagation characteristics in the process of propagation, and show inertia in macroscopic objects. When the spin of an electron is determined, because another electron is not disturbed by external disturbance, and the spin of the electron remains in its original state. In other words, the spin is in the opposite state. This is by no means as a kind of over-distanced action, but caused by the wave keeping its original characteristics in the process of propagation, so it is a misunderstanding about the over distance action.

Photons are 1-dimensional. If we use experimental means to make the B and a photon close enough to the point C (Fig. 2.3.12 and Fig. 2.3.13), and two photons are coaxial, then photon A will affect another photon B. One photon constrains another photon (or one photon measures another photon), which is experimentally called two photons in entangled states. Because the vacuum strain has the physical property of restoring the flat state, the strain of the two photons will be in the opposite state, so entangled photons are opposite to spins. When the two photons A and B are separated, if the spin state of photon A is determined without external interference, it can be determined that the spin state of photon B is opposite to that of photon A.

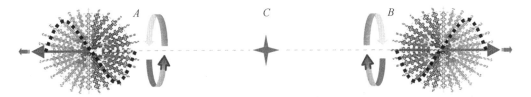

Fig. 2.3.12 The vacuum coaxial correlation of dual photon

Logically, after a couple divorces, no matter how far they are, once one of them is found, the other one is determined. This confirmation process has nothing to do with superluminal speed.

Uncertainty, the measurement results show that low-dimensional space cannot constrain high-dimensional objects. For example, when we rou the dice, the dice is a 3-dimensional cube, but when it is thrown on the floor, which is 2-dimensional. In this process, we use 2-dimensional floor to measure 3-dimensional dice. 2-dimensional space cannot constrain 3-dimensional objects, which inevitably leads to uncertainty. But the objective reality that we have in our experimental means to use two different dimensions of quantum interaction to achieve detection. The uncertainty confusion caused by this dimensional asymmetry is misinterpreted as the

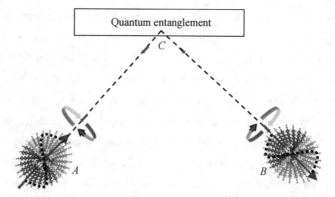

Fig. 2. 3. 13　The reflex coaxial correlation of dual photon

quantum field, which is completely uncertain. It can even change at will, such as Schrodinger cat. If we can find the constraint space of the same dimension as the Schrodinger cat, we can determine the life and death status of the Schrodinger cat.

2.4　The expression of photon strain in observational space

2.4.1　Non-proliferation of photons longitudinal waves

We note that photon is longitudinal wave. But it does not diffuse to form a spherical wave like sound. It is always in the form of particles. This is because the photon has fiber structure. Let's look at the ξ^1 axis of fibers (Fig. 2. 4. 1). When O moves along the ξ^1 axis to reach the tensile compression limit $\mathscr{H}/2$ (i. e. photon forming condition), the compression limit is not reached except on the ξ^1 axis, although the fiber deforms outside the ξ^1 axis, and the compression limit is only reached on the ξ^1 axis, while the propagation condition is satisfied to form longitudinal wave. Vacuum cannot form outward diffusion wave and lose energy outside the ξ^1 axis, photons only propagate along the ξ^1 axis direction, and will not form outward radiation spherical wave, thus maintaining the properties of photons propagating particles. With the photon center O as the origin, the deformation occurs in the spherical region of R radius. When O leaves, the vacuum of the fiber restores to a flat state without any energy. Therefore, all the energy of the

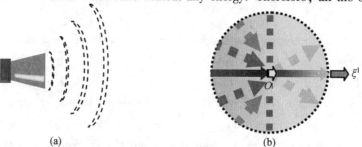

Fig. 2. 4. 1　Particle nature of photons

quantum field is determined by the O point, which is called the center point of the photon.

2.4.2 Energy momentum wave characteristics

Let's discuss $\varphi_{\text{4- momentum}}(x) = \exp[-i[\varepsilon_{\mu\mu}](\xi)]$ in detail. In the intrinsic space of quantum field, there is

$$\varphi(\xi) = \exp[-i\varepsilon_{\mu\nu}(\xi)\xi^\alpha] \Rightarrow \varphi(x) = \exp[-i\varepsilon_{\alpha\beta}(\xi)L^\alpha_\nu x^\nu] \quad (2.4.1)$$

$$L^\alpha_\nu = \begin{bmatrix} \frac{1}{\sqrt{1-u^2/c^2}} & \frac{u/c}{\sqrt{1-u^2/c^2}} & 0 & 0 \\ 0 & 0 & 0 & 0 \\ 0 & 0 & 0 & 0 \\ 0 & 0 & \frac{u/c}{\sqrt{1-u^2/c^2}} & \frac{1}{\sqrt{1-u^2/c^2}} \end{bmatrix} \quad (2.4.2)$$

Then, $x^3 = \frac{\xi^3 - ut}{\sqrt{1-u^2/c^2}}; x^1 = \xi^1; x^2 = \xi^2; x^0 = \frac{\xi^0 - u\xi^1/c}{\sqrt{1-u^2/c^2}}$. ξ^μ is a coordinate of a synchronous motion reference frame. The photon is stationary in this coordinate system. x^μ is the observation coordinate system, and the photon moving at the speed of light in this coordinate system is $u = c$. Therefore, the coordinate system changes from the intrinsic space to the observational space which is equivalent to the change from static state to the speed of light c. In this transformation process, coordinates are not expressed in the direction of motion and time dimension, which means that there is no-observability in these two dimensions. This produces a strange effect that the momentum and energy of the intrinsic quantum strain in the observed space along the direction of motion which cannot be expressed.

$$\phi_\gamma(x) = \exp(-i)\left(\begin{bmatrix} k_0 & 0 & 0 & 0 \\ 0 & 0 & 0 & 0 \\ 0 & 0 & 0 & 0 \\ 0 & 0 & 0 & -k_3 \end{bmatrix}\begin{bmatrix} \xi^0 & 0 & 0 & 0 \\ 0 & 0 & 0 & 0 \\ 0 & 0 & 0 & 0 \\ 0 & 0 & 0 & \xi^3 \end{bmatrix} + \begin{bmatrix} 0 & 0 & 0 & 0 \\ 0 & -k_1 & 0 & 0 \\ 0 & 0 & -k_2 & 0 \\ 0 & 0 & 0 & 0 \end{bmatrix}\begin{bmatrix} 0 & 0 & 0 & 0 \\ 0 & x^1 & 0 & 0 \\ 0 & 0 & x^2 & 0 \\ 0 & 0 & 0 & 0 \end{bmatrix}\right) \quad (2.4.3)$$

2.4.3 Photon transverse fluctuation and probability wave

We can get Lorentz normality condition directly by 4-dimensional geometric properties of vacuum and the law of vacuum strain conservation.

$$\left.\begin{array}{r}\Box A_\mu = 0 \\ A_{\mu,\mu} = 0\end{array}\right\} \Rightarrow A_\mu = e_\mu e^{ik_\nu x^\nu}, \ k^\mu k_\mu = -\frac{m_0^2 c^2}{\hbar} \Rightarrow m_0 = 0, \ k^\mu e_\mu = \eta^{\mu\nu} k_\mu e_\mu = 0 \quad (2.4.4)$$

Electromagnetic waves are transverse wave and photons have zero static mass. We further analyze the physical properties of photons from the perspective of vacuum strain.

1. Photon transverse wave

Photon sphere can be regarded as a string of length Φ which spread along the direction ξ^3 in the simplified structure. Fibre deformation in the direction ξ^3 is the largest, the surface $\xi^1\xi^2$ are perpendicular in the propagation direction ξ^3, $A_{/\!/}(\xi) = \exp[i(k_3\xi^3 - k_0\xi^0)]$. It is the intrinsic field function of photons, and there is no-observability in the observation space.

When photons are emitted by electrons, the photon center on $\xi^1\xi^2$ plane is inevitably disturbed by the electron spin. The center of the string is disturbed and the vibration occurs on the surface $\xi^1\xi^2$, thus constituting the transverse probability wave of photons. Vibration direction is polarization direction of the photon. Therefore, photons have transverse and longitudinal waves. The center of the photon string propagation along a straight line is at the speed of light and vibrates horizontally at the same time, and $p_1 = \partial u_1/\partial x_1, p_2 = \partial u_2/\partial x_2 \ll p_3$. So the locus of photon centers is formation of a sine curve. At the peak, its phase velocity is slowest, center appears most likely. Transverse wave and longitudinal wave synthesis

$$A(\xi,x) = A_\perp(x)A_{/\!/}(\xi) = \exp[i(k_1x^1 + k_2x^2 + k_3\xi^3 - k_0\xi^0)] \quad (2.4.5)$$

$$\varepsilon_0 = k_0, \quad \varepsilon_i = -k_i$$

k_1x^1 and k_2x^2 make the photon propagation exist transverse vibration in the 2-dimensional surface $\xi^1\xi^2$. The photon center of vibration in 4 directions is k_μ. Basis vector is $\varepsilon_{k\mu}$. Considering the amplitude, the intrinsic field function is

$$A(\xi) = \sum_{\lambda=i} \varepsilon_{k\lambda} (A e^{i(k_1x^1+k_2x^2+k_3\xi^3-k_0\xi^0)} + \bar{A} e^{-i(k_1x^1+k_2x^2+k_3\xi^3-k_0\xi^0)}) \quad (i = 0,1,2,3)$$

(2.4.6)

The photon center of 1-dimensional string along a straight line forward propagation at speed of light, while it's vibrate in vertical that the locus of photon centers to form a sinusoidal curve. At the peak, the speed is the slowest, center appears most likely, which constitutes photon probability $|A(x)|^2$ and probability wave function $A(x)$.

$$A(t,x) = A e^{-i(px-Et)} \quad (2.4.7)$$

$\varphi(x)$ is $A_{4-\text{momentum}}(x) = \exp\{[\varepsilon_{\mu\mu}(\xi)](x)\}$. Describing the occurring possibility of photon center, A is the amplitude of wave function. The terms of the longitudinal wave cannot be observable.

2. Probability wave function of the photon longitudinal wave

The concept of the wave function derives from the probability amplitude (Fig. 2.4.2). However, the photon longitudinal wave probability amplitude cannot be observed, thus using probability wave concept to describe the longitudinal photons is not applicable.

For longitudinal wave in the intrinsic space, there are two independent longitudinal polarization states, i.e. along the time direction and momentum direction. The polar coordinate system is established in the intrinsic space, and the momentum along the propagation direction is k_3, which satisfies the following orthogonal conditions:

Chapter 2 Structure of the Photons

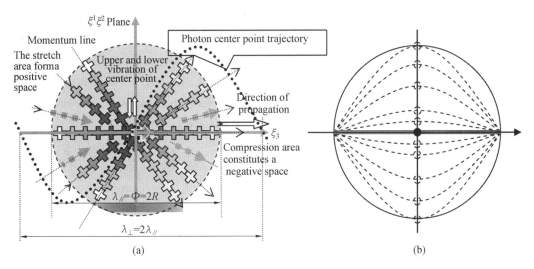

Fig. 2.4.2 Photonic fiber structure

$$\varepsilon_{k\lambda}\varepsilon_{k\lambda'} = \delta_{\lambda\lambda'} \tag{2.4.8}$$

Because the longitudinal character of $A_{/\!/}$, there are longitudinal conditions as follows.

$$k \cdot \varepsilon_{k\lambda'} = k_3 , \quad k_0\tau - k\xi = 0 \tag{2.4.9}$$

For longitudinal photon:

$$\Box A_{/\!/}(\xi) = 0 \tag{2.4.10}$$

Available photon mass shell relations:

$$k^2 = |k|^2 - E_k^2 = -m_0^2 = 0 \tag{2.4.11}$$

Photon rest mass is zero. Take $\hbar = c = 1$, and the single photon energy $E_k = \omega_k = |k|$. The arbitrary solutions of equation $\Box A_{/\!/}(\xi) = 0$ are expressed as a linear combination, namely the longitudinal photon wave function

$$A_{/\!/}(\xi) = \sum_{k,\lambda = i}(a_{k\lambda}A_{k\lambda}^{(+)}(\xi) + a_{k\lambda}^{*}A_{k\lambda}^{(-)}(\xi)) \quad (i,k = 0,3) \tag{2.4.12}$$

Here, $a_{k\lambda}$ and $a_{k\lambda}^{*}$ are the expansion coefficients. Because in the longitudinal and abscissa coordinates, the epsilon klambda can be used as the third and fourth axes base vectors. From the view of the uncertainty relation $\Delta\lambda \cdot \Delta p \geqslant h/2$, the longitudinal wave length is $\Phi = \lambda/2$. P-wave cannot be observed in the outer space of quantum field, and thus join $\theta(R-x)$, when $x > R$, $\theta(R-x) = 0$; when $x < R$, $\theta(R-x) = 1$. In this way, the intrinsic space can be expressed by observation space.

$$A_{/\!/}(x) = \frac{1}{\sqrt{V}}\sum \frac{1}{\sqrt{2|k|}}\sum_{\lambda = i}\varepsilon_{k\lambda}\theta(R-x)(a_{k\lambda}e^{ik\cdot x} + \bar{a}_{k\lambda}e^{-ik\cdot x}) \quad (i = 0,3) \tag{2.4.13}$$

The above representation is a longitudinal photon, which is in a virtual photon state because of no-observability.

3. The relationship between photon probability wave function and photon intrinsic function

If the physical propertiesin intrinsic space can be expressed in the macroscopic observation, then has the basic condition to expand. The description of in the intrinsic space or in observation space is differences. In the intrinsic space $A(\xi) = \exp[i(k_i\xi^i - k_0\xi^0)]$, intrinsic momentum $k = k_i e^i$, and e^i is the basis vector in intrinsic space. For the intrinsic spatial expansion to the observation space, 4-momentum k is unchanged. Considering the coincidence of internal space and observation space coordinate frame, the relationship between photon probabilistic wave function and photon intrinsic function shows $2\Phi = \lambda, 2\Phi_0 = T$.

$$\xi = x/2 \tag{2.4.14}$$

$$A(\xi) = u\exp\left[i\left(\frac{\pi}{T}\xi^0 - \frac{\pi}{\lambda_i}\xi^i\right)\right] \tag{2.4.15}$$

Photon probability wave function:

$$A_\perp(x) = u\exp\left[i\left(\frac{2\pi}{T}t - \frac{2\pi}{\lambda_i}x^i\right)\right] \tag{2.4.16}$$

Intrinsic 2-dimensional simplified wave function and photon probability wave function has exactly the same form. Intrinsic spatial frame can be extended to the particle space frame (Fig. 2.4.3), $\xi^\mu = x^\mu(\xi)$, and $x^\mu(\xi)$ is abbreviated as x^μ. 4-dimensional wave vector in intrinsic space and 4-dimensional wave vector in degrees of freedom space is strictly equivalent, and the equivalence is with longitudinal wave vector.

$$k = \frac{\pi}{\Phi} \equiv \frac{2\pi}{\lambda}, \quad k_0 = \frac{\pi}{\Phi_0} \equiv \frac{2\pi}{T} \tag{2.4.17}$$

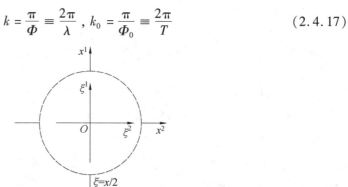

Fig. 2.4.3 Intrinsic space frame can be extended to particle space frame

4. The relationship between the photon longitudinal and transverse waves

Look at the fiber field in observation space (Fig. 2.4.4), the photon move along x-axis (Fig. 2.4.4). The transverse vibration of photon center constitutes a transverse wave. Necessary to consider the vibration of photon center would be contrary to the principle of the speed of light cannot exceed, the answer is negative. Because dimensions are independent respectively. We do a simple experiment. The intersection of n beams does not affect each other. This experiment illustrates the same point along x-direction and y-direction, and its propagation velocity are independent respectively, this is the independent transmission principle of vacuum. According to this principle, we can regard photon sphere centers vibration along y-shaft (nature of vibration is the wave propagation), which does not affect the propagation of x, namely the speed of photon sphere along x-axis is still c and on y-axis that the maximum speed of photon center is c. The minimum is zero, and photon center trajectory is

$$v = c\sqrt{1 + \cos^2 kx} \qquad (2.4.18)$$

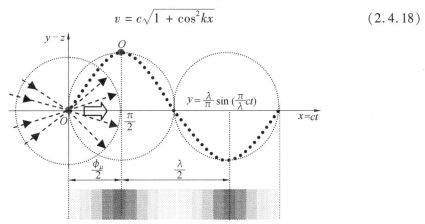

Fig. 2.4.4 Traiectory and probability of photon center

The motion of photon center in y axis is simple harmonic vibration. When photon center achieve the photon spherical shell, its speed is zero. Pass x axis, and the speed is c, therefore $v_y = c \cdot \cos kx$, where $k = \pi/\lambda_\perp$. Along the axis direction, photon center always moves at the speed of light, which is not affected by the y-direction, so $v_x = c$. The movement of the center trajectory meet the cosine function $\cos kx$, in this composite trajectory, center line speed are equal to: when $kx = n\pi$, pass x-axis, v is a maximum value, and phase velocity is $\sqrt{2}c$ that phase velocity is maximum where probability minimum; when $n = (2n + 1)\pi/2$, the speed is maximum, and phase velocity is c; phase velocity is minimum on the peaks where probability is maximum.

Here, make a note: the 4-momentum of photon did not change and overall deformation of the photon still is h. Photons do not fluctuate horizontally until being disturbed, while the results of transverse wave characteristics are disturbed. The original longitudinal photon has no probability wave, and longitudinal photon is beyond observation, which we call the virtual photon. Process of photon Generation itself is a kind of disturbance, the measurement itself is also a kind of disturbance, which is also gives the fluctuation of photon. Only transverse photons have characteristics of probability wave, which can be observed.

$A_{//}(\xi)$ and $A_\perp(\xi)$ are the two sides of a coin. See $A_{//}(\xi)$ in the intrinsic space, and see $A_\perp(\xi)$ out of the intrinsic space.

5. Geometric image of uncertainty principle

Can we determine location of photon center? When we detect quantum measurements in a given direction, the position from the surface to the center of the photosphere is R. Considering that another quantum field also has internal space, measuring distance is $\Delta x \geqslant R$. The essence of measurement is to detect target particles by probe photons. The diameter of the probe photon sphere is Φ, and the momentum is Δp. So $\Delta p \cdot \Phi = h$. $\Delta \Phi$ is the distance between two center of quantum field. Δx is the distance of causes a disturbance, and it's the distance between probe photon center and the target particle center, $\Delta x \geq \Phi$. The probe photon can be a real photon or can be a virtual photon. Here, $\Delta \Phi \geq \Phi$. If $\Delta \Phi \leq \Phi$, it will lose the space which virtual photons pass through.

$$\Delta p \cdot \Delta \Phi \geq \frac{h}{2} \qquad (2.4.19)$$

This is known as the uncertainty relation (Fig. 2.4.5). When a probe photon ball touch another quantum field, so the quantum center position cannot be accurately detected, the detection limit of experimental measuring uses a quantum field to sense another quantum field.

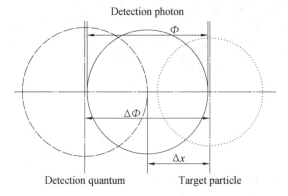

Fig. 2.4.5 Uncertainty principle

6. The description of photon transverse wave in quantum space

In the experiment, we can classify photon by polarizer and filter out the photon, whose polarization direction is orthogonal. As a result, we believe that the photon has a transverse wave characteristics, the polarization surface is perpendicular to the propagate direction k.

Photons are measured by a detector, which detects a single photon and generates electrical signals (Fig. 2.4.6). The detector can only measure the total energy and momentum of photons, but cannot measure the components p_1, p_2, p_3. This is because the detector cannot enter the photon interior space. Limited by the experimental conditions, the detector cannot enter the photon interior space. Based on the existing experimental conditions, we have established the physical image of photons. Photon has wave-particle duality and photon is transverse wave, and the momentum k is in the direction of propagation. In the other direction the momentum is zero. According to the present physical image, we can see that $|k| = k_3$.

From the view of the field deformation, $A_\mu(\xi) = h\exp(ik_\mu \xi^\mu)$, k_μ moving along ξ^μ produced $A_\mu(\xi)$. Experiments in observation space show that only transverse wave have no longitudinal waves:

$$\begin{cases} k_3 \neq 0, A_3(x) = 0 \text{ (Along the direction of propagation, i.e. longitudinal wave direction)} \\ k_1 = k_2 = 0, A_1(x) \neq 0, A_2(x) \neq 0 \text{ (Transverse wave direction)} \end{cases}$$

$$(2.4.20a)$$

Chapter 2 Structure of the Photons

Photon entering the detector generates an electrical signal

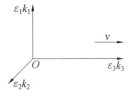

Fig. 2.4.6 Polar coordinate system

$$A(t,x) = A^0(t,x)\varepsilon_{(0)} - A^1(t,x)\varepsilon_{(1)} - A^2(t,x)\varepsilon_{(2)} - A^3(t,x)\varepsilon_{(3)}$$

Vector potential $A_1(x)$ and $A_2(x)$ are perpendicular to the direction k of light propagation and the direction of vector potential is the direction of light polarization, which satisfies the following requirements:

$$k \cdot \varepsilon_{(1)} = k \cdot \varepsilon_{(2)} = 0 \qquad (2.4.20b)$$

This theory is contrary to the principle of wave generation by vacuum strain $u(\xi) = \exp(k\xi)$. This description is well consistent with experimental observations that show the form of theory description is determined by experimental observation. Obviously, there is a huge difference between the internal space and the observation space. The photon transverse waves meet

$$\Box A_t(x) = 0 \qquad (2.4.21)$$

Any solution of the equation is expressed as the following linear combination:

$$A_t(x) = \sum_{k,\lambda=1}^{2}(a_{k\lambda}A_{k\lambda}^{(+)}(x) + a_{k\lambda}^{*}A_{k\lambda}^{(-)}(x)) = \frac{1}{\sqrt{V}}\sum\frac{1}{\sqrt{2|k|}}\sum_{\lambda=1}^{2}\varepsilon_{k\lambda}(a_{k\lambda}e^{-ik\cdot x} + \bar{a}_{k\lambda}e^{ik\cdot x})$$

$$(2.4.22)$$

Among them, $a_{k\lambda}$ and $a_{k\lambda}^{*}$ are expansion coefficients $|k| = \omega_k$. Because $\varepsilon_{k\lambda}$ can be used as the first and the second axis base vector in vertical and horizontal coordinate system, so the above equation in the vertical and horizontal coordinate system is written as

$$A_{k\lambda}^{(+)}(x) = \frac{1}{\sqrt{V}}\frac{1}{\sqrt{2|k|}}\varepsilon_{k\lambda}e^{-ik\cdot x} \quad (\lambda = 1,2) \qquad (2.4.23a)$$

$$A_{k\lambda}^{(-)}(x) = \frac{1}{\sqrt{V}}\frac{1}{\sqrt{2|k|}}\varepsilon_{k\lambda}e^{ik\cdot x} \quad (\lambda = 1,2) \qquad (2.4.23b)$$

Here, $A_{k\lambda}^{(+)}(x)$ and $A_{k\lambda}^{(-)}(x)$ are positive and negative frequency plane wave solutions; $\frac{1}{\sqrt{V}}\frac{1}{\sqrt{2|k|}}$ is the normalization constant; $\varepsilon_{k\lambda}$ is the unit vector in the direction $A_{k\lambda}^{(+)}(x)$. $\lambda = 1,2$ are two transversely polarized states of photons. Because the two polarized states are independent of each other, the following orthogonal conditions are satisfied.

$$\varepsilon_{k\lambda}\varepsilon_{k\lambda'} = \delta_{\lambda\lambda'} \qquad (2.4.24)$$

$A_t(x)$ is the expression of physical fluctuation of photon intrinsic vacuum strain in observation space. The idea that photons have only two transverse polarizations is obviously limited by the experimental conditions.

7. Photon probability wave function

Why the photon has the wave-particle duality? Because photon has both longitudinal and transverse waves and non-proliferation of photon longitudinal waves which performance as particle, while the transverse wave shows that the phase velocity of photon center has periodic speed changing. Observations show that photons have the properties of probability waves, and photons have two properties at the same time, which are defined as the wave-particle duality of photons.

In fact, only probability waves can be observed. The wave function is very simple. It does not involve the intrinsic structure of photons, but only describes the probability distribution of photon centers. We observed that the intensity of light in a region is the probability distribution of the photon center. In other words, only the shear wave can be observed. Expression of probability of photons is given by $\phi = a_k e^{i(\omega \cdot t - kx)}$. Photon of the opposite direction of motion with dual structure is $\phi^* = \bar{a}_k e^{-i(\omega \cdot t - kx)}$, and a_k is the amplitude of probability wave. Photon of movement in the opposite direction has antithesis structure. Photon wave function is

$$A_\perp(x) = \frac{1}{\sqrt{V}} \sum \frac{1}{\sqrt{2|k|}} \sum_{\lambda=1}^{2} \varepsilon_{k\lambda} (a_{k\lambda} e^{ikx} + \bar{a}_{k\lambda} e^{-ikx})$$

Contrast to the photon longitudinal wave function

$$A_{/\!/}(x) = \frac{1}{\sqrt{V}} \sum \frac{1}{\sqrt{2|k|}} \sum_{\lambda=i} \varepsilon_{k\lambda} \theta(R - x) (a_{k\lambda} e^{ikx} + \bar{a}_{k\lambda} e^{-ikx}) \quad (i = 0, 3)$$

The above two formulas are combined as

$$A(x) = \frac{1}{\sqrt{V}} \sum \frac{1}{\sqrt{2|k|}} \sum_{\lambda=0}^{3} [\varepsilon_{k\lambda(\lambda=1,2)} + \varepsilon_{k\lambda(\lambda=0,3)} \theta(R - x)](a_{k\lambda} e^{ikx} + \bar{a}_{k\lambda} e^{-ikx})$$

(2.4.25)

This is a complete representation of the photon wave function. It can be seen from the above equation, when $x \leq R$, the limit by the uncertainty principle $\Delta p \cdot R \geq h$ that longitudinal wave is without observability, it's virtual photon. When $x \geq R$, since it is impossible to detect the interior of photons, the P-wave has no observable effect, so the θ-function is used to delete the P-wave term. For experimental observers, only photons are retained as shear-wave terms.

2.4.4 The degree of freedom of photon spin

1. Spin angular momentum

Photon spin is the rotational strain of electric flux line field in photon interior space. Rotational strain $\omega_{ij}^B = \frac{\partial A_i}{\partial \xi_j} - \frac{\partial A_j}{\partial \xi_i}$ constitutes photon magnetic field which is a spin. ω_{12} is positive rotation, and ω_{21} reverse rotation. We know by the fiber structure of photon $A_1(\xi), A_2(\xi) \ll A_3(\xi)$. $A_3(\xi)$ reaches the limit of deformation in the ξ_3 axis. When a substance is deformed,

Chapter 2 Structure of the Photons

and the deformation again becomes relatively more difficult, so in the direction of propagation cannot add a new strain, while ω_{13}, ω_{31}, ω_{23}, ω_{32} does not exist, and there is only ω_{12} and ω_{21}. So photons have only two spin states: left-handed and right-handed. Left and right spin angular momentum correspond to positive and negative spin angular momentum respectively.

We have to repeat that spin waves do not cause the photons to rotate, but they produce magnetic field effects. Although the spin of the photon is the same in the front and back hemispheres, the photon is composed of positive and negative power lines. The magnetic field generated is equal in size and opposite in direction, and the superposition is 0.

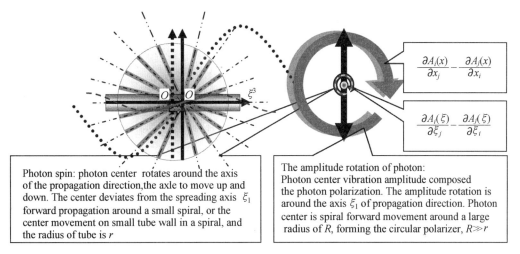

Fig. 2.4.7 Photon spin and amplitude spin
(Photon center is the synthesis of the two types of movement)

The photons with spiral precession and the total strain of the spin wave is $u_S = \oint_{2\pi r} du(\xi) = \hbar$. Momentum $P_S = \partial u_{rS}/\partial S$, while $S = \theta \cdot r$ is the arc length. Photon center spiral precession, set r as radius of photon center to rotate around the helix. When $S = 2\pi r$, the circle locus of photon center is wavelength $\lambda = 2\pi r$ that constitutes a single standing wave (Fig. 2.4.7), which is low-energy stable state. Therefore rotation waves meet $2\pi r \times P_S = h$, then

$$r \times P_S = \frac{h}{2\pi} = \hbar \qquad (2.4.26)$$

This is the photon spin angular momentum.

In quantum mechanics, photon spin angular momentum is $\pm 1\hbar$. Among them, the left is positive and the right is negative which corresponds to quantum states $S_z = + \hbar$ and $S_z = - \hbar$ (Fig. 2.4.8)

$$|L\rangle = \frac{1}{\sqrt{2}} \begin{bmatrix} 1 \\ i \\ 0 \end{bmatrix} e^{i(kz-\omega t)}$$

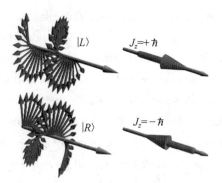

Fig. 2.4.8 Photon spin state

$$|R\rangle = \frac{1}{\sqrt{2}} \begin{bmatrix} 1 \\ -i \\ 0 \end{bmatrix} e^{i(kz-\omega t)} \qquad (2.4.27)$$

2. Photon spin in particle space

Photon spin should have three polarization states. However, due to electromagnetic waves are transverse wave, photons actually have only two independent transverse polarization states. The polarization vector $\varepsilon_{k\lambda}$ ($\lambda = 1, 2$) is as the amplitude of the photon spin wave function. Due to the horizontal of electromagnetic wave, the third axis basis vectors ε_{k3} in intrinsic space vertical horizontal coordinates is photon polarization vector, the $\varepsilon_{k\sigma}$ ($\sigma = 1, 2, 3$) constitutes a set of common eigenfunctions which are "complete", orthogonal and normalized by $\hat{s}^2$ and $\hat{s}_3$.

$$\varepsilon_{k\sigma} \cdot \varepsilon_{k\sigma'} = \delta_{\sigma\sigma'} \qquad (\sigma, \sigma' = 1, 2, 3) \qquad (2.4.28)$$

$\varepsilon_{k\sigma}$ is a 3-dimensional vector, can be expressed as the following columns matrix $\varepsilon_{k\sigma} = \begin{bmatrix} \varepsilon_{k\sigma}^1 \\ \varepsilon_{k\sigma}^2 \\ \varepsilon_{k\sigma}^3 \end{bmatrix}$, $\sigma = 1, 2, 3$. Accordingly, amplitude of spin component operator $\hat{s}_j$ ($j = 1, 2, 3$) is 3×3 matrix:

$$\hat{s}_j = \begin{bmatrix} s_j^{11} & s_j^{12} & s_j^{13} \\ s_j^{21} & s_j^{22} & s_j^{23} \\ s_j^{31} & s_j^{32} & s_j^{33} \end{bmatrix}$$

For example, when $j = 1$, having

$$\hat{s}_1 \varepsilon_{k\sigma} = \hat{s}_1 \begin{bmatrix} \varepsilon_{k\sigma}^1 \\ \varepsilon_{k\sigma}^2 \\ \varepsilon_{k\sigma}^3 \end{bmatrix} = \begin{bmatrix} s_j^{11} \varepsilon_{k\sigma}^1 + s_j^{12} \varepsilon_{k\sigma}^2 + s_j^{13} \varepsilon_{k\sigma}^3 \\ s_j^{21} \varepsilon_{k\sigma}^1 + s_j^{22} \varepsilon_{k\sigma}^2 + s_j^{23} \varepsilon_{k\sigma}^3 \\ s_j^{31} \varepsilon_{k\sigma}^1 + s_j^{32} \varepsilon_{k\sigma}^2 + s_j^{33} \varepsilon_{k\sigma}^3 \end{bmatrix}$$

According to quantum mechanics, the operator $\hat{s}_1$, $\hat{s}_2$ and $\hat{s}_3$ matrix are as follows.

$$\hat{s}_1 = \begin{bmatrix} 0 & 0 & 0 \\ 0 & 0 & -i \\ 0 & -i & 0 \end{bmatrix}, \hat{s}_2 = \begin{bmatrix} 0 & 0 & i \\ 0 & 0 & 0 \\ -i & 0 & 0 \end{bmatrix}, \hat{s}_3 = \begin{bmatrix} 0 & -i & 0 \\ i & 0 & 0 \\ 0 & 0 & 0 \end{bmatrix} \qquad (2.4.29)$$

Chapter 2 Structure of the Photons

According to the above three additive, it is available that

$$\hat{s}^2 = \begin{bmatrix} 2 & 0 & 0 \\ 0 & 2 & 0 \\ 0 & 0 & 2 \end{bmatrix} \quad (2.4.30)$$

Matrix form of operator $\hat{s}^2$ shows that photon amplitude of spin is 1.

If the field $A_\mu(\xi)$ rotates around ξ_3 axis turn to γ, then the point ξ_1 change into ξ_1''.

$$\begin{bmatrix} \xi_1' \\ \xi_2' \\ \xi_3' \end{bmatrix} = \begin{bmatrix} \cos\gamma & -\sin\gamma & 0 \\ \sin\gamma & \cos\gamma & 0 \\ 0 & 0 & 1 \end{bmatrix} \begin{bmatrix} \xi_1 \\ \xi_2 \\ \xi_3 \end{bmatrix}$$

Similarly, field rotates around ξ_1-axial rotation angle is α, and rotates around ξ_2-axial rotation angle is β. The rotation matrices are respectively as

$$\begin{bmatrix} 1 & 0 & 0 \\ 0 & \cos\alpha & -\sin\alpha \\ 0 & \sin\alpha & \cos\alpha \end{bmatrix}, \quad \begin{bmatrix} \cos\beta & 0 & \sin\beta \\ 0 & 1 & 0 \\ -\sin\beta & 0 & \cos\beta \end{bmatrix}$$

General rotation matrix are the above three matrices. According to $(\theta^1, \theta^2, \theta^3) = (\alpha, \beta, \gamma)$, when θ^i is infinitesimal, general rotation matrix can be written as

$$(a_j^i) = \begin{bmatrix} 1 & -\theta^3 & \theta^2 \\ \theta^3 & 1 & -\theta^1 \\ -\theta^2 & \theta^1 & 1 \end{bmatrix} = 1 - iJ_k\theta^k = e^{-iJ_k\theta^k} \quad (2.4.31)$$

Here, J_i is the following matrix:

$$J_1 = \begin{bmatrix} 0 & 0 & 0 \\ 0 & 0 & -i \\ 0 & -i & 0 \end{bmatrix}, \quad J_2 = \begin{bmatrix} 0 & 0 & i \\ 0 & 0 & 0 \\ -i & 0 & 0 \end{bmatrix}, \quad J_3 = \begin{bmatrix} 0 & -i & 0 \\ i & 0 & 0 \\ 0 & 0 & 0 \end{bmatrix}$$

They are generating the rotation operator of three dimensional coordinate spaces $J_i = \hat{s}_i$. In vertical horizontal coordinate system, the amplitude of spin third component operator $\hat{s}_3$ is with helicity operator h consistent.

$$\hat{s}_3 = \hat{s} \cdot \varepsilon_{k3} = \hat{s} \cdot k/|k| = h. \quad (2.4.32)$$

Photon rest mass is zero. Spin state is Lorentz invariant using the h eigenvalue to describe. It should be noted that the rest mass of photon is zero, and photon amplitude of spin angular momentum and orbital angular momentum exists in the same Minkonwski space. There is no additional "spin space".

For photon polarization states, in vertical horizontal coordinate system, ε_{k3} is expressed as

$$\varepsilon_{k3} = \begin{bmatrix} 0 \\ 0 \\ 1 \end{bmatrix} = e^0 = e^3 \quad (2.4.33)$$

Therefore, $h = \varepsilon_{k3} = \begin{bmatrix} 0 \\ 0 \\ 0 \end{bmatrix} = 0 \times \begin{bmatrix} 0 \\ 0 \\ 1 \end{bmatrix}$; ε_{k3} is description of $h=0$ polarization states. For $\varepsilon_{k\lambda}$ ($\lambda = 1,2$), $h = \pm 1$ polarization state should be get. Use eigenvalue equations $h\varepsilon_{k1} = \varepsilon_{k1}$, $h\varepsilon_{k2} = -\varepsilon_{k2}$ and formula (2.4.32), it is not difficult to get

$$\varepsilon_{k1} = \frac{1}{\sqrt{2}}\begin{bmatrix} 1 \\ i \\ 0 \end{bmatrix} = \frac{1}{\sqrt{2}}(e^1 + ie^2) = e^+ \ , \ \varepsilon_{k2} = \frac{1}{\sqrt{2}}\begin{bmatrix} 1 \\ -i \\ 0 \end{bmatrix} = \frac{1}{\sqrt{2}}(e^1 - ie^2) = e^- \quad (2.4.34)$$

(Right circular polarization state)   (Left circular polarization states)

According to the horizontal condition (2.4.28), $h=0$, and longitudinal polarization state is automatically ruled out, while photon has only two transverse polarization states. The three physical significances of the eigenvectors can be understood as a vector harmonic propagating along $k \ // \ \varepsilon_{k3}$, and it is expanded by these three eigenvectors

$$A(\xi,t) = (a_1\varepsilon_{k1} + a_2\varepsilon_{k2} + a_3\varepsilon_{k3})e^{-i(\omega t - k\xi_3)} = (a_+ e^+ + a_- e^- + a_0 e^0)e^{-i(\omega t - k\xi_3)} \quad (2.4.35)$$

From the e^+ in (2.3.34), it can be seen that the e^2 phase lags behind the e^1 phase $\pi/2$, and e^- is just the opposite. J_3 projection in propagation direction is $s = +1$, and its component a_+ is right circular wave. Similarly, $s = -1$, whose component a_- is left circular wave, and $s = 0$, whose component a_0 is longitudinal wave.

Photon spin is the intrinsic property of photon, which has no direct observability in experiment. But because of the existence of spin, photon has the magnetic field of positive and negative superposition state, and photon has the property of electromagnetic wave.

2.4.5 Photon orbital angular momentum

1. Photon spatial displacement function in the observation space

The photon propagation is along z-axis direction (Fig. 2.4.9), and vector potential $A_r(t,r)$ is

$$A_r(t,r) = A_x(t,r)\varepsilon_x + A_y(t,r)\varepsilon_y + A_z(t,r)\varepsilon_z \quad (2.4.36)$$

Here, $\varepsilon_x, \varepsilon_y, \varepsilon_z$ are polarization vectors; A_x, A_y, A_z are perpendicular to each other, meet

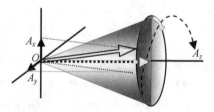

Fig. 2.4.9   Vector potential $A(r,t)$ is the amount of rotation

$$A_r(t,r) = \sqrt{A_x^2(t,r) + A_y^2(t,r) + A_z^2(t,r)}$$

Here, $A_{xy} = \sqrt{A_x^2 + A_y^2}$, A_z could not be observed. $A_r(t,r)$ is an invariant. $A_x(t,r)$ and $A_y(t,r)$ are two components to meet

Chapter 2 Structure of the Photons

$$\begin{cases} A_x = A_r\sin\theta\cos\Phi = A_{xy}\cos\Phi \\ A_y = A_r\sin\theta\sin\Phi A_r = A_{xy}\sin\Phi \end{cases} \quad (2.4.37)$$

Here, Φ is the argument of vector potential at the point. Vector potential $A_r(t,r)$ is an amount of rotation in 3-dimensional space.

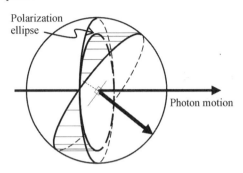

Fig. 2.4.10 Elliptical polarization

Point oscillations in xy 2-dimensional plane, speed v_x and v_y are between $0 - v_{max}$ (Fig. 2.4.11). The speed is zero, and is most likely measured. The speed v_{max} is least likely measured. Trajectory of photon center in 2-dimensional surface is oval ring.

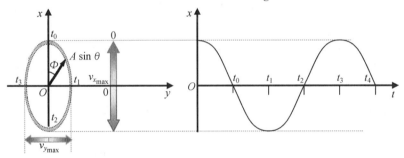

Fig. 2.4.11 Photon spin

Center of single photon is propagation along z direction and at the same time this point elliptical rotation along the z direction, it is the spiral precession. As time increases, $+\omega t$ (the amplitude of phase argument) increases 2π argument. The center rotation a cycle corresponding the photon center along z direction propagation a cycle, and it can be seen as photons forward propagation along z axis. On the contrary, $-\omega t$ is the photon along z axis conversely propagation.

On the other hand, the linearly polarized light in either direction can be decomposed into a left-handed circularly polarized light A_L and a circularly polarized light A_R.

$$\begin{cases} A_L = Ae^{+i\Phi} = A_x + iA_y \\ A_R = Ae^{-i\Phi} = A_x - iA_y \end{cases} \quad (2.4.38)$$

Write $\exp(-i\omega t)$ as the time factor of wave function, when the phase A_L and A_R are $\Phi-\omega t$ and $\Phi+\omega t$. As time t increases, photon center spiral precession is along z axis, the same phase point of argument increase (Left) or decrease (Right). In cycle $T = 2\pi/\omega$, vector potential turn 2π argument.

With rotation of the vector potential, we further give photons spin angular momentum in an expressive sense. By the analogy with spherical harmonic functions, there are

$$\begin{cases} A_L \sim Y_{1+1} \\ A_R \sim Y_{1-1} \end{cases} \quad (2.4.39)$$

Rotating nature of the vector potential similar spherical harmonic function $Y_{lm}, l = 1$. Therefore, this analogy gets photon with $l = 1$ spin angular momentum. This conclusion is universal. The spin angular momentum $l = 1$ for any particle is described by a vector wave function. For left and right circularly polarized photon, spin angular momentum projection on z axis respectively are $+\hbar$ and $-\hbar$.

The angular momentum of optical orbit is caused by the rotation of shear wave amplitude. The strain of rotation of shear wave amplitude $A_i(x)$ is $B_{ij} = \dfrac{\partial A_i(x)}{\partial x_j} - \dfrac{\partial A_j(x)}{\partial x_i}$.

The equation (2.4.27) is the two circularly polarized states in the vertical plane in observation space.

So, the light is only two kinds of polarized forms: left-handed and right-handed. Left and right are correspond to positive and negative spin angular momentum respectively.

Left and right rotation polarized state of photons is the eigenstate of the spin component in the moving direction (Fig. 2.4.10). The eigenstate is ± 1. Linearly polarized light is the superposition state by left-rotated and right-rotated polarized light. So, the linearly polarized light is not the eigenstate of the spin component in the moving direction, and the spin component of the motion direction is not determined. Thus, it has no the determination spin component in the motion direction. In addition, the mean value of the spin component of the linearly polarized light in the direction of motion is 0.

The following is the description of the intrinsic field function of photon at external observation space.

For experimental observations, single photon is just a highlight on the photographic film, which cannot show intrinsic wave $A_{//}(\xi, t) = (a_1 \varepsilon_{k1} + a_2 \varepsilon_{k2} + a_3 \varepsilon_{k3}) e^{-i(\omega t - k\xi_3)}$ effect of the photon. So we mistakenly think that photonic structures are the point. Many highlights show the probability of transverse wave, and only transverse wave shows the characteristics of photon fluctuations. And longitudinal wave is redundant. Photon spin is an intrinsic spin wave, which is not observable, but the amplitude rotation of polarized light is observable, which shows circular or elliptical polarization of left or right rotation. We unified the spin of photons and the rotation of shear wave amplitude as $A(\xi) + A(x)$.

$$A(x) + A(\xi) = \frac{1}{\sqrt{V}} \sum \frac{1}{\sqrt{2|k|}} \sum_{\lambda=0}^{3} [\varepsilon_{k\lambda(\lambda=1,2)} + \varepsilon_{k\lambda(\lambda=0,3)} \theta(R-x)] \cdot$$
$$[a_{k\lambda}(e^{ikx} + e^{ik\xi}) + \bar{a}_{k\lambda}(e^{-ikx} + e^{-ik\xi})] \quad (2.4.40)$$

Chapter 2 Structure of the Photons

This describes the information inside the photon. The photon has the characteristic of longitudinal photon and electromagnetic field. Deleting the intrinsic wave functions which is no-observability, we can get the expression, and we can also get the expression of wave functions which is consistent with the experiment.

$$A_t(x) = \sum_{k,\lambda=1}^{2} [a_{k\lambda} A_{k\lambda}^{(+)}(x) + a_{k\lambda}^* A_{k\lambda}^{(-)}(x)] = \frac{1}{\sqrt{V}} \sum_k \frac{1}{\sqrt{2|k|}} \sum_{\lambda=1}^{2} \varepsilon_{k\lambda} (a_{k\lambda} e^{-ikx} + \bar{a}_{k\lambda} e^{ikx})$$

(2.4.41)

This is the wave function of light.

2.5 Complete field functions of free photons

All these characteristics of photons are caused by elastic strain, so we can use a concise formula of quantum field wave function to express the intrinsic space properties of photon.

$$\phi_\gamma(\xi) = \exp\{-i[\varepsilon_{\alpha\beta}](\xi)\}$$ (2.5.1)

The quantum field contains all the information. Since the intrinsic space has no-observability in the observation space, for convenience, we mix the information in the observation space with the intrinsic information of the quantum field and express it by a unified wave function, which is written as

$$\varphi_\gamma(X) = \exp(-i)\{[\varepsilon_{\alpha\beta}](\xi,x)\} = \exp(-i)\begin{bmatrix}\varepsilon_0 & 0 & 0 & 0\\ 0 & 0 & 0 & 0\\ 0 & 0 & 0 & 0\\ 0 & 0 & 0 & 0\end{bmatrix}\underset{\gamma_0}{\sim}\begin{bmatrix}\xi_0 & 0 & 0 & 0\\ 0 & 0 & 0 & 0\\ 0 & 0 & 0 & 0\\ 0 & 0 & 0 & 0\end{bmatrix} +$$
$$\underbrace{}_{(1)\ \text{Energy of photon } E_0}$$

$$\underbrace{\begin{bmatrix}0 & 0 & 0 & 0\\ 0 & 0 & 0 & 0\\ 0 & 0 & 0 & 0\\ 0 & 0 & 0 & \varepsilon_{33}\end{bmatrix}\underset{\gamma_3}{\sim}\begin{bmatrix}0 & 0 & 0 & 0\\ 0 & 0 & 0 & 0\\ 0 & 0 & 0 & 0\\ 0 & 0 & 0 & \xi_3\end{bmatrix}}_{\substack{(11)\ \text{Photon intrinsic momentum,}\\ \text{confined to the intrinsic space}}} +$$

$$\underbrace{\begin{bmatrix}0 & 0 & 0 & 0\\ 0 & \Delta\varepsilon_{11} & 0 & 0\\ 0 & 0 & 0 & 0\\ 0 & 0 & 0 & 0\end{bmatrix}\underset{\gamma_1}{\sim}\begin{bmatrix}0 & 0 & 0 & 0\\ 0 & x_1 & 0 & 0\\ 0 & 0 & 0 & 0\\ 0 & 0 & 0 & 0\end{bmatrix}}_{(12)\ \text{Momentum } p_1,\ p_1\gamma_1} +$$

$$\underbrace{\begin{bmatrix}0 & 0 & 0 & 0\\ 0 & 0 & 0 & 0\\ 0 & 0 & \Delta\varepsilon_{22} & 0\\ 0 & 0 & 0 & 0\end{bmatrix}\underset{\gamma_2}{\sim}\begin{bmatrix}0 & 0 & 0 & 0\\ 0 & 0 & 0 & 0\\ 0 & 0 & x_2 & 0\\ 0 & 0 & 0 & 0\end{bmatrix}}_{(13)\ \text{Momentum } p_2,\ p_2\gamma_2} +$$

$$\begin{bmatrix} 0 & 0 & 0 & 0 \\ 0 & 0 & \gamma_{12} & \gamma_{13} \\ 0 & \gamma_{21} & 0 & \gamma_{23} \\ 0 & \gamma_{31} & \gamma_{32} & 0 \end{bmatrix} (1 \pm \gamma_5) \begin{bmatrix} 0 & 0 & 0 & 0 \\ 0 & 0 & \xi_1 & \xi_1 \\ 0 & \xi_2 & 0 & \xi_2 \\ 0 & \xi_3 & \xi_3 & 0 \end{bmatrix} +$$

$\underbrace{\qquad\qquad\qquad\qquad\qquad\qquad\qquad\qquad\qquad\qquad}_{\text{Hoop strain constitutes weak field of photon}\Leftrightarrow(1+\gamma_5)}$

(3) Photon weak field

$$\begin{bmatrix} 0 & 0 & 0 & 0 \\ 0 & 0 & \omega_{12} & \omega_{13} \\ 0 & \omega_{21} & 0 & \omega_{23} \\ 0 & \omega_{31} & \omega_{32} & 0 \end{bmatrix} \check{\Sigma}_i \begin{bmatrix} 0 & 0 & 0 & 0 \\ 0 & 0 & \xi_1 & \xi_1 \\ 0 & \xi_2 & 0 & \xi_2 \\ 0 & \xi_3 & \xi_3 & 0 \end{bmatrix} +$$

(4) $\hbar_1$, spin of photon

(4) Rotational strain in internal space

$$\begin{bmatrix} 0 & 0 & 0 & 0 \\ 0 & 0 & \Omega_{12} & \Omega_{13} \\ 0 & \Omega_{21} & 0 & \Omega_{23} \\ 0 & \Omega_{31} & \Omega_{32} & 0 \end{bmatrix} \check{\Sigma}_i \begin{bmatrix} 0 & 0 & 0 & 0 \\ 0 & 0 & x_1 & x_1 \\ 0 & x_2 & 0 & x_2 \\ 0 & x_3 & x_3 & 0 \end{bmatrix} +$$

(4) Photon spin angular momentum in observable space

(4) Rotational strain in observable space

$$\begin{bmatrix} 0 & \gamma_{01} & \gamma_{02} & \gamma_{03} \\ \gamma_{10} & 0 & 0 & 0 \\ \gamma_{20} & 0 & 0 & 0 \\ \gamma_{30} & 0 & 0 & 0 \end{bmatrix} \check{I} \begin{bmatrix} 0 & \xi_0 & \xi_0 & \xi_0 \\ \xi_1 & 0 & 0 & 0 \\ \xi_2 & 0 & 0 & 0 \\ \xi_3 & 0 & 0 & 0 \end{bmatrix} + \begin{bmatrix} 0 & \omega_{01} & \omega_{02} & \omega_{03} \\ \omega_{10} & 0 & 0 & 0 \\ \omega_{20} & 0 & 0 & 0 \\ \omega_{30} & 0 & 0 & 0 \end{bmatrix} \check{I} \begin{bmatrix} 0 & \xi_0 & \xi_0 & \xi_0 \\ \xi_1 & 0 & 0 & 0 \\ \xi_2 & 0 & 0 & 0 \\ \xi_3 & 0 & 0 & 0 \end{bmatrix} +$$

(5) Intrinsic shear 4-momentum flow   (6) Intrinsic 4-momentum flow of Spin

$[\varepsilon_{i0}]$ or $[\varepsilon_{0i}]$ flux of lepton intrinsic 4-momentum through the surface ξ^i. The flux is limited in intrinsic space.

$$\frac{1}{2} \begin{pmatrix} A_{0,0}(\xi) & 0 & 0 & 0 \\ 0 & A_{1,1}(\xi) & 0 & 0 \\ 0 & 0 & A_{2,2}(\xi) & 0 \\ 0 & 0 & 0 & A_{3,3}(\xi) \end{pmatrix} \check{I} \begin{bmatrix} \xi_0 & 0 & 0 & 0 \\ 0 & \xi_1 & 0 & 0 \\ 0 & 0 & \xi_2 & 0 \\ 0 & 0 & 0 & \xi_3 \end{bmatrix} -$$

Intrinsic electromagnetic four vector field of photon $e_\gamma/2$

Intrinsic anti-electromagnetic four vector of photon $e_\gamma/2$

$$\frac{1}{2} \begin{pmatrix} A_{0,0}(\xi) & 0 & 0 & 0 \\ 0 & A_{1,1}(\xi) & 0 & 0 \\ 0 & 0 & A_{2,2}(\xi) & 0 \\ 0 & 0 & 0 & A_{3,3}(\xi) \end{pmatrix} \check{I} \begin{bmatrix} \xi_0 & 0 & 0 & 0 \\ 0 & \xi_1 & 0 & 0 \\ 0 & 0 & \xi_2 & 0 \\ 0 & 0 & 0 & \xi_3 \end{bmatrix} +$$

Intrinsic anti-electromagnetic four vector of photon $e_\gamma/2$

$$\frac{1}{2}\begin{bmatrix} 0 & E_{01} & E_{02} & E_{03} \\ E_{10} & 0 & 0 & 0 \\ E_{20} & 0 & 0 & 0 \\ E_{30} & 0 & 0 & 0 \end{bmatrix}\check{I}\begin{bmatrix} 0 & \xi_0 & \xi_0 & \xi_0 \\ \xi_1 & 0 & 0 & 0 \\ \xi_2 & 0 & 0 & 0 \\ \xi_3 & 0 & 0 & 0 \end{bmatrix} - \frac{1}{2}\begin{bmatrix} 0 & E_{01} & E_{02} & E_{03} \\ E_{10} & 0 & 0 & 0 \\ E_{20} & 0 & 0 & 0 \\ E_{30} & 0 & 0 & 0 \end{bmatrix}\check{I}\begin{bmatrix} 0 & \xi_0 & \xi_0 & \xi_0 \\ \xi_1 & 0 & 0 & 0 \\ \xi_2 & 0 & 0 & 0 \\ \xi_3 & 0 & 0 & 0 \end{bmatrix} +$$

$\underbrace{\qquad\text{Front half electric field } E \qquad\qquad\qquad\qquad\qquad\qquad \text{Seeond half electric field } E \qquad\qquad}$
(15) Electric field in electric flux line formation area, no electric field in observable space

$$\frac{1}{2}\begin{bmatrix} 0 & 0 & 0 & 0 \\ 0 & 0 & \omega_{12}^B & \omega_{13}^B \\ 0 & \omega_{21}^B & 0 & \omega_{23}^B \\ 0 & \omega_{31}^B & \omega_{32}^B & 0 \end{bmatrix}\check{\Sigma}_i\begin{bmatrix} 0 & 0 & 0 & 0 \\ 0 & 0 & \xi_1 & \xi_1 \\ 0 & \xi_2 & 0 & \xi_2 \\ 0 & \xi_3 & \xi_3 & 0 \end{bmatrix}^T - \frac{1}{2}\begin{bmatrix} 0 & 0 & 0 & 0 \\ 0 & 0 & \omega_{12}^B & \omega_{13}^B \\ 0 & \omega_{21}^B & 0 & \omega_{23}^B \\ 0 & \omega_{31}^B & \omega_{32}^B & 0 \end{bmatrix}\check{\Sigma}_i\begin{bmatrix} 0 & 0 & 0 & 0 \\ 0 & 0 & \xi_1 & \xi_1 \\ 0 & \xi_2 & 0 & \xi_2 \\ 0 & \xi_3 & \xi_3 & 0 \end{bmatrix} +$$

$\underbrace{\text{First half photon magnetic line in internal formation area } B/2, \quad \text{second half photon reverse magnetic line in internal formation area } -B/2}$
(10) Intrinsic magnetic have no effect in observable space

$$\frac{1}{2}\begin{bmatrix} 0 & E_{01} & E_{02} & E_{03} \\ E_{10} & 0 & 0 & 0 \\ E_{20} & 0 & 0 & 0 \\ E_{30} & 0 & 0 & 0 \end{bmatrix}\check{I}\begin{bmatrix} 0 & x_0 & x_0 & x_0 \\ x_1 & 0 & 0 & 0 \\ x_2 & 0 & 0 & 0 \\ x_3 & 0 & 0 & 0 \end{bmatrix} - \frac{1}{2}\begin{bmatrix} 0 & E_{01} & E_{02} & E_{03} \\ E_{10} & 0 & 0 & 0 \\ E_{20} & 0 & 0 & 0 \\ E_{30} & 0 & 0 & 0 \end{bmatrix}\check{I}\begin{bmatrix} 0 & x_0 & x_0 & x_0 \\ x_1 & 0 & 0 & 0 \\ x_2 & 0 & 0 & 0 \\ x_3 & 0 & 0 & 0 \end{bmatrix} +$$

$\underbrace{\qquad\text{Front half electric field } E \qquad\qquad\qquad\qquad\qquad\qquad \text{second half electric field } E \qquad\qquad}$
(15) No electric field in observable space

$$\frac{1}{2}\begin{bmatrix} 0 & 0 & 0 & 0 \\ 0 & 0 & B_{12} & B_{13} \\ 0 & B_{21} & 0 & B_{23} \\ 0 & B_{31} & B_{32} & 0 \end{bmatrix}\check{I}\begin{bmatrix} 0 & 0 & 0 & 0 \\ 0 & 0 & x_1 & x_1 \\ 0 & x_2 & 0 & x_2 \\ 0 & x_3 & x_3 & 0 \end{bmatrix} - \frac{1}{2}\begin{bmatrix} 0 & 0 & 0 & 0 \\ 0 & 0 & B_{12} & B_{13} \\ 0 & B_{21} & 0 & B_{23} \\ 0 & B_{31} & B_{32} & 0 \end{bmatrix}\check{I}\begin{bmatrix} 0 & 0 & 0 & 0 \\ 0 & 0 & x_1 & x_1 \\ 0 & x_2 & 0 & x_2 \\ 0 & x_3 & x_3 & 0 \end{bmatrix}$$

$\underbrace{\qquad\text{First half photon magnetic field } B \qquad\qquad\qquad\qquad \text{second half photon magnetic field } B \qquad}$
(16) No magnetic field in observable space

(2.5.2)

The above is simplified as

$$\varphi_\gamma(X) = \exp(-\mathrm{i})\{[\varepsilon]_n X_n\} \qquad (2.5.3)$$

Each strain constitutes a kind of freedom, which forms a generalized multi-dimensional space. The generalized degree of freedom is introduced into the metric g_{mn}, satisfying $X_m = g_{mn}X^n = (X^0, -X)$.

$$\varphi_r(X) = \exp\{-\mathrm{i}[\varepsilon]_n g_{mn} X^n\} \qquad (2.5.4)$$

2.5.1 Eigenvalue operator of photons

Based on the above discussions, we give eigenoperators of all photon physical quantities (Table 2.5.1).

Table 2.5.1 The relationship between the free particle intrinsic space and freedom of space

Items	Intrinsic strain	Degree of freedom coordinates	Eigenvalue acquisition operator
Momentum p_1	$p_1 = \begin{bmatrix} 0 & 0 & 0 & 0 \\ 0 & \Delta\varepsilon_{11} & 0 & 0 \\ 0 & 0 & 0 & 0 \\ 0 & 0 & 0 & 0 \end{bmatrix}$	$X_1 = \begin{bmatrix} 0 & 0 & 0 & 0 \\ 0 & x_1 & 0 & 0 \\ 0 & 0 & 0 & 0 \\ 0 & 0 & 0 & 0 \end{bmatrix}$	Spatial character γ_1 $\Delta\hat{\varepsilon}_1 = \hat{p}_1 = -i\hbar\dfrac{\partial}{\partial X_1} = \hat{K}_1$ Eigenvalue: p_1
Momentum p_2	$p_2 = \begin{bmatrix} 0 & 0 & 0 & 0 \\ 0 & 0 & 0 & 0 \\ 0 & 0 & \Delta\varepsilon_{22} & 0 \\ 0 & 0 & 0 & 0 \end{bmatrix}$	$X_2 = \begin{bmatrix} 0 & 0 & 0 & 0 \\ 0 & 0 & 0 & 0 \\ 0 & 0 & x_2 & 0 \\ 0 & 0 & 0 & 0 \end{bmatrix}$	Spatial character γ_2 $\Delta\hat{\varepsilon}_2 = \hat{p}_2 = -i\hbar\dfrac{\partial}{\partial X_2} = \hat{K}_2$ Eigenvalue: p_2
Momentum p_3	$p_3 = \begin{bmatrix} 0 & 0 & 0 & 0 \\ 0 & 0 & 0 & 0 \\ 0 & 0 & 0 & 0 \\ 0 & 0 & 0 & \varepsilon_{33} \end{bmatrix}$	$X_3 = \begin{bmatrix} 0 & 0 & 0 & 0 \\ 0 & 0 & 0 & 0 \\ 0 & 0 & 0 & 0 \\ 0 & 0 & 0 & \xi_3 \end{bmatrix}$	Spatial character γ_3 $\Delta\hat{\varepsilon}_3 = \hat{p}_3 - i\hbar\dfrac{\partial}{\partial X_3} = \hat{K}_3$ Eigenvalue: p_3, photon radial mass, photon dynamic mass
Kinetic energy E_0	$E_0 = \begin{bmatrix} \varepsilon_0 + \Delta\varepsilon_0 & 0 & 0 & 0 \\ 0 & 0 & 0 & 0 \\ 0 & 0 & 0 & 0 \\ 0 & 0 & 0 & 0 \end{bmatrix}$	$X_0 = \begin{bmatrix} x_0 & 0 & 0 & 0 \\ 0 & 0 & 0 & 0 \\ 0 & 0 & 0 & 0 \\ 0 & 0 & 0 & 0 \end{bmatrix}$	Spatial character I $\Delta\hat{\varepsilon}_0 = \Delta\hat{E} = -i\hbar\dfrac{\partial}{\partial X_0}$ Eigenvalue: E
Spin $\pm S_1$	$s_1 = \begin{bmatrix} 0 & 0 & 0 & 0 \\ 0 & 0 & 0 & 0 \\ 0 & 0 & 0 & \omega_{23} \\ 0 & 0 & \omega_{32} & 0 \end{bmatrix}$	$X_5 = \begin{bmatrix} 0 & 0 & 0 & 0 \\ 0 & 0 & 0 & 0 \\ 0 & 0 & 0 & \xi_2 \\ 0 & 0 & \xi_3 & 0 \end{bmatrix}$	Spatial character Σ_1 $\hat{\omega}_{[23]} = \hat{S}_1 = -i\hbar\dfrac{\partial}{\partial X_5}$ Eigenvalue: $\pm\hbar/2$
$\pm S_2$	$s_2 = \begin{bmatrix} 0 & 0 & 0 & 0 \\ 0 & 0 & 0 & \omega_{13} \\ 0 & 0 & 0 & 0 \\ 0 & \omega_{31} & 0 & 0 \end{bmatrix}$	$X_6 = \begin{bmatrix} 0 & 0 & 0 & 0 \\ 0 & 0 & 0 & \xi_1 \\ 0 & 0 & 0 & 0 \\ 0 & \xi_3 & 0 & 0 \end{bmatrix}$	Spatial character Σ_2 $\hat{\omega}_{[13]} = \hat{S}_2 = -i\hbar\dfrac{\partial}{\partial X_6}$ Eigenvalue: $\pm\hbar/2$
$\pm S_3$	$s_3 = \begin{bmatrix} 0 & 0 & 0 & 0 \\ 0 & 0 & \omega_{12} & 0 \\ 0 & \omega_{21} & 0 & 0 \\ 0 & 0 & 0 & 0 \end{bmatrix}$	$X_7 = \begin{bmatrix} 0 & 0 & 0 & 0 \\ 0 & 0 & \xi_1 & 0 \\ 0 & \xi_2 & 0 & 0 \\ 0 & 0 & 0 & 0 \end{bmatrix}$	Spatial character Σ_3 $\hat{\omega}_{[12]} = \hat{S}_3 = -i\hbar\dfrac{\partial}{\partial X_7}$ Eigenvalue: $\pm\hbar/2$

Chapter 2 Structure of the Photons

Continued Table 2.5.1

Items	Intrinsic strain	Degree of freedom coordinates	Eigenvalue acquisition operator
Weak interaction field $\gamma_{\mu\nu}(\xi)$	$K_8 = \pm \dfrac{1}{2}\begin{bmatrix} 0 & \gamma_{01} & \gamma_{02} & \gamma_{03} \\ \gamma_{10} & 0 & \gamma_{12} & \gamma_{13} \\ \gamma_{20} & \gamma_{21} & 0 & \gamma_{23} \\ \gamma_{30} & \gamma_{31} & \gamma_{32} & 0 \end{bmatrix}$	$X_8 = \begin{bmatrix} 0 & \xi_0 & \xi_0 & \xi_0 \\ \xi_1 & 0 & \xi_1 & \xi_1 \\ \xi_2 & \xi_2 & 0 & \xi_2 \\ \xi_3 & \xi_3 & \xi_3 & 0 \end{bmatrix}$	$\hat{\gamma}_{\mu\nu} = -i\hbar\dfrac{\partial}{\partial X_8}$ Semi-space $(1+\gamma_5)$ Photon weak field, photon circumferential mass
Intrinsic 4-momentum of charge	$\partial A_\mu/\partial \xi_\mu =$ $\pm \dfrac{1}{2}\begin{bmatrix} \partial A_0/\partial \xi_0 & 0 & 0 & 0 \\ 0 & \partial A_1/\partial \xi_1 & 0 & 0 \\ 0 & 0 & \partial A_2/\partial \xi_2 & 0 \\ 0 & 0 & 0 & \partial A_3/\partial \xi_3 \end{bmatrix}$ It has been included in the intrinsic four momentum of the electron	$X_9 = \begin{bmatrix} \xi_0 & 0 & 0 & 0 \\ 0 & \xi_1 & 0 & 0 \\ 0 & 0 & \xi_2 & 0 \\ 0 & 0 & 0 & \xi_3 \end{bmatrix}$	Photon 4-vector strain Spatial character I $\dfrac{\partial A_\mu}{\partial \xi_\mu} = -i\hbar\dfrac{\partial}{\partial X_9}$ Photon pure electromagnetic mass
Electronic internal electric field	$E(\xi) = \pm \dfrac{1}{2}\begin{bmatrix} 0 & \omega_{01}^B & \omega_{02}^B & \omega_{03}^B \\ \omega_{10}^B & 0 & 0 & 0 \\ \omega_{20}^B & 0 & 0 & 0 \\ \omega_{30}^B & 0 & 0 & 0 \end{bmatrix}$	$X_{10} = \begin{bmatrix} 0 & \xi_0 & \xi_0 & \xi_0 \\ \xi_1 & 0 & 0 & 0 \\ \xi_2 & 0 & 0 & 0 \\ \xi_3 & 0 & 0 & 0 \end{bmatrix}$	Internal electric field Spatial character I $\hat{\omega}_{[0i]}^B = -i\hbar\dfrac{\partial}{\partial X_{10}}$
Electronic internal magnetic field	$B(\xi) = \pm \dfrac{1}{2}\begin{bmatrix} 0 & 0 & 0 & 0 \\ 0 & 0 & \omega_{12}^B & \omega_{13}^B \\ 0 & \omega_{21}^B & 0 & \omega_{23}^B \\ 0 & \omega_{31}^B & \omega_{32}^B & 0 \end{bmatrix}$	$X_{11} = \begin{bmatrix} 0 & 0 & 0 & 0 \\ 0 & 0 & \xi_1 & \xi_1 \\ 0 & \xi_2 & 0 & \xi_2 \\ 0 & \xi_3 & \xi_3 & 0 \end{bmatrix}$	Spin magnetic moment, eigenvalue of positive and negative magnetic moment: 0 $\hat{\omega}_{[ij]}^B = -i\hbar\dfrac{\partial}{\partial X_{11}} = \hat{K}_{11}$
Electric field in observation space	$E(x) = \pm \dfrac{1}{2}\begin{bmatrix} 0 & E_{01} & E_{02} & E_{03} \\ E_{10} & 0 & 0 & 0 \\ E_{20} & 0 & 0 & 0 \\ E_{30} & 0 & 0 & 0 \end{bmatrix}$	$X_{12} = \begin{bmatrix} 0 & x_0 & x_0 & x_0 \\ x_1 & 0 & 0 & 0 \\ x_2 & 0 & 0 & 0 \\ x_3 & 0 & 0 & 0 \end{bmatrix}$	Multi-electron electric field $\hat{E}_{0i} = -i\dfrac{\partial}{\partial X_{12}}$
Magnetic field in observation space	$B(x) = \pm \dfrac{1}{2}\begin{bmatrix} 0 & 0 & 0 & 0 \\ 0 & 0 & B_{12} & B_{13} \\ 0 & B_{21} & 0 & B_{23} \\ 0 & B_{31} & B_{32} & 0 \end{bmatrix}$	$X_{13} = \begin{bmatrix} 0 & 0 & 0 & 0 \\ 0 & 0 & x_1 & x_1 \\ 0 & x_2 & 0 & x_2 \\ 0 & x_3 & x_3 & 0 \end{bmatrix}$	Multi-electron magnetic field $\hat{B}_{[ij]} = -i\dfrac{\partial}{\partial X_{13}}$
	All the internal structures are superimposed on one point, and form a spherically symmetric gravitational charge.		

2.5.2 Physics image of photon wave-particle duality

We know that photons have wave-particle duality (Fig. 2.5.1). The intrinsic space of photons forms a spherical field, and the vibration of the photon center is confined to the spherical shell, forming a probability wave. Any disturbance will change the position of the center of the spherical wave and the shape of the spherical wave. If the edge of the spherical wave is squeezed and any deformation occurs, it will disturb the spherical wave.

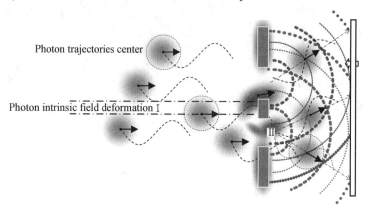

Fig. 2.5.1 The nature of photon wave-particle duality

Then the spherical wave is divided into two parts, one is spherical wave without center point, and the other is spherical wave with center point. The photon separates two parts interfere with each other when they pass through double slits and produce interference fringes. In the whole interference process, the spherical wave has no energy loss and does not scatter. The total deformation of the sphere is h. The center of the spherical wave will not split, and the spherical wave will merge into one after passing through two slits (Fig. 2.5.1). It should be noted that the position of the spherical wave center is ultimately determined by the screen spot position.

Let's look at the single slit diffraction of photons. There is a slit in A. The width of the slit is approximately the same as the amplitude of photons, which can cause interference (without considering slit II). In front of A, all photons have the same momentum and energy, and the photon center is in a simple harmonic vibration state, but because of different phases, no fringes are formed. When a photon passes through slot I of A, it will produce perturbation, and the perturbation positions of all photons are the same, so that the photon can get a unified phase through slot I.

In addition, the perturbation of photons changes the original direction of motion. Because of the low energy, the perturbation is elastic. It does not change the original momentum of photons, but only changes the direction. Therefore, photons form single slit diffraction fringes on screen B through the first slit. The slit itself is an interference source. The narrower the slit is, the larger the elastic deformation of photon sphere is, the stronger the interference is, and the

Chapter 2 Structure of the Photons

wider the scattering angle is. In Figure 2.5.1, there is a schematic of the photon, dashed ring expression photon intrinsic spatial boll, and the dashed sine curve representation photon center-locus. We can use dashed lines to indicate that the peak (thick dashed lines) is the most probabilistic, while the thin dashed lines are the least probabilistic. As we can understand above, double slit diffraction is formed by a single photon, which is part of its own wave and part of its own interference. These special properties come from spherical waves.

Photon center has a strict and clear propagation path, so Einstein is right. If we can know all boundary conditions of the internal vacuum and background vacuum, we can accurately determine the photon center position and velocity. Its trajectory can be accurately described by the wave function. But on the other hand, photon center locus cannot measure, and boundary conditions are unknown. Because the most sophisticated tools which we used to perceive the target photon is the probe electronic, it contacts target photon then its center position was given lethal perturbations, so that we don't know the track of photon center, and we will never not be found to the more sophisticated tools than the photon. We had to use the way of statistical to represent the photon behavior, so Bohr is right. The trajectory of the measured particles can not be verified by experiments. Because the perturbation caused by measurement is that the position and momentum of the target particle cannot be determined simultaneously, we will be in a dilemma. So, we extend the measurement space to infinity, and the probability of finding particles in a sufficient range must be 1, that is

$$\int_{-\infty}^{+\infty} |\psi(x,t)|^2 dx = 1 \qquad (2.5.5)$$

The integral area $\int_{-\infty}^{+\infty}$ is the presentation of this dilemma.

2.5.3 Quantization of electromagnetic field

The classical electromagnetic wave is a continuous wave composed of low energy photon spherical waves. The quantization of electromagnetic field is the restoration of this physicalimage. The classical electromagnetic wave is decomposed into many plane waves with different frequency and different wave vector. A plane wave (k, ω_k) is as an electromagnetic oscillator[①]. Introduce the normalized volume $V = l^3$, and assume the field on the edge of V to meet the periodic boundary conditions. The components of the wave vector k are taken as

$$k_j = \frac{2\pi}{l} n_j \quad (j = 1, 2, 3) \qquad (2.5.6)$$

In this way, the solution of the field equation $\Box A_t(x) = 0$ can be expressed as the sum of superposition series of the plane wave solutions.

---

① Discussions on the quantized electromagnetic field can be found in any textbook about quantum electrodynamics, in which more detailed discussions exist.

$$A_t(x,t) = \sum_k \{c_k(t)e^{ikx} + c_k^*(t)e^{-ikx}\} \tag{2.5.7}$$

Here, $c_k(t)$ is a horizontal vector, which is orthogonal to k that is the vibration of (k, ω_k). Electromagnetic field energy in normalized volume V is

$$E = \frac{1}{2}\int_V (|E|^2 + |B|^2)\,dV = H \text{ (Hamiltonian)} \tag{2.5.8}$$

A harmonic (k, ω_k, λ) is in the state $\Phi_{k\lambda}^{(n)}(Q_{k\lambda})$ in the quantum field, the quantum electromagnetic oscillator two quantization (using Coulomb specification), and the energy is

$$E_{k\lambda}^{(n)} = \left(N_{k\lambda} + \frac{1}{2}\right)\hbar\omega_k$$

These harmonics contain $N_{k\lambda}$ photons, and their energy is $\hbar\omega_k$. Momentum is $\hbar k$ and the polarization is λ. Here, the state of the harmonics (k, ω_k, λ) is completely determined by the $N_{k\lambda}$ photons. Therefore, the right vector $|n_{k\lambda}\rangle$ is used to describe the state of the wave vector.

The state vector of the entire quantum field is

$$\prod_k \prod_{\lambda=1}^{2} |n_{k\lambda}\rangle \text{ (There is no interaction between photons)} \tag{2.5.9}$$

The corresponding field energy is

$$E = \sum_{k,\lambda}^{2}\left(N_{k\lambda} + \frac{1}{2}\right)\hbar\omega_k \tag{2.5.10}$$

Particularly, when the quantum field is in the ground state (vacuum state), there is

$$|0\rangle = \prod_k |0_{k1}\rangle|0_{k2}\rangle \tag{2.5.11}$$

The photon number $n_{k\lambda} = 0$, and the wave is stopped, but the quantum field is still there, and

$$E_0 = \sum_k \hbar\omega_k \to \infty \tag{2.5.12}$$

The view of quantum field theory is that there are still the quantum electromagnetic fields in the space of no electromagnetic wave, and just this field is in vacuum state. Because E_0 cannot be measured, so make $E_0 = 0$, that regard vacuum as zero background, and all physical processes and physical measurements are carried out in this background. The particle number operator is Hermitian operator.

$$N_{k\lambda} = a_{k\lambda}^+ a_{k\lambda} \tag{2.5.13}$$

This is the number of particles in the quantum field. It is the eigenvalue $N_{k\lambda}$ of the photon number with (k, ω_k, λ). In the representation of the number of particles (Using $\{|n_{k\lambda}\rangle\}$ as base coordinates):

$$N_{k\lambda} = \begin{bmatrix} 0 & 0 & 0 & 0 \\ 0 & 1 & 0 & 0 \\ 0 & 0 & 2 & 0 \\ 0 & 0 & 0 & \ddots \end{bmatrix} \tag{2.5.14}$$

$a_{k\lambda}$ is the photon annihilation operator, and $a_{k\lambda}^+$ photon creation operator, i.e.

$$N_{k\lambda} a_{k\lambda} |n_{k\lambda}\rangle = (n_{k\lambda} - 1) a_{k\lambda} |n_{k\lambda}\rangle \tag{2.5.15}$$

Chapter 2 Structure of the Photons

$$N_{k\lambda} a_{k\lambda}^+ \mid n_{k\lambda} \rangle = (n_{k\lambda} + 1) a_{k\lambda}^+ \mid n_{k\lambda} \rangle \quad (2.5.16)$$

$$a_{k\lambda} \mid n_{k\lambda} \rangle = \sqrt{n_{k\lambda}} \mid n_{k\lambda} - 1 \rangle \quad (2.5.17)$$

$$a_{k\lambda}^+ \mid n_{k\lambda} \rangle = \sqrt{n_{k\lambda} + 1} \mid n_{k\lambda} + 1 \rangle \quad (2.5.18)$$

Here, $a_{k\lambda}(t) = a_{k\lambda} e^{-i\omega_k t}$; $a_{k\lambda}^+(t) = a_{k\lambda}^+ e^{i\omega_k t}$. Existing particle production and annihilation phenomenon in the relativistic field are

$$a_{k\lambda}^+ \mid 0_{k\lambda} \rangle = \mid 1_{k\lambda} \rangle \quad \text{(Particles produced in vacuum)} \quad (2.5.19\text{ a})$$

$$a_{k\lambda} \mid 1_{k\lambda} \rangle = \mid 0_{k\lambda} \rangle \quad \text{(Particles disappeared in the vacuum)} \quad (2.5.19\text{ b})$$

$$a_{k\lambda} \mid 0_{k\lambda} \rangle = 0 \quad \text{(Vacuum cannot disappear again)} \quad (2.5.20)$$

In quantum electrodynamics theory, Lorentz condition is applied to the state function $A_\mu(x)$ of field, and after the second quantization, $A_\mu(x)$ is field operator, so quantum field $A_\mu(x)$ Lorentz condition should limit to the state vector (physical state). Lorentz condition is changed to the following formula.

$$\partial_\mu A_\mu(x) \mid a \rangle = 0 \quad (2.5.21)$$

Among them, $\mid a \rangle$ is arbitrary state vector of the quantization electromagnetic field. Lorentz condition in momentum space is

$$(a_{k3} - i a_{k4}) \mid a \rangle = 0 \quad \text{(For any } k\text{)} \quad (2.5.22)$$

That is to say, for any physical state, the scalar photon and the longitudinal photon exist simultaneously. There is no single independent scalar photon or longitudinal photon. The sum of the contributions of longitudinal photon and the time photon is zero, so that from the effect of physical observation, there is only transverse photons without longitudinal photons and time photons in the quantum electromagnetic field.

2.5.4 The physical significance of operator commutation relations

Coordinate x^i and momentum p_i can be used instead by the coordinate operator $\hat{x}^i$ and momentum operator $\hat{p}_i$. $\hat{p}_i$ is the differential operator, and $\hat{x}^i$ act on the wave function is multiplied, and for any wave function ψ, there are

$$\hat{x}^i \hat{p}_i \psi = \frac{\hbar}{i} x \frac{\partial \psi}{\partial x_i}; \quad \hat{p}_i \hat{x}^i \psi = \frac{\hbar}{i} x^i \frac{\partial}{\partial x^i}(x^i \psi) = \frac{\hbar}{i} x^i \frac{\partial}{\partial x^i} + \frac{\hbar}{i} \psi$$

The two results are not the same, and

$$\hat{x}^i \hat{p}_i \psi - \hat{p}_i \hat{x}^i \psi = i\hbar \psi \quad (2.5.23)$$

In the quantum field, λ is defined as the coordinate space of the tangent vector, and p is the momentum space of the tangent vectors (Fig. 2.5.2), d/dx integral curve is $x(\lambda)$, and integral curve of $d/d\lambda$ is $\lambda(k)$.

$$\lambda = x \frac{\partial}{\partial \lambda} = x e_x = \frac{d\lambda}{dk} \frac{\partial}{\partial \lambda} = \frac{d}{dk} \quad (2.5.24)$$

$$p = p \frac{\partial}{\partial x} = p e_p = \frac{dx}{d\lambda} \frac{\partial}{\partial x} = \frac{d}{d\lambda} \quad (2.5.25)$$

$$x^i(B) - x^i(A) = [e^{\varepsilon d/d\lambda}, e^{\varepsilon d/dk}]x^i|_p$$

$$[e^{\varepsilon d/d\lambda}, e^{\varepsilon d/dx}]x^i = [1 + \varepsilon \frac{d}{d\lambda} + \frac{1}{2}\varepsilon^2 \frac{d^2}{d\lambda^2} + 0(\varepsilon^3), 1 + \varepsilon \frac{d}{dk} + \frac{1}{2}\varepsilon^2 \frac{d^2}{dk^2} + 0(\varepsilon^3)]x^i$$

$$x^i(B) - x^i(A) = \{\varepsilon^2[\lambda, p] + 0(\varepsilon^3)\}x^i|_p \qquad (2.5.26)$$

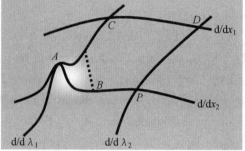

Fig. 2.5.2 The physical significance of commutation relations

λ, p and p consist the incomplete open part of the quadrilateral (just $\hbar$), which is total deformation of quantum field of space-time. Because $\hbar$ quantum exists, it has to take a detour along $d/d\lambda_1$ and $d/d\lambda_2$, which is exactly the photonic field in vacuum with the maximum deformation $\hbar$, with using the Lie bracket. You can deduct the background field deformation, so there is such formulation $[x^i, p_i] \equiv i\hbar$, which can be applied to a broader scope.

2.5.5 Localization and spatiotemporal discontinuity of quantum field

Our knowledge of quantum fields comes from detection methods, which use probing particles to detect a target particle. Unfortunately, this detection method is the only one. We use test particles to detect the target particles. Once a particle is detected, we determine its position and momentum. Every point particle has its corresponding coordinates in space-time, which are obviously discontinuous. Because of the uniqueness of cognitive mode, we believe that the space-time of quantum field is local and discontinuous. Particular attention should be paid to the fact that particles that have not been detected exist objectively. Because their positions are not disturbed, their coordinate positions are accurate and their trajectories are continuous. The spatiotemporal discontinuity of quantum fields is only measured (Fig. 2.5.3).

In the process of detecting target particles by probe particles, the detection is accomplished by exchanging momentum (virtual photons are exchanged between particles), and the momentum and coordinates of virtual photons satisfy the uncertainty relationship ($\Delta x \Delta p \geqslant h/4\pi$).

1. Physical Image of $[x^i, p_i] \equiv i\hbar$

The non-commutative relation of quantum field is essentially caused by the exchange of virtual photon momentum in the measurement process. There is an undetected quantum field at a point P in space-time. If we want to determine the position and momentum of the quantum field ψ, there are only two ways: ①we suppose the position of the undetected quantum field is x_i,

Chapter 2 Structure of the Photons

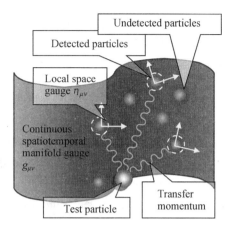

Fig. 2.5.3 The existence of quantum fields

and we input the detection momentum p_i to verify the hypothesis; ②we input the detection momentum p_i, and then determine the position x_i. The physical image of the first way is to set the position of x_i and then input a detection momentum p_i, which means that the detection particle has not yet emitted virtual photons. The physical image of the second way is to input a momentum p_i first and then determine x_i, which means that the virtual photons have been transmitted. So $\hat{x}^i \hat{p}_i \psi - \hat{p}_i \hat{x}^i \psi = i\hbar \psi$, and the difference between the two methods is $1\hbar$, while the quantum field is noncommutative.

2. Quantum field gauge

For quantum field, the bending of intrinsic quantum field is not observable. From the point of view of measurement, the position and momentum of quantum field are detected by detecting particles. The bending of space-time has no effect on the detection. For any detected quantum field, it is a point particle, so the space describing the point particle is the same. In other words, the space of the quantum field is flat, and the metric $g_{\mu\nu}$ degenerates to $\eta_{\mu\nu}$.

2.5.6 Basic physical images of quantum mechanics

The essence of space-time dimension is constraints. Dimension determination comes from measurement, and the essence of measurement lies in constraints, and the essence of constraints lies in determination. Dimension is the physical effect of vacuum strain in measurement.

The lepton and photon are simplified as point models, and the physical images of probability waves are obtained. Probability wave is the basic physical image of quantum mechanics. The axiom of quantum mechanics is the cornerstone of the establishment of quantum mechanics. Here we will re-understand the axiom of quantum mechanics from the viewpoint of field.

1. Axiom Ⅰ of Quantum Mechanics

For a system consisting of point particles, its state is completely determined by a complex value (wave) function $\Psi(x,t)$. We interpret it as probability density amplitude

$|\Psi(x,t)|^2 dx$. It denotes the probability that the particle will be in the closed range x to $x+dx$ when the position of the particle is measured at time t. Since the total probability must be 1, we have a normalization condition:

$$\int_{-\infty}^{\infty} |\Psi(x,t)|^2 dx = 1 \qquad (2.5.27)$$

We also need to re-understand the uncertainty principle. Due to the limitations of human measurement tools and means, we cannot measure the trajectory of the electronic. If it is necessary, the measurement is only the simple method of detecting particles, which can disturb the measured electrons. The macroscopic manifestation is the position and momentum of the measured electrons, which cannot be determined at the same time. This effect is defined as the uncertainty principle.

2. Axiom II of quantum mechanics

The possible states of the system mentioned in Axiom I constitute a set $L_2(-\infty,\infty)$, which consists of all normalized square integrable functions, which are defined on $(-\infty,\infty)$.

On the basis of Axiom I, the properties of wave functions are further defined, so Axiom II can get the very important theorem of line overlap immediately. The arbitrary normalized linear combination of the two possible states of the system at time t is a possible state of the system at time t. Let $\Psi_1(x,t)$ and $\Psi_2(x,t)$ be two possible kinematic states, so if c_1 and c_2 are two numbers in the complex field (constants independent of position x and time t), and satisfy

$$|c_1|^2 + |c_2|^2 = 1 \qquad (2.5.28)$$

Then

$$\Psi(x,t) = c_1\Psi_1(x,t) + c_2\Psi_2(x,t) \qquad (2.5.29)$$

It is also a possible state of kinematics. According to the principle of superposition of states (Fig. 2.5.4), the probability density of particles appearing at a point P on the screen is

$$|\Psi| = |c_1\Psi_1 + c_2\Psi_2|^2 = (c_1^*\Psi_1^* + c_2^*\Psi_2^*)(c_1\Psi_1 + c_2\Psi_2)$$
$$= |c_1\Psi_1|^2 + |c_2\Psi_2|^2 + c_1^*c_2\Psi_1^*\Psi_2 + c_1c_2^*\Psi_1\Psi_2^*$$

On the right side of the equation, the first term is the probability density of the particle passing through the upper slit at point P, and the second term is the probability density of the particle passing through the lower slit at point P, then the third and fourth term is the interference term of Ψ_1 and Ψ_2. The interference term is due to the perturbation caused by photon selectivity.

The particle's trajectory makes the particle have the characteristic of probability wave. Lepton central point trajectory is a 1-dimensional curve. For two propagating particles, the probability of collision between two leptons and lepton central point is almost zero at low energy, and the two trajectory curves are independent of each other. The independence also applies to n. The existence of independence makes the principle of linear superposition of states of wave functions exist from a statistical point of view.

When a lepton with the same momentum propagates from a source i in different directions,

Chapter 2   Structure of the Photons

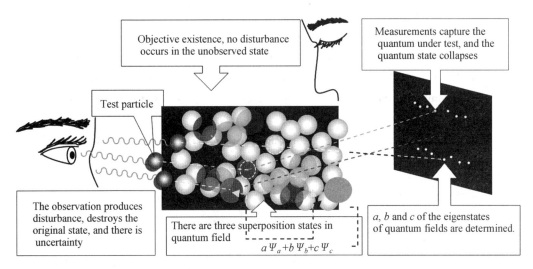

Fig. 2.5.4   Physical images of quantum field of basic assumptions

the set of wave trajectories at the lepton center points constitutes a probability wave function Ψ_i. The number and size of elements in the set c_i denotes the strength of probability wave function Ψ_i and constitutes $c_i \Psi_i$. It is noteworthy that the wave field can also be extended to the probability wave function $c_i \Psi_i$, where c_i is a relatively small quantity.

De Broglie wave is the basic element of probability wave. It can be seen from the propagation property of wave that the fluctuation is independent of each other, or orthogonal to each other. Therefore, the probability wave functions Ψ_i and Ψ_j emitted by any two sources are orthogonal to each other. Each wave function Ψ_i constitutes a dimension coordinate base, and then Ψ_i and Ψ_j constitute an n-dimensional orthogonal space. It consists of different wave vectors and different sources of de Broglie waves, and n-dimensional orthogonal space is Hilbert space. For $\Psi_1, \Psi_2, \cdots, \Psi_i, \cdots, \Psi_n$, there is

$$\Psi = c_1 \Psi_1 + c_2 \Psi_2 + \cdots + c_n \Psi_n + \cdots = \sum_n c_n \Psi_n$$

We can understand Ψ as a vector in n-dimensional orthogonal space, and $\Psi_1, \Psi_2, \cdots, \Psi_i, \cdots, \Psi_n$ as n-dimensional orthogonal coordinate frame, $c_1, c_2, \cdots, c_i, \cdots, c_n$ as the corresponding projection value of Ψ vector on n-dimensional orthogonal coordinate frame (Fig. 2.5.5). When a lepton with the same momentum propagates from a source i in different directions, the set of wave trajectories of the lepton center points constitutes a probability wave function Ψ_i of leptons. The set is also called state function.

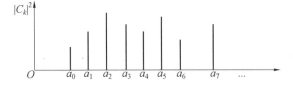

Fig. 2.5.5   Probability spectrum

3. Axiom III of quantum mechanics

All possible information about the system is contained in the wave function and can be extracted by a suitable set of linear self-adjoint operators. If the system is in state $\Psi(x,t)$, the probability of a_k is $|C_k|^2$ by measuring the physical quantity represented by operand A, where a_k is the eigenvalue of A corresponding to eigenfunction $\Phi_k(x)$ (in the discrete spectrum).

$$A\Phi_k(x) = a_k\Phi_k(x) \qquad (2.5.30)$$

Here, C_k is the coefficient when $\Psi(x)$ is expanded by the eigenfunction of A at time t.

$$\Psi(x,t) = \sum_l C_l\Phi_l(x) + \int C(\lambda)\Phi(\lambda,x)d\lambda \qquad (2.5.31)$$

Similarly, the probability of measuring A to obtain the values from $a(\lambda)$ to $a(\lambda+d\lambda)$ in the continuous spectrum is $|C(\lambda)|^2$, where $a(\lambda)$ is the eigenvalue of A corresponding to the irregular eigenfunction $\Phi(\lambda,x)$.

$$A\Phi(\lambda,x) = a(\lambda)\Phi(\lambda,x) \qquad (2.5.32)$$

Here, $C(\lambda)$ is the expansion coefficient of $\Phi(\lambda,x)$ at time t.

In Axiom III, all relevant information is included in the wave function, which can be obtained from the intrinsic structure of photons and leptons.

For a large number of simultaneous measurements of identical systems, the average or expected value A is given by the following formula.

$$\langle A \rangle = \sum |C_k|^2 a_k + \int |C(\lambda)|^2 a(\lambda) d\lambda = \int_{-\infty}^{\infty} \Psi^*(x,t) A \Psi(x,t) dx \qquad (2.5.33)$$

4. Axiom IV of quantum mechanics

Similar to the Hamilton form of classical mechanics, the self-adjoint operator functions of the following basic position operators and momentum operators are represented in quantum mechanics.

$$\hat{x} = x, \hat{p} = -i\hbar\partial/\partial x$$

It has a complete set of eigenfunctions (i.e. observables).

5. Axiom V of quantum mechanics

The dynamics of quantum system is determined by Schrodinger equation

$$H\psi(x,t) = i\hbar\partial\psi(x,t)/\partial t$$

This is the natural conclusion of vacuum strain conservation.

Chapter 3 The Structure of Lepton

According to the standard model theory, we know that the universe consists of leptons and quarks. To ascertain the lepton's structure is essential. Based on vacuum, the goal of this chapter is established the image of the intrinsic structure of particle which coincidence experimental observations. Leptons are composed by $e, \mu, \tau, \nu_e, \nu_\mu, \nu_\tau$ and their anti-particle. Lepton spin is 1/2, it does not participate in strong effect, but other forces are involved. Leptons are divided into two classes: lepton e, μ, τ have the rest mass and 1 unit charge. Neutrinon ν_e, ν_μ, ν_τ have no charge and rest mass. Anti-lepton and lepton are duality particles, and have opposite charge. The nature is viewed in the table as below.

Table 3.0.1 Lepton

charge	lepton	spin	I	II	III	Mass
−1	1	$\hbar/2$	e	μ	τ	have
0	1	$\hbar/2$	ν_e	ν_μ	ν_τ	no

Table 3.0.2 Anti-lepton

charge	lepton	spin	I	II	III
+1	−1	$\hbar/2$	e^+	μ^+	τ^+
0	−1	$\hbar/2$	$\bar{\nu}_e$	$\bar{\nu}_\mu$	$\bar{\nu}_\tau$

The strain form of vacuum determines the intrinsic structure of electrons, which can explain the inherent degrees of freedom of leptons, such as mass, charge, spin, etc. It involves the following aspects: ①the formation of intrinsic structure of electrons; ②causes of formation of electric flux line and properties of electric and magnetic fields; ③causes of formation of the electron spin; ④causes of formation of the mass; ⑤electronic wave-particle duality.

Leptons are the vacuum local strain, and strain contains all the information of leptons. The strain tensor of the leptons is composed of two parts. First part is the internal strain, and the second part is the electric flux line strain tensor caused by fiber shift.

$$[\varepsilon]_{\text{Lepton}} = \underbrace{\left[\frac{\partial u_\mu}{\partial \xi_\nu}\right]}_{\substack{\text{Microcosmic space-time} \\ \text{rupture region, internal strain}}} + \underbrace{\left[\frac{\partial A_\mu}{\partial x_\nu}\right]}_{\substack{\text{The fiber field in the observational} \\ \text{space, electric flux line strain}}}$$

$$= \underbrace{\varepsilon_{\mu\nu}(\xi)}_{\text{4-momentum}} + \frac{1}{2}\underbrace{\gamma_{\mu\nu}(\xi)}_{\substack{\text{Weakly} \\ \text{acting strain}}} + \frac{1}{2}\underbrace{\omega_{\mu\nu}(\xi)}_{\text{Spin}} + \underbrace{\Delta E_{\mu\nu}(x)}_{\substack{\text{Electromagnetic field of} \\ \text{scalar and vector potential}}} + \frac{1}{2}\underbrace{\Omega_{\mu\nu}(x)}_{\text{Electromagnetic field}}$$

Our task in this chapter is to analyze the strain of lepton one by one.

3.1 The strain $\varepsilon_{\mu\nu}$ of lepton

3.1.1 Formation of electron fiber strain field

Now let's study how electron pairs are produced in vacuum. The excitation of electron pairs originates from local disturbances in vacuum.

1. Electron(e^-)

Considering that the vacuum region is stimulated(Fig. 3.1.1), a basic unit flies away at the initial position and leaves a hole, which would have an impact on the surrounding vacuum. Similar to the bathtub filled with water, and when we pulled out the cork of bottom suddenly appeared in a hollow will form a convergent flow vortex around the hole. If a hole suddenly appears in the vacuum, the vacuum will rotate and converge to fill the hole. This effect is electronic. The deformation region around the basic unit hole is defined as the intrinsic space of electrons, and the overall effect is called electrons. Electrons have "spin" and "convergence" structures.

2. Positron(e^+)

Vacuum being stimulated and a basic unit from the original position fly away(Fig. 3.1.1), this basic unit of vacuum will compress it around the vacuum, and resulting antispin source forms an antielectron. Positron has the structure of spin source.

A basic unit is excited to produce pairs of electrons.

3. Production of fiber structure of single electron

We can ignore the spin effect. When a small area disapears, a basic unit of vacuum will form a hole which resulting in deformation zone where is near the lepton center. Further analyzed as follows.

(1) Radial strain.

Because of the vacuum lost a basic unit to form a hole, surrounding the vacuum extreme bend so that the vacuum along direction r reach the limit of compression, still cannot meet the curvature then makes the vacuum produce rupture form gap(the gap make the electronic structure with fibrosis), the vacuum will fill this gap and basic unit of vacuum will be towards the cavity occurs in a tiny movement Δr, and the closely related basic unit of vacuum also occur a tiny movement, this process continued down will extend to infinity in this direction to form an electric flux line ξ^i(Fig. 3.1.1(a)). Positron compression strain is opposite to electron tension strain. The positron center surplus a basic unit, outward compression form positive electric flux line in the vacuum (Fig. 3.1.1(b)), and electric flux line is no longer Faraday conceived the image, but the real physical reality.

(2) The spherical symmetric strain in θ, φ direction.

Using spherical coordinate system to examine the formation of the electron (Fig. 3.1.2(a)),

Chapter 3 The Structure of Lepton

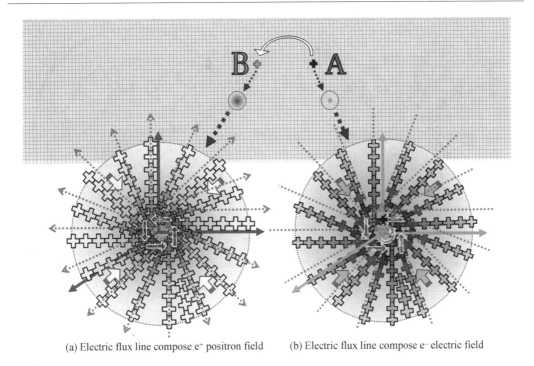

(a) Electric flux line compose e⁺ positron field   (b) Electric flux line compose e⁻ electric field

Fig. 3.1.1 Schematic of electronic vacuum excitation

except the strain which along the radial direction, also exist the strain which along the direction of θ and φ. From the static analysis of vacuum in Chapter 1 that the spherically symmetric the deformation ε_r in the radial direction r and deformation along the direction θ, φ which are: $\varepsilon_r = \dfrac{\partial u_r}{\partial r}$; $\varepsilon_\vartheta = \varepsilon_\varphi = \dfrac{u_r}{r}$. Shear strain in the case of spherical symmetry is $\gamma_{\vartheta\varphi} = \dfrac{1}{2}\left(\dfrac{\partial u_\vartheta}{\partial \varphi} + \dfrac{\partial u_\varphi}{\partial \vartheta}\right) = \varepsilon_\vartheta = \varepsilon_\varphi$, and constitute a weak field. Filling gap effect result in vacuum rupturein direction θ, φ constitutes tructure of fiber. That is caused by the vacuum particle structure, which is different to the continuous medium.

Consider a radius R of sphere in a spherical coordinate, along the radial expansion ΔR, then the radius of the sphere into $R + \Delta R$, spherical geodesic length from $2\pi R$ into $2\pi(R + \Delta R)$. Because of the basic unit only has weak elastic, so that the original spherical geodesic need $2\pi R$ basic unit of vacuum covered, after the expansion it needs $2\pi(R + \Delta R)$ basic unit of vacuumcan be covered, then there are $2\pi\Delta R$ holes of basic unit of vacuum which need new basic unit to fill, and it cracks. The total amount of fill is $2\pi R$ basic unit of vacuum. Lotus-like structure is shown in Fig. 3.1.2. n is the number of vacuum basic units needed to fill the gap. Starting point of filling hole is the initial end of electric flux line (Rays), which is the starting point of the crack, along the direction θ, φ was pulled out of cracks at the surface of the front-end along the radial and θ, φ direction have reached the tensile limit. The basic unit have been squeezed along the radial direction, and stretching in the direction θ, φ, the basic unit of front end of electric flux line (e⁻ electric flux line) reached the deformation limit. The

The Vacuum Super-unified Theory

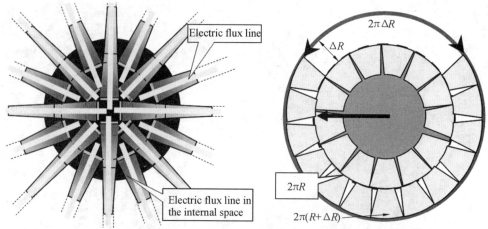

(a) The basic unit of vacuum on electric flux line front surface deformation limit was reached

(b) A basic unit along the radial the amount of change is ΔR, the amount of change in $\theta\varphi$ direction is $2\pi\Delta R$

Fig. 3.1.2 Fiber structure of mass charge and the deformation along θ, ϕ direction
(A basic unit along the radial the amount of change is ΔR, and the amount of change in θ, ϕ direction is $2\pi\Delta R$)

ring strain of single fiber is $\mathscr{H}_{circular}$. The discussions correspond to antileptons.

For lepton, when being stretched $\Delta R/2$ along the radial direction, then the radius of sphere is $R-\Delta R$, perimeter contracted into $2\pi(R-\Delta R)$. Similarly, there are $2\pi\Delta R$ holes of basic unit of vacuum which need new basic unit to fill. The front face of the basic unit of electric flux line has reached the ultimate tensile along the radial direction θ and φ. The deformation of basic unit of vacuum meet $\varepsilon_1 = \varepsilon_2 = \varepsilon_3$, the normal direction of principal strain surface along r, the radial strain of the whole single fiber is $\mathscr{H}_{Radial}$.

The initial end of the electric flux line along ξ_1, ξ_2, ξ_3 dimension of direction to achieve the deformation limit, this effect also makes time to become slowly, time and space the total deformation $\mathscr{H}/2$ ($\mathscr{H} = \mathscr{H}_{Radial} + \mathscr{H}_{circular}$). Therefore, the circumferential strain is independent of the type of charged leptons. The volume of electronic fibers only needs to consider the length, just like photonic fibers. In the establishment of strain equation, only radial deformation is considered, and no circumferential deformation is considered, which can greatly simplify the problem.

3.1.2 The single fiber deformation of intrinsic field of electron

We establish a coordinate system with O as the coordinate origin that is the center of the sphere of lepton and this coordinate space is the intrinsic space of leptons (Fig. 3.1.3). We will discuss issue on this basis.

1. A single internal area electric flux line within electronic intrinsic space

First look at ξ^1 dimensional displacement function of lepton field (Fig. 3.1.4). If force to a basic unit join a 1-dimensional flat vacuum, then AB 1-dimensional vacuum will loss of a basic unit, so the basic unit of vacuums of both sides were stretched by point $\bar{O}$, and form stretch

Chapter 3 The Structure of Lepton

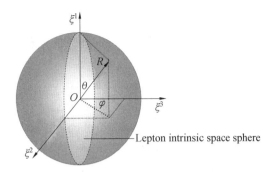

Fig. 3.1.3 The intrinsic space of lepton

lepton. Adding a basic unit that make the basic unit of vacuums of both sides of point O are compressed, forming compression lepton field. This deformation will not disrupt the dimension properties of the central area of the lepton. In the interval AB, the vacuum is in the tensile state. If the infinity approaches the central point $\bar{O}$, the vacuum reaches the tensile limit. In the interval AB, the vacuum is in the compression state. If the infinity approaches the central point O, the vacuum reaches the compression limit.

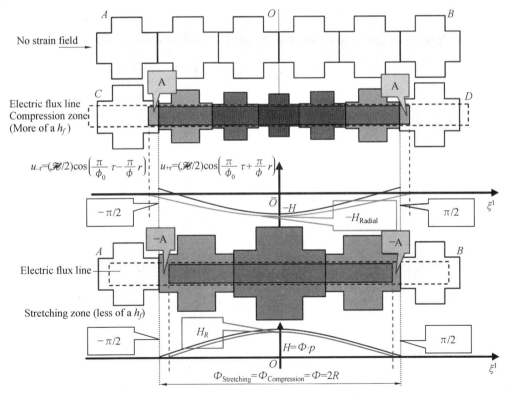

Fig. 3.1.4 A dimension of lepton field

(Single electric flux line total strain is $\mathscr{H}$, and h_f is the lepton field total strain. Since leptons with the electric flux line structure, then $\hbar_f = \varPhi \cdot Np$)

2. Intrinsic displacement function

AB is intrinsic diameter of stretch lepton, denoted by Φ_{Stretch}, similar to the tensile zone of the photon ξ_1-dimensional situation $A\bar{O} \cdot p = \mathscr{H}/2$, $\Phi_{\text{Stretching}} \cdot p = \mathscr{H}$. In compression zone $\Phi_{\text{Compression}} \cdot p = \mathscr{H}$, while

$$\Phi_{0\text{Stretching}} c \equiv \Phi_{0\text{Compression}} c \equiv \Phi \quad (3.1.1)$$

Fiber contraction $\mathscr{H}$ along the "sink" the direction. In tensile field, along the "source" of the direction expansion H, here, $\mathscr{H} \ll \Phi$. The spatial scale is length of photon propagation path. Time scale is efficiency of photon propagation, the compression zone and tension zone will reduce the efficiency of photon propagation and the effect are same, so in the two dual space meet $\Phi_{0\text{Stretch}} c \equiv \Phi_{0\text{Compression}} c \equiv \Phi$, Φ_0 –particle formation time, so the dual particle formation time strictly equal $\Phi_{0\text{Stretch}} \equiv \Phi_{0\text{Compression}}$, therefore, the tension zone is the same as the compression zone.

$$\Phi \cdot p = \mathscr{H} \quad (3.1.2)$$

We can draw a schematic of the vacuum strain. The vacuum from plane state to deformation state:

$$\xi^\mu \to \xi'^\mu$$

Here, ξ^μ is the 4-dimensional coordinates of deformation state of vacuum, and the displacement components is

$$u^\mu = \xi'^\mu - \xi^\mu \quad (3.1.3)$$

Time items:

$$u^0(\tau) = \left(\frac{\mathscr{H}}{2}\right) \cos\left(\frac{\pi}{\Phi_0}\xi_0\right) = \left(\frac{\mathscr{H}}{2}\right) \cos(k^0 \xi_0) \quad (3.1.4a)$$

Space items:

$$u(r) = \left(\frac{\mathscr{H}}{2}\right) \cos\left(\frac{\pi}{\Phi}r\right) = \left(\frac{\mathscr{H}}{2}\right) \cos(k^1 \xi_1 i + k^2 \xi_2 j + k^3 \xi_3 k) \quad (3.1.4b)$$

All strain to produce wave in the intrinsic space, and all the wave are transmitted at the speed of light, therefore intrinsic space frame to meet

$$(c d\xi^0)^2 - (d\xi^1)^2 - (d\xi^2)^2 - (d\xi^3)^2 = 0 \quad (3.1.5)$$

Use Minkowski space Gauge, it can be written as

$$u^\mu(\xi) = \left(\frac{\mathscr{H}}{2}\right) \cos(k_0 \xi_0 + k^1 \xi_1 i + k^2 \xi_2 j + k^3 \xi_3 k) \quad (3.1.6)$$

Here, $0 \leq \xi_0 \leq \Phi_0$; $0 \leq \xi_i \leq \Phi_i (i=1,2,3)$; $k^\mu = \frac{\pi}{\Phi^\mu} (\mu = 0,1,2,3)$. Displacement components of the vacuum deformation $u_\mu(u_0, u_1, u_2, u_3)$ called the displacement function of intrinsic field of the quantum field.

3.1.3 Lepton intrinsic strain [$\varepsilon_{\mu\nu}$]

Lepton intrinsic strain in the spherical coordinate system is

Chapter 3 The Structure of Lepton

$$[\varepsilon](\tau, r, \theta) = \left\{ \begin{bmatrix} \varepsilon_\tau & 0 & 0 & 0 \\ 0 & \varepsilon_r & 0 & 0 \\ 0 & 0 & \varepsilon_\theta & 0 \\ 0 & 0 & 0 & \varepsilon \end{bmatrix} + \begin{bmatrix} 0 & \gamma_{\tau r} & \gamma_{\tau \theta} & \gamma_\tau \\ \gamma_{r\tau} & 0 & \gamma_{r\theta} & \gamma_r \\ \gamma_{\theta\tau} & \gamma_{\theta r} & 0 & \gamma_\theta \\ \gamma_\tau & \gamma_r & \gamma_\theta & 0 \end{bmatrix} \right\} \quad (3.1.7)$$

(1) 4-momentum of electron   (2) Weak field of electron, spin, charge

It contains all information of lepton. Considering spherically symmetric radial strain (Fig. 3.1.5), there are

$$u_r = H_r \cos \frac{\pi r}{2R} \quad (0 \leqslant r \leqslant R) \quad (3.1.8a)$$

Taking account of spin

$$u_\phi = H_\phi \cos \frac{\phi}{2}, 0 \leqslant \phi \leqslant 2\pi ; \quad u_\theta = H_\theta \sin \theta \quad (-\pi/2 \leqslant \theta \leqslant \pi/2) \quad (3.1.8b)$$

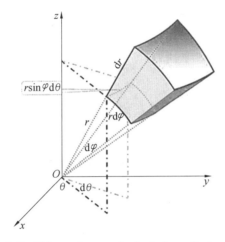

Fig. 3.1.5 Volume element of spherical coordinate system

Here, H is coefficient. It's maximum deformation, and vacuum is isotropic, then $H_r = H_\theta = H_\varphi = H$.

$$\varepsilon_0 = \sqrt{\varepsilon_1^2 + \varepsilon_2^2 + \varepsilon_3^2}$$

$$\varepsilon_r = \frac{\partial u_r}{\partial r} = -H \frac{\pi}{\Phi} \sin \frac{\pi r}{\Phi}$$

$$\varepsilon_\phi = \frac{1}{r} \frac{\partial u_\phi}{\partial \phi} + \frac{u_r}{r} = \frac{H}{r} \cos \frac{\pi r}{\Phi} - \frac{1}{2} \frac{H}{r} \sin \frac{\phi}{2} = \varepsilon_{\phi\phi} + \Delta \varepsilon_\phi \quad (3.1.9)$$

$$\varepsilon_\theta = \frac{1}{r \sin \phi} \frac{\partial u_\theta}{\partial \theta} + \frac{u_r}{r} + \frac{\cot \phi}{r} u_\phi = \frac{H}{r} \cos \frac{\pi r}{\Phi} + \frac{H}{r} \left(\cot \phi \cos \frac{\phi}{2} + \frac{\cos \theta}{\sin \phi} \right) = \varepsilon_{\theta\theta} + \Delta \varepsilon_\theta$$

Note: the repeated lower index represents the spherically symmetric strain.

$$\begin{cases} \gamma_{r\phi} = \dfrac{1}{r}\dfrac{\partial u_r}{\partial \phi} + \dfrac{\partial u_\phi}{\partial r} - \dfrac{u_\phi}{r} = -\dfrac{H}{r}\cos\dfrac{\phi}{2} \\ \gamma_{\phi\theta} = \dfrac{1}{r\sin\phi}\dfrac{\partial u_\phi}{\partial \theta} + \dfrac{1}{r}\dfrac{\partial u_\theta}{\partial \phi} - \dfrac{\cot\phi}{r}u_\theta = -\dfrac{H}{r}\cot\phi\sin\theta \\ \gamma_{\theta r} = \dfrac{\partial u_\theta}{\partial r} + \dfrac{1}{r\sin\phi}\dfrac{\partial u_r}{\partial \theta} - \dfrac{u_\theta}{r} = -\dfrac{H}{r}\sin\theta \end{cases} \quad (3.1.10)$$

$$dV = r^2\sin\phi\, dr\, d\theta\, d\phi \quad (3.1.11)$$

The total amount of deformation of electrons (including spin, Fig. 3.1.6)

$$I = \int_0^\pi d\phi \int_0^{2\pi} d\theta \int_0^{r(\theta,\phi)} F(r,\theta,\phi) r^2 \sin\phi\, dr$$

In spherical coordinates, the strain function $\dfrac{\partial u(r)}{\partial r} = \dfrac{e\pi H}{8\pi r^2 \phi}\sin\dfrac{\pi r}{\phi}\cdot\dfrac{r}{r}$ is spherically symmetric, has nothing to do with θ,ϕ. Lepton 4-dimensional total of internal area electric flux line deformation rate is

$$F(r) = \dfrac{eH}{8r^2\phi}\cdot\sin\left(\dfrac{\pi r}{\phi}\right) \quad (3.1.12)$$

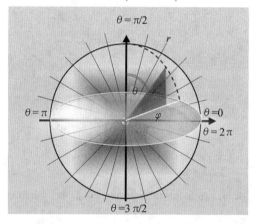

Fig. 3.1.6 2-dimensional lepton field

The total amount of space-time deformation h_r is

$$u_\theta = H\cos\dfrac{\theta}{2},\; 0 \leqslant \theta \leqslant 2\pi$$

$$u_\phi = H\sin\phi,\; -\pi/2 \leqslant \theta \leqslant \pi/2$$

$$F(r,\theta,\phi) = \dfrac{e\pi}{4\pi r^2\phi}\cdot H\sin\left(\dfrac{\pi r}{\phi}\right)\cdot H\cos\dfrac{\theta}{2}\cdot H\sin\phi$$

$$h_f = \int_0^\pi \sin\phi\, d\phi \int_0^{2\pi}\cos\dfrac{\theta}{2}\sin\theta\, d\theta \int_0^R \dfrac{e\pi}{4\pi r^2\phi}H^3\sin\left(\dfrac{\pi r}{\phi}\right)r^2\, dr$$

$$= 2\cdot\dfrac{8}{3}\cdot\dfrac{e}{4\pi}H^3\int_0^R -\sin\left(\dfrac{\pi r}{\phi}\right)d\left(\dfrac{\pi r}{\phi}\right) = \dfrac{4}{3}e\mathcal{H}$$

The 3-dimensional deformation is attributed to one dimensional deformation. Set $H^3 = \dfrac{\mathcal{H}}{2}$,

then
$$h_f = \frac{4}{3}e\mathscr{H} = 2h \qquad (3.1.13)$$

Lepton intrinsicstrain within the Cartesian coordinate system:

$$[\varepsilon_{\mu\nu}](\xi) = \left\{ \begin{bmatrix} \varepsilon_0 & 0 & 0 & 0 \\ 0 & \varepsilon_1 & 0 & 0 \\ 0 & 0 & \varepsilon_2 & 0 \\ 0 & 0 & 0 & \varepsilon_3 \end{bmatrix} + \begin{bmatrix} 0 & \gamma_{01} & \gamma_{02} & \gamma_{03} \\ \gamma_{10} & 0 & \gamma_{12} & \gamma_{r3} \\ \gamma_{20} & \gamma_{21} & 0 & \gamma_{13} \\ \gamma_{30} & \gamma_{31} & \gamma_{32} & 0 \end{bmatrix} + \begin{bmatrix} 0 & \omega_{10} & \omega_{02} & \omega_{03} \\ \omega_{10} & 0 & \omega_{12} & \omega_{13} \\ \omega_{20} & \omega_{21} & 0 & \omega_{23} \\ \omega_{30} & \omega_{31} & \omega_{32} & 0 \end{bmatrix} \right\} \cdot$$

4-momentum of electron   Weak field of electron   Electromagnetic field of electron, spin

$$\begin{pmatrix} \xi_0 & \xi_0 & \xi_0 & \xi_0 \\ \xi_1 & \xi_1 & \xi_1 & \xi_1 \\ \xi_2 & \xi_2 & \xi_2 & \xi_2 \\ \xi_3 & \xi_3 & \xi_3 & \xi_3 \end{pmatrix} \qquad (3.1.14)$$

$$\begin{cases} u_x = H\cos\dfrac{\pi r}{\varPhi}\sin\theta\sin\phi + H\cos\dfrac{\theta}{2}\cos\theta\sin\phi - H\sin\phi\sin\phi \\ u_x = H\cos\dfrac{\pi r}{\varPhi}\sin\theta\cos\phi + H\cos\dfrac{\theta}{2}\cos\theta\cos\phi + H\cos\phi\sin\phi \\ u_z = H\cos\dfrac{\pi r}{\varPhi}\dfrac{z}{r} - H\cos\dfrac{\theta}{2}\cos\phi \end{cases} \qquad (3.1.15)$$

$$\begin{cases} r = \sqrt{x^2+y^2+z^2} \\ \theta = \arccos(z/r) \\ \phi = \arctan(y/x) \end{cases}, \quad \begin{cases} x = r\sin\theta\cos\phi \\ y = r\sin\theta\sin\phi \\ z = r\cos\theta \end{cases} \qquad (3.1.16)$$

$$\begin{cases} u_x = H\cos\dfrac{\pi r}{\varPhi}\dfrac{y}{r} + H\cos\dfrac{\theta}{2}\dfrac{x}{r} - H\sin\phi\sin\phi \\ u_x = H\cos\dfrac{\pi r}{\varPhi}\dfrac{x}{r} + H\dfrac{z}{r}\cos\dfrac{\theta}{2}\cos\phi + H\cos\phi\sin\phi \\ u_z = H\cos\dfrac{\pi r}{\varPhi}\dfrac{z}{r} - H\cos\dfrac{\theta}{2}\cos\phi \end{cases} \qquad (3.1.17)$$

3.1.4 The intrinsic principal strain $\varepsilon_{\mu\mu}$ of the lepton

Now we discuss the intrinsic principal strain of the leptonin the spherical coordinate system:

$$[\varepsilon](\tau,r,\theta,\phi) = \begin{bmatrix} \varepsilon_\tau & 0 & 0 & 0 \\ 0 & \varepsilon_{rr} & 0 & 0 \\ 0 & 0 & \varepsilon_{\theta\theta} & 0 \\ 0 & 0 & 0 & \varepsilon_{\phi\phi} \end{bmatrix} \begin{pmatrix} c\tau & 0 & 0 & 0 \\ 0 & r & 0 & 0 \\ 0 & 0 & \theta & 0 \\ 0 & 0 & 0 & \phi \end{pmatrix} +$$

(1)

$$\begin{bmatrix} 0 & 0 & 0 & 0 \\ 0 & 0 & 0 & 0 \\ 0 & 0 & \Delta\varepsilon_\theta & 0 \\ 0 & 0 & 0 & \Delta\varepsilon_\phi \end{bmatrix} \begin{pmatrix} 0 & 0 & 0 & 0 \\ 0 & 0 & 0 & 0 \\ 0 & 0 & \theta & 0 \\ 0 & 0 & 0 & \phi \end{pmatrix} \quad (3.1.18)$$

$$\begin{cases} \varepsilon_0 = \sqrt{\varepsilon_1^2 + \varepsilon_2^2 + \varepsilon_3^2} \,;\, \varepsilon_{rr} = \dfrac{\partial u_r}{\partial r} = - H \dfrac{\pi}{\Phi}\sin\dfrac{\pi r}{\Phi} \\ \varepsilon_{\phi\phi} = \dfrac{H}{r}\cos\dfrac{\pi r}{\Phi} \,;\, \varepsilon_{\theta\theta} = \dfrac{H}{r}\cos\dfrac{\pi r}{\Phi} \end{cases} \quad (3.1.19)$$

Leptons have fibre structure, and shear strain fibers form the intrinsic mass field of leptons. We call strained fibers in the inner space the electron flux lines in the inner region. Now consider the specific situation of an electron flux line.

1. Intrinsic strain and electric flux line strain of electrons flux line in generating regions

Considering the characteristics of quantum field vacuum fibrosis, $\varepsilon_{\theta\theta}, \varepsilon_{\varphi\varphi}$ can be incorporated into a single fiber.

$$\frac{\mathscr{H}}{R} = \frac{\mathscr{H}_{Radial}}{R} + \frac{\mathscr{H}_{Ring}}{R} = |\bar{\varepsilon}_r| + |\bar{\varepsilon}_\theta + \bar{\varepsilon}_\phi| \quad (3.1.20)$$

The circumferential strain ε_θ, ε_φ of a single fiber is incorporated into a single fiber strain $u(r) = \dfrac{\mathscr{H}}{2}\cos\left(\dfrac{\pi}{\Phi}r\right)$ in the intrinsic space. The dimension shifting Δu_μ of a single electric flux line is Δr.

$0 \leqslant r \leqslant r_0$ is electric flux line generating regions in internal space. This region is a spherical region, constituting a mass charge. Electric flux line number is $e(r) = k \cdot 4\pi r^2$ in generating regions, when the spherical smaller, then the fiber number fewer (Fig. 3.1.7), electric flux line are formed in a small area of internal space. Set the total number of lines is e, internal area electric flux line is proportional to the surface area. When the surface is $4\pi r_0^2$, the electric flux line are all formed has e. The number of electric flux lines passing through a sphere per unit area is internal area electric flux line intensity which is constant.

$$P = e\frac{4\pi r^2}{4\pi r_0^2}p = e\frac{r^2}{r_0^2}p, \quad E_{\text{Internal space}} = e\frac{4\pi r^2}{4\pi r_0^2}\Delta r_B = e\frac{r^2}{r_0^2}\Delta r_B \quad (3.1.21)$$

Here, Δr_B is single fiber offset. See Fig. 3.1.2. In the fiber formation zone, when $r \to 0$, while $E \to 0$. It is different with the classical theory, here E is not infinite. Obviously, Coulomb's Law is only applicable in the range $r > r_0$. It should be noted here r_0 is a small amount which much smaller than the radius of the intrinsic space. From view of the experimental, the radius is estimated $r_0 < 10^{-16}$ cm, when less than a certain value, Coulomb theorem will be ineffective. The deformation fiber within intrinsic space which meet

$$u_\mu = \left(\frac{\mathscr{H}}{2}\right)\cos(k^\mu \xi_\mu), \quad \Delta u_\mu = \Delta r_B = \text{constant} \quad (3.1.22)$$

Here, u_μ is the displacement function of single electric flux line. Consider space term in a closed

Chapter 3 The Structure of Lepton

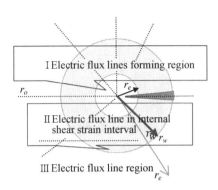

Fig. 3.1.7 Regional division of fiber strain field

spherical ($4\pi r^2$), the total number of single-electron fiber is e. $u_{0\sim r_0}(r)/S_{ball}$ is deformation distribution of the electric flux line in forming region.

$$\mathscr{U}_{0\sim r_0} = \frac{u_{0\sim r_0}(r)}{S_{ball}} = e\frac{r^2}{r_0^2}\frac{\mathscr{H}}{2}\cos\left(\frac{\pi}{\phi_\mu}r_\mu\right)\cdot\frac{r}{r}\,;\,\mathscr{E}_{0\sim r_0} = \frac{\Delta u_{0\sim r_0}(r)}{S_{ball}} = e\frac{r^2}{r_0^2}\Delta r_B \quad (3.1.23)$$

Above is lepton intrinsic field function, $e(r^2/r_0^2)$ is the wave amplitude of fiber formation region. $\mathscr{U}$ is fiber strain strength, and $\mathscr{E}$ is electric field intensity.

2. Internal space electric flux line

$r_0 \leqslant r \leqslant R$ is the scope of internal spaceelectric flux line, and the total deformation is h_e

$$u(\xi) = eu_\mu(\xi)\,;\,h_e = e\Delta r_B \quad (3.1.24)$$

The number of electric flux lines per unit erea of internal space (deformation fiber density) is

$$\mathscr{U}_{r_0\sim R}(\xi) = \frac{u_{r_0\sim R}(\xi)}{S_{ball}} = \frac{e}{4\pi r^2}\frac{\mathscr{H}}{2}\cos\left(\frac{\pi}{\phi_\mu}\xi_\mu\right)\,;\,\mathscr{E}_{r_0\sim R}(\xi) = \frac{\Delta u_{r_0\sim R}(\xi)}{S_{ball}} = \frac{e\Delta r_B}{4\pi r^2} \quad (3.1.25)$$

Here, $r_0 \sim R$ is weak force zone. Lepton intrinsic internal spaceelectric flux line was closed in the radius of the sphere. To consider the whole sphere, the total deformation is

$$u(\xi) = 4\pi r^2\frac{e}{4\pi r^2}\frac{\mathscr{H}}{2}\cos\left(\frac{\pi}{\phi_\mu}\xi_\mu\right) = \frac{e\mathscr{H}}{2}\cos\left(\frac{\pi}{\phi_\mu}\xi_\mu\right) = h_f\cos k^\mu\xi_\mu \quad (3.1.26)$$

Here, $k^\mu = k_\mu$. This is the displacement function of the lepton intrinsic field.

3. The size of a basic unit of vacuum

Field the basic unit is extremely small, basic unit vacuumis excited that produce a pair of dual electronic e^+e^-, lepton field strain is restricted to the intrinsic space (tensile state lepton has a basic unit hole), and finally the field strain almost can fill in basic unit hole, size of basic unit of vacuum equal to basic unit hole, so total deformation h_f should be equal to basic unit hole, i. e.

$$V_0 - \Delta V_0 = h_f\,;\quad \Delta V_0 = h_e = e\Delta r_B \quad (3.1.27)$$

h_f is a new constant, V_0 is a basic unit size, call h_f as a basic unit of vacuum. ΔV_0 is the total dimension deviation of electric flux line, it is a very small quantity (Fig. 3.4.1). The intrinsic-momentum of the single-electron is $\bar{p}_\mu = h_f/R_\mu$. The shear strain area of electric flux line constitute the mass charge, the mass charge bending the background space-time to form the gravitational field of the second order strain, bending split background fibrosis vacuum to form the electrostatic field.

Because electrons have fiber structure in the internal space of electrons ($r<R$), there is intrinsic electromagnetic field. The strain tensor of electromagnetic field is

$$[A_{\mu,\mu}](\xi) + \frac{1}{2}([\Omega_{\mu 0}] + [\Omega_{ij}])(\xi) \qquad (3.1.28)$$

Here, $A_{\mu,\mu}$ is intrinsic electromagnetic four vectors; $\Omega_{\mu 0}$ is intrinsic magnetic field; Ω_{ij} is intrinsic electric field.

4. Region of electrostatic field electric flux line

The deformation is within intrinsic space. Outside the intrinsic space ($r \geqslant R$, R is the radius of electronic intrinsic space), the first-order vacuum deformation is zero, i.e. $u_\mu(x) = 0$.

$$e\Delta r = A = \text{const}, \quad \partial u_\mu(x)/\partial x_\mu = 0 \qquad (3.1.29)$$

Here, u_r does not vary with r changes, the strain is zero. The electric flux line is without influence on space-time. The dimension mobile effect of electric flux line makes the electric flux line is penetrate the intrinsic spacesurface into the observation space that is the electrostatic field. The electric flux lines are rays, and the radius interval of electrostatic field is $r = 0 \sim \infty$.

$$\mathscr{U}_{R\sim\infty}(\xi) = \frac{u_{R\sim\infty}(\xi)}{S_{\text{ball}}} = 0 \; ; \; \mathscr{E}_{R\sim\infty}(\xi) = \frac{\Delta u_{R\sim\infty}(\xi)}{S_{\text{ball}}} = \frac{e\Delta r_B}{4\pi r^2} \qquad (3.1.30)$$

5. The intrinsic diameter of the electron Φ_e

According to $\bar{p}_e = \dfrac{h_f}{\Phi_e} = \dfrac{2h}{\Phi_e}$; $\bar{p}_e = m_e c$, it can be obtained $\Phi_e = \dfrac{2h}{m_e c}$, by $h = 6.626\,069\,3(11) \times 10^{-34}$ J·s, and the mass of the electron $m_e = 9.109\,382\,15(45) \times 10^{-31}$ that it into the above equation. It can be calculated as follows.

$$\Phi_e = \frac{2 \times 6.626\,069\,3 \times 10^{-34}}{9.8 \times 9.109\,382\,15 \times 10^{-31} \times 2.997\,924\,58 \times 10^8}$$

$$= 0.495\,165\,375\,648\,9 \times 10^{-13}$$

$$= 4.951\,65 \times 10^{-12}(\text{m})$$

It is two times the wavelength of the electron Compton, that is $\Phi_e = 2\lambda_c$. The intrinsic diameter and the present radius of the electron (0.090 880 914 (40) Fermi) are two different concepts. Now we know that the electron radius is the radius of a tiny spherical with charge.

3.2 Lepton mass

Mass is an unsolved mystery in current physics. Now we have to solve problems of mass. Substances have mass, as large as the universe as small as quarks. From the point of view of

Chapter 3 The Structure of Lepton

vacuum strain, mass is easy to understand. We need to understand the propagation characteristics of lepton first.

Elementary particle mass. The propagation effect of vacuum strain of quantum field in a certain direction is the mass of particles in that direction. The projection of vacuum strain in the direction of motion determines the magnitude of mass in the direction. Fermions with static mass always propagate along the direction of minimum strain.

3.2.1 Lepton propagation characteristics

The essence of electronic motion is that the vortex waves of "source" and "sink" propagate from one point to another. Since it is a wave, why can electrons propagate at any speed?

Review the propagation of photons (Fig. 3.2.1(a)). Point O forward moving with time t pass, and the vacuum deformation to reach the reaching the maximum $\mathscr{H}/2$, then recovery, eventually returning to initial location. Point O is reach $\mathscr{H}/2$ to form wave, the propagation velocity is c. Only consider the space deformation, and defines the basic unit of deformation of spatial and temporal is $\mathscr{H}_\xi$.

$$c = \frac{\mathscr{H}_\xi}{\tau_h} \qquad (3.2.1)$$

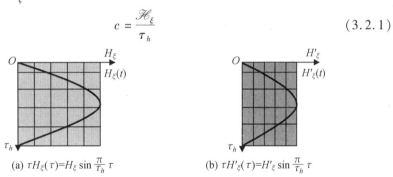

(a) $\tau H_\xi(\tau) = H_\xi \sin \frac{\pi}{\tau_h} \tau$ (b) $\tau H'_\xi(\tau) = H'_\xi \sin \frac{\pi}{\tau_h} \tau$

Fig. 3.2.1 Measurement covariant of quantum wave

The displacement function with time is

$$\mathscr{H}_\xi(\tau) = \mathscr{H}_\xi \sin \frac{\pi}{\tau_h}\tau \quad (0 \leqslant \tau \leqslant \tau_h) \qquad (3.2.2)$$

It's nothing to do with the photon frequency and wavelength. When the entire background field is compressedor stretched (Fig. 3.2.1(b)), $\mathscr{H}_\xi$ is into $\mathscr{H}'_\xi$. Because of vacuum strain, the vacuum elasticity decreases, and the maximum space-time deformation decreases to $\mathscr{H}'_\xi$, i.e. vacuum hardening and propagation speed decreases $c' = \mathscr{H}'_\xi/\tau_h$. τ_h is the time that is a point from a free state deformation to the maximum then back to the initial state in the vacuum, it's the strain cycle time of basic unit. The elasticity of a basic unit vacuum has a common characteristic with the vibration of a simple pendulum, that is, the period τ_h of vibration is a constant independent of the amplitude (Fig. 3.2.2). τ_h is defined as the smallest unit time in the micro world. It is Planck time in theory of vacuum superunified field, which is one of the most basic invariants. Here, $\int_0^{\Phi_0} d\tau = \Phi_0, d\tau \equiv \tau_h$.

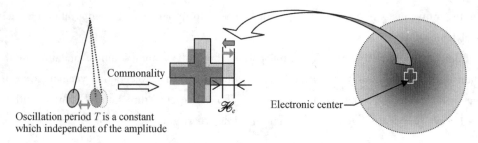

Fig. 3.2.2 Electronic transmission properties

Point I of simple harmonic motion vibrates continuously because of inertia, while the basic unit of vacuum vibrates only once (Fig. 3.2.1), because the vacuum itself is only an elastic medium and has no inertia.

The center of lepton determines the propagating nature of lepton. The center of the electron reaches its limit. Leptons propagate along x, taking into account the 1-dimensional strain of the x axis. The strain $\Delta\varepsilon(x)$ applied externally satisfies the principle of strain superposition $\mathcal{H} \equiv \int_a^b [\,|\varepsilon_0(x)| + |\Delta\varepsilon(x)|\,] \mathrm{d}x$. The central point is always in the strain limit state. $\dfrac{\mathcal{H}}{b-a} = p_f = p_0 + \Delta p = |\bar{\varepsilon}_0(x)| + |\Delta\bar{\varepsilon}(x)|$. Applying external strain $\Delta\varepsilon(x)$ will result in a smaller lepton strain range. Therefore, any external vacuum strain can satisfy the propagation conditions. The strain of the lepton center reaches the limit value, which enables the lepton to move at any speed. In addition, when the lepton accelerates its motion, it will produce superluminal effect, radiate electromagnetic waves and gravitational waves.

Lepton propagation is determined by the averagestrain of lepton is the average deformation of electrons in the direction of propagation, Φ_0 is the intrinsic diameter of electrons. The subscript V indicates the direction of propagation. The propagation velocity increases with the increase of deformation $\Delta\overline{\mathcal{H}}_{eV}$. The electron has intrinsic deformation $\mathcal{H}_{e0}$, the propagation is the vacuum deformation superposition. The condition of wave propagation is $|\mathcal{H}| = |\mathcal{H}_{e0}| + |\Delta\overline{\mathcal{H}}_{eV}|$. The vacuum deformation of photon is $|\mathcal{H}_\gamma| = |\mathcal{H}|$, and $|\Delta\overline{\mathcal{H}}_{eV}| > 0$, then $|\Delta\overline{\mathcal{H}}_{eV}| < |\mathcal{H}_\gamma|$. $V_e = \Delta\overline{\mathcal{H}}_{eV}/\tau_h$. The vibration period of the basic unit vacuum from deformation to natural state is τ_h. Because the Lepton vacuum deformation $\mathcal{H}_{e0}$ exists in advance, $\Delta\overline{\mathcal{H}}_{eV}$ will never exceed the deformation of free photon in $\mathcal{H}$ space-time, so the electron propagation ability is not as good as photon, and the propagation speed is $V_e < c$. That's why photons always travel at the speed of light. Fermions can travel at any speed less than c.

The propagation speed of leptons is

$$V \equiv \frac{\Delta\overline{\mathcal{H}}_e}{\tau_h} \quad (3.2.3)$$

3.2.2 Inertial mass of Fermions

There are two kinds of mass, inertial mass and gravitational mass (Fig. 3.2.3). Consider

Chapter 3 The Structure of Lepton

two macroscopic objects, which have no friction and are static. If we use the same force to act on two objects, according to $F = ma$, we know that the acceleration of heavy objects is small, that of light objects is large, and that under the same force, the velocities of two objects are different. Relatively speaking, the inertia mass of low-speed object is large, while that of high-speed object is small. When momentum (i.e. strain) is applied to an object, the process of action is expressed as force. Here, $F = dp/dt$; $dp = dv \cdot m$. The inertial mass of elementary particles is the embodiment of the propagation effect of fermions in the observation space. By $h_f = P_f \Phi$ known that the static momentum of particle is greater its diameter is smaller.

$$F = ma \qquad (3.2.4)$$

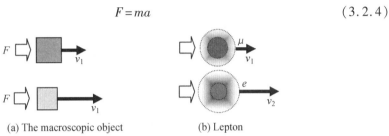

(a) The macroscopic object (b) Lepton

Fig. 3.2.3 Inertial mass

① The intrinsic momentum of charged leptons is $p_f = m_0 c = \dfrac{h_f}{\Phi} = \varepsilon_L$.

Here, ε_L is the intrinsic strain of lepton spherical symmetry; Φ is the intrinsic diameter of lepton.

② Momentum P_f of lepton moving at velocity v in observation space is $P_f = \dfrac{1}{2} m_0 v$.

③ Applied the same external momentum to different lepton field will be get different speeds, by momentum conservation to know

$$P = v_1 m_1 = v_2 m_2 = \cdots \qquad (3.2.5)$$

The macroscopic effect of inertia mass is shown by the propagation state of quantum field. We give the definition of Fermion inertia mass that the static Fermion intrinsic strain is not zero and the central point strain reaches the limit, so that Fermion can propagate at any speed less than the speed of light. The average value of all internal strains of particles determines the propagation performance. This macroscopic propagation effect is defined as Fermion Inertial Mass.

It should be noted that the mass of elementary particles has nothing to do with Higgs particles.

Arbitrary form of strain will reduce the transmission effect and produce mass effect. The electron all intrinsic static are closed internal space which represents the static energy and rest mass of particles. Mixed coordinates are used in observation space.

$$\varphi(\xi) = \exp\left[i\left(\underbrace{\begin{bmatrix} \varepsilon_{00} & 0 & 0 & 0 \\ 0 & 0 & 0 & 0 \\ 0 & 0 & 0 & 0 \\ 0 & 0 & 0 & 0 \end{bmatrix}}_{E_0} \underbrace{\begin{bmatrix} \xi^0 & 0 & 0 & 0 \\ 0 & 0 & 0 & 0 \\ 0 & 0 & 0 & 0 \\ 0 & 0 & 0 & 0 \end{bmatrix}}_{X_0} + \underbrace{\begin{bmatrix} 0 & \varepsilon_{01} & \varepsilon_{02} & \varepsilon_{03} \\ \varepsilon_{10} & \varepsilon_{11} & \varepsilon_{12} & \varepsilon_{13} \\ \varepsilon_{20} & \varepsilon_{21} & \varepsilon_{22} & \varepsilon_{23} \\ \varepsilon_{30} & \varepsilon_{31} & \varepsilon_{32} & \varepsilon_{33} \end{bmatrix}}_{m_0 c} \underbrace{\begin{bmatrix} 0 & \xi_0 & \xi_0 & \xi_0 \\ \xi_1 & \xi_1 & \xi_1 & \xi_1 \\ \xi_2 & \xi_2 & \xi_2 & \xi_2 \\ \xi_3 & \xi_3 & \xi_3 & \xi_3 \end{bmatrix}}_{X_4}\right)\right]$$

$$= \exp\left[i(E_0 X_0 + m_0 X_4)\right] \tag{3.2.6}$$

Radial strain mass m_{0R} and shear strain mass m_{0S} can keep electrons at rest.

$$m_0 c \equiv \begin{bmatrix} 0 & \varepsilon_{01} & \varepsilon_{02} & \varepsilon_{03} \\ \varepsilon_{10} & \varepsilon_{11} & \varepsilon_{12} & \varepsilon_{13} \\ \varepsilon_{20} & \varepsilon_{21} & \varepsilon_{22} & \varepsilon_{23} \\ \varepsilon_{30} & \varepsilon_{31} & \varepsilon_{32} & \varepsilon_{33} \end{bmatrix}$$

Contain

$$m_{0R} c \equiv \begin{bmatrix} 0 & 0 & 0 & 0 \\ 0 & \varepsilon_1 & 0 & 0 \\ 0 & 0 & \varepsilon_2 & 0 \\ 0 & 0 & 0 & \varepsilon_3 \end{bmatrix} \ ; \ m_{0c} c \equiv \begin{bmatrix} 0 & \gamma_{01} & \gamma_{02} & \gamma_{03} \\ \gamma_{10} & \gamma_{11} & \gamma_{12} & \gamma_{13} \\ \gamma_{20} & \gamma_{21} & \gamma_{22} & \gamma_{23} \\ \gamma_{30} & \gamma_{31} & \gamma_{32} & \gamma_{33} \end{bmatrix} \tag{3.2.7}$$

This strain has 3-dimensional characteristics. The pure spin strain constitutes a neutrino with 2-dimensional characteristics. Photons have 1-dimensional characteristics, only dynamic mass, no static mass, and only 3-dimensional strain has static quantity, which we will discuss later. The 3-dimensional mass of electron satisfies $P_m = m_0 c$, $P_m \phi = h_f$. Electronic intrinsic diameter is ϕ. $\xi_1^2 + \xi_1^2 + \xi_1^2 = (\phi/2)^2$.

3.2.3 Orthogonal of lepton mass and momentum

When photons are coupled with electrons, electrons gain kinetic energy and propagate in the direction of momentum. The strain of electrons along the direction of motion is momentum. In motion, the intrinsic radius R_0 of space along the direction x of motion is contraction R_x (Fig. 3.2.4), which conforms to special relativity.

$$R_x = R_0 \sqrt{1 - (v_x/c)^2} \tag{3.2.8}$$

Lepton motion causes the contraction of internal space. Considering a single particle, see formula (3.2.8), the momentum function of electric flux line in intrinsic space is

$$p_0(r) = -\frac{e\mathscr{H}\pi r}{8\pi r^3 \Phi} \sin\frac{\pi}{\Phi} r \quad (0 \leqslant r \leqslant R)$$

The direction x of movement of particles along the intrinsic space, then we should consider the value R of the projection in the direction

$$R = R_0 \cos\varphi \sin\theta$$

Lepton movement caused by contraction of the intrinsic space to meet special theory of relativity:

$$p_0 \cdot R_0 \cos\varphi \sin\theta = p \cdot R_0 \cos\varphi \sin\theta \cdot \gamma$$
$$\left| -\frac{e\mathcal{H}\pi r}{8\pi r^3 \Phi} \sin\frac{\pi}{\Phi} r \right| \cdot 2R_0 \cos\varphi \sin\theta = h \quad (3.2.9)$$

Photonelectric flux line part of the Planck's constant is

$$p_0 \cdot \Phi_0 = p \cdot \Phi\gamma = h \; ; \; m_0 c \cdot \Phi_0 = h = mc \cdot \Phi\gamma \quad (3.2.10)$$

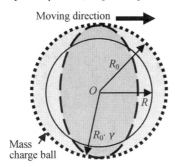

Fig. 3.2.4 Moving leads to quantum field intrinsic deformation

$\varphi=0$, $\theta=\pi/2$, the intrinsic space direction $\xi /\!/ x$ same as v, $\gamma = 1/\sqrt{1-(v/c)^2}$, $\Phi = \Phi_0 \gamma$. In order to simplify the problem, we averaged to R along the direction of motion.

$$R = R_0 \sqrt{1-(v/c)^2} \, \frac{1}{\pi}\int_{\frac{\pi}{2}}^{\frac{\pi}{2}} \cos\varphi \mathrm{d}\varphi \, \frac{1}{\pi}\int_0^{\pi} \sin\theta \mathrm{d}\theta = \frac{4R_0}{\pi^2 \gamma}$$

$$R = \frac{4R_0}{\pi^2}\sqrt{1-(v/c)^2} = R_{0x}\sqrt{1-(v/c)^2}$$

Take $R_{0x} = 4R_0/\pi^2$.

$$mc \cdot R = h_f = m_0 \gamma c \cdot R_{0x} \quad (3.2.11)$$

Here, Φ_0 is the internal space diameter of static electrons. $\Phi_0 \geqslant \Phi$. The internal space diameter of static electrons contraction along the direction x of movement is

$$\Phi_0 = \Phi_0/\gamma \; ; p = p_0 \gamma \; ; m = \gamma m_0 \quad (3.2.12)$$

The intrinsic mass m_{0x} satisfies the following relationships in the observation space:

$$m_x = m_{x0}/\sqrt{1-(v/c)^2}$$

Particle mass-energy relation is $E/c = \sqrt{p^2 + m_0^2 c^2}$ the relationship can be obtained by field strain it's proved as follows:

$$R_x m_x = h/2 = R_{0x} m_{0x} \quad (3.2.13)$$

Static intrinsic momentum function is

$$p_m(r) = -\frac{e\mathcal{H}\pi r}{8\pi r^3 \Phi}\sin\frac{\pi}{\Phi} R_m \quad (3.2.14)$$

Here, R_m is Static intrinsic space radius. The function of the increment of intrinsic momentum in motion is

$$\Delta p(r) = -\frac{e\mathcal{H}\pi r}{8\pi r^3 \Phi}\sin\frac{\pi}{\Phi}R_{\Delta p} \quad (3.2.15)$$

Total intrinsic momentum function in direction x:

$$p(r) = \frac{\partial u_r}{\partial r} = -\frac{e\mathcal{H}\pi r}{8\pi r^3 \Phi}\sin\frac{\pi}{\Phi}R, R = \Phi/2 \text{ (total length of strain area)}$$

$$p = \bar{p}(r) = -\mathcal{H}\frac{\pi}{\Phi}$$

The relationship between the total momentum and rest mass can be obtained by

$$p(r) = -\frac{e\mathcal{H}\pi r}{8\pi r^3 \Phi}\sin\left[\frac{\pi}{\Phi}(R_m + R_{\Delta p})\right]$$

$$= -\frac{e\mathcal{H}\pi r}{8\pi r^3 \Phi}\left[\sin\left(\frac{\pi}{\Phi}R_m\right)\cos\left(\frac{\pi}{\Phi}R_{\Delta p}\right) + \sin\left(\frac{\pi}{\Phi}R_{\Delta p}\right)\cos\left(\frac{\pi}{\Phi}R_{\Delta p}\right)\right]$$

Here, $\Phi/2 = R_m + R_{\Delta p}$.

$$R_{\Delta p} = \Phi/2 - R_m; R_m = \Phi/2 - R_{\Delta p} \quad (3.2.16)$$

$$p(r) = -\frac{e\mathcal{H}\pi r}{8\pi r^3 \Phi}\left[\sin\left(\frac{\pi}{\Phi}R_m\right)\cos\frac{\pi}{\Phi}\left(\frac{\Phi}{2} - R_m\right) + \sin\left(\frac{\pi}{\Phi}R_{\Delta p}\right)\cos\frac{\pi}{\Phi}\left(\frac{\Phi}{2} - R_{\Delta p}\right)\right]$$

$$p(r) = -\frac{e\mathcal{H}\pi r}{8\pi r^3 \Phi}\left[\sin^2\left(\frac{\pi}{\Phi}R_m\right)\sin^2\left(\frac{\pi}{\Phi}R_{\Delta p}\right)\right]$$

Multiplication the factor: $-\frac{e\mathcal{H}\pi r}{8\pi r^3 \Phi}$, by $R_m = \Phi/2 - R_{\Delta p}$ get

$$\left(-\frac{e\mathcal{H}\pi r}{8\pi r^3 \Phi}\right)^2 = \left[-\frac{e\mathcal{H}\pi r}{8\pi r^3 \Phi}\sin\left(\frac{\pi}{\Phi}R_m\right)\right]^2 + \left[-\frac{e\mathcal{H}\pi r}{8\pi r^3 \Phi}\sin\left(\frac{\pi}{\Phi}R_{\Delta p}\right)\right]^2 \quad (3.2.17)$$

$$p^2 = p_m^2 + \Delta p^2, \text{ i. e. }: m_0^2 c^2 + p^2 = (E/c)^2$$

$$E/c = \sqrt{p^2 + m_0^2 c^2} \quad (3.2.18)$$

This is known as the mass-energy relation in relativistic quantum mechanics of particles. This is the intrinsic properties of particles. According to the special theory of relativity

$$p^\mu = (p^0, p^1, p^2, p^3) = \left(\frac{E}{c}, p_x, p_y, p_z\right) \quad (3.2.19)$$

$$\sum_{\mu=0}^{3} p_\mu p^\mu = p_\mu p^\mu = \frac{E^2}{c^2} - p \cdot p \equiv m_0^2 c^2 \quad (3.2.20)$$

In the vacuum unified theory, the essence of mass is the strain effect, essence of static mass and dynamic mass is same, just state of observe are different.

Lepton and longitudinal photon coupling makes the intrinsic space $\xi_\alpha(\alpha = 0, 1, 2, 3)$ and the space of freedom $x_\mu(\mu = 0, 1, 2, 3)$ Lorentz transformation between the two reference frames to meet

$$\Delta\xi_\mu = \gamma\Delta x_\mu \quad (\mu = 0, 1, 2, 3) \quad (3.2.21)$$

The intrinsic space frame ξ_α transform to space frame x_μ, the particle mass to meet the Lorentz transformation: $m = \gamma\Delta m_0$, the total energy of the particle:

$$E = \gamma mc = \frac{m_0 c}{\sqrt{1 - (v/c)^2}} = m_0 c^2 + \frac{1}{2}m_0 v^2 + \cdots \quad (3.2.22)$$

$E_0 = m_0 c^2$ in intrinsic space-time; $\Delta E = m_0 v^2 / 2$ in observation space-time.

Particle momentum $p^v = \dfrac{m_0 \dot{x}^v}{\sqrt{-\dot{x}^\mu \dot{x}_\mu}}$, $\dot{p}^v = 0$.

Kinetic energy of the leptons: $E = \dfrac{p^2}{2m_E}$, momentum: $p = \sqrt{2m_c E}$.

3.3 The lepton spin strain [ω_{ij}]

3.3.1 Virtual photon-spin wave propagation in the lepton intrinsic space-time

1. Generation of lepton spin

Why the electronics will have to spin? Here we give the reason. Vacuum strain is always propagation in the form of wave. The spherical wave propagates toward the center in Fermion's interior space, and the wavefront is frozen at the regional center. Without spin waves, leptons are unstable. The spin wave is enclosed in the inner space of spherical shell radius R, which constitutes an intrinsic degree of freedom of leptons.

As we can see from the previous discussion, there was the vacuum medium of 1 free-state basic unit in the center of e^+, and 1 basic unit hole-state in the center of e^-. The hole stretches the surrounding vacuum medium to the deformation limit, and a convergent wave propagating toward the center will be formed in the intrinsic space (Fig. 3.3.1). The eccentricity of spin wavecan not be accurately concentrated in the center; the spin wave axis is the longitudinal photon axis. The spin wave rotates along the ring of radius R and never stops. This is like the water in the bathtub, when we pulled out of the cork of the bottom suddenly, then appeared a hole, water will be formed around a hollow flow swirl. When the vacuumappears a basic unit hole, the deformation vacuum around the hole achieves the propagation condition, will form waves of whirlpool, it is the electron Spin wave. On the contrary, if the vacuum has 1 free-state basic unit, it will form a wave that spread from the center outward, which is the anti-electron Spin wave. The electronic spin wave effect is defined as the electron spin. Electron spin angular momentum is $\hbar/2$.

The lepton with the spin is a more complex, but we can be decomposed into two parts.

(1) Static lepton field. Because of the gains and losses a h_f, the lepton form a local internal area electric flux line (i.e. static electric field).

(2) The Spin wave infinitely approaches the center of the lepton. We can understand the spin waves as longitudinal photons (virtual photons) rotating around the lepton center, which can be called virtual photon-spin wave. Virtual photon-spin wave is 1 virtual photon rotating state in the inner space of lepton.

2. Virtual photon-spin wave propagation in the lepton intrinsic space-time

The concept existence in the classical physics, that is obviously wrong if the electron spin

Fig. 3.3.1 Convergent wave propagating toward the center
(When we pulled out the cork on the bottom suddenly, there appeared a hole, the water will flow to form a swirl around the hole)

as mechanical rotation, because if we envisaged electronic is a ball which uniformly distributed charge, the classical electronic radius $r_c = e^2/mc^2 = 2.8 \times 10^{-13}$ cm, and his moment to reach 1 Bohr magnet, the rotating speed of surface to exceed the speed of light. Because it is easy to estimate

$$p \sim \Delta p \sim \hbar/r_c = \frac{\hbar mc^2}{c^2} = mc/\alpha \quad (\alpha = e^2/\hbar c \approx 1/137)$$

So the speed $v \sim 137c$ is much faster than the speed of light. The problem is serious, which makes the spin has no physical image. From the view of the vacuum, this problem will be solved.

Lepton strain is restricted within the intrinsic spherical shell (Fig. 3.3.2), the radius is less than R, the lepton internal area electric flux line with strain in intrinsic space, internal area electric flux line itself does not rotate, rotating is virtual photon-spin wave, and the fiber structure field deformation will reduce propagation properties, and the internal time will slow down. Virtual photon-spin waves propagate slowly in the inner space of leptons $v = \cos\frac{\pi}{2R}(R - r) \cdot c$ (R is intrinsic spatial radius). Near the center $r \to 0$, the propagation speed of the virtual photon-spin wave center tends to zero. Therefore spin does not superluminal. If we can small to enter the electronic intrinsic space, it will happen as follows.

Time dilation:

$$t = t_0/\cos\frac{\pi}{2R}(R - r) \qquad (3.3.1\text{a})$$

Scales reduced

$$l = l_0 \cdot \cos\frac{\pi}{2R}(R - r) \qquad (3.3.1\text{b})$$

The measured speed of light remains c is an invariant.

Set $k = \cos\frac{\pi}{2R}(R - r)$ is space compression coefficient, electron intrinsic space of edge length is $2\pi R = ct_0$. The propagation velocity of virtual photon-spin waves in the internal space

Chapter 3 The Structure of Lepton

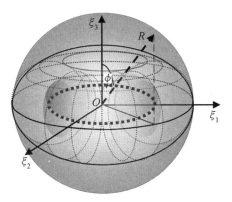

Fig. 3.3.2   Spin intrinsic structure

is $c \cdot 2\pi(R \cdot k) = (c \cdot k) t_0$. In other words, the measurement result of intrinsic spatial at outer edge is same as at $r \to 0$. The total momentum of virtual photon-spin waves is constant, independent of r.

3. Deformation of space-time of spin quantum field

The spin is a longitudinal photon rotation propagation surrounded spin axis. The volume integral of the photon strain in the ring tire is the total virtual photon-spin wave deformation is h. We can get the total deformation of internal area electric flux line contained the virtual photon-spin wave ring tire

$$u_S = \int_{-\pi/2}^{\pi/2} \varepsilon_\varphi \sin\varphi \, d\varphi \cdot \int_0^{2\pi} \varepsilon_\theta d\theta \int_0^R p_{S\gamma}(r) r^2 dr = h$$

The virtual photon-spin wave has the structure of circular tire which can be simplified as a ring. Spin total deformation is h.

The longitudinal photon momentum $p_s = h/\Phi_\gamma$, when the longitudinal photon rotates around the ring axis, the length of the ring $2\pi r$ is the original diameter Φ_γ.

$$p_s = h/2\pi r \qquad (3.3.2)$$

This is the momentum of virtual photon-spin waves.

3.3.2 Spin angular momentum

We simplify the central region in the inner space of lepton(Fig. 3.3.3(b)). Set r is average radius of rotation of center of virtual photon-spin wave, $r < R, r \geqslant 0$, O is center of lepton intrinsic space. Special attention to the lepton intrinsic space relative to the virtual photon-spin wave is stationary, virtual photon-spin waves propagating in the intrinsic space.

The electron spin value is $\hbar/2$. The virtual photon-spin wave center orbit as shown by dotted lines in Fig. 3.3.3(a), photon center path into a ring, r is radius of the fluctuation track, ring length is $2\pi r$, when the wavelength of virtual photon-spin wave $\lambda = 2\pi r$ is low-energy stable state of the photon. Along the circular direction, the total deformation of the entire ring photons is h, thus the virtual photon-spin wave meet

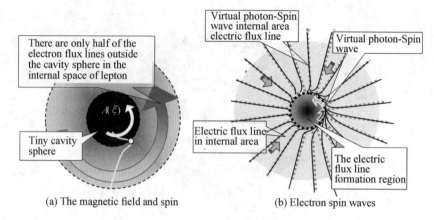

(a) The magnetic field and spin   (b) Electron spin waves

Fig. 3.3.3   Electron spin waves

$$2\pi |r \times p_\gamma| = h,$$

$$|r \times p_S| = rp_S \sin(r\hat{\ }p_S) \frac{h}{2\pi} = \hbar \quad (3.3.3)$$

Here, $r \perp p_S$. The spin angular momentum of lepton is constituted by the spin-orbital angular momentum of the virtual photon-spin wave.

Due to virtual photon-spin wave infinitely approaches the center of the lepton, the virtual photon-spin wave propagation along a lepton center basic unit of vacuum hole a ball surface, region of ball hole is t no vacuum medium, no fiber field exists, and thus outside the hole ball have the internal area electric flux line, virtual photon-spin wave total internal area electric flux line only half of the original, that is $S = \hbar/2$.

From the perspective of observation, we can not observe from the inside to the outside, but from the outside to the inside (Fig. 3.3.4). The center of the lepton is a tiny cavity sphere, and there is no electron flux line. In this way, there are only half of the electron flux lines outside the cavity sphere in the internal space of lepton. For our experiment, the spin angular momentum is $\hbar/2$.

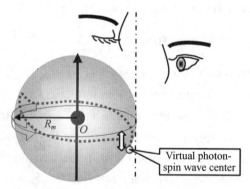

Fig. 3.3.4   Probability standing waves of spin

3.3.3 Rotation strain constitute the spin momentum

Total deformation of virtual photon-spin wave space is

$$u_S = \oint_l \Delta u \, dl = \iint_S (\operatorname{curl} \Delta u) \cdot dS = \iint_S \omega_{21} \cdot dS = h \tag{3.3.4}$$

Here, $\operatorname{curl} \Delta u = \dfrac{\partial u_2}{\partial \xi^1} - \dfrac{\partial u_1}{\partial \xi^2} = \omega_{12}$. Thus, the space-time deformation of virtual photon-spin wave and the virtual photon-spin wave radius of rotation unrelated. Shear strain as shown in Fig. 3.3.5. The rotational strain is $\omega_{21} = \dfrac{1}{2}\left(\dfrac{\partial u}{\partial \xi_2} - \dfrac{\partial v}{\partial \xi_1}\right) = \dfrac{1}{2}[\beta - \alpha]$. Considering the half space effect, the corresponding deformation amount $u(\beta - \alpha) = \dfrac{h}{2}\left(\cos\dfrac{\beta}{2} - \cos\dfrac{\alpha}{2}\right)$, $u_{Se} = \int_0^{2\pi} u(\beta - \alpha)/2 = h/2$, Considering the half space effect, the corresponding average value of deformation $\bar{\omega}_{21} = \dfrac{\int_0^{2\pi} u(\beta - \alpha)/2}{\oint_l \Delta u(\theta) \, dl} = \dfrac{1}{2}$.

The space-time strain of virtual photon-spin waves can be used to shear strain expressed by

$$\omega = \frac{1}{2} \nabla \times S = \frac{1}{2} \begin{vmatrix} i & j & k \\ \dfrac{\partial}{\partial \xi_1} & \dfrac{\partial}{\partial \xi_2} & \dfrac{\partial}{\partial \xi_3} \\ u_1 & u_2 & u_3 \end{vmatrix}$$

The spin-strain of vacuum (i.e. the spin momentum field) is

$$S_{\mu\nu} = \iint_S \omega_{\mu\nu} \cdot dS = \iint_S \left(\frac{\partial u_\mu(\xi)}{\partial \xi^\nu} - \frac{\partial u_\nu(\xi)}{\partial \xi^\mu}\right) \cdot dS = h/2 \quad (\mu, \nu = 0,1,2,3) \tag{3.3.5}$$

(1) High angular momentum.

The centerof photon virtual photon-spin wave along the rim of the electric flux line generating area around the center point of fermion spread form the fermion spin angular momentum $\hbar/2$. When there are second photon virtual photon-spin waves, the center points leave the edge of the electric flux line generating area, then the complete virtual photon-spin waves can be observed in the intrinsic shear strain region for experimental observations. The orbital angular momentum of photons is $\hbar$ that is also the fermion of angular momentum. In this case, the total angular momentum of fermion is $3\hbar/2$. For higher angular momentum, when the virtual photon-spin wave moves around the center, the virtual photon-spin waves have corresponding energy levels. The more virtual photon-spin waves, the greater the angular momentum. If there are n virtual photon-spin waves, then the spin angular momentum of the fermion is $(2n-1)\hbar/2$ (Fig. 3.3.5).

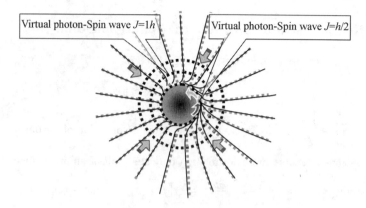

Fig. 3.3.5 High spin angular momentum of fermion

(2) The direction of the spin angular momentum.

Lepton movement from the lepton center coupling with photon center result of coupling make the trajectory of lepton center like photon with fluctuation and spin (Fig. 3.3.6). On this basis, the wave vector of lepton center is k_x, k_y, k_z. The field strain $\varepsilon_x > \varepsilon_z, \varepsilon_y > \varepsilon_z$. Momentum distribution is largest in in x-axis or y-axis and minimum the z-axis. The smaller the vacuum elastic strain is, the easier the vacuum deformation is. The vacuum occurrence deformation is the easiest in ground state. If another vacuum strain is applied, it will occur on the z-axis, which shows that leptons move along the z-axis, and the z-axis constitutes the axis of spin angular momentum.

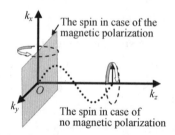

Fig. 3.3.6 Electronic center trajectory and spin orientation

In the case, non-magnetic polarization, the projection of the spin angular momentum on the x-axis are $\pm\hbar/2$ mixed state, projection value relative to the x-axis is 0. $[\langle \hbar/2 | + \langle -\hbar/2 |]/2 = 0$. Spin lepton can be seen as a small magnetic needle. If we want to observe the spin of particles, we can use external magnetic field to turn the spin lepton axis from z to x. In the case of magnetic field polarization that spin split, spin direction pointing to the magnetic field, the value in the projection of x-axis perpendicular to the direction of propagation are $+\hbar/2$ or $-\hbar/2$. It should be noted that the natural spin angular momentum points to the propagation direction, but the measurement results lead to the spin angular momentum perpendicular to the propagation direction.

Chapter 3 The Structure of Lepton

3.4 The intrinsic structure of lepton electric flux line

3.4.1 Electric flux line influence on the space-time

We need to consider that the electric flux line has dimensions independence. Vacuum has 3-dimensional, when ξ^1-dimensional move a small amount Δr does not change the nature of ξ^2, ξ^3 dimensional they are independent of each other. Small "shuttle" does not change the nature of remainder 2-dimensional and the vacuum has no effect, so the existence of the photon has no effect on the surrounding space-time.

We should attention to the fact that Δr_B has nothing to do with the inner structure of lepton. In other words, Δr_B is the residual additional produced by the intrinsic strain of lepton. Any fiber will produce a residual additional effect of the fiber shifting Δr_B in vacuum (Fig. 3.4.1). For e, μ, τ, the total fiber shifting he is equal. The total number of electric flux lines is n_e.

$$h_e = n_e \Delta r_B \qquad (3.4.1)$$

e, μ, τ have the same charge, which is called charge homogeneity. For a single electric flux line, no matter how long the electric flux line is, the electric flux line is shifting Δr_B along the r. The existence of electric flux line does not cause changes in the background space-time, so the electric flux line themselves are massless. In other words, in a strong electric field, the clock will not slow down.

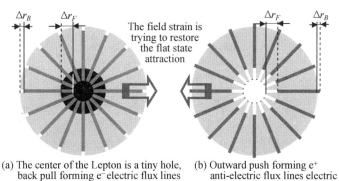

(a) The center of the Lepton is a tiny hole, back pull forming e⁻ electric flux lines which extends to infinity

(b) Outward push forming e⁺ anti-electric flux lines electric field extends to infinity

Fig. 3.4.1 Positive and negative electorn structure of electric flux line

There are the following relationships:

$$h_e \subset h_{f(\text{Radial})} \subset h_f, \quad h_f = h_{f(\text{Radial})} + h_{f(\text{Circumferential})}; \quad h_e < h_{f(\text{Radial})} < h_f \qquad (3.4.2)$$

h_f is enclosed in the inner space. Both shear strain $\gamma_{\mu\nu}$ and spin strain $\omega_{\mu\nu}$ belong to circumferential strain. h_e exists in the observational space.

According to $|\Delta r_{\max}| = |\Delta r_{F\max}| - |\Delta r_{B\max}|$, We known the radial deformation of single fiber.

$$\begin{cases} |\Delta r_{B\max}| = \text{const} \\ \mathcal{H}_{\text{Radial}}(r) = |\Delta r_{F\max}|\cos\dfrac{\pi r}{2R} - |\Delta r_{B\max}|\cos\dfrac{\pi r}{2R} \end{cases} \quad (3.4.3)$$

Set:

$$\begin{cases} u_{r\max} \equiv \mathcal{H}_{\text{Radial}} = u_{rF\max} - u_{rB\max} \\ u_{rF} \equiv |\Delta r_{F\max}|\cos\dfrac{\pi r}{2R} = u_{rF\max}\cos\dfrac{\pi r}{2R} \end{cases} \quad (3.4.4a)$$

$$u_{rB} \equiv |\Delta r_{B\max}|\cos\dfrac{\pi r}{2R} = u_{rB\max}\cos\dfrac{\pi r}{2R} \quad (3.4.4b)$$

Basic unit of vacuum arrangement constituted a fibrosis which compression or stretching in $\xi^1\xi^2\xi^3$ direction where near the compression zone, and circumferential deformation has nature of spherically symmetric, the deformation in radial direction exist directing. Consider ξ^1-axis through the center of the lepton, so the vacuum along ξ^1 to move forward an tiny quantity ε_r (Fig. 3.4.2), which does not change the 3-dimensional features of basic unit. And it closely linked to the basic unit of vacuum also to move forward a small amount, so it continues in ξ^1 to form an electric flux line. In the compression zone is formed compression electric flux line, the formation of tensile electric flux line in tensile area.

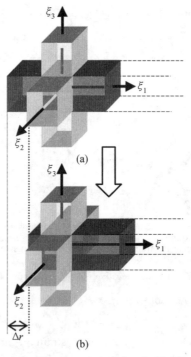

Fig. 3.4.2 The formation of electric flux line

3.4.2 Intrinsic vector potential $\mathscr{A}_{i,i}$ and intrinsic scalar potential $\mathscr{A}_{0,0}$

$$[\varepsilon](\tau,r,\theta,\varphi) = \begin{bmatrix} \varepsilon_\tau & 0 & 0 & 0 \\ 0 & \varepsilon_{rr} & 0 & 0 \\ 0 & 0 & 0 & 0 \\ 0 & 0 & 0 & 0 \end{bmatrix} + \begin{bmatrix} \varepsilon_0 & 0 & 0 & 0 \\ 0 & 0 & 0 & 0 \\ 0 & 0 & \varepsilon_{\theta\theta} & 0 \\ 0 & 0 & 0 & \varepsilon_{\phi\phi} \end{bmatrix} + \begin{bmatrix} 0 & 0 & 0 & 0 \\ 0 & 0 & 0 & 0 \\ 0 & 0 & \Delta\varepsilon_\theta & 0 \\ 0 & 0 & 0 & \Delta\varepsilon_\phi \end{bmatrix}$$

(1) Expression of spherically symmetric strain fiber structure   (2) Spherically symmetric circumferential strain   (3) Non spherically symmetric circumferential strain

(3.4.5)

$$\varepsilon(\tau,r,\theta,\varphi) = \begin{bmatrix} \varepsilon_\tau & 0 & 0 & 0 \\ 0 & \varepsilon_r & 0 & 0 \\ 0 & 0 & 0 & 0 \\ 0 & 0 & 0 & 0 \end{bmatrix} \begin{matrix} \text{Using cartesian} \\ \text{coordinates} \\ \Rightarrow \\ \begin{cases} \xi_1 = r\sin\theta\cos\theta \\ \xi_2 = r\sin\theta\sin\varphi \\ \xi_3 = r\cos\theta \end{cases} \end{matrix} \quad \varepsilon_r(\xi) = \begin{bmatrix} \varepsilon_{r0} & 0 & 0 & 0 \\ 0 & \varepsilon_{r1} & 0 & 0 \\ 0 & 0 & \varepsilon_{r2} & 0 \\ 0 & 0 & 0 & \varepsilon_{r3} \end{bmatrix}$$

(3.4.6)

That is

$$\varepsilon_{\mu\mu} = \begin{bmatrix} du_{r0}/cd\tau & 0 & 0 & 0 \\ 0 & du_{r1}/d\xi_1 & 0 & 0 \\ 0 & 0 & du_{r2}/d\xi_2 & 0 \\ 0 & 0 & 0 & du_{r3}/d\xi_3 \end{bmatrix}$$

$$= \begin{bmatrix} du_{rF0}/cd\tau & 0 & 0 & 0 \\ 0 & du_{rF1}/d\xi_1 & 0 & 0 \\ 0 & 0 & du_{rF2}/d\xi_2 & 0 \\ 0 & 0 & 0 & du_{rF3}/d\xi_3 \end{bmatrix} - \begin{bmatrix} du_{rB0}/cd\tau & 0 & 0 & 0 \\ 0 & du_{rB1}/d\xi_1 & 0 & 0 \\ 0 & 0 & du_{rB2}/d\xi_2 & 0 \\ 0 & 0 & 0 & du_{rB3}/d\xi_3 \end{bmatrix}$$

$$\underbrace{\qquad\qquad\qquad\qquad\qquad\qquad\qquad\qquad\qquad\qquad\qquad}_{\text{The strain of the electrostatic field}}$$

$$\varepsilon_{\mu\nu} = \varepsilon_{\mu\nu}^F - \varepsilon_{\mu\nu}^B \qquad (3.4.7)$$

Here, $\partial_i u_r = \partial_i(u_{rF} - u_{rB}) = \mathscr{A}_{i,i}(\xi)$ is the spatial strain function which is called the intrinsic vectorpotential, $\partial_0(u_{rF0} - u_{rB0}) = \mathscr{A}_{0,0}(\xi)$ is called intrinsic scalar potential. $u_r = (u_{rF} - u_{rB}) = \mathscr{A}_i(\xi)$ is the spatial displacement function and $u_{r0} = (u_{rF0} - u_{rB0}) = \mathscr{A}_0(\xi)$ is the time displacement function. They are the radial displacement function $A(\xi)$.

$$u_r = \mathscr{A}_\mu(\xi) = \begin{bmatrix} \mathscr{A}_0(\xi) & 0 & 0 & 0 \\ 0 & \mathscr{A}_1(\xi) & 0 & 0 \\ 0 & 0 & \mathscr{A}_2(\xi) & 0 \\ 0 & 0 & 0 & \mathscr{A}_3(\xi) \end{bmatrix}$$

$$\mathcal{A}_{\mu,\mu}(\xi) = \begin{bmatrix} \mathcal{A}_{0,0}(\xi) & 0 & 0 & 0 \\ 0 & \mathcal{A}_{1,1}(\xi) & 0 & 0 \\ 0 & 0 & \mathcal{A}_{2,2}(\xi) & 0 \\ 0 & 0 & 0 & \mathcal{A}_{3,3}(\xi) \end{bmatrix} \quad (3.4.8)$$

Scalar potential   Vector potential

The radial displacement function $\mathcal{A}(\xi)$ in intrinsic space ($0 \leqslant r \leqslant R$). This constitutes the mass of the electron's electromagnetic field.

$$\mathcal{A}(r) = e\left(\frac{u_r}{2}\right)\cos\left(\frac{\pi}{\Phi}r\right) \ (0 \leqslant r \leqslant R); r = \sqrt{(\xi^1)^2 + (\xi^2)^2 + (\xi^3)^2} \quad (3.4.9a)$$

$$\mathcal{A}_0(\xi^0) = e\left(\frac{u_r}{2}\right)\cos\left(\frac{\pi}{\Phi_0}\xi^0\right) \ (0 \leqslant \xi^0 \leqslant R_0) \quad (3.4.9b)$$

This part deformation generated electromagnetic field.

3.4.3 Electronic magnetic field lines

The electric flux line is the fiber micro-offset caused by the fiber field strain of the lepton.

$$\varepsilon_r = \frac{\partial u_{rB}}{\partial r} = -u_{rB\max}\frac{\pi}{\Phi}\sin\frac{\pi r}{\Phi}$$

$$\varepsilon_\phi = \frac{u_{rB\max}}{r}\cos\frac{\pi r}{\Phi} - \frac{1}{2}\frac{u_{rB\max}}{r}\sin\frac{\varphi}{2} \quad (3.4.10)$$

$$\varepsilon_\theta = \frac{u_{rB\max}}{r}\cos\frac{\pi r}{\Phi} + \frac{u_{rB\max}}{r}\left(\cot\phi\cos\frac{\varphi}{2} + \frac{\cos\theta}{\sin\varphi}\right)$$

The displacement function of electric flux line is transformed from spherical coordinate system to Cartesian coordinate system.

$$\begin{cases} A_2 = u_{rB\max}\cos\frac{\pi r}{\Phi}\frac{y}{r} + u_{rB\max}\cos\frac{\theta}{2}\frac{x}{r} - u_{rB\max}\sin\varphi\sin\varphi \\ A_1 = u_{rB\max}\cos\frac{\pi r}{\Phi}\frac{x}{r} + u_{rB\max}\frac{z}{r}\cos\frac{\theta}{2}\cos\varphi + u_{rB\max}\cos\varphi\sin\varphi \\ A_3 = u_{rB\max}\cos\frac{\pi r}{\Phi}\frac{z}{r} - u_{rB\max}\cos\frac{\theta}{2}\cos\varphi \\ A_0 = \sqrt{A_1^2 + A_2^2 + A_3^2} \end{cases} \quad (3.4.11)$$

What needs to be distinguished is that $u_{rB} = A_i(\xi)$ is the spatial displacement function of electromagnetic field and $u_{rB0} = A_0(\xi)$ is the time displacement function of electromagnetic field. $\partial_i u_{rB} = A_{i,i}(\xi)$ is the intrinsic vectorpotential of electromagnetic field, and $\partial_0 u_{rB0} = A_{0,0}(\xi)$ is intrinsic scalar potential of electromagnetic field, which can be obtained

Chapter 3 The Structure of Lepton

$$\left[\frac{\partial A_\mu(\xi)}{\partial \xi_\nu}\right] = \left\{ \begin{bmatrix} \varepsilon^B_{00} & 0 & 0 & 0 \\ 0 & \varepsilon^B_{11} & 0 & 0 \\ 0 & 0 & \varepsilon^B_{22} & 0 \\ 0 & 0 & 0 & \varepsilon^B_{33} \end{bmatrix} + \frac{1}{2}\begin{bmatrix} 0 & \gamma^B_{01} & \gamma^B_{02} & \gamma^B_{03} \\ \gamma^B_{10} & 0 & \gamma^B_{12} & \gamma^B_{13} \\ \gamma^B_{20} & \gamma^B_{21} & 0 & \gamma^B_{23} \\ \gamma^B_{30} & \gamma^B_{31} & \gamma^B_{32} & 0 \end{bmatrix} + \frac{1}{2}\begin{bmatrix} 0 & \omega^B_{01} & \omega^B_{02} & \omega^B_{03} \\ \omega^B_{10} & 0 & \omega^B_{12} & \omega^B_{13} \\ \omega^B_{20} & \omega^B_{21} & 0 & \omega^B_{23} \\ \omega^B_{30} & \omega^B_{31} & \omega^B_{32} & 0 \end{bmatrix} \right\}$$

(1) Intrinsic 4-momentum of electromagnetic field, that is vector and scalar potentials

(2) Intrinsic electric flux line shear strain of electron

(3) Intrinsic electromagnetic field of electron

(3.4.12)

Here, B denotes the strain at back end of the electric flux line; F denotes the strain at front end of the electric flux line.

1. Magnetic line of force from a tiny split of the dimension

Why are the electric and magnetic fields always together like a shadow? Why are the magnetic field lines always closed? Does magnetic monopole exist? We will answer these questions. Now let's analyze it one by one.

(1) The intrinsic 4-momentum of the electromagnetic field.

$$\begin{bmatrix} \varepsilon^B_{00} & 0 & 0 & 0 \\ 0 & \varepsilon^B_{11} & 0 & 0 \\ 0 & 0 & \varepsilon^B_{22} & 0 \\ 0 & 0 & 0 & \varepsilon^B_{33} \end{bmatrix} = \begin{bmatrix} du_{rB0}/cd\tau & 0 & 0 & 0 \\ 0 & du_{rB1}/d\xi_1 & 0 & 0 \\ 0 & 0 & du_{rB2}/d\xi_2 & 0 \\ 0 & 0 & 0 & du_{rB3}/d\xi_3 \end{bmatrix}$$

(3.4.13)

It is the intrinsic 4-momentum of the pure electrostatic field, which is contained in the intrinsic 4-momentum $\varepsilon_{\mu\mu}$ of lepton $\varepsilon_{\mu\mu} = \varepsilon^F_{\mu\mu} - \varepsilon^B_{\mu\mu}$. In experiments, it can not be peeled off and has no independent observability. The intrinsic 4-momentum $\varepsilon_{\mu\mu}$ of lepton corresponds to $\varepsilon(\tau,r)$ in the spherical coordinate system. The intrinsic energy momentum of the electric field is identified as the mass of the charge or the mass of the charged electromagnetic field.

It is worth mentioning that $\varepsilon^F_{\mu\nu}$ can only be observed from the interior of the lepton center (Fig. 3.4.1), but it is obvious that it is impossible. Therefore, as an intrinsic characteristic, there is no observability. But $\varepsilon^B_{\mu\nu}$ is observed from outside to inside, so the electric flux line strain characteristics are observable. Therefore, $\varepsilon^F_{\mu\mu}$ and $\varepsilon^B_{\mu\mu}$ do not have the physical meaning of existence independently, but only retain the $\varepsilon_{\mu\mu}$.

(2) The shear strain $[\gamma^B_{\mu\nu}]$ of electron intrinsic electric flux line.

The intrinsic shear strain $\gamma^B_{\mu\nu}$ of the pure electrostatic field is contained in the intrinsic shear strain $\gamma_{\mu\nu}$ of the lepton $\gamma_{\mu\nu} = \gamma^F_{\mu\nu} - \gamma^B_{\mu\nu}$. In the experiment, it can not be divestied from it, and there is no independent observability. The shear strain of the electron intrinsic electric flux line corresponds to $\varepsilon(\tau,\theta,\varphi)$ in the spherical coordinate system, which is regarded as the circumferential mass of charge, that is, the naked mass of the charge. Similarly, $\gamma^F_{\mu\nu}$ and $\gamma^B_{\mu\nu}$ have no independent physical meaning. Only $\gamma_{\mu\nu}$ is retained.

(3) The intrinsic electromagnetic field of lepton is composed of rotational strain $[\omega^B_{\mu\nu}]$ of electric flux line.

$$[\omega^B_{\mu\nu}](\xi) = \begin{pmatrix} 0 & B_3 & -B_2 & -E_1 \\ -B_3 & 0 & B_1 & -E_2 \\ B_2 & -B_1 & 0 & -E_3 \\ E_1 & E_2 & E_3 & 0 \end{pmatrix} \begin{pmatrix} 0 & \xi^0 & \xi^0 & \xi^0 \\ \xi^1 & 0 & \xi^1 & \xi^1 \\ \xi^2 & \xi^2 & 0 & \xi^2 \\ \xi^3 & \xi^3 & \xi^3 & 0 \end{pmatrix} \quad (3.4.14)$$

$[\omega_{ij}]$ is the rotational strain of the electric flux line in the internal space of electron. In the elastic deformation region, the dimensions are independent, but there is an influence between the vacuum dimensions when the vacuum deformation reaches its limit (Fig. 3.4.3), which is caused by a slight origin offset of the coordinate system. The starting end of the electric flux line is to reach the deformation limit state, and the starting end is the mass element. The new strain applied at the starting end of the electric flux line will result in a 3-dimensional origin offset.

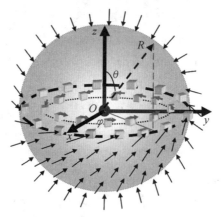

Fig. 3.4.3 Intrinsic structure of spin quantum field
(Each small cube represents a basic unit of
deformation limit. It's the starting end of
the electric flux line)

2. Closed magnetic field lines

Consider the effect of virtual photon-spin waves on the starting end of electric flux lines (Fig. 3.4.4). From the Z-dimensional point to the tangential direction of virtual photon-spin wave, there is a small origin offset (Fig. 3.4.5). Z-direction deformation has an effect on Y-direction, resulting in small offset of Y. The internal space of electrons is half space, and the spatial dimension has only one direction, so the Z-dimensional offset of the same lepton field can only be one direction (Fig. 3.4.5(b)). The Z-dimension fibers produced by small offset are different from electric flux lines. There is no basic vacuum loss in Z-dimension. Therefore, when the Z-axis is offset by a tiny "ε", there will inevitably be "holes" (Fig. 3.4.5(c)). It must be filled by adjacent basic unit vacuum, and finally a closed magnetic line will be formed, this ensures that no defects occur in the vacuum.

Chapter 3 The Structure of Lepton

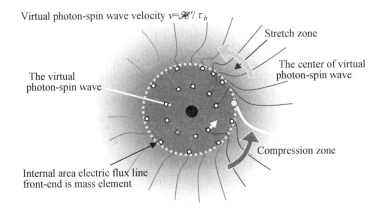

Fig. 3.4.4 virtual photon-spin wave on the surface of spherical shell in the generating region
(The half of spin-waves in the quality ball, there is no observable. Spin wave was enclosed in the sphere that propagation makes mass element fluctuations, a slight deformation along the propagation direction; the deformation caused by the tiny dimensions of split, formation the magnetic field lines)

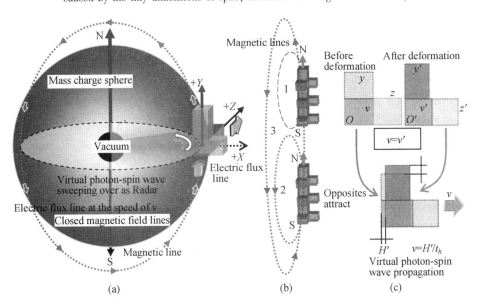

Fig. 3.4.5 Formation of magnetic electric flux line

3. The magnetic properties

In half space, spin has only one point (Fig. 3.4.5(a)). If the Z dimension shifts upward, then B shifts upward, so A and B have attraction in order to restore the natural state of vacuum, which is magnetic force. The flux line is coupled by the shortest path. When two magnets meet, the path of the magnetic line changes directly from path 1 to coupling through path 2, and path 3 forms a closed magnetic line. The spin is reversed, the encounter of the bottom magnetic lines of B and A will result in the superposition of vacuum deformation and repulsion. It is worth mentioning that the image of the magnetic field of vacuum negate the existence of magnetic monopoles. Static magnetic field is a rotary passive field div $B=0$, we introduce a vector potential A in the usual way, so that $B=\text{curl } A$.

The magnetic field is defined as in the new theory. During the movement of the ray starting point of electric flux line, the direction of motion is perpendicular to the electric flux line. The two dimensions produces tiny offset and form a closed structure of the flux line, defined as the magnetic lines. Vacuum of the magnetic lines is defined as the magnetic field, magnetic lines with 1-directional, is vector field, represented by the symbol B.

4. The propagation velocity of the virtual photon-spin wave

When X, Y and Z reach the deformation limit, that is

$$u_y(\text{Compression}) = u_z(\text{Compression}) = u_x(\text{Stretch}) = \mathscr{H} \quad (3.4.15)$$

This is the basic condition of dimension offset. For unit field O', vacuum is in the limit strain state, and vacuum is incompressible and liquid. The properties of liquids can be used to describe the basic unit vacuum in a strain limit state in a certain dimension. The three dimensions are interrelated. Y-deformation Δu_y is transfer to Z-dimension to meet

$$\Delta u_y = \Delta u_z \quad (3.4.16)$$

Y-deformation is proportional to Z-dimensional deformation and magnetic force.

The maximum deformation of free photon space-time is h. The virtual photon-spin wave is analogous to radar. Just as a radar wave sweeps through a circular screen, a photon's virtual photon-spin wave sweeps through a mass element. In the swept position, the space-time deformation approaches the deformation limit, the virtual photon-spin wave background hardens in space-time, the virtual photon-spin wave deformation from h to h', $h'<h$. Lepton field intrinsic 4-momentum (that is the rest mass) is bigger. Set up single fiber deformation of virtual photon-spin wave is $\mathscr{H}$, the virtual photon-spin wave propagation velocity is $v = \mathscr{H}/t_h$, and virtual photon-spin wave momentum is $p = h'/2\pi R$. The hardening of background space results in a slower propagation rate, so virtual photon-spin waves do not exceed the speed of light.

5. The spinor of 3-dimensional rotational strain constitutes a magnetic field

$$\text{curl } A = \left(\frac{\partial A_3}{\partial \xi_2} - \frac{\partial A_2}{\partial \xi_3}\right) i + \left(\frac{\partial A_1}{\partial \xi_3} - \frac{\partial A_3}{\partial \xi_1}\right) j + \left(\frac{\partial A_2}{\partial x_1} - \frac{\partial A_1}{\partial \xi_2}\right) k \quad (3.4.17)$$

$$\text{curl } A = B_1 i + B_2 j + B_3 k = B$$

Consider the space rotation generated magnetic field

$$F_{ij} = \partial_j A_i - \partial_i A_j \quad (3.4.18)$$

It's called electromagnetic fields.

$$F_{ij} = \begin{bmatrix} 0 & 0 & 0 & 0 \\ 0 & 0 & -B_3 & B_2 \\ 0 & B_3 & 0 & -B_1 \\ 0 & -B_2 & B_1 & 0 \end{bmatrix} \quad (3.4.19)$$

Can also be writtenas $F_{\mu\nu} = -F_{\nu\mu} = \partial_\mu A_\nu - \partial_\nu A_\mu$, called electromagnetic field tensor.

The rotation of time and space inside lepton produces electric field $F_{0\mu}$:

Chapter 3 The Structure of Lepton

$$E_i = \frac{\partial A(\xi^i)}{\partial t} - \nabla A(\xi^0) = F_{0\mu} \qquad (3.4.20)$$

The rotation of time and space in the fibre field is not what we usually understand. The electric flux line does not rotate, but the virtual photon-spin wave propagates around the electron center and sweeps through the power line, causing the rotation between time and space of the intrinsic fiber field of the electron. [ω_{ij}] The lepton electric flux line space dimensions are rotating generated magnetic field and spin.

6. Magnetic moment

For the lepton, the interaction force of field is proportional to the field deformation rate, so the virtual photon-spin wave caused magnetic force $F_B = e \cdot H'/H$. Torque is $F \times r$, and then the magnetic moment is (for example as electronic)

$$\mu_e = F_B \times r = e \cdot \frac{H'}{H} \times r = e \cdot \frac{H'/t_h}{H/t_h} \cdot \frac{m_e}{m_e} \times r = e \cdot \frac{p_e}{m_e \cdot c} \times r, \text{and } p_e \times r = \frac{1}{2}\hbar$$

$$\mu_e = \frac{e}{2m_e c}\hbar \qquad (3.4.21)$$

This is known as the expression of the magnetic moment of electron.

7. The rotation effect between time and space constitute electric flux line

Dimension move caused a "rotation" between time and space. The rotation can be expressed with spinor field. In electrodynamics, the definition is

$$F_{0\mu} = -F_{\mu 0} = iE_\mu \qquad (3.4.22)$$

The rotation effect between time and space constitute electricfield (Fig. 3.4.6):

$$\omega_{01} = \frac{\partial u_{e0}}{\partial \xi^1} - \frac{\partial u_{e1}}{\partial \tau} = \alpha - \beta = F_{01}$$

$$\omega_{0i} = \frac{\partial A_0(\xi)}{\partial \xi^i} - \frac{\partial A_i(\xi)}{\partial \tau} = F_{0i}(\xi) = -F_{i0}(\xi) \qquad (3.4.23a)$$

$$\omega_{ji} = \frac{\partial A_j(\xi)}{\partial \xi^i} - \frac{\partial A_i(\xi)}{\partial \xi^j} = F_{ji}(\xi) = -F_{ij}(\xi) \qquad (3.4.23b)$$

$F_{0\nu}$ indicates that existence of a rotation between time and space. Such rotation produces an electric field. Electrostatic field electric flux line in the vacuum unified theory: $F_{0\mu}$ is positive

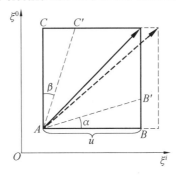

Fig. 3.4.6 Time and space rotation produces an electric field

electric field; $F_{\mu 0}$ is negative electric field. Visible, rotating in different directions are resulting in positive and negative electric field.

Electric field strain is expressed as: $iE = \begin{pmatrix} 0 & \omega_{01} & \omega_{02} & \omega_{03} \\ \omega_{10} & 0 & 0 & 0 \\ \omega_{20} & 0 & 0 & 0 \\ \omega_{30} & 0 & 0 & 0 \end{pmatrix}$ ($\nabla \cdot E = \rho/\varepsilon_0$)

Constitute a charge by distribution of strain electric flux line "source" or "convergence". ρ is corresponding charge e.

8. Intrinsic degree of freedom of lepton electromagnetic field

$$\varphi_{EB}(\xi) = \exp\left[i \left(\begin{bmatrix} \partial_0 A_0 & B_3 & -B_2 & -iE_1 \\ -B_3 & -\partial_1 A_1 & B_1 & -iE_2 \\ B_2 & -B_1 & -\partial_2 A_2 & -iE_3 \\ iE_1 & iE_2 & iE_3 & -\partial_3 A_3 \end{bmatrix} \begin{pmatrix} \xi^0 & \xi^0 & \xi^0 & \xi^0 \\ \xi^1 & \xi^1 & \xi^1 & \xi^1 \\ \xi^2 & \xi^2 & \xi^2 & \xi^2 \\ \xi^3 & \xi^3 & \xi^3 & \xi^3 \end{pmatrix} \right) \right] \quad (3.4.24)$$

$$= \exp\left[i \underbrace{\left(\begin{bmatrix} \partial_0 A_0 & 0 & 0 & 0 \\ 0 & -\partial_1 A_1 & 0 & 0 \\ 0 & 0 & -\partial_2 A_2 & 0 \\ 0 & 0 & 0 & -\partial_3 A_3 \end{bmatrix} \begin{pmatrix} \xi^0 & 0 & 0 & 0 \\ 0 & \xi^1 & 0 & 0 \\ 0 & 0 & \xi^2 & 0 \\ 0 & 0 & 0 & \xi^3 \end{pmatrix} \right)}_{(1)\text{Electromagnetic intrinsic 4-momentum}} + \underbrace{\begin{bmatrix} 0 & B_3 & -B_2 & -iE_1 \\ -B_3 & 0 & B_1 & -iE_2 \\ B_2 & -B_1 & 0 & -iE_3 \\ iE_1 & iE_2 & iE_3 & 0 \end{bmatrix} \begin{pmatrix} 0 & \xi^0 & \xi^0 & \xi^0 \\ \xi^1 & 0 & \xi^1 & \xi^1 \\ \xi^2 & \xi^2 & 0 & \xi^2 \\ \xi^3 & \xi^3 & \xi^3 & 0 \end{pmatrix}}_{(2)\text{Electromagnetic field in intrinsic space-time}} \right] \quad (3.4.25)$$

Pure electromagnetic energy and pure electromagnetic mass are included in the charged lepton static energy and static mass, no independent performance. The independent contribution of pure electromagnetic energy and pure electromagnetic mass in the experiment is no-observability. For experimental observations $m_{\text{electron}} = m_{\text{electromagnetic mass}} + m_{\text{bare mass}}$.

The mass effect of electromagnetic field is radial strain of electron. Electric field fibers have no mass effect. The mass effect of magnetic field is caused by rotational strain of electron and the rotational strain corresponds to a virtual photon-spin wave. Magnetic fibers have no mass effect, so the mass of the magnetic field is the angular momentum of virtual photon-spin waves.

Electric flux line and magneticlines penetrate the internal space, the lepton electromagnetic field can be observed in observation space, $F_{\mu\nu}(\xi) \to F_{\mu\nu}(x)$. The magnetic field $F_{ij}(\xi)$ are generated by the spin of the lepton charge. The electromagnetic field $F_{\mu\nu}(x)$ is present in the macroscopic observation space.

Chapter 3   The Structure of Lepton

9. The structure image of the electron

Lepton virtual photon-spin waves with magnetic fields in the 10^{-16} cm range. e^+ center is a basic unit of free state, which constitutes the "source" of scattering wave. e^- center is a hole of basic unit and constitutes the "sink" of convergent wave (Fig. 3.4.7).

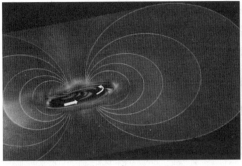 

(a) Electronic structure   (b) Structure of electronic center

Fig. 3.4.7   Electronic structure image

Lepton is a superimposed wave of virtual photon-spin wave and convergent wave. Leptons have mass, charge, spin and spin magnetic moments. The electric flux line of an electron has the characteristics of radiation, forming the electromagnetic field outside the electron.

All these structures are caused by vacuum strain and can be clearly explained by strain theory.

3.5   The characteristics outside the intrinsic space of lepton electric flux line field

3.5.1   Simplification of fiber field with intrinsic structure

It is impossible to know the internal structure of leptons for the current physical experimental observations. Lepton is a point-like particle.

On the other hand, the experimental observation can not distinguish the fiber structure from the electromagnetic field. Therefore, we can simplify the fiber structure of the electromagnetic field. It is considered that the electromagnetic field is distributed continuously in time and space. The intensity of the electromagnetic field is inversely proportional to the square of the distance.

From the point of view of strain, in the internal electronic space (Fig. 3.5.1), the total electric flux line movement is $h_e = e\Delta u_B$. It is re-understood that the total strain of electromagnetic field in the whole observation time and space is h_e. The internal space of the electron is simplified to a point in the observation space. This point is often set as the origin of the coordinates.

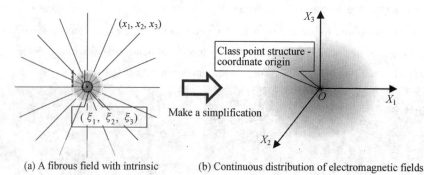

(a) A fibrous field with intrinsic   (b) Continuous distribution of electromagnetic fields

Fig. 3.5.1 The fiber field with intrinsic structure is simplified as point structure continuous electromagnetic field

1. The basic characteristic of electric field in observational space

The fiber structure of the vacuum is known as the electric field. Macroscopic electronic regarded as a point, we know from the front that electric flux line with a directional, and thus electric flux line is a vector field in vacuum and this field are defined as the electric field, and used to E represent. Electric flux line can be expressed by field theory (Fig. 3.5.2). The electric flux line from intrinsic space extension to the external particle space, x^μ is the observation space. We know vector field $A = A(x^1, x^2, x^3)$ by basic theory, beyond R_e, $M(x^1, x^2, x^3)$ as an arbitrary point on vector line, its radius vector is

$$r = x^1 i + x^2 j + x^3 k \qquad (3.5.1)$$

Then the differential is

$$dr = dx^1 i + dx^2 j + dx^3 k \qquad (3.5.2)$$

This is tangent with vector lines at points M. According to vector lines defined it will be collinear with the vector field

$$A = Ax^1 i + Ax^2 j + Ax^3 k \qquad (3.5.3)$$

Therefore

$$\frac{dx^1}{Ax^1} = \frac{dx^2}{Ax^2} = \frac{dx^3}{Ax^3} \qquad (3.5.4)$$

This is the differential equation satisfied by the vector line. The solution is getting the vector line family. The electric flux lines are isotropic. $|Ax^1| = |Ax^2| = |Ax^2|$ Have the same nature, set Ax^1, Ax^2, Ax^3. A is field parameter, there are

$$\frac{dx^1}{x^1} = \frac{dx^2}{x^2} = \frac{dx^3}{x^3} \qquad (3.5.5)$$

This is the mathematical expression of the electric flux line.

2. The electric field intensity

Set up the number of electric flux line through the unit surface S area is the field density (Fig. 3.5.2). It's called the electric field intensity, indicated by E. The total number of electric flux line is e (Experimental measurements to determine electric flux line numbere by a

Chapter 3 The Structure of Lepton

test charge), the total area of closure ball $S_{\text{Ball}} = 4\pi r^2$, then at r the field intensity is

$$E = k\frac{e}{r^3}r \qquad (3.5.6)$$

Here, $k = 1/4\pi\varepsilon_0$ is characterization the intensity of the interaction of the electrostatic field. $r \geqslant R_e$, R_e is radius of electric flux line generating zone. Do not consider internal structure of point charge, from the view of macro, E is a source irrotational field.

$$\text{div } E = \frac{\rho}{\varepsilon_0} \qquad (3.5.7)$$

$$\text{rot } E = 0 \qquad (3.5.8)$$

Point charge of electron constitutes by the electric flux line.

3. Positive and negative characteristics of the electric field

Electric field is the fibrosis vacuum with electric flux line. The electric field has a positive electric field and negative electric field two categories. When positive electric flux line field meet the anti-electric flux line that vacuum has a trend to restore the flat state then both trying to combine together, on the contrary, both repel each other (Fig. 3.4.1). Here special attention that to fill the vacuumhole volume determines the number of electric flux line of e^-. Due to the loss of the vacuum is a basic unit, electric flux line amount is a fixed number. Because the vacuum defect of $-e$ is 1 basic unit ($+e$ is 1 more basic unit), so the total number of electric flux lines is a constant.

4. The charge cloud of vacuum intensification

Fiber offset of quantum field can constitute anelectric flux line. For the small movement of each basic unit on a single electric flux line, if there is a strain that aggravates such "offset", it can be understood as inducing the formation of weak e^+e^- pairs in vacuum (Fig. 3.5.3), thus forming an electronic cloud.

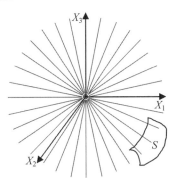

Fig. 3.5.2 The electric flux line of electrostatic field

5. Vacuum basic unit size

The cavitysize of an electron determines by the number of electric flux line (Fig. 3.1.1). Setting the total number of electric flux line is e, each electric flux line volume change is $\mathcal{H}/2$ which to fill the vacuum gap and field. So the total amount is

$$e\mathcal{H}4/3 = h_f \qquad (3.5.9)$$

h_f is the void volume of vacuum, this size is a basic unit vacuum. Lepton mass is determined by the deformation of intrinsic spatial field. A basic unit of vacuum constitutes a basic "point" in vacuum. 1 h_f excited state is the simplest and most stable. For the Fermi field, the lepton center lost to only 1 basic unit, so the total number of internal area electric flux line has been decided.

6. Pure electromagnetic mass

A total electric flux lines is 1 basic chargee which has nothing to do with the intrinsic four momentums of the charged Fermions (Fig. 3.4.1). $\Delta r \equiv \mathcal{H}_{Radial} = $ const. Consider having electric flux lines, so the total deformation of lepton charge is $e\mathcal{H}_{Radial}4/3 = h_{f(Radial)}$, $h_{f(Radial)} < h_f$ is constant that indicates all charged lepton with the same charge. Here, $h_{f(Radial)} = h_f - h_{f(Ring)}$. $h_{f(Radial)}$ is total radial deformation; $h_{f(Ring)}$ is total ring deformation. The electrostatic field constitute by electric flux line. Pure electromagnetic mass is

$$m_{electromagnetic\ mass} = h_{f(Radial)}/\phi c \qquad (3.5.10)$$

7. Baremass

Baremass is the mass effect of the ring strain.

$$m_{bare\ mass} = h_{f(Ring)}/\phi c \qquad (3.5.11)$$

For the experiment, because the detection cannot enter the intrinsic space of particle, the radial strain and circumferential strain cannot separate, electromagnetic mass and bare mass exist only in theory.

8. Definition of vacuum charge

According to the definition of the electric charge $div E = \lim_{\Delta V \to 0} \oint_S E \cdot dS/\Delta V$, we redefine a volume, which is the total vacuum deformation of charged fermions is h_f. Charge in vacuum is defined as

$$div\ e = k\frac{\pm e}{h_f} \qquad (3.5.12)$$

k is the electric influence coefficient to vacuum.

9. Vector potential A and scalar potential A_0 in observation space

From the observation space, the total offset of the electronic flux line is $h_e = e\Delta r_B$. Because h_e is only the "offset" of fibers, not strain, it has no effect on space-time. On the other hand, there is no intrinsic structure in the simplified model, so it is regarded as a point. The distribution of the electronic flux line caused by the "total offset Δr_B" of the "electronic fiber" is the distribution of the radiation (Fig. 3.5.4). The distribution of "source" or "sink" constitutes an electric charge. We stipulate that the total vacuum variable of charge is h_e. Here, the power line is used to distribute h_e in the macroscopic 4-dimensional space, and is the macroscopic radius $r \gg R_{in}$ (internal space radius), so that the energy e/ε_0 of the charge is much less than the static mass energy $m_e c^2$. That is $E_e = e/\varepsilon_0 \ll m_0 c^2$.

Chapter 3 The Structure of Lepton

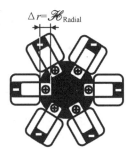

Fig. 3.5.3   Electronic vacuum intensification

Considering the distribution of h_e in the sphere space, the $E(r)$ of the electric flux lines distribution function is satisfied. Electric field intensity is

$$E(r) = \frac{h_e}{4\pi r^2} = \frac{n_e \Delta r_B}{4\pi r^2} = k\frac{e}{4\pi r^2} \quad (3.5.13)$$

This is Coulomb's Law, $E = \frac{e}{4\pi\varepsilon_0 r^3}r$. The distance from charge is r, the potential energy Ke/r. Scalar potential A_0 has the physical meaning that it is the potential energy of 1 unit charge when it is moved from a reference point to a specified point $A_{0,0}(x,y,z) = \frac{e}{4\pi\varepsilon_0}\frac{1}{r}$. In classical electromagnetic theory, vector potential and scalar potential at r' in vacuum are

$$\phi(r,t) = \int\frac{\rho(r',t')}{4\pi\varepsilon_0 R}dv' \Rightarrow \text{Energy}; \ A_{i,i}(r,t) = \int\frac{j(r',t')}{4\pi\varepsilon_0 c^2 R}dv' \Rightarrow \text{Momentum}$$

$$(3.5.14)$$

Here, $R = |r - r'|$, $t' = t - R/v$, $v = 1/\sqrt{\mu_0\varepsilon_0} = c$. Take $r=0$, $t=0$. The Coulomb's Law describes the electrostatic properties of point charges e.

When $r \to 0$, we get the conclusion that the electric field strength of the electron is $E \to \infty$, which is obviously unreasonable, because we do not consider $r \leq r_0$ (the electric flux line generation region). The electric field originates from the charge $\nabla \cdot E = 4\pi\rho$, and the magnetic field (Fig. 3.5.4) is the passive field $\nabla \cdot B = 0$. The "curl" of four dimensional vectors forms an electric field.

$$F^{0\mu} = -F^{\mu 0}; \ F^{0\mu} = \frac{1}{2}(\frac{\partial A}{\partial \tau} - \nabla A^0)$$

10. Vacuum dielectric constant

The existence of mass charge will bend the background space-time, and the curved space-time is the gravitational field. The existence of charged particles makes the electric flux line space curved, and the curved electric flux line space is electromagnetic field. Electromagnetic fields and gravitation are very similar. The deformation of fiber structure in the intrinsic space of the lepton is very large, and the large deformation is enclosed in the intrinsic space. However, this deformation will have a small effect on the background fibers, which will make the background field deform h_e.

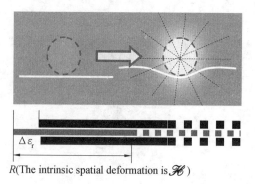

R(The intrinsic spatial deformation is $\mathcal{H}$)

Fig. 3.5.4 Fiber structure charges bend the background fiber field space to form electromagnetic field

Unlike mass in the gravitational field, the most basic charge in the universe is $\pm e$. Therefore, the degree of bending of the background fiber field is strictly equal in the vacuum environment. But if a medium exists and the background field becomes hard, it is relatively difficult to bend the background fiber field. For observation, the electromagnetic field becomes weak. The coefficient to measure the effect of bending electric field is $1/\mu$ ($1/\mu_0$ in vacuum is the maximum), the coefficient to measure the effect of bending magnetic field is $1/\varepsilon$ ($1/\varepsilon_0$ in vacuum is the maximum). The propagation speed of electromagnetic field is $c = 1/\sqrt{\mu_0 \varepsilon_0}$ which is the speed of light, and $1/\sqrt{\mu \varepsilon}$ is the speed in the corresponding medium. The total energy of the single electron electric field is e/ε_0 and the total static momentum is $e/c\varepsilon_0$. h_e is considered to be an effect transmitted from intrinsic space, or field deformation observed outside intrinsic space. Vacuum displacement h_e is related to the "soft" and "hard" of the background space-tim, so the effect of outward transmission is related to the background medium, and the coefficient related to the medium is expressed by ε. For vacuum medium: $\varepsilon_0 = \dfrac{1}{4\pi (3)^2 \times 10^9}$ F/m.

11. Electronic homosexuality

For the Fermi field, the vacuum lost by the lepton center is only 1 basic unit h_f. As a result, the total number of electric flux lines is determined. The total amount of electric flux lines in the Fermi field is 1 basic charge e, which is independent of the intrinsic four momentum of the Fermi field. Therefore, the essence of the basic charge is caused by the gain and loss of 1 basic unit vacuum. This indicates that all stable fermions have the same basic unit charge e.

12. The difference of the photon and lepton

Photons and leptons have electron flux line structure, and the strain of electron flux line is limited in intrinsic space. The number electric flux line are same. Photon electric flux line caused by the dimensions of the offset, and thus its length extends to infinity. Photon has no gain or loss 1 basic unit vacuum. Number of positive electric flux line strictly equal negative electric flux line, thus no charge. Photonic structure is similar to half a positron and half an anti-electron combination, lepton was gain or loss 1 basic unit vacuum and existence of 1 charge,

Chapter 3 The Structure of Lepton

its electric flux line caused by the fiber offset which length is infinity.

The intrinsic space of leptons is spherically symmetric and has 4-momentum at rest, the intrinsic energy $E = m_0 c^2$ to meet $m_0 c^2 \cdot \tau = m_0 c \cdot R = h_f$. The intrinsic momentum has not directly observable effects. Leptons have wave-particle duality $h = E_e/\nu = \lambda p_e$, ν, λ is frequency and wavelength of Compton probability wave. Photon intrinsic structure is axial symmetry and can be simplified to 1-dimensional. The intrinsic energy and momentum of photons can be directly observed. Photon no rest mass and only dynamic 4-momentum. The energy and momentum of photon are meets $E \cdot \Delta t = h = p \cdot \lambda$.

3.5.2 Lepton Maxwell equations

The intrinsic degrees of freedomcan be directly showing in observation space $F_{\mu\nu}(x)$ is the electromagnetic field which originated in the charge is asource $\nabla \cdot E = 4\pi\rho$. The magnetic field is passive field, with the nature $\nabla \cdot B = 0$. Known by the electric flux line model, the magnetic field caused by the rotation of the electric field. Electric and magnetic fields orthogonal to each other, the relationship between electric field and magnetic field expressed by the Maxwell equations. Maxwell's standard form as follows:

$$\nabla \cdot E = 4\pi\rho \qquad (3.5.15)$$

$$\nabla \times B - \frac{1}{c}\frac{\partial E}{\partial t} = \frac{4\pi}{c}J \qquad (3.5.16)$$

$$\nabla \cdot B = 0 \qquad (3.5.17)$$

$$\nabla \times E + \frac{1}{c}\frac{\partial B}{\partial t} = 0 \qquad (3.5.18)$$

Here, equation(3.5.15) gives the form of electric flux line distribution, which determines the form of Coulomb's theorem. Also indicate that the existence of the monopole charge. Equation (3.5.17) is denying existence of magnetic monopoles. Equation(3.5.18) expression mechanism of the magnetic field of lepton and there was the spin magnetic moment. $\nabla \times E$ Vortex vector it's caused by the spin. Equation(3.5.16) is no direct representation lepton nature of the electric flux line in intrinsic space, but expressed the macroscopic nature of electric flux line field. Equation (3.5.15),(3.5.17) and (3.5.18) not only expressed the nature of lepton electric flux line field in the intrinsic space, but also expressed the macro nature of electric flux line outside the intrinsic space.

3.6 Estimate of electromagnetic coupling constants

In the standard model, the interaction between particles is transmitted by exchanging intermediate bosons, and the interaction intensity of each force field is expressed by the coupling constant. Coupling constant is the basic constant in the present theory, which can not be calculated. But in the new theory, we can estimate the coupling constants according to the characteristics of the field flux lines of electrons and photons.

3.6.1 The physical image of the electromagnetic coupling constant

The physical image of the coupling constants shown in Fig. 3.6.1, it shows how the possibility of virtual photon and electron interaction when the virtual photon (only consider the longitudinal wave) into the electronic coupling spherical $4\pi R_e^2$. We know that in vacuum unified theory, the lepton rest mass is inversely proportional to the size of the intrinsic space. Assumptions photon and electron coupling constant 1, it indicates that as long as the photon into electronic coupling space (R_e is the radius of electronic coupling spherical space), the photon will interact with electronic and occurs deflection. In other words, the photon only to enter the electronic coupling space photon be possible to the radiation virtual photons (longitudinal photon) act on electronic, if this probability is 1, indicating that as long as hit the target, must be interaction. The truth is that the photon and electron coupling is a small $\alpha_e = 1/137$.

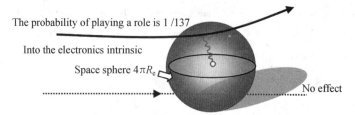

Fig. 3.6.1 The physical image of coupling constant

$$\text{Oscillation amplitude} = 1 + \alpha_e(\ ) + \alpha_e^2(\ ) + \cdots \tag{3.6.1}$$

The nature of coupling constants is interaction probability. Now look at a typical example, consider Thomson scattering (Fig. 3.6.2). For long-wavelength photons, the cross-section is

$$\sigma_{TH} = \frac{2}{3}\alpha(4\pi R_e^2) \tag{3.6.2}$$

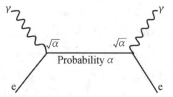

Fig. 3.6.2 Feynman diagrams of Thomson scattering

Here, R_e is Compton wavelength of electron (natural units), $R_e = \frac{\hbar}{m_e c} = \frac{1}{m_e}$ is intrinsic space radius of lepton. $\alpha = g_e = 1/137$ is electromagnetic coupling constant.

Why would α be so small? Now we can analyze this problem from the point of view of vacuum. This is because if a single electron satisfies the following conditions, electromagnetic coupling can occur.

(1) It must exist in the electron flux lines channel.

(2) Photon-electron coupling through the electron flux lines.

(3) Photon flux line and electric flux line are connected to form the coupling channels, there is an optimal path, which exists in the best transmission angle conducive to momentum (longitudinal photon).

3.6.2 Consider the case of electrons

We look at the issue from the perspective of the electric flux line. Set up the photons from left to right into the electronic interior space (Fig. 3.6.3). The essence of photon and electron interaction is the photon center and electronic center coupling pass through by the electric flux line achieve energy exchange then interactions occur, such as Thomson scattering. When the photon center enters the inner space of the electron, the interaction may not be realized. For a single electron, the interaction occurs after satisfying the above three coupling conditions.

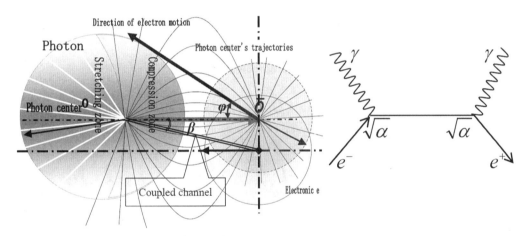

Fig. 3.6.3 Electromagnetic interaction

We can estimate the probability of coupling between lepton centers and photon centers via electron flux lines. Positive and negative charge exchange photon achieves mutual attractive force while the same charge exchange photon to realize mutual repulsion forces (Fig. 3.6.4).

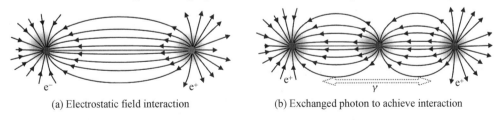

(a) Electrostatic field interaction (b) Exchanged photon to achieve interaction

Fig. 3.6.4 Electronic and photonic electric flux line coupling

(1) Electric flux line to lock the electronic center and the center of the photon, it is clear that the right hemisphere of electron cannot be locked photon center, while existing the

possibility only electric flux line within the left hemisphere can lock the photon center, so that only 1/2 electric flux line is possible.

(2) For the electric flux lines, the basic unit of vacuumis 3-dimensional, which only 1-dimensional vacuum medium can form the electric flux lines, so the photon center and the lepton center coupling channel is not 3-dimensional while only 1-dimensional. When the 3-dimensional participate in coupling transfer, then the channel width is 1, so that the channel width is 1/3.

(3) In these electric flux lines, the photon center and the lepton center is connecting by anelectric flux line which consisting an angle φ with center electric flux line. The smaller the value of φ is, the more favorable the photon transfer momentum is. Therefore, considering all possible coupling angles φ, the coupling average of the left hemisphere is estimated as follows.

The electric flux line length is $R_r(r,\varphi,\theta) = r\sin\theta\cos\varphi$ insidea electron intrinsic space (Fig. 3.6.5). Considering the hemisphere, the photon and electronic coupling electric flux line distribution is

$$e(\theta,\varphi) = e_N \sin\theta\cos\varphi \qquad (3.6.3)$$

Here, e_N is the number of the electric flux line. The electric flux line distribution is spherically symmetric which unrelated with θ,φ, that is $e(\theta,\varphi) = e_N$.

$$\text{Coupling ratio} = \frac{e_N \sin\theta\cos\varphi}{e_N} = \sin\theta\cos\varphi \qquad (3.6.4)$$

There is also a factor will affect the success rate of photon and electron coupling, that is the angle of relative motion between each other at coupled instant, the angle is $\alpha = V_e V_\gamma$ (Fig. 3.6.3), when $\alpha = 0$ electron and photon most easy to establish coupling relationship. The above factors, the overall coupling rate is also multiplied a factor$\cos\alpha$, that is

$$\sin\theta\cos\varphi\cos\alpha \qquad (3.6.5)$$

To movement of electrons and photons has volatility, the coupling of the success rate is a statistical process. The easiest way is to take the average, then the average coupling rate

$$\frac{1}{\pi}\int_0^\pi \sin\theta d\theta \frac{1}{\pi}\int_0^\pi \cos\varphi d\varphi \frac{2}{\pi}\int_0^{\pi/2}\sin\alpha d\alpha = \frac{8}{\pi^3} \qquad (3.6.6)$$

The coupling probability for a single electronicelectric flux line coupler locking a photon center is

$$\sqrt{\alpha_e} = \frac{1}{2} \cdot \frac{1}{3} \cdot \frac{8}{\pi^3} = \frac{4}{3\pi^3} \qquad (3.6.7)$$

3.6.3 Consider the case of photon

Photon is similar to electron. Photon power lines consist of stretching and compressing electric flux line. The photon flux lines and the electron flux lines are coupled to form a 1-dimensional channel to transmit the longitudinal photons between photons and leptons, thus realizing the 4-momentum transfer between photons and electrons (Fig. 3.6.3). If only consider the photon, the photon from the left into the right, only consider half of vacuum it will be possible

Chapter 3 The Structure of Lepton

to form 1-dimensional channel. The photon flux lines number and the electron flux lines are same, but the distribution is not the same, specifically consider the following. The strain of photon center near the surface which vertical to the direction of propagation is very small, and the electric flux line strain is zero which in photon center intersects the vertical surface yz. This difference makes the flux lines of photon coupling smaller than that of electron. Therefore, we have to consider the spatial distribution of the electric flux line, so it is necessary to seek the coupling rate of photon.

The photon of the face yz no deformation (Fig. 3.6.5), the electric flux line r in the surface yz is zero, considering the surface yz, when $\theta = \pi/2$ along the direction x the space distribution should meet

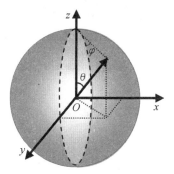

Fig. 3.6.5 Photon spherical coordinates

$$R_r(r,\varphi) = r\cos\varphi$$

Consider the direction θ, when $\theta=0$ no electric flux line, in spherical coordinates

$$R_r(r,\varphi,\theta) = r\sin\theta\cos\varphi$$

Electric flux line distribution of photons:

$$e(\varphi,\theta) = e_N \sin\theta\cos\varphi$$

Coupling required the distribution of electric flux line is

$$e(\varphi,\theta) = e_N \sin\theta\cos\varphi$$

Can be obtained

$$\text{coupling ratio} = \frac{e_N \sin\theta\cos\varphi}{e_N \sin\theta\cos\varphi} = 1 \qquad (3.6.8)$$

Comprehensive consideration, get

$$\sqrt{\alpha_r} = \frac{1}{2} \times \frac{1}{3} \times 1 = \frac{1}{6} \qquad (3.6.9)$$

Photon and electron coupling that is photonic and electronic forming a 1-dimensional channel probability product, so the photoelectric coupling constants (i.e. probability) is

$$\alpha = \sqrt{\alpha_e} \cdot \sqrt{\alpha_\gamma} = \frac{4}{3\pi^3} \cdot \frac{1}{6} \approx \frac{1}{139.5} \qquad (3.6.10)$$

This is slightly smaller than the measured value $\alpha = 1/137$. Considering the second-order effect of the interaction between virtual photons and virtual electrons, see (3.6.1), then

$$\alpha \approx 1/139.5 + (1/139.5)^2 \approx 1/138.6$$

If higher order effects are considered, the coupling constants will increase further. In addition, the calculation here is an ideal value, which is the maximum distance of the flux line coupling value, and the value should be the lower limit of the coupling constant. For the experiment, this distance is much closer. The effect of running coupling constant should be considered. When the interaction distance is reduced, the coupling parameters will increase slightly. After considering many factors, the calculation of the coupling probability of photon-electron flux line will be more reliable.

3.6.4 Close condition

The previous calculation does not consider close condition. In the case of large momentum, particle interactions become shorter, more electric flux line intersect, and resulting in the probability of interaction increases and coupling constant become larger. $\alpha(Q^2) \equiv \dfrac{e^2(Q^2)}{4\pi}$ is the running coupling constant. In the large limit $Q^2 \equiv -q^2$, $\alpha(Q^2) = \alpha_0 \cdot \left[1 - \dfrac{\alpha_0}{3\pi}\ln\left(\dfrac{Q^2}{M^2}\right)\right]^{-1}$, the running coupling constant $\alpha(Q^2)$ describes the dependency relationship that the two charged particles coupling and the distance between them. Here, Q is momentum, M is particles mass.

3.6.5 The leptons point model form quantum mechanics

Lepton and photon simplified to the point model can obtain the physical image of the probability wave. The probability wave constitutes the basic physics image of quantum mechanics.

For a system composed by point particle, its state is completely determined by a complex function $\psi(x,t)$. We interpreted it as the probability density amplitude, that is

$$|\Psi(x,t)|^2 dx$$

That measuring the particle position at time t, the particle is located $x = -\infty$ to $x = \infty$, all probability must be 1 in this closed interval, so we have the normalization condition

$$\int_{-\infty}^{\infty} |\Psi(x,t)|^2 dx = 1 \qquad (3.6.11)$$

Lepton has wave-particle duality. After the lepton is disturbed, the center of the lepton makes a simple harmonic vibration perpendicular to the propagation direction r. If there is no disturbance to the lepton center, the central orbit is a straight line. Of course, this possibility does not exist, because the emission of electrons from the atom is a disturbance. The trajectory of electronic center becomes cosine or sinusoidal curve (It can also be considered that the photon center and the electron center are the same after the photon and the electron are coupled, so the electron has the fluctuation). The propagation speed of lepton center is the slowest on the peak (positive and negative peaks) and the fastest on the axis. From a statistical point of view,

Chapter 3 The Structure of Lepton

the probability of the top is the largest, and the probability of the propagation axis is the smallest. This is the probability transverse wave.

1. Single slit diffraction

Electrons have wave-particle duality. Now we look at the electronic single-slit diffraction. A crack in the A, the slit size is roughly close the fluctuation amplitude of electron to will be able to form a disturbance. If the propagation speed of electrons is equal, the probability of shear wave frequency is equal. Because of the different phases, the fringes can not be formed. When the electron passes through the slot I of the baffle, it gives the electron a new disturbance, and the position of the disturbance is the same for all electrons. The electrons get the same phase from the slot I. In addition, because the electrons are disturbed, the original direction of motion is changed. Because of the low energy, the disturbance is elastic, and only the direction is changed without changing the original electron momentum. Therefore, after passing through slit I, a single slit diffraction fringe appears on the screen.

For electronic single-slit experiment (Fig. 3.6.6), the seam itself is a disturbance source, the smaller the slit width, the larger the quantum field deformation, the more intense the disturbance, the wider the angle of the corresponding scattering. Can be obtaining the uncertainty relation by relationship between the single slit width and the momentum distribution: $\Delta x \cdot \Delta p \geqslant \hbar/2$.

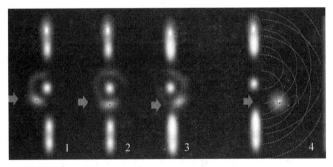

Fig. 3.6.6 Single electron double slit interference

2. Single electron double slit interference

Electron propagation is produced by lepton-photon coupling. The double slit itself constitutes two disturbance sources (Fig. 3.6.6). The intrinsic strain field of the electron contacts with the shadow field around the double slit. The intrinsic field of the electron deforms, and the central point of the electron may enter I or II slit after being disturbed, as for which slit to enter, it is determined by the state of electronic deformation. The fibre field inside the electron is spherical. If the spherical wave is squeezed and deformed, it will disturb the spherical wave, and any disturbance will change the position of the center point. When a spherical wave passes through a double slit, the distance between the two slits is similar to the diameter of the spherical wave, and the spherical wave is destroyed by the extrusion deformation, the spherical wave is divided into two parts. At the same time, the spherical wave passes through the double slit, and the interference between one part of the wave and the other part of the wave forms the single electron double slit interference. The electrons pass through the double slits and merge into complete electrons. A new phase is obtained with no change in amplitude.

The center of electronics itself has a strict and clear communication path, Einstein was right. If we can know all the boundary conditions, we will be able to accurately determine the position and velocity of the electronic center, its trajectory can be accurately described. But on the other hand, the trajectory of the center of the lepton cannot be measured, and thus the boundary conditions are unknown. Because the most sophisticated tools that we used to perceive the worlds are photons (or electrons). When this clumsy tool is contact with the measured electrons will make the measured electron's position has been fatal disturbances, so that we cannot know the whereabouts of electrons, and we never find a more sophisticated tool than electronic photon. In this desperation, we are represent the electronic behavior had to use the statistical method. Measured particle with track, if we do not measure particles we will be in a state of ignorance, and measure particles that although we know the position of the particle at time t. But the measurement disturbance makes the position of particles unknown, they satisfy the uncertainty relation $\Delta x \cdot \Delta p \geqslant \hbar/2$. Einstein and Bohr are correct, so we are in a dilemma. We introduce the probability wave to describe the behavior of the electrons, used $|\psi(x,t)|^2$ to indicate the size of the probability amplitude. We measure that space becomes infinite, and the probability of discovery must be 1. When the particle is in enough space, that is $\int_{-\infty}^{+\infty} |\psi(x,t)|^2 dx = 1$. Integrating range $\int_{-\infty}^{+\infty}$ is the presentation of this dilemma. Figuratively speaking, the electronic as elusive Monkey Sun Wukong, but Sun Wukong always cannot escape out of the palm of Buddha. Integrating range $\int_{-\infty}^{+\infty}$ is the palm of Buddha.

3.7 The complete field function of lepton

3.7.1 The complete field function of free lepton

Based on the previous discussion, we need to make a brief summary. Leptons have total deformed h_f, charge and electromagnetic field, stationary mass and energy, spin angular momentum $\hbar/2$ and 4-momentum. All these characteristics are caused by the vacuum elastic strain, so we can use a simple wave function of quantum field to express the intrinsic properties of electrons.

$$\varphi(\xi) = \exp\{-i[\varepsilon_{\alpha\beta}](\xi)\} \qquad (3.7.1)$$

Dual quantum field wave function

$$\varphi^+(\xi) = \exp\{i[\varepsilon_{\alpha\beta}](\xi)\}$$

The wave function of quantum field contains all the information. Since the intrinsic space

Chapter 3　The Structure of Lepton

has no observability, for convenience, we mix the information in the observation space with the intrinsic information of the quantum field and express it by a unified wave function, which is written as

$$\varphi_e(X) = \exp(-i)\{[\varepsilon_{\alpha\beta}](\xi,x)\}$$

$$= \exp(-i)\left\{\begin{bmatrix} \varepsilon_0 & 0 & 0 & 0 \\ 0 & 0 & 0 & 0 \\ 0 & 0 & 0 & 0 \\ 0 & 0 & 0 & 0 \end{bmatrix} \breve{I} \begin{pmatrix} \xi_0 & 0 & 0 & 0 \\ 0 & 0 & 0 & 0 \\ 0 & 0 & 0 & 0 \\ 0 & 0 & 0 & 0 \end{pmatrix} + \right.$$

$$\underbrace{}_{\text{(1) Rest energy of electron } E_0}$$

$$\begin{bmatrix} 0 & 0 & 0 & 0 \\ 0 & \varepsilon_{11} & 0 & 0 \\ 0 & 0 & \varepsilon_{22} & 0 \\ 0 & 0 & 0 & \varepsilon_{33} \end{bmatrix} \breve{\gamma}_4 \begin{pmatrix} 0 & 0 & 0 & 0 \\ 0 & \xi_1 & 0 & 0 \\ 0 & 0 & \xi_2 & 0 \\ 0 & 0 & 0 & \xi_3 \end{pmatrix} +$$

$$\underbrace{}_{\text{(2) Rest intrinsic momentum of electric field of electron}}$$

Rest Radial mass m_{0R}

Fiber radial strain, mass charge

$$\begin{bmatrix} 0 & 0 & 0 & 0 \\ 0 & 0 & \gamma_{12} & \gamma_{13} \\ 0 & \gamma_{21} & 0 & \gamma_{23} \\ 0 & \gamma_{31} & \gamma_{32} & 0 \end{bmatrix} (1-\breve{\gamma}_5) \begin{pmatrix} 0 & 0 & 0 & 0 \\ 0 & 0 & \xi_1 & \xi_1 \\ 0 & \xi_2 & 0 & \xi_2 \\ 0 & \xi_3 & \xi_3 & 0 \end{pmatrix} +$$

$$\underbrace{}_{\text{(3) Shear strain}}$$

Hoop strain constitution weak field of electron$\Leftrightarrow(1+\gamma_5)$.

Circumferential mass M_{0C}, Bare mass $M_{0B}=M_{0R}+M_{0C}+M_{0S}$

$$\begin{bmatrix} 0 & 0 & 0 & 0 \\ 0 & 0 & \delta(i-1)\omega_{12} & \delta(i-2)\omega_{13} \\ 0 & \delta(i-1)\omega_{21} & 0 & \delta(i-3)\omega_{23} \\ 0 & \delta(i-2)\omega_{31} & \delta(i-3)\omega_{32} & 0 \end{bmatrix} \breve{\Sigma}_4 \begin{pmatrix} 0 & 0 & 0 & 0 \\ 0 & 0 & \xi_1 & \xi_1 \\ 0 & \xi_2 & 0 & \xi_2 \\ 0 & \xi_3 & \xi_3 & 0 \end{pmatrix} +$$

$$\underbrace{}_{\text{(4) Spin strain, spin angular momentum } \hbar/2}$$

(4) Rotational strain. spin photon mass M_{0S}

$$\begin{bmatrix} 0 & \gamma_{01} & \gamma_{02} & \gamma_{03} \\ \gamma_{10} & 0 & 0 & 0 \\ \gamma_{20} & 0 & 0 & 0 \\ \gamma_{30} & 0 & 0 & 0 \end{bmatrix} \breve{I} \begin{pmatrix} 0 & \xi_0 & \xi_0 & \xi_0 \\ \xi_1 & 0 & 0 & 0 \\ \xi_2 & 0 & 0 & 0 \\ \xi_3 & 0 & 0 & 0 \end{pmatrix} + \begin{bmatrix} 0 & \omega_{01} & \omega_{02} & \omega_{03} \\ \omega_{10} & 0 & 0 & 0 \\ \omega_{20} & 0 & 0 & 0 \\ \omega_{30} & 0 & 0 & 0 \end{bmatrix} \breve{I} \begin{pmatrix} 0 & \xi_0 & \xi_0 & \xi_0 \\ \xi_1 & 0 & 0 & 0 \\ \xi_2 & 0 & 0 & 0 \\ \xi_3 & 0 & 0 & 0 \end{pmatrix} +$$

(5) Intrinsic shear 4-momentum flow　　　　　(6) Intrinsic 4-momentum flow of spin

$$\underbrace{}_{\text{Hoopstrain constitution wave of weak field}} \quad \underbrace{}_{\text{The fluctuation caused by the rotational strain}}$$

$[\varepsilon_{i0}]$ or $[\varepsilon_{0i}]$ flux of lepton intrinsic 4-momentum through the surface ξ^i, the flux is limited in intrinsic space

$$\underbrace{\begin{pmatrix} A_{0,0}(\xi) & 0 & 0 & 0 \\ 0 & A_{1,1}(\xi) & 0 & 0 \\ 0 & 0 & A_{2,2}(\xi) & 0 \\ 0 & 0 & 0 & A_{3,3}(\xi) \end{pmatrix} \check{I} \begin{pmatrix} \xi_0 & 0 & 0 & 0 \\ 0 & \xi_1 & 0 & 0 \\ 0 & 0 & \xi_2 & 0 \\ 0 & 0 & 0 & \xi_3 \end{pmatrix}}_{\text{Intrinsic fiber radial shift strain}} +$$

(8) Without observability. Intrinsic electric charge. The intrinsic space is the generation area of the electric flux line. $A_\mu(\xi)$ i.e., intrinsic electromagnetic four vector

$$\underbrace{\begin{bmatrix} 0 & \Omega_{01} & \Omega_{02} & \Omega_{03} \\ \Omega_{10} & 0 & 0 & 0 \\ \Omega_{20} & 0 & 0 & 0 \\ \Omega_{30} & 0 & 0 & 0 \end{bmatrix}}_{\text{Intrinsic electrostatic tensor } E} \check{I} \begin{pmatrix} 0 & \xi_0 & \xi_0 & \xi_0 \\ \xi_1 & 0 & 0 & 0 \\ \xi_2 & 0 & 0 & 0 \\ \xi_3 & 0 & 0 & 0 \end{pmatrix} +$$

(9) Intrinsic electromagnetic tensor, fiber shuttling. Electric field mass merging in radial strain

$$\underbrace{\begin{bmatrix} 0 & 0 & 0 & 0 \\ 0 & 0 & \delta(i-1)\Omega_{12} & \delta(i-2)\Omega_{13} \\ 0 & \delta(i+1)\Omega_{21} & 0 & \delta(i+3)\Omega_{23} \\ 0 & \delta(i+2)\Omega_{31} & \delta(i+3)\Omega_{32} & 0 \end{bmatrix}}_{\text{Magnetic tensor } B} \check{\Sigma} \begin{pmatrix} 0 & 0 & 0 & 0 \\ 0 & 0 & \xi_1 & \xi_1 \\ 0 & \xi_2 & 0 & \xi_2 \\ 0 & \xi_3 & \xi_3 & 0 \end{pmatrix}$$

(10) Intrinsic rotational strain of electric flux line, forming an electron Intrinsic magnetic field

$$\underbrace{\frac{1}{2} \begin{bmatrix} 0 & \omega_{01}^H & \omega_{02}^H & \omega_{03}^H \\ \omega_{10}^H & 0 & 0 & 0 \\ \omega_{20}^H & 0 & 0 & 0 \\ \omega_{30}^H & 0 & 0 & 0 \end{bmatrix} \check{I} \begin{pmatrix} 0 & \xi_1 & \xi_2 & \xi_3 \\ \xi_0 & 0 & 0 & 0 \\ \xi_0 & 0 & 0 & 0 \\ \xi_0 & 0 & 0 & 0 \end{pmatrix} + \frac{1}{2} \begin{bmatrix} 0 & 0 & 0 & 0 \\ 0 & 0 & \omega_{12}^H & \omega_{13}^H \\ 0 & \omega_{21} & 0 & \omega_{23}^H \\ 0 & \omega_{31}^H & \omega_{32}^H & 0 \end{bmatrix} \check{\Sigma} \begin{pmatrix} 0 & 0 & 0 & 0 \\ 0 & 0 & \xi_1 & \xi_1 \\ 0 & \xi_2 & 0 & \xi_2 \\ 0 & \xi_3 & \xi_3 & 0 \end{pmatrix}}_{\text{High angular momentum} \Leftrightarrow (n+1)\hbar/2, \text{ spin degree of freedom}} +$$

(7) Rotational strain, spin photon mass

$$\underbrace{\begin{pmatrix} A_{0,0} & 0 & 0 & 0 \\ 0 & A_{1,1} & 0 & 0 \\ 0 & 0 & A_{2,2} & 0 \\ 0 & 0 & 0 & A_{3,3} \end{pmatrix} \check{I} \begin{pmatrix} x_0 & 0 & 0 & 0 \\ 0 & x_1 & 0 & 0 \\ 0 & 0 & x_2 & 0 \\ 0 & 0 & 0 & x_3 \end{pmatrix}}_{\text{Vector and scalar potentials. Momentum and rest energy of electric field of electron } E_e} +$$

(8) Electromagnetic rest mass M_{0E}, without observability, fiber radial shift strain electric charge, $A_\mu(\xi)$ (i.e. intrinsic electromagnetic four vector); $M_0 = M_{0E} + M_{0B}$

Chapter 3 The Structure of Lepton

$$\frac{1}{2}\begin{bmatrix} 0 & \Omega_{01} & \Omega_{02} & \Omega_{03} \\ \Omega_{10} & 0 & 0 & 0 \\ \Omega_{20} & 0 & 0 & 0 \\ \Omega_{30} & 0 & 0 & 0 \end{bmatrix} \widetilde{I} \begin{pmatrix} 0 & x_0 & x_0 & x_0 \\ x_1 & 0 & 0 & 0 \\ x_2 & 0 & 0 & 0 \\ x_3 & 0 & 0 & 0 \end{pmatrix} +$$

$\underbrace{\qquad\qquad\qquad\text{Electrostatic tensor } E \qquad\qquad\qquad}$

(9) Electromagnetic tensor, fiber shuttling, have no effect of mass, electric field mass merging in radial strain

$$\frac{1}{2}\begin{bmatrix} 0 & 0 & 0 & 0 \\ 0 & 0 & \delta(i-1)\Omega_{12} & \delta(i-2)\Omega_{13} \\ 0 & \delta(i+1)\Omega_{21} & 0 & \delta(i+3)\Omega_{23} \\ 0 & \delta(i+2)\Omega_{31} & \delta(i+3)\Omega_{32} & 0 \end{bmatrix} \widetilde{\Sigma} \begin{pmatrix} 0 & 0 & 0 & 0 \\ 0 & 0 & x_1 & x_1 \\ 0 & x_2 & 0 & x_2 \\ 0 & x_3 & x_3 & 0 \end{pmatrix} +$$

$\underbrace{\qquad\qquad\qquad\text{Magnetic tensor } B \qquad\qquad\qquad}$

(10) Rotational strain of electric magnetic line, forming an electron magnetic field. It shows the existence of electron spin magnetic moment μ_S

$$\begin{bmatrix} \Delta\varepsilon_0 & 0 & 0 & 0 \\ 0 & 0 & 0 & 0 \\ 0 & 0 & 0 & 0 \\ 0 & 0 & 0 & 0 \end{bmatrix} \widetilde{\gamma_0} \begin{pmatrix} x_0 & 0 & 0 & 0 \\ 0 & 0 & 0 & 0 \\ 0 & 0 & 0 & 0 \\ 0 & 0 & 0 & 0 \end{pmatrix} +$$

$\underbrace{\qquad\qquad\text{(11) Kinetic energy } E\qquad\qquad}$

$$\begin{bmatrix} 0 & 0 & 0 & 0 \\ 0 & \Delta\varepsilon_{11} & 0 & 0 \\ 0 & 0 & 0 & 0 \\ 0 & 0 & 0 & 0 \end{bmatrix} \widetilde{\gamma_1} \begin{pmatrix} 0 & 0 & 0 & 0 \\ 0 & x_1 & 0 & 0 \\ 0 & 0 & 0 & 0 \\ 0 & 0 & 0 & 0 \end{pmatrix} +$$

$\underbrace{\qquad\qquad\text{(12) Momentum } p_1, \text{space } \gamma_1\qquad\qquad}$

$$\begin{bmatrix} 0 & 0 & 0 & 0 \\ 0 & 0 & 0 & 0 \\ 0 & 0 & \Delta\varepsilon_{22} & 0 \\ 0 & 0 & 0 & 0 \end{bmatrix} \widetilde{\gamma_2} \begin{pmatrix} 0 & 0 & 0 & 0 \\ 0 & 0 & 0 & 0 \\ 0 & 0 & x_2 & 0 \\ 0 & 0 & 0 & 0 \end{pmatrix} +$$

$\underbrace{\qquad\qquad\text{(13) Momentum } p_2, \text{space } \gamma_2\qquad\qquad}$

$$\begin{bmatrix} 0 & 0 & 0 & 0 \\ 0 & 0 & 0 & 0 \\ 0 & 0 & 0 & 0 \\ 0 & 0 & 0 & \Delta\varepsilon_{33} \end{bmatrix} \widetilde{\gamma_3} \begin{pmatrix} 0 & 0 & 0 & 0 \\ 0 & 0 & 0 & 0 \\ 0 & 0 & 0 & 0 \\ 0 & 0 & 0 & x_3 \end{pmatrix} \quad (i=\pm 1,\pm 2,\pm 3) \qquad (3.7.2)$$

$\underbrace{\qquad\qquad\text{(14) Momentum } p_3, \text{space } \gamma_3\qquad\qquad}$

Here, introducing the δ function, there is 1 virtual photon-spin wave in the intrinsic space, usually the lepton propagates along the spin axis ξ^3, $i=1$. The above is simplified as

$$\varphi_e(X) = \exp\mathrm{i}\{[\varepsilon]_n X_n\} \qquad (3.7.3)$$

Each strain is regarded as a degree of freedom. These degrees of freedom form a generalized multidimensional space with a metric of g_{mn}, satisfying $X_m = g_{mn}X^n = (X_0, -X)$

$$\varphi_e(X) = \exp\left[\mathrm{i}([\varepsilon]_n g_{mn} X^n)\right] \qquad (3.7.4)$$

3.7.2 Eigen value equation of free lepton

We discussed the intrinsic properties of particles. Now let's describe the global properties of leptons. The field function of spin particle is expressed by $\psi(x,\xi)$.

Table 3.7.1 The relationship between the free particle intrinsic space and freedom of space

Items	Intrinsic strain	Degree of freedom coordinates	Eigenvalue acquisition operator
Static Energy E_0	$K_0 = \begin{bmatrix} \varepsilon_0 & 0 & 0 & 0 \\ 0 & 0 & 0 & 0 \\ 0 & 0 & 0 & 0 \\ 0 & 0 & 0 & 0 \end{bmatrix}$	$X_{0I} = \check{I} \begin{pmatrix} \xi_0 & 0 & 0 & 0 \\ 0 & 0 & 0 & 0 \\ 0 & 0 & 0 & 0 \\ 0 & 0 & 0 & 0 \end{pmatrix}$	$\hat{\varepsilon}_0 = -\mathrm{i}\dfrac{\partial}{\partial X_0}$ Eigenvalue: E_0
Momentum p_1	$K_1 = \begin{bmatrix} 0 & 0 & 0 & 0 \\ 0 & \Delta\varepsilon_{11} & 0 & 0 \\ 0 & 0 & 0 & 0 \\ 0 & 0 & 0 & 0 \end{bmatrix}$	$X_1 = \check{\gamma}_1 \begin{pmatrix} 0 & 0 & 0 & 0 \\ 0 & x_1 & 0 & 0 \\ 0 & 0 & 0 & 0 \\ 0 & 0 & 0 & 0 \end{pmatrix}$	$\Delta\hat{\varepsilon}_1 = \hat{p}_1 = -\mathrm{i}\dfrac{\partial}{\partial X_1}$ Eigenvalue: p_1
Momentum p_2	$K_2 = \begin{bmatrix} 0 & 0 & 0 & 0 \\ 0 & 0 & 0 & 0 \\ 0 & 0 & \Delta\varepsilon_{22} & 0 \\ 0 & 0 & 0 & 0 \end{bmatrix}$,	$X_2 = \check{\gamma}_2 \begin{pmatrix} 0 & 0 & 0 & 0 \\ 0 & 0 & 0 & 0 \\ 0 & 0 & x_2 & 0 \\ 0 & 0 & 0 & 0 \end{pmatrix}$	$\Delta\hat{\varepsilon}_2 = \hat{p}_2 - \mathrm{i}\dfrac{\partial}{\partial X_2}$ Eigenvalue: p_2
Momentum p_3	$K_3 = \begin{bmatrix} 0 & 0 & 0 & 0 \\ 0 & 0 & 0 & 0 \\ 0 & 0 & 0 & 0 \\ 0 & 0 & 0 & \Delta\varepsilon_{33} \end{bmatrix}$	$X_3 = \check{\gamma}_3 \begin{pmatrix} 0 & 0 & 0 & 0 \\ 0 & 0 & 0 & 0 \\ 0 & 0 & 0 & 0 \\ 0 & 0 & 0 & x_3 \end{pmatrix}$	$\Delta\hat{\varepsilon}_3 = \hat{p}_3 = -\mathrm{i}\dfrac{\partial}{\partial X_3}$ Eigenvalue: p_3
Kinetic Energy ΔE	$\Delta K_0 = \begin{bmatrix} \Delta\varepsilon_0 & 0 & 0 & 0 \\ 0 & 0 & 0 & 0 \\ 0 & 0 & 0 & 0 \\ 0 & 0 & 0 & 0 \end{bmatrix}$	$X_0 = \check{I} \begin{pmatrix} x_0 & 0 & 0 & 0 \\ 0 & 0 & 0 & 0 \\ 0 & 0 & 0 & 0 \\ 0 & 0 & 0 & 0 \end{pmatrix}$	$\Delta\hat{\varepsilon}_0 = \Delta\hat{E} = -\mathrm{i}\dfrac{\partial}{\partial X_0}$ Eigenvalue: ΔE
Mass m_0	$K_4 = \begin{bmatrix} 0 & 0 & 0 & 0 \\ 0 & \varepsilon_{11} & 0 & 0 \\ 0 & 0 & \varepsilon_{22} & 0 \\ 0 & 0 & 0 & \varepsilon_{33} \end{bmatrix}$	$X_4 = \check{\gamma}_4 \begin{pmatrix} 0 & 0 & 0 & 0 \\ 0 & \xi_1 & 0 & 0 \\ 0 & 0 & \xi_2 & 0 \\ 0 & 0 & 0 & \xi_3 \end{pmatrix}$	$\hat{\varepsilon}_{ii} = \hat{m}_0 = -\mathrm{i}\dfrac{\partial}{\partial X_4}$ Eigenvalue: m_0

Chapter 3 The Structure of Lepton

Continued Table 3.7.1

Item	Intrinsic strain	Degree of freedom coordinates	Eigenvalue acquisition operator
Spin $\pm S_1$	$K_5 = \begin{bmatrix} 0 & 0 & 0 & 0 \\ 0 & 0 & 0 & 0 \\ 0 & 0 & 0 & \omega_{23} \\ 0 & 0 & \omega_{32} & 0 \end{bmatrix}$	$X_{S1} = \breve{\Sigma}_1 \begin{pmatrix} 0 & 0 & 0 & 0 \\ 0 & 0 & 0 & 0 \\ 0 & 0 & 0 & \xi_2 \\ 0 & 0 & \xi_3 & 0 \end{pmatrix}$	$\hat{\omega}_{[23]} = -i\dfrac{\partial}{\partial X_5}$ Eigenvalue: $\pm\omega_1$
$\pm S_2$	$K_6 = \begin{bmatrix} 0 & 0 & 0 & 0 \\ 0 & 0 & 0 & \omega_{13} \\ 0 & 0 & 0 & 0 \\ 0 & \omega_{31} & 0 & 0 \end{bmatrix}$	$X_6 = \breve{\Sigma}_2 \begin{pmatrix} 0 & 0 & 0 & 0 \\ 0 & 0 & 0 & \xi_1 \\ 0 & 0 & 0 & 0 \\ 0 & \xi_3 & 0 & 0 \end{pmatrix}$	$\hat{\omega}_{[13]} = -i\dfrac{\partial}{\partial X_6}$ Eigenvalue: ω_2
$\pm S_3$	$K_7 = \begin{bmatrix} 0 & 0 & 0 & 0 \\ 0 & 0 & \omega_{12} & 0 \\ 0 & \omega_{21} & 0 & 0 \\ 0 & 0 & 0 & 0 \end{bmatrix}$	$X_7 = \breve{\Sigma}_3 \begin{pmatrix} 0 & 0 & 0 & 0 \\ 0 & 0 & \xi_1 & 0 \\ 0 & \xi_2 & 0 & 0 \\ 0 & 0 & 0 & 0 \end{pmatrix}$	$\hat{\omega}_{[12]} = -i\dfrac{\partial}{\partial X_7}$ Eigenvalue: $\pm\omega_3$
ω_{0i}, ω_{i0}	$K_{8.1} = \begin{bmatrix} 0 & \omega_{01} & \omega_{02} & \omega_{03} \\ \omega_{10} & 0 & 0 & 0 \\ \omega_{20} & 0 & 0 & 0 \\ \omega_{30} & 0 & 0 & 0 \end{bmatrix}$	$X_{8.1} = \breve{I} \begin{pmatrix} 0 & \xi_0 & \xi_0 & \xi_0 \\ \xi_1 & 0 & 0 & 0 \\ \xi_2 & 0 & 0 & 0 \\ \xi_3 & 0 & 0 & 0 \end{pmatrix}$	Intrinsic spin four momentum flow $\hat{\omega}_{[0i]} = -i\dfrac{\partial}{\partial X_{8.1}}$
Weak interaction field $\gamma_{\mu\nu}(\xi)$	$K_9 = \begin{bmatrix} 0 & \gamma_{01} & \gamma_{02} & \gamma_{03} \\ \gamma_{10} & 0 & \gamma_{12} & \gamma_{13} \\ \gamma_{20} & \gamma_{21} & 0 & \gamma_{23} \\ \gamma_{30} & \gamma_{31} & \gamma_{32} & 0 \end{bmatrix}$	$X_9 = (1 \pm \breve{\gamma}_5) \begin{pmatrix} 0 & \xi_0 & \xi_0 & \xi_0 \\ \xi_1 & 0 & \xi_1 & \xi_1 \\ \xi_2 & \xi_2 & 0 & \xi_2 \\ \xi_3 & \xi_3 & \xi_3 & 0 \end{pmatrix}$	$\hat{\gamma}_{\mu\nu} = -i\dfrac{\partial}{\partial X_9}$ Half-space, Eigenvalue: m_{0Bare} Weak charge
Intrinsic vector and scalar potentials of charge	$K_{10} = \begin{bmatrix} \partial A_0/\partial \xi_0 & 0 & 0 & 0 \\ 0 & \partial A_1/\partial \xi_1 & 0 & 0 \\ 0 & 0 & \partial A_2/\partial \xi_2 & 0 \\ 0 & 0 & 0 & \partial A_3/\partial \xi_3 \end{bmatrix}$	$X_{10} = \breve{I} \begin{pmatrix} \xi_0 & 0 & 0 & 0 \\ 0 & \xi_1 & 0 & 0 \\ 0 & 0 & \xi_2 & 0 \\ 0 & 0 & 0 & \xi_3 \end{pmatrix}$	Intrinsic four vector, Eigenvalue: $A_{\mu,\mu}$ $\dfrac{\partial A_\mu}{\partial \xi_\mu} = -i\dfrac{\partial}{\partial X_{10}}$ No observability
Electronic internal electric field	$K_{8.2} = E(\xi) = \begin{bmatrix} 0 & \omega_{01}^B & \omega_{02}^B & \omega_{03}^B \\ \omega_{10}^B & 0 & 0 & 0 \\ \omega_{20}^B & 0 & 0 & 0 \\ \omega_{30}^B & 0 & 0 & 0 \end{bmatrix}$	$X_{8.2} = \breve{I} \begin{pmatrix} 0 & \xi_0 & \xi_0 & \xi_0 \\ \xi_1 & 0 & 0 & 0 \\ \xi_2 & 0 & 0 & 0 \\ \xi_3 & 0 & 0 & 0 \end{pmatrix}$	Eigen value of internal electric field of charge: e $\hat{\omega}_{[0i]}^B = -i\dfrac{\partial}{\partial X_{8-2}}$

Continued Table 3.7.1

Item	Intrinsic strain	Degree of freedom coordinates	Eigenvalue acquisition operator
Electronic internal magnetic field	$K_{11} = B(\xi) = \begin{bmatrix} 0 & 0 & 0 & 0 \\ 0 & 0 & \omega^B_{12} & \omega^B_{13} \\ 0 & \omega^B_{21} & 0 & \omega^B_{23} \\ 0 & \omega^B_{31} & \omega^B_{32} & 0 \end{bmatrix}$	$X_{11} = \check{Y} \begin{pmatrix} 0 & 0 & 0 & 0 \\ 0 & 0 & \xi_2 & \xi_2 \\ 0 & \xi_2 & 0 & \xi_3 \\ 0 & \xi_3 & \xi_3 & 0 \end{pmatrix}$	Spin magnetic moment, Eigenvalue: $\mu B = eh/4\pi m_e$ $\hat{\omega}^B_{[ij]} = -i\dfrac{\partial}{\partial X_{11}}$
Intrinsic electronic and magnetic field	$K_{12} = \begin{bmatrix} 0 & \omega^H_{01} & \omega^H_{02} & \omega^H_{03} \\ \omega^H_{10} & 0 & \omega^H_{12} & \omega^H_{13} \\ \omega^H_{20} & \omega^H_{21} & 0 & \omega^H_{23} \\ \omega^H_{30} & \omega^H_{31} & \omega^H_{32} & 0 \end{bmatrix}$	$X_{12} = \check{\Sigma} \begin{pmatrix} 0 & \xi_0 & \xi_0 & \xi_0 \\ \xi_1 & 0 & \xi_1 & \xi_1 \\ \xi_2 & \xi_2 & 0 & \xi_2 \\ \xi_3 & \xi_3 & \xi_3 & 0 \end{pmatrix}$	High angular momentum, Eigenvalue: n $\hat{\omega}^H_{[\mu\nu]} = -i\dfrac{\partial}{\partial X_{12}}$
Electric field in observation space	$K_{13} = \begin{bmatrix} 0 & E_{01} & E_{02} & E_{03} \\ E_{10} & 0 & 0 & 0 \\ E_{20} & 0 & 0 & 0 \\ E_{30} & 0 & 0 & 0 \end{bmatrix}$	$X_{13} = \check{Y} \begin{pmatrix} 0 & x_0 & x_0 & x_0 \\ x_1 & 0 & 0 & 0 \\ x_2 & 0 & 0 & 0 \\ x_3 & 0 & 0 & 0 \end{pmatrix}$	Multi electron electric field $\hat{E}_{0i} = -i\dfrac{\partial}{\partial X_{13}}$
Magnetic field in observation space	$K_{14} = \begin{bmatrix} 0 & 0 & 0 & 0 \\ 0 & 0 & B_{12} & B_{13} \\ 0 & B_{21} & 0 & B_{23} \\ 0 & B_{31} & B_{32} & 0 \end{bmatrix}$	$X_{14} = \check{Y} \begin{pmatrix} 0 & 0 & 0 & 0 \\ 0 & 0 & x_1 & x_1 \\ 0 & x_2 & 0 & x_2 \\ 0 & x_3 & x_3 & 0 \end{pmatrix}$	Multi electron magnetic field $\hat{B}_{[ij]} = -i\dfrac{\partial}{\partial X_{13}}$

(All the internal structures are superimposed on one point, form a spherically symmetric gravitational charge.)

Lepton generalized field functions containing complete information:

$$\varphi_{\text{lepton}}(X) = \exp i[K_\mu X^\nu g_{\nu\mu}] \tag{3.7.5}$$

Generalized field equations of lepton.

The establishment of equations is first to find the conserved quantities of eigenvalues. The main methods are as follows.

(1) We use energy momentum conservation to find the eigenvalue equation.

(2) Using the conservation of angular momentum, we find the eigenvalue equation.

(3) Find the eigenvalue equation by using stress balance.

Chapter 3 The Structure of Lepton

3.8 Several examples of generalized field equation of free particle

3.8.1 Example 1: Lepton and photon coupling constitutes a de Broglie wave

Photon-lepton coupling is the central overlap of photons and leptons. Let's first look at the physical meaning of photons in observational space (Fig. 3.8.1).

$$\varphi_\gamma(x) = \exp i \left[\begin{pmatrix} E_0 & 0 & 0 & 0 \\ 0 & 0 & 0 & 0 \\ 0 & 0 & 0 & 0 \\ 0 & 0 & 0 & 0 \end{pmatrix} \begin{pmatrix} x^0 & 0 & 0 & 0 \\ 0 & 0 & 0 & 0 \\ 0 & 0 & 0 & 0 \\ 0 & 0 & 0 & 0 \end{pmatrix} - \begin{pmatrix} 0 & 0 & 0 & 0 \\ 0 & p_1 & 0 & 0 \\ 0 & 0 & p_2 & 0 \\ 0 & 0 & 0 & p_3 \end{pmatrix} \begin{pmatrix} 0 & 0 & 0 & 0 \\ 0 & x^1 & 0 & 0 \\ 0 & 0 & x^2 & 0 \\ 0 & 0 & 0 & \xi^3 \end{pmatrix} \right]$$

(3.8.1)

Total energy term $\hbar k_0$ Longitudinal momentum of photons ($\hbar k_3$)

Photons transverse momentum which   it cause photons propagate along the k_3 direction

causing the photons has the fluctuation

Static electronic intrinsicfield function $\varphi_e(\xi) = \exp i[\varepsilon_{\mu\mu}](\xi)$, expressions are as follows:

$$\varphi_{0e}(\xi) = \exp i \left[\begin{pmatrix} E_0 & 0 & 0 & 0 \\ 0 & 0 & 0 & 0 \\ 0 & 0 & 0 & 0 \\ 0 & 0 & 0 & 0 \end{pmatrix} \begin{pmatrix} \xi^0 & 0 & 0 & 0 \\ 0 & 0 & 0 & 0 \\ 0 & 0 & 0 & 0 \\ 0 & 0 & 0 & 0 \end{pmatrix} - \begin{pmatrix} 0 & 0 & 0 & 0 \\ 0 & p_0/\sqrt{3} & 0 & 0 \\ 0 & 0 & p_0/\sqrt{3} & 0 \\ 0 & 0 & 0 & p_0/\sqrt{3} \end{pmatrix} \begin{pmatrix} 0 & 0 & 0 & 0 \\ 0 & \xi^1 & 0 & 0 \\ 0 & 0 & \xi^2 & 0 \\ 0 & 0 & 0 & \xi^3 \end{pmatrix} \right]$$

Static energy   rest mass

(3.8.2)

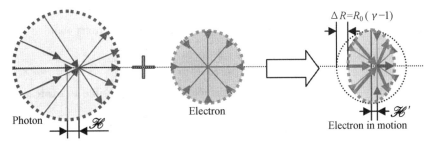

Fig. 3.8.1 Photon and static lepton coupling constitutes the movement electron

Meet $Tr[\varepsilon_{\mu\mu}] = 0$, that is $E_0 = m_0 c^2$. Static lepton and the photon are coupling:

$$\varphi(x) = \varphi_\gamma(x)\varphi_{0e}(\xi)$$

$$= \exp i \left\{ \begin{pmatrix} E_\gamma + E_0 & 0 & 0 & 0 \\ 0 & 0 & 0 & 0 \\ 0 & 0 & 0 & 0 \\ 0 & 0 & 0 & 0 \end{pmatrix} \begin{pmatrix} x^0 + \xi^0 & 0 & 0 & 0 \\ 0 & 0 & 0 & 0 \\ 0 & 0 & 0 & 0 \\ 0 & 0 & 0 & 0 \end{pmatrix} - \right.$$

$$\begin{pmatrix} 0 & 0 & 0 & 0 \\ 0 & p_0/3 & 0 & 0 \\ 0 & 0 & p_0/3 & 0 \\ 0 & 0 & 0 & p_0/3 \end{pmatrix} \begin{pmatrix} 0 & 0 & 0 & 0 \\ 0 & \xi^1 & 0 & 0 \\ 0 & 0 & \xi^2 & 0 \\ 0 & 0 & 0 & \xi^3 \end{pmatrix} -$$

$$\begin{pmatrix} 0 & 0 & 0 & 0 \\ 0 & p_1 & 0 & 0 \\ 0 & 0 & 0 & 0 \\ 0 & 0 & 0 & 0 \end{pmatrix} \begin{pmatrix} 0 & 0 & 0 & 0 \\ 0 & x^1 & 0 & 0 \\ 0 & 0 & 0 & 0 \\ 0 & 0 & 0 & 0 \end{pmatrix} - \begin{pmatrix} 0 & 0 & 0 & 0 \\ 0 & 0 & 0 & 0 \\ 0 & 0 & p_2 & 0 \\ 0 & 0 & 0 & 0 \end{pmatrix} \begin{pmatrix} 0 & 0 & 0 & 0 \\ 0 & 0 & 0 & 0 \\ 0 & 0 & x^2 & 0 \\ 0 & 0 & 0 & 0 \end{pmatrix} -$$

$$\begin{pmatrix} 0 & 0 & 0 & 0 \\ 0 & 0 & 0 & 0 \\ 0 & 0 & 0 & 0 \\ 0 & 0 & 0 & p_0 \end{pmatrix} \begin{pmatrix} 0 & 0 & 0 & 0 \\ 0 & 0 & 0 & 0 \\ 0 & 0 & 0 & 0 \\ 0 & 0 & 0 & x^3 \end{pmatrix} \Bigg\} \quad (3.8.3)$$

Here

$$X^0 = \begin{pmatrix} x^0 + \xi^0 & 0 & 0 & 0 \\ 0 & 0 & 0 & 0 \\ 0 & 0 & 0 & 0 \\ 0 & 0 & 0 & 0 \end{pmatrix}; \begin{pmatrix} 0 & 0 & 0 & 0 \\ 0 & p_0/3 & 0 & 0 \\ 0 & 0 & p_0/3 & 0 \\ 0 & 0 & 0 & p_0/3 \end{pmatrix} = m_0 c;$$

$$X^4 = \begin{pmatrix} 0 & 0 & 0 & 0 \\ 0 & \xi^1 & 0 & 0 \\ 0 & 0 & \xi^2 & 0 \\ 0 & 0 & 0 & \xi^3 \end{pmatrix} \quad (3.8.4)$$

$$\varphi_e(x) = \exp i\{[(E_0(\xi) + \Delta E(x))X^0 - p_1 X^1 - p_2 X^2 - p_3 X^3 - m_0 X^4]\} \quad (3.8.5)$$

Here, $E_0(\xi)$ is electronic static energy; $\Delta E(x)$ is electron kinetic energy.

Electrons with momentum are considered to be successful electrons coupled with photons. In our customary way:

$$\varphi_e(x) = \exp[i(Et - px - m_0 X_m)] \quad (3.8.6)$$

Since the mass is an intrinsic property of leptons, it is impossible to express it in the observational space, so the above formula is further written as

$$\psi_e(x) = u\exp[i(Et - px)]$$

This is the wave function that we know very well. Here, $E = E_\gamma + m_0 c^2 = m_0 v^2/2 + m_0 c^2$, the photon energy is complete absorbed by electron, $E_\gamma = m_0 v^2/2 = \hbar k_0$ is completely converted to the electron momentum.

1. Lagrangian quantities of Lagrangian quantum field

Lagrangian quantum field has a very clear physical meaning in vacuum field theory. The strain of vacuum field exists in the form of wave, and there are two forms of wave. One is the fluctuation that is enclosed in the intrinsic space. The other is the wave that can be measured in the observation space. The energy state of the observed wave in the observation space is

Chapter 3 The Structure of Lepton

Laplace quantity. According to this principle, the Lagrangian quantum field can be obtained only when the quantum field operator is confined to the observation space. Therefore, the Lagrangian quantum field has a very simple expression.

$$\mathscr{L} = i\gamma^{\mu} \partial_{\mu} \psi(X) \tag{3.8.7}$$

Field operator ∂_{μ} action in X space.

2. Electronics movement

The electron momentum is considered as static lepton coupling with photon effect. Low-energy momentum particles are described by Schrodinger equation. The propagation of lepton field is the coupling of lepton and photon. When the photon center enters the inner space of the electron, total deformation h of photon are compressed into h' become smaller and the propagation speed c into v (the electron motion velocity). The strain superposition makes the propagation characteristics of leptons worse, that is, the dynamic mass increases.

(1) After the coupling that lepton center and photon center is the same point and lepton electric flux line and electric flux line of photon is coupled into a same electric flux line which number is same with single lepton field.

(2) Coupling does not change the properties of photons. The propagation characteristics of photons coupled with free-state photons are the same. The trajectory of electron propagation is sinusoidal wave (Fig. 3.8.2).

$$\Psi_{\perp} = \cos[k \cdot r - k_0 t] \tag{3.8.8}$$

$$k = \frac{2\pi}{\lambda} n = \frac{\pi}{\phi_i}, k_0 = \frac{2\pi}{t_\lambda} = \frac{\pi}{\phi_0} = \omega \ ; \ p = \frac{h}{\lambda} n = \hbar k, \ E = h\nu = \hbar\omega \tag{3.8.9}$$

Here, $\Psi = e^{i(k \cdot r - \omega \cdot t)}$. Set A is the amplitude of wave.

$$\Psi = A\exp\{i[k \cdot r - k_0 t]\} = A\exp\{i[p \cdot r - Et]/\hbar\} \tag{3.8.10}$$

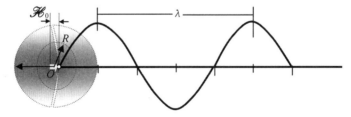

Fig. 3.8.2 The propagation wave trajectory of lepton center

This is the lepton transverse wave, i.e. de Broglie wave. $E = \frac{p^2}{2m_E}$, and de Broglie wavelength is $\lambda = \frac{h}{p} = \frac{h}{\sqrt{2m_e E}}$.

3.8.2 Example 2: The field equation of the real scalar field

Real scalar particles have the simplest spherically symmetric strain and can be directly written as

$$\phi_{\text{Real scalar}}(\xi) = \exp\left[i\begin{pmatrix}\varepsilon_{00} & 0 & 0 & 0\\ 0 & 0 & 0 & 0\\ 0 & 0 & 0 & 0\\ 0 & 0 & 0 & 0\end{pmatrix}\begin{pmatrix}\xi_0\\ 0\\ 0\\ 0\end{pmatrix} + \begin{pmatrix}0 & 0 & 0 & 0\\ 0 & \varepsilon_{11} & 0 & 0\\ 0 & 0 & \varepsilon_{22} & 0\\ 0 & 0 & 0 & \varepsilon_{33}\end{pmatrix}\begin{pmatrix}0\\ \xi_1\\ \xi_2\\ \xi_3\end{pmatrix}\right]$$

(3.8.11)

The intrinsic property of spherically symmetric strain is expressed as scalar field, which is expressed as a mass point in the observation space. Since the intrinsic space is a servo coordinate system, therefore, the kinetic energy and momentum of real scalar particles can not be expressed in the internal coordinate space. So the field function is written in a mixed expression which contains two coordinate systems.

$$\phi_{\text{Real scalar}}(\xi,x) = \exp\left\{i\left[\underbrace{\begin{pmatrix}\varepsilon_{00} & 0 & 0 & 0\\ 0 & 0 & 0 & 0\\ 0 & 0 & 0 & 0\\ 0 & 0 & 0 & 0\end{pmatrix}\begin{pmatrix}\xi_0 & 0 & 0 & 0\\ 0 & 0 & 0 & 0\\ 0 & 0 & 0 & 0\\ 0 & 0 & 0 & 0\end{pmatrix}}_{\text{Intrinsic energy } E_0 = m_0c^2} + \underbrace{\begin{pmatrix}\Delta\varepsilon_{00} & 0 & 0 & 0\\ 0 & 0 & 0 & 0\\ 0 & 0 & 0 & 0\\ 0 & 0 & 0 & 0\end{pmatrix}\begin{pmatrix}x_0 & 0 & 0 & 0\\ 0 & 0 & 0 & 0\\ 0 & 0 & 0 & 0\\ 0 & 0 & 0 & 0\end{pmatrix}}_{\text{Momentum } \Delta E = pc} + \right.\right.$$

$$\begin{pmatrix}0 & 0 & 0 & 0\\ 0 & \varepsilon_{11} & 0 & 0\\ 0 & 0 & \varepsilon_{22} & 0\\ 0 & 0 & 0 & \varepsilon_{33}\end{pmatrix}\begin{pmatrix}0 & 0 & 0 & 0\\ 0 & \xi_1 & 0 & 0\\ 0 & 0 & \xi_2 & 0\\ 0 & 0 & 0 & \xi_3\end{pmatrix} + \begin{pmatrix}0 & 0 & 0 & 0\\ 0 & \Delta\varepsilon_{11} & 0 & 0\\ 0 & 0 & 0 & 0\\ 0 & 0 & 0 & 0\end{pmatrix}\begin{pmatrix}0 & 0 & 0 & 0\\ 0 & x_1 & 0 & 0\\ 0 & 0 & 0 & 0\\ 0 & 0 & 0 & 0\end{pmatrix} +$$

$$\left.\left.\begin{pmatrix}0 & 0 & 0 & 0\\ 0 & 0 & 0 & 0\\ 0 & 0 & \Delta\varepsilon_{22} & 0\\ 0 & 0 & 0 & 0\end{pmatrix}\begin{pmatrix}0 & 0 & 0 & 0\\ 0 & 0 & 0 & 0\\ 0 & 0 & x_2 & 0\\ 0 & 0 & 0 & 0\end{pmatrix} + \begin{pmatrix}0 & 0 & 0 & 0\\ 0 & 0 & 0 & 0\\ 0 & 0 & 0 & 0\\ 0 & 0 & 0 & \Delta\varepsilon_{33}\end{pmatrix}\begin{pmatrix}0 & 0 & 0 & 0\\ 0 & 0 & 0 & 0\\ 0 & 0 & 0 & 0\\ 0 & 0 & 0 & x_3\end{pmatrix}\right]\right\}$$

(3.8.12)

This is a complete real scalar field function that contains all the information of the real scalar field, which describes the internal information of the scalar particles and the information in the observational space. The space-time strain satisfies the following equation.

$$\varepsilon_{00}^2 - \varepsilon_{11}^2 - \varepsilon_{22}^2 - \varepsilon_{33}^2 = 0 \tag{3.8.13}$$

That is the conservation of energy. Real scalar field wave function is

$$\phi(x) = u\exp i[P_0X^0 - p_1X^1 - p_2X^2 - p_3X^3 - m_0X^4] \tag{3.8.14}$$

The operator $\gamma = \gamma_1 + \gamma_2 + \gamma_3$, $\gamma_4 \equiv \gamma_M$ be understood as the base vector of degrees of freedom of mass, $I = \gamma_E$ be understood as base vector of degrees of freedom of energy. Set the existence the operator corresponding relationship as following:

$$X_0 = \breve{\gamma}_0\begin{pmatrix}x_0 & 0 & 0 & 0\\ 0 & 0 & 0 & 0\\ 0 & 0 & 0 & 0\\ 0 & 0 & 0 & 0\end{pmatrix}, X_1 = \breve{\gamma}_1\begin{pmatrix}0 & 0 & 0 & 0\\ 0 & x_1 & 0 & 0\\ 0 & 0 & 0 & 0\\ 0 & 0 & 0 & 0\end{pmatrix}, X_2 = \breve{\gamma}_2\begin{pmatrix}0 & 0 & 0 & 0\\ 0 & 0 & 0 & 0\\ 0 & 0 & x_2 & 0\\ 0 & 0 & 0 & 0\end{pmatrix},$$

Chapter 3 The Structure of Lepton

$$X_3 = \breve{\gamma}_3 \begin{pmatrix} 0 & 0 & 0 & 0 \\ 0 & 0 & 0 & 0 \\ 0 & 0 & 0 & 0 \\ 0 & 0 & 0 & x_3 \end{pmatrix} ; X_4 = \breve{\gamma}_4 \begin{pmatrix} x_0 & 0 & 0 & 0 \\ 0 & \xi_1 & 0 & 0 \\ 0 & 0 & \xi_2 & 0 \\ 0 & 0 & 0 & \xi_3 \end{pmatrix} \quad (3.8.15)$$

$$\phi_{\text{Real scalar}}(\xi, x) = \phi_{\text{Real scalar}}(X) = \exp i[K_\mu X^\mu g_{\nu\mu}] \ ; \ \hat{E} = i\frac{\partial}{\partial X_0} \ ;$$

$$\hat{p}_i = -i\frac{\partial}{\partial X_i} \ ; \ \hat{m} = -i\frac{\partial}{\partial X_4} \quad (3.8.16)$$

Energy term:

$$\hat{m}\phi_{\text{Real scalar}}(X) = m_0 c^2 \phi_{\text{Real scalar}}(X) \ ; \ \Delta\hat{E}\phi_{\text{Real scalar}}(X) = pc\phi_{\text{Real scalar}}(X) \quad (3.8.17)$$

Then

$$i\partial_0 \phi_{\text{Real scalar}}(X) = \sqrt{m_0^2 c^4 + c^2 p^2} \phi_{\text{Real scalar}}(X) \quad (3.8.18)$$

It is described by the square of the operator. Considering the conservation of energy, get

$$(\partial_0^2 - \partial_1^2 - \partial_2^2 - \partial_3^2 - \partial_4^2) \phi_{\text{Real scalar}}(X) = 0$$

That is

$$(\partial_\mu \partial^\mu g_{\mu\nu}) \phi_{\text{Real scalar}}(X) = 0 \quad (3.8.19)$$

Flat space: $g_{\mu\nu} = \eta_{\mu\nu}$. This is the generalized Klein–Gordon Equation.

3.8.3 Example3: Free particle Dirac Equation

The operator

$$\hat{p}_\mu = -i\frac{\partial}{\partial X_\mu} \quad (\mu = 0,1,2,3,4) \quad (3.8.20)$$

Mass is an independent intrinsic degree of freedom, which can be written as

$$(i\gamma^\mu \partial_\mu)\psi = 0 \quad (3.8.21)$$

The definition the $\overline{\psi}$ is Dirac conjugate field.

$$\overline{\psi} = \psi^+ \gamma^0 \quad (3.8.22)$$

Dirac Equation (3.8.21), using the Hermitian conjugate and right multiplication γ^0, can obtain $\overline{\psi}$, Dirac Equation can be written as

$$\overline{\psi}(i\gamma^v \partial_v - i\gamma^m \partial_m) = \overline{\psi}(i\gamma^v \overleftarrow{\partial_v} + \hat{m}) = 0 \quad (3.8.23)$$

That is

$$\overline{\psi}(i\gamma^v \overleftarrow{\partial_v}) = 0$$

Here the arrow indicates that it acts on function $\overline{\psi}$ in front of it. Here, $b\overleftarrow{\partial_\mu} \equiv -\partial_\mu b$. Dirac field ψ should be four components, can be expressed as

$$\psi = \begin{pmatrix} \psi_0 \\ \psi_1 \\ \psi_2 \\ \psi_3 \end{pmatrix} = (\psi_0 \ \psi_1 \ \psi_2 \ \psi_3)^T \quad (3.8.24)$$

Where the superscript T indicates the matrix transpose. Thus, equation (3.8.21) actually contains four equations:

$$(i\gamma^\mu \partial_\mu + i\gamma^4 \partial_4)_{\alpha\beta}\psi_\beta = 0.$$
$$(i\gamma^\mu \partial_\mu)_{\alpha\beta}\psi_\beta = 0, \mu = 0,1,2,3,4 \qquad (3.8.25)$$

1. Free electrons corresponding static electronic solution

Observation of physical quantity represent by bilinear spinor. From the point of view of physics, the physical fact corresponding to Lorentz invariant is that the change of lepton background space does not change the motion characteristics of lepton physics, that is, Dirac Equation does not change under Lorentz transformation.

Consider a free electron and its corresponding static solution of electronic, lepton in rest state, we use a spinoramplitude $w^1(0)$ to expandvarious state space of lepton. Particle in its own reference system, Dirac Equation solutions can be divided into two categories according to the positive and negative energy: positive energy solutions ψ^1, ψ^2 and negative energy solutions ψ^3, ψ^4. Positive solution or negative energy solutions are divided into two spin states: ψ^1, ψ^3 description the space projection of spin angular momentum in the third axis is $\hbar/2$ state; ψ^2, ψ^4 describing the spin angular momentum projection in the third axis of the space is the $-\hbar/2$ state. Lepton wave function is φ, and thus the total wave function can be written as

$$\psi^r(x) = w^r(0) e^{-(i\varepsilon_r mc^2/\hbar)t}; \quad \varepsilon_r = \begin{cases} +1 & (r=1,2) \\ -1 & (r=3,4) \end{cases} \qquad (3.8.26)$$

Spinor amplitude $w^r(0)$ is characterization the classification of the lepton and spin state.

$$\psi^{1,2}(0) = \begin{bmatrix} 1 \\ 1 \\ 0 \\ 0 \end{bmatrix} e^{-i(\frac{mc^2}{\hbar}t - \frac{mc}{\hbar}x)} ; \quad \psi^{3,4}(0) = \begin{bmatrix} 0 \\ 0 \\ 1 \\ 1 \end{bmatrix} e^{i(\frac{mc^2}{\hbar}t - \frac{mc}{\hbar}x)} \qquad (3.8.27)$$

Lepton spin state: $\pm 1/2$, Mass: $+m$ Anti-lepton spin states $\pm 1/2$, Mass: $-m$

2. Free particles solution of the Dirac Equation

The Hamiltonian of a free particle is

$$\hat{H} = c\hat{\alpha} \cdot \hat{p} + \hat{\beta} mc^2 \qquad (3.8.28)$$

Momentum operator $\hat{p} = (\hat{p}^1, \hat{p}^2, \hat{p}^3) = -i\hbar\left\{\frac{\partial}{\partial x^1}, \frac{\partial}{\partial x^2}, \frac{\partial}{\partial x^3}\right\}$, first of all, operator $\hat{p}$ and $\hat{H}$ is commutation

$$[\hat{p}, \hat{H}] = 0 \qquad (3.8.29)$$

Conservation of momentum of free Dirac particles. To solve the Dirac Equation of free particle:

$$\psi(x,t) = \psi(x)\exp(-\frac{i}{\hbar}Et) \qquad (3.8.30)$$

The steady-state Schördinger equation is obtained by separating (3.8.23) variables.

$$(c\hat{\alpha} \cdot \hat{p} + \hat{\beta}mc^2)\psi(x) = E\psi(x) \qquad (3.8.31)$$

Further, since the matrix α and β can be written in Pauli matrix σ and matrix is the elements of 2×2 the matrix, therefore, 4 component spinor wave function can be decomposed into two spinor wave function φ and χ.

$$\psi = \begin{bmatrix} \psi_1 \\ \psi_2 \\ \psi_3 \\ \psi_4 \end{bmatrix} = \begin{pmatrix} \varphi \\ \chi \end{pmatrix} \qquad (3.8.32)$$

Take it into (3.8.31), get

$$E\begin{pmatrix} \varphi \\ \chi \end{pmatrix} = c \begin{pmatrix} 0 & \sigma \\ \sigma & 0 \end{pmatrix} \hat{p} \begin{pmatrix} \varphi \\ \chi \end{pmatrix} + mc^2 \begin{pmatrix} 1 & 0 \\ 1 & -1 \end{pmatrix} \begin{pmatrix} \varphi \\ \chi \end{pmatrix} \qquad (3.8.33)$$

That is

$$\begin{cases} E\varphi = c\hat{\sigma} \cdot \hat{p}\chi + mc^2\varphi \\ E\chi = c\hat{\sigma} \cdot \hat{p}\varphi - mc^2\chi \end{cases} \qquad (3.8.34)$$

Due to the free Dirac particle momentum conservation, it has identified momentum state

$$\begin{pmatrix} \varphi \\ \chi \end{pmatrix} = \begin{pmatrix} \varphi_0 \\ \chi_0 \end{pmatrix} \exp\left(\frac{i}{\hbar} p \cdot x\right) \qquad (3.8.35)$$

Take (3.8.35) into (3.8.34), get

$$\begin{cases} (E - mc^2)\varphi - c\hat{\sigma} \cdot \hat{p}\chi_0 = 0 \\ -c\hat{\sigma} \cdot \hat{p}\varphi_0 + (E + mc^2)\chi_0 = 0 \end{cases} \qquad (3.8.36)$$

The non-zero solution condition of equation (3.8.36) is that the determinant of coefficient is zero.

$$\begin{vmatrix} E - mc^2 & -c\hat{\sigma} \cdot \hat{p} \\ -c\hat{\sigma} & E + mc^2 \end{vmatrix} = 0 \qquad (3.8.37)$$

Touse the Pauli matrix formula

$$(\sigma \cdot A)(\sigma \cdot B) = A \cdot B + i\sigma \cdot (A \times B) \qquad (3.8.38)$$

We can solve (3.8.37):

$$E = \pm\sqrt{c^2p^2 + m^2c^4} \equiv \lambda\sqrt{c^2p^2 + m^2c^4} \equiv \lambda E_p \qquad (3.8.39)$$

Equation (3.8.39) $\lambda = \pm 1$ is positive or negative respectively corresponding to the Dirac Equation positive energy solution and negative solution. Substitution of $E = \lambda E_p$ into (3.8.36), get

$$\chi_0 = \frac{c\hat{\sigma} \cdot \hat{p}}{E + mc^2}\chi_0; \quad \varphi_0 = \frac{c\hat{\sigma} \cdot \hat{p}}{E + mc^2}\varphi_0 \qquad (3.8.40)$$

If given φ_0, we canget χ_0 and φ_0 from the normalization condition. Take $\varphi_0 = \begin{pmatrix} u_1 \\ u_2 \end{pmatrix}$, the normalized condition is

$$\varphi_0^+ \varphi_0 = u_1^* u_1 + u_2^* u_2 = 1 \qquad (3.8.41)$$

u_1, u_2 is a constant. Therefore, the solution of the free Dirac Equation is

$$\psi_{p\lambda}(x,t) = N_\lambda \begin{pmatrix} \varphi_0 \\ \dfrac{c\hat{\sigma}\cdot\hat{p}}{mc^2 + \lambda\sqrt{c^2p^2 + m^2c^4}}\varphi_0 \end{pmatrix} \dfrac{\exp\left[\dfrac{i}{\hbar}(p\cdot x - \lambda E_p t)\right]}{(2\pi\hbar)^{3/2}} \qquad (3.8.42)$$

The above, $\lambda = \pm 1$, N_λ is the normalized constant, by

$$\int \psi_{p\lambda}^+(x,t)\psi_{p'\lambda'}(x,t)d^3x = \delta_{\lambda\lambda'}\delta(p - p') \qquad (3.8.43)$$

The determinable result is

$$N_\lambda = \sqrt{\dfrac{m_0c^2 + \lambda E_p}{2\lambda E_p}} \qquad (3.8.44)$$

(1) Angular momentum.

The magic of the Dirac Equation is automatically included in the lepton spin characteristics. Although the Dirac Equation does not give the cause of spin, from the point of view of vacuum dynamics, we can understand the cause of spin. Here we find the reason for spin mathematically.

Orbital angular momentum $L = r \times p$ is not a good quantum number. Here $\hat{L}$ and $\hat{H}$ is not commutation. Although the Dirac Equation does not give the cause of spin, from the point of view of vacuum dynamics, we can understand the cause of spin. Here we find the reason for spin mathematically. The proof is as follows.

$$\dfrac{d\hat{L}_x}{dt} = \dfrac{1}{i\hbar}[\hat{L}_x, \hat{H}] = \dfrac{c}{i\hbar}[\hat{L}_x, \hat{\alpha}_x\hat{p}_x + \hat{\alpha}_y\hat{p}_y + \hat{\alpha}_z\hat{p}_z]$$

$$= \dfrac{c}{i\hbar}\{\hat{\alpha}_x[\hat{L}_x, \hat{p}_x] + \hat{\alpha}_y[\hat{L}_x, \hat{p}_y] + \hat{\alpha}_z[\hat{L}_x, \hat{p}_z]\}$$

$$= c[\hat{\alpha}_y\hat{p}_z - \hat{\alpha}_z\hat{p}_y] = c(\hat{\alpha}\times\hat{p})_x \neq 0$$

That is

$$\dfrac{d\hat{L}}{dt} = c(\hat{\alpha}\times\hat{p}) \qquad (3.8.45)$$

Orbital angular momentum is not conserved. This means that the Dirac particle must exists the intrinsic angular momentum. Its orbital angular momentum and intrinsic angular momentum to form the angular momentum is conserved quantity. In order to find the intrinsic angular momentum, a matrix is defined.

$$\Sigma \equiv \begin{pmatrix} \sigma & 0 \\ 0 & \sigma \end{pmatrix} \text{ or } \Sigma_i = \begin{pmatrix} \sigma_i & 0 \\ 0 & \sigma_i \end{pmatrix}, (i = x,y,z) \qquad (3.8.46)$$

Chapter 3 The Structure of Lepton

The commutation relation can be proved by using Pauli operator directly.

$$[\Sigma,\beta] = 0, \{\Sigma_i,\alpha_i\} \quad (3.8.47)$$

$$[\Sigma_x,\alpha_y] = 2i\alpha_z, [\Sigma_y,\alpha_z] = 2i\alpha_x, [\Sigma_z,\alpha_x] = 2i\alpha_y \quad (3.8.48)$$

The commutation relations of operator Σ and H is

$$[\Sigma_x, H] = [\Sigma_x, c\alpha \cdot p] = c[\Sigma_x, \alpha_y p_y + \alpha_z p_z]$$
$$= 2ic(\alpha_z p_y + \alpha_y p_z) = -2ic(\alpha \times p)_x \quad (3.8.49)$$

$$\left[\frac{\hbar}{2}\Sigma, H\right] = -i\hbar c(\alpha \times p) \quad (3.8.50)$$

The total angular momentum J by (3.8.45) and (3.8.50) is

$$J = L + \frac{\hbar}{2}\Sigma \quad (3.8.51)$$

$$\frac{dJ}{dt} = 0 \quad (3.8.52)$$

It is concluded that the intrinsic angular momentum S of Dirac particles is

$$S = \frac{\hbar}{2}\Sigma \quad (3.8.53)$$

S and the orbital angular momentum compose the total angular momentum $J=L+S$ is conservation. S is spin angular momentum. Therefore, spin quantum number are automatically included in the Dirac Equation.

(2) Helicity operator.

The corresponding of helicity operator Λ_S is the projection of spin operator S in the momentum direction.

$$\Lambda_S = S\frac{p}{|p|} = \frac{\hbar}{2}\Sigma \cdot \frac{p}{|p|} \quad (3.8.54)$$

The operator to meet

$$[\Lambda_S, H] = 0 \quad (3.8.55)$$

$$[\Lambda_S, p] = 0 \quad (3.8.56)$$

The following is the representation of the matrix. Without loss of generality, choose the direction p is the direction z, that is $p = \{0,0,p\}$, so by the (3.8.54) get

$$\hat{\Lambda}_S = S_z = \frac{\hbar}{2}\Sigma_z = \frac{\hbar}{2}\begin{bmatrix}\sigma_z & 0 \\ 0 & \sigma_z\end{bmatrix} = \frac{\hbar}{2}\begin{bmatrix}1 & 0 & 0 & 0 \\ 0 & -1 & 0 & 0 \\ 0 & 0 & -1 & 0 \\ 0 & 0 & 0 & -1\end{bmatrix} \quad (3.8.57)$$

The $\hat{\Lambda}_S$ eigenvalues is $\pm\hbar/2$.

$$\begin{pmatrix}u_1 \\ 0\end{pmatrix}, \begin{pmatrix}u_{-1} \\ 0\end{pmatrix}, \begin{pmatrix}0 \\ u_1\end{pmatrix}, \begin{pmatrix}0 \\ u_{-1}\end{pmatrix} \quad (3.8.58)$$

Among them:

$$u_1 = \begin{pmatrix}1 \\ 0\end{pmatrix}, u_{-1} = \begin{pmatrix}0 \\ -1\end{pmatrix} \quad (3.8.59)$$

This is the eigenvector of σ_z. Free Dirac particle wave function can be classified by positive and negative energy or positive and negative helicity.

$$\Psi_{p_z,\lambda,\sigma} = N_\lambda \begin{pmatrix} u_\sigma \\ \dfrac{c\sigma_z p}{mc^2 + \lambda E_p} u_\sigma \end{pmatrix} \exp\left[\frac{i}{\hbar}(p_z z - \lambda E_p t)\right] \quad (\lambda = \pm 1, \sigma = \pm 1)$$

(3.8.60)

From the solution of Dirac field equation, it can be seen that all physical quantities observed in observation space are wave functions. In the new theory, all information is contained in the field function, while in the Dirac wave function, the intrinsic physical quantities of leptons are only four momentum, the amplitude of lepton wave function contains spin, and all other information is erased.

3.8.4 The energy and momentum of the Dirac field

We can derive Lagrange by the principle of action. From the free Dirac Equation

$$i\gamma^\mu \partial_\mu \psi(x) = 0 \qquad (3.8.61)$$

Left to multiply $\overline{\delta\psi}$ to structure all the Lagrangian integral which among t_1 and t_2:

$$\int_{t_1}^{t_2} d^4 x \overline{\delta\psi}(x)(i\gamma^\mu \partial_\mu)\psi(x) = \delta \int_{t_1}^{t_2} d^4 x \overline{\psi}(x)(i\gamma^\mu \partial_\mu)\psi(x) = 0 \qquad (3.8.62)$$

Get the Lagrangian density

$$\mathscr{L} = \overline{\psi}(i\gamma^\mu \partial_\mu - m)\psi = i\psi_a^* \partial_0 \psi_a + \psi_b^*(i\gamma^i_{ba}\partial_i - m\delta_{ba})\psi_a \qquad (3.8.63)$$

Back to our familiar expressions

$$\mathscr{L} = i(\overline{\psi}\gamma^\nu \partial_\nu \psi) - m\overline{\psi}\psi \qquad (3.8.64)$$

Hamiltonian density of Dirac field is

$$\mathscr{H} = \pi\dot{\Psi} - L = \psi^+(-i\alpha \cdot \nabla + \beta m)\psi = \psi^+ i\frac{\partial}{\partial t}\psi \qquad (3.8.65)$$

$\mathscr{H}$ is consistent with the non-relativistic form. Conservation laws of energy, momentum and angular momentum can be automatically obtained from translation invariance and Lorentz invariance, and can be calculated.

$$\mathscr{F}^{\nu\mu} = i\overline{\psi}\gamma^\nu \frac{\partial}{\partial x_\mu}\psi \qquad (3.8.66)$$

By

$$H = \int \mathscr{P}^{00} d^3 x = \int \psi^+(-i\alpha \cdot \nabla + \beta m)\psi d^3 x = \int \psi^*(-i\alpha \cdot \nabla + i\beta\partial_M)\psi d^3 x$$

$$= \int \psi^+(-i\delta_{\mu\nu}\gamma^\mu \partial_\nu)\psi d^3 x$$

As well as

$$P = \int \psi^+(-i\nabla)\psi d^3 x \qquad (3.8.67)$$

A vailable angular momentum density $\mathscr{U}^{\mu\nu\lambda}$ and angular momentum of conservation $M^{\nu\mu}$:

$$\mathcal{U}^{\mu\nu\lambda} = i\bar{\psi}\gamma^{\mu}\left(x^{\nu}\frac{\partial}{\partial x_{\lambda}} - x^{\lambda}\frac{\partial}{\partial x_{\nu}} + \Sigma^{\nu\lambda}\right)\psi \qquad (3.8.68)$$

$$M^{\nu\lambda} = \int d^{3}x\,\mathcal{U}^{0\nu\lambda}$$

Here, $\Sigma^{\nu\mu} = \frac{1}{4}[\gamma^{\nu}, \gamma^{\mu}]$ is the spinor rotation matrix which under Lorentz transform. The last term is added spin angular momentum, especially the spatial component.

$$J \equiv (M^{23}, M^{31}, M^{12}) = \int d^{3}x\psi^{+}\left(r \times \frac{1}{i}\nabla + \frac{1}{2}\sigma\right)\psi \qquad (3.8.69)$$

Easy to see, $\mathscr{L}$ is invariant under the Poincare transformation, therefore, according to Noether theorem that momentum and angular momentum are conserved quantity, and using the equations of motion (mean $\mathscr{L}=0$) received Dirac field energy momentum tensor is

$$T^{\mu\nu} = \frac{i}{2}(\bar{\psi}\gamma^{\nu}\overleftrightarrow{\partial}_{\nu}\psi) \qquad (3.8.70)$$

Thus, the energy and momentum respectively is

$$H = p^{0} = \frac{i}{2}\int\psi^{*}\overleftrightarrow{\partial}_{0}\psi d^{3}x \qquad (3.8.71a)$$

$$p = \frac{i}{2}\int\psi^{*}\overleftrightarrow{\nabla}\psi d^{3}x \qquad (3.8.71b)$$

Lagrangian density (3.6.38) and the following internal transformation invariance:

$$\psi(x) \to \psi'(x) = e^{-i\alpha\omega}\psi(x), \psi^{+}(x) \to \psi'^{+}(x) = e^{i\alpha\omega}\psi^{+}(x) \qquad (3.8.72)$$

ω is infinitesimal parameters, which is independent of x. Therefore, corresponding to the presence of conserved currents and the conserved charge, respectively are

$$J^{\mu} = e\bar{\psi}\gamma^{\mu}\psi \qquad (3.8.73)$$

$$Q = e\int\psi^{+}\psi d^{3}x \qquad (3.8.74)$$

Lepton is Fermi, so Dirac field obey the Pauli's exclusion principle, and meet Fermi-Dirac statistics. It also sees other quantum field books.

3.9 Neutrino

This section is based on the nature of vacuum to discuss the structure of the neutrino and explain the various properties of the neutrino.

3.9.1 Neutrino intrinsic space composition

Neutrino is an independent virtual photon-spin wave which isconstituted by a pure spin state photon. The internal structure is shown in Fig.3.9.1 and Fig.3.9.3.

The structure of neutrinos is similar to that of tornadoes. The central point of virtual photon-spin wave propagates around the spin axis near the ring radius R. This virtual photon-spin

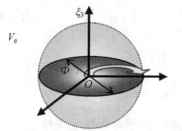 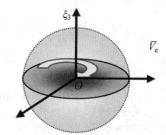

(a) Virtual photon-spin waves spread outward from the center f neutrino stretcher surface $\xi^I\xi^I$

(b) Virtual photon-spin waves spread inward to the center f neutrino compressed surface $\xi^I\xi^I$

Fig. 3.9.1 Neutrino structure

wave center is spread in the 2-dimensional surface of the disc-shaped, virtual photon-spin waves spread inward or outward of relative to the center of lepton, but neutrinos have no charge, no electromagnetic field. The experiment can distinguish the neutrinos are sinistral or dextral, left and right rotation is degeneration.

Here, neutrinos sinistral and dextral are corresponds to the virtual photon-spin wave of inward and outward. The 2-dimensional lepton field formed by inward compression is defined as a dextral neutrino. Out the formation of stretch 2-dimensional lepton field is the left-neutrino. Therefore, there are only two kinds of neutrinos, R neutrino and L neutrino. Neutrino and electronic are different, the electron both have virtual photon-spin wave and the electrostatic field, while the neutrino is a pure state of virtual photon-spin waves, or a pure state of virtual photon-spin wave, and no charge. Due to the virtual photon-spin wave center spread in the approximate 2-dimensional surface of the disks, neutrino internal space can be simplified as a 2-dimensional circular surface (Fig. 3.9.2). The intrinsic space-time frame $\xi^\alpha, \alpha=0,1,2$. Neutrino radius is R. Neutrino virtual photon-spin waves are similar to electron virtual photon-spin waves, 1/2 electric flux line within the ring, no-observed effect, 1/2 outside the ring, the spin angular momentum is $\hbar/2$.

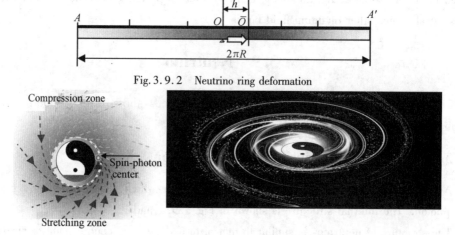

Fig. 3.9.2 Neutrino ring deformation

Fig. 3.9.3 Neutrino spin waves

Chapter 3   The Structure of Lepton

3.9.2   Neutrino intrinsic field function

The virtual photon-spin wave propagation around the ring which total deformation of space-time is h. As shown in Fig. 3.9.4 and Fig. 3.9.5. If the virtual photon-spin wave propagation inward or outward, then, the virtual photon-spin wave is compressed or stretched into the vacuum in a ring on a 2-dimensional surface to form a 2-dimensional antilepton or lepton.

Fig. 3.9.4   Relationship between spin and speed

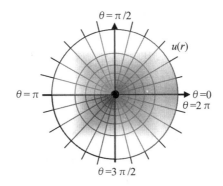

Fig. 3.9.5   2-dimensional lepton field

Consider $u(r, \theta)$ including radial r deformation and θ circumferential deformation. We use cylindrical coordinates to describe the neutrino, while the static neutrino has only two dimensional structure, then $z=0$.

Strain expressions in cylindrical coordinates:

$$\varepsilon_r = \frac{\partial u_r}{\partial r} \; ; \; \varepsilon_\theta = \frac{1}{r} \cdot \frac{\partial u_\theta}{\partial \theta} + \frac{u_r}{r} \tag{3.9.1}$$

$$\gamma_{\theta r} = \gamma_{r\theta} = \frac{1}{2}\left(\frac{\partial u_\theta}{\partial r} + \frac{1}{r} \cdot \frac{\partial u_r}{\partial \theta} - \frac{u_\theta}{r}\right) \tag{3.9.2}$$

$$\omega_{\theta r} = \omega_{r\theta} = \frac{1}{2}\left(\frac{\partial u_\theta}{\partial r} + \frac{1}{r} \cdot \frac{\partial u_r}{\partial \theta} + \frac{u_\theta}{r}\right) \tag{3.9.3}$$

The radial strain of fiber field can be seen from the previous discussions.

$$u_r = \cos\left(\frac{\pi r}{2R}\right) \quad (0 < r \leqslant R) \tag{3.9.4}$$

$\theta = 0$ is the ultimate compression zone, $\theta = 2\pi$ is the ultimate tensile zone. $u_\theta = u(r)\cos\left(\frac{\theta}{2}\right), 0 \leqslant \theta \leqslant 2\pi$. Considering that the outer edge of the intrinsic space is zero, the center is the largest.

According to the boundary conditions, the simplest strain function can be constructed as

$$u_\theta = \cos\frac{\pi r}{2R}\cos\frac{\theta}{2} \quad (0 < r \leq R, 0 \leq \theta \leq 2\pi) \tag{3.9.5}$$

$$\begin{cases} \varepsilon_r = -\dfrac{\Phi}{\pi}\sin\dfrac{\pi r}{2R}\,;\ \varepsilon_0 = \dfrac{1}{c}\sqrt{\varepsilon_r^{\,2}+\varepsilon_\theta^{\,2}} \\[4pt] \varepsilon_\theta = \dfrac{1}{r}\cos\dfrac{\pi r}{2R} - \dfrac{1}{2r}\sin\dfrac{\theta}{2}\cos\dfrac{\pi r}{2R} \\[4pt] \gamma_{r\theta} = -\dfrac{\pi}{4R}\sin\dfrac{\pi r}{2R}\cos\dfrac{\theta}{2} - \dfrac{1}{2r}\cos\dfrac{\theta}{2}\cos\dfrac{\pi r}{2R} \\[4pt] \omega_{r\theta} = -\dfrac{\pi}{4R}\sin\dfrac{\pi r}{2R}\cos\dfrac{\theta}{2} + \dfrac{1}{2r}\cos\dfrac{\theta}{2}\cos\dfrac{\pi r}{2R} \end{cases} \tag{3.9.6}$$

The neutrino field function is only 3-dimensional structure.

$$\varphi_\nu(\xi) = \exp(-i)\left\{ \underbrace{\begin{bmatrix} \varepsilon_0 & 0 & 0 & 0 \\ 0 & 0 & 0 & 0 \\ 0 & 0 & 0 & 0 \\ 0 & 0 & 0 & \varepsilon_3 \end{bmatrix}\begin{pmatrix} \xi_0 & 0 & 0 & 0 \\ 0 & 0 & 0 & 0 \\ 0 & 0 & 0 & 0 \\ 0 & 0 & 0 & x_3 \end{pmatrix}}_{\text{Total Energy and Longitudinal momentum}} + \right.$$

The strain structure similar virtual photon. Neutrinos propagate at the speed of light and transmit momentum

$$\underbrace{\begin{bmatrix} 0 & \varepsilon_{01} & \varepsilon_{02} & 0 \\ \varepsilon_{10} & 0 & 0 & 0 \\ \varepsilon_{20} & 0 & 0 & 0 \\ 0 & 0 & 0 & 0 \end{bmatrix}\begin{pmatrix} 0 & \xi_0 & \xi_0 & 0 \\ \xi_1 & 0 & 0 & 0 \\ \xi_2 & 0 & 0 & 0 \\ 0 & 0 & 0 & 0 \end{pmatrix}}_{\text{Energy momentum flow}} +$$

$$\underbrace{\begin{bmatrix} 0 & 0 & 0 & 0 \\ 0 & \varepsilon_1 & 0 & 0 \\ 0 & 0 & \varepsilon_2 & 0 \\ 0 & 0 & 0 & 0 \end{bmatrix}\begin{pmatrix} 0 & 0 & 0 & 0 \\ 0 & \xi & 0 & 0 \\ 0 & 0 & \xi_2 & 0 \\ 0 & 0 & 0 & 0 \end{pmatrix}}_{\text{2-dimensional momentum}} +$$

Constitute 2-dimensional weak field radial strain part

$$\frac{1}{2}\underbrace{\begin{bmatrix} 0 & 0 & 0 & 0 \\ 0 & 0 & \gamma_{12} & 0 \\ 0 & \gamma_{21} & 0 & 0 \\ 0 & 0 & 0 & 0 \end{bmatrix}\begin{pmatrix} 0 & 0 & 0 & 0 \\ 0 & 0 & \xi_1 & 0 \\ 0 & \xi_2 & 0 & 0 \\ 0 & 0 & 0 & 0 \end{pmatrix}}_{\text{2-dimensional circumferential shear mass}} +$$

Constitute 2-dimensional weak field shear strain part

$$\frac{1}{2}\underbrace{\begin{bmatrix} 0 & 0 & 0 & 0 \\ 0 & 0 & \omega_{12} & 0 \\ 0 & 0 & 0 & 0 \\ 0 & 0 & 0 & 0 \end{bmatrix}\begin{pmatrix} 0 & 0 & 0 & 0 \\ 0 & 0 & 0 & 0 \\ 0 & \xi_2 & 0 & 0 \\ 0 & 0 & 0 & 0 \end{pmatrix}}_{\text{Spin angular momentum along the direction of motion } x_3} +$$

Constitute two dimensional weak field spin strain part

Chapter 3 The Structure of Lepton

$$\frac{1}{2}\underbrace{\begin{bmatrix}0 & 0 & 0 & 0\\ 0 & 0 & 0 & 0\\ 0 & -\omega_{12} & 0 & 0\\ 0 & 0 & 0 & 0\end{bmatrix}}_{\substack{\text{Spin angular momentum is opposite}\\ \text{to the direction of motion } x_3}}\begin{pmatrix}0 & 0 & 0 & 0\\ 0 & 0 & \xi_1 & 0\\ 0 & 0 & 0 & 0\\ 0 & 0 & 0 & 0\end{pmatrix} \right\} \qquad (3.9.7)$$

We know that most intrinsic strains can not be directly shown in the observation space.

3.9.3 Angle momentum of virtual photon-spin wave

The neutrino is a pure state of virtual photon-spin waves it rotational spread around the spin axis. Neutrino also has a nature of half-space. Set r is the average radius of rotation of the virtual photon-spin wave, $r<R$, $r\geqslant 0$, O is center of lepton intrinsic space, when the wavelength of the virtual photon-spin wave $\lambda=2\pi r$ it's in low-energy steady state. virtual photon-spin waves meet $2\pi r\times P=h$, namely $r\times P=h/2\pi$ is the spin angular momentum. Because the nature of half space of neutrinos, the outside of the virtual photon-spin wave is observed, the inside of the virtual photon-spin wave cannot be observed. Spin angular momentum is $S=\hbar/2$. Neutrino spin surface no deformation which perpendicular transmission direction, when neutrino in the dissemination, the field strain always occurred in the direction of minimum deformation, and thus the neutrino is always propagation parallel the direction of spin angular momentum, as shown in Fig. 3.9.5. Lepton motion is caused by the coupling of lepton and photon. The coupling is the smallest deformation along the direction of lepton, so the motion always follows the z-axis. As neutrino field in this direction has no deformation which can travel at the speed of light. 2-dimensional lepton center achieve deformation limit, and thus the surface xy has the rest mass. Because there is no vacuum strain in z-direction, it has no 3-dimensional rest mass.

$$\varphi_\nu(\omega)=\exp\left\{-\frac{\mathrm{i}}{2}\underbrace{\begin{bmatrix}0 & 0 & 0 & 0\\ 0 & 0 & \omega_{12} & 0\\ 0 & -\omega_{12} & 0 & 0\\ 0 & 0 & 0 & 0\end{bmatrix}}_{\substack{\text{Spin angular momentum along}\\ \text{the direction of motion } x_3}}(1-\breve{\gamma}_5)\begin{pmatrix}0 & 0 & 0 & 0\\ 0 & 0 & \xi_1 & 0\\ 0 & \xi_2 & 0 & 0\\ 0 & 0 & 0 & 0\end{pmatrix}\right\} \qquad (3.9.8)$$

virtual photon-spin wave constitutes neutrino, and the virtual photon-spin wave has three energy states corresponding to ν_e, ν_μ and ν_τ, respectively. Because the lowest energy state is the most stable state, ν_μ and ν_τ should release energy into the lower state ν_e.

Neutrino is virtual photon-spin waves. The virtual photon-spin wave inward is right-handed anti-neutrino with antihalf-space. Neutrino virtual photon-spin wave outward corresponds to L-neutrino with positive half-space. This essence has nothing to do with left or right of the virtual photon-spin wave. The neutrino is not charged, the laboratory is unable to distinguish the spin direction. The nature of the half-space of neutrino is understood as the virtual photon-spin wave of the left and right in the quantum field theory, this can easily lead to misunderstanding.

$$h = \int_0^{2\pi R} \omega_{12} d(\theta R). \quad \omega_{12} = \frac{1}{2}\left(\frac{\partial u_1}{\partial \xi_2} - \frac{\partial u_2}{\partial \xi_1}\right) \quad (\text{Set}: \Omega_3 = \omega_{12})$$

$$\varphi_\nu(X) = \begin{pmatrix} \nu_e \\ \nu_\mu \\ \nu_\tau \end{pmatrix} \exp\left\{-i\left[E_0 x_0 \pm \frac{\Omega_3}{2} x_3 + \begin{bmatrix} 0 & 0 & 0 & 0 \\ 0 & p_1 & \gamma_{12} & 0 \\ 0 & -\gamma_{12} & p_2 & 0 \\ 0 & 0 & 0 & 0 \end{bmatrix} (1-\breve{\gamma}_5) \begin{pmatrix} 0 & 0 & 0 & 0 \\ 0 & \xi_1 & \xi_1 & 0 \\ 0 & \xi_2 & \xi_2 & 0 \\ 0 & 0 & 0 & 0 \end{pmatrix}\right]\right\}$$

(3.9.9a)

$$\phi_{\bar{\nu}}(X) = \begin{pmatrix} \bar{\nu}_e \\ \bar{\nu}_\mu \\ \bar{\nu}_\tau \end{pmatrix} \exp\left\{i\left[E_0 x_0 \pm \frac{\Omega_3}{2} x_3 + \begin{bmatrix} 0 & 0 & 0 & 0 \\ 0 & p_1 & \gamma_{12} & 0 \\ 0 & -\gamma_{12} & p_2 & 0 \\ 0 & 0 & 0 & 0 \end{bmatrix} (1+\breve{\gamma}_5) \begin{pmatrix} 0 & 0 & 0 & 0 \\ 0 & \xi_1 & \xi_1 & 0 \\ 0 & \xi_2 & \xi_2 & 0 \\ 0 & 0 & 0 & 0 \end{pmatrix}\right]\right\}$$

(3.9.9b)

This is the neutrino andanti-neutrino wave functions. Each neutrino wave function has three fluctuation amplitudes, which correspond to three neutrinos respectively.

3.9.4 The neutrino rest mass

That is intrinsic 4-momentum of neutrino. The internal space $(\xi_0, \xi_1, \xi_2, \xi_3)$ is extended to coordinate space (x_0, x_1, x_2, x_3). For experimental observation, neutrinos propagate at the speed of light. It can be understood as the coupling of the static neutrino and the photon:

$$\varphi_{\nu(E,m,p)}(\xi, x) = \exp i\left\{\begin{bmatrix} \varepsilon_{00} & 0 & 0 & 0 \\ 0 & \varepsilon_{11} & 0 & 0 \\ 0 & 0 & \varepsilon_{22} & 0 \\ 0 & 0 & 0 & 0 \end{bmatrix}\begin{pmatrix} \xi_0 & 0 & 0 & 0 \\ 0 & \xi_1 & 0 & 0 \\ 0 & 0 & \xi_2 & 0 \\ 0 & 0 & 0 & 0 \end{pmatrix} + \begin{bmatrix} \Delta\varepsilon_{00} & 0 & 0 & 0 \\ 0 & 0 & 0 & 0 \\ 0 & 0 & 0 & 0 \\ 0 & 0 & 0 & \Delta\varepsilon_{33} \end{bmatrix}\begin{pmatrix} \xi_0 & 0 & 0 & 0 \\ 0 & 0 & 0 & 0 \\ 0 & 0 & 0 & 0 \\ 0 & 0 & 0 & x_3 \end{pmatrix}\right\}$$

(3.9.10)

Neutrino in $x_3 = z$ shaft having a longitudinal photon effect, satisfies the following equation:

$$\frac{\partial^2 \varphi_\nu}{\partial z^2} - \frac{1}{c^2}\frac{\partial \varphi_\nu}{\partial t^2} = 0$$

Neutrino along the z-axis moving at the speed of light (for the observer, the helicity operator is introduced to determine the direction of propagation). On the other hand, the neutrino having

Chapter 3 The Structure of Lepton

2-dimensional static mass in the surface which perpendicular to the direction of propagation:

$$\frac{1}{c^2}\left(\frac{\partial^2 \varphi_v}{\partial x^2} + \frac{\partial \varphi_v}{\partial y^2}\right) = \varepsilon_x^2 + \varepsilon_y^2 = m_{xy}^2 \tag{3.9.11}$$

This mass has no properties of 3-dimension, therefore along the propagation z-axis direction has no observable effect. Neutrinos have unique properties. They can travel along the direction of propagation at the speed of light without rest mass. From the view of experimental, the neutrinos will take away some energy, the energy is $\varepsilon_0 = E_s$, E_s is the spin energy. Consider only observable physical properties in observation space.

$$\varphi_{v(E,p)}(X) = \exp\{i[(E_0 + \Delta E)X_0 + pX_1]\} \tag{3.9.12}$$

3.9.5 Neutrino oscillation

The two dimensional properties of neutrinos make it move at the speed of light. We can also understand that the motion of the speed of light is static neutrinos coupled with and photons. Due to the intrinsic level change cycle which leads to the intrinsic diameter of neutrino periodically changes, transformation between tensile limit and compression limit, this changes make intrinsic time change $\phi_\alpha P_\alpha = \phi_{0\alpha} P_{0\alpha}$. Longitudinal strain is transformed into 2-dimensional transverse strain (energy conservation $E_v = E_S + E_\gamma$). Neutrinos can only beat the speed of light or at rest in vacuum theory ($u_v = 0$; $u_v = c$).

Part of the neutrino into other types of neutrinos, called "neutrino oscillation". Why are the different types of neutrinos going to change? We now know that neutrino intrinsic space is 2-dimensional ball, virtual photon-spin wave spreading in a 2D circle, there are three level state, corresponding to the three radii neutrino $R_\tau < R_\mu < R_e$. The same relationship exists between anti neutrinos and the wavelength of virtual photon-spin wave $\lambda = 2\pi R_\alpha$. For the neutrino, the virtual photon-spin wave can be in the three energy states transition, similar to the atomic energy level of the hydrogen atom, the neutrino is the superposition of three states $|v\rangle = |v_e\rangle + |v_\mu\rangle + |v_\tau\rangle$. The energy levels are affected by the background space. The propagation perturbations are not the same in the vacuumor atmospheric, the specific expressions are as follows.

$$v_\alpha = U_{\alpha 1} v_1 + U_{\alpha 2} v_2 + U_{\alpha 3} v_3 \tag{3.9.13}$$

Among them $\alpha = e, \mu, \tau$.

Neutrinos with energy E are generated from the reactor and are lost due to oscillation after the propagation distance L. The probability of neutrinos losing due to oscillation is

$$P_{\text{react}} = \sin^2 2\theta_{13} \sin^2(\Delta m_{31}^2 L / 4\hbar c E)$$

Neutrino oscillation exists in the process of neutrino propagation in vacuum, and a neutrino with a fixed flavor will be transformed into other flavors.

$$P_{\text{sur}} \approx 1 - \underbrace{\sin^2(2\theta_{13})}_{\text{Amplitude}} \cdot \underbrace{\sin^2(1.27 \Delta E_{31}^2 \cdot \overbrace{L}^{\text{Flight distance/Neutrno energy}} / E)}_{\text{Oscillation frequency}} \tag{3.9.14}$$

We use 1, 2, 3 to mark these with a fixed energy the neutrino: e, μ, τ, so different angles θ on behalf of the relationship between different types of neutrinos. For example, θ_{12} is related to the mixing angle between ν_e and ν_τ, usually called the solar neutrino, and the test results are

$$\Delta m_{12}^2 = 8.0^{-5} \times 10 \text{ eV}^2; \sin^2(2\theta_{12}) = 0.85; \theta_{12} = 33°$$

$$\Delta m_{32}^2 \approx \Delta m_{31}^2 = (2.32) \times 10^{-3} \text{eV}^2; \sin^2(2\theta_{23}) = 0.97; \theta_{23} = 45°$$

The mixing angles are the basic physical constants, it could not be obtained by calculation. Solar neutrino oscillation and atmospheric neutrino oscillation experiments clearly tell us, θ_{12} and θ_{23} are relatively large, and θ_{12} is very small. China's Daya Bay neutrino experiment 2012 get $\sin^2(2\theta_{13}) = 0.092 \pm 0.017$.

Now the theoretical explanation is that neutrinos are produced by the flavor eigenstate and have three different "flavors". Each flavor eigenstate neutrino is composed of three types of neutrino mass eigenstates. Because of the quantum coherence of the mass eigenstate, one flavor neutrino oscillates into another flavor neutrino during its propagation. In quantum field theory, neutrinos propagate at the speed of light. The laboratory can judge the validity of vacuum neutrino theory by identifying the neutrino velocity equal to c.

3.9.6 Calculation of weak mixing angle

2-dimensional vacuum strain can explain neutrino oscillation simply and clearly.

Neutrinos have 2-dimensional mass. The mass along the propagation direction is no-observability, so for the experimental observations, using 3-dimensional space compression factor γ to determine the mass of change is no longer applicable. The change of neutrino mass m is the vacuum periodic strain effect caused by the fluctuation of the virtual photon-spin wave center around the spin axis region. The closer the virtual photon-spin wave center is to the axis, the greater the strain is near the axis. As the neutrinos propagate along the spin axis, the increase of strain in the axis region will reduce the propagation efficiency of neutrinos, showing the mass effect, propagation speed is less than c. Neutrino oscillation can lead to periodic variation of propagation speed. When the virtual photon-spin wave reaches its peak inward, the radius R is the smallest, and the strain near the axis region is the largest, indicating that the 2-dimensional mass increases, when the virtual photon-spin wave reaches its peak outward, the radius R is the largest and the strain in the axis region is the smallest, indicating that the 2-dimensional mass decreases.

The photon is a transverse wave, while propagating along a ring with radius $(R+3r)$, it still retains the characteristics of transverse wave (Fig. 3.9.6). When it vibrates inward, the amplitude is in $\langle \nu_\tau |$ state, while when it vibrates outward, the amplitude is in $\langle \nu_e |$ state. When it propagates along a ring with radius $(R+r)$, inward is $\langle \nu_\tau |$, and outward is $\langle \nu_\mu |$

Chapter 3 The Structure of Lepton

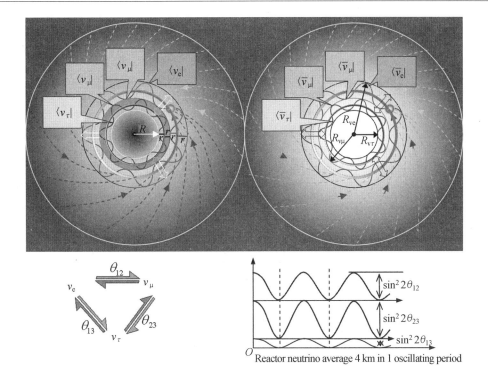

Fig. 3.9.6 The energy levels of neutrino oscillation

state. The neutrino oscillation is a pulsation in the ring, that is, the radius of the ring appears periodic fluctuation. It is important to note that on the compressed 2-dimensional surface, time slows down, and the central point of photons propagating in the ring travels much slower than the speed of light, which leads to a longer oscillation period. The essence of oscillation is the transformation of kinetic energy and potential energy. For measurement, we can only measure kinetic energy, which will lead to confusion about energy loss. Because kinetic energy is transformed into potential energy, the smaller the radius, the larger the potential energy, the larger the mass, the larger the radius, the larger the kinetic energy and the smaller the mass.

$$R \propto 1/U; \quad R \propto \Delta E \tag{3.9.15}$$

$$U_{\nu\tau} > U_{\nu\mu} > U_{\nu e} \tag{3.9.16}$$

From the measurement of neutrino oscillation in S_1, S_4 two rings were measured to be ν_e and ν_τ, in S_2, S_3 two rings were ν_μ, to calculate the probability of neutrino change into another neutrino need to consider the following two factors:

(1) The area of the ring is bigger, the possibility of being detected is higher. Neutrino oscillation is from one ring to another, if two dimensional ring area is not the same, The probabilities of neutrinos detected in a 2-dimensional torus is changed, so the probability of being detected by proportional to the area of the ring $U_i \propto S_i$.

(2) Because neutrinos have only 2-dimensional structure, 2-dimensional space is

compressed or stretched, the stronger the more inward strain space, space compression by γ expression, in the strong strain zone, time slows down, the probability will be detected becomes large $U_i \propto \gamma_i$.

After considering the above two factors, we can calculate the probability that a neutrino into another neutrino.

$$\sin^2(2\theta_{12}) = \frac{S_1}{S_2}\frac{\gamma_e}{\gamma_\mu} = \frac{\pi(R+4r)^2 - \pi(R+3r)^2}{\pi(R+3r)^2 - \pi(R+2r)^2}\frac{\sqrt{1-[0/(R+4r)]^2}}{\sqrt{1-[2r/(R+4r)]^2}}$$

(3.9.17)

$$\sin^2(2\theta_{23}) = \frac{S_2}{S_3}\frac{\gamma_\mu}{\gamma_\tau} = \frac{\pi(R+2r)^2 - \pi(R+r)^2}{\pi(R+r)^2 - \pi R^2}\frac{\sqrt{1-[2r/(R+4r)]^2}}{\sqrt{1-[4r/(R+4r)]^2}}$$

(3.9.18)

Take $r = 1.62$.

$$\sin^2(2\theta_{12}) = \frac{2+5r}{2+7r}\frac{1}{\sqrt{1-[2r/(R+4r)]^2}} = 0.84 \quad \text{(Experimental value 0.86)}$$

$$\sin^2(2\theta_{23}) = \frac{2+r}{2+3r}\frac{\sqrt{1-[2r/(R+4r)]^2}}{\sqrt{1-[4r/(R+4r)]^2}} = 0.95 \quad \text{(Experimental value 0.97)}$$

According to the theory of $\sin^2(2\theta_{12})$ is zero, but considering the quantum fluctuations of the tunneling effect, the fluctuation of low energy electron neutrinos can cross the barrier ν_μ into the S_4 torus, as $\nu_e \rightarrow \nu_\tau$. Tunneling probability is very low, so $\sin^2(2\theta_{12})$ tends to 0, but not equal to 0.

3.9.7 The energy limit of photons and the length of Planck

The quantum vacuum deformation satisfies $h_f = 2h$. Because the microcosmic deformation of space-time has the property of fibrosis, h_f can be simplified as a length, in this way, the length of the vacuum basic unit is $2h$. For photon, when the energy reaches the maximum limit, there are only two vacuum basic units. One stretch to the limit, the other is compressed to the limit, so the shortest wavelength of the photon $\Phi_{min} = 4h/K_{GP}(K_{GP} = 1)$. The maximum momentum of a photon $p_{max} = h/\phi_{min}$, maximum energy $E_{max} = h/t_{min}$, $ct_{min} = \Phi_{min} = 4h$, t_{min} is the shortest time in the real world that has physical meaning. Φ_{min} is the shortest distance with physical meaning in the real world. We can calculate the energy of the limit photon according to a known photon energy formula. $E_{max} = E_\gamma \lambda_\gamma / 4h$, λ_γ is the measure wavelength of photons, and E_γ is the energy of measurable photon.

The limit fermion is the heaviest particle in the physical world, no charge, having a spin. Bending the limit photon into a circle constitutes the ultimate Fermi particle. The circumference of the circle is $4h$, that is $2\pi R = 4h$. The radius of the limit particle is $R = h/2\pi$. According

Chapter 3 The Structure of Lepton

to the definition of particle mass $m = h_f/2Rc$, $m_{max} = \pi h_f/hc$ can be obtained. Interestingly, the structure of this particle is a standard Taiji structure (Fig. 3.9.7). The particle is the basic particle that makes up a black hole.

Fig. 3.9.7 Bending the limit photon into a circle constitutes the limit scalar particle

Chapter 4 Hadron Structure

In this chapter, the static intrinsic structure of hadrons is established by the nature of vacuum, and the strong interaction is studied from the point of view of vacuum field①. The basic unit of vacuum will cause splitting dimension constitute quarks with string. Because the strong interaction is extremely complex, the discussion in this chapter is limited to the nature of quarks, the hadrons static intrinsic structure and understanding basic physical images of strong interaction.

The simplest fundamental particles with rest mass are electrons and quarks. Internal structure of leptons and quarks are same, and there are also three generations, the form of wave functions ϕ of free quarks and free electron are same. Quark is different from lepton, it comes from dimension division of basic unit which makes the quark with string, and therefore there is no free quarks. A basic unit dimension cannot be divided constitutes quark confinement and existing physical mechanism of colorless singlet state. The following will specifically introduce the hadrons string model image in vacuum.

The mathematical description of the new theory is still based on the mathematical model of strain. The vacuum breakdown is aggravated by strong interaction. In addition, the dimension is not a complete 3-dimensional space in the hadron interior space, which hinders the accurate description of the theory.

The relationship between lepton and quark mass and mass generation cannot be directly estimated. Considering that the essence of mass comes from vacuum strain and vacuum rupture, any change of vacuum will produce mass effect. So at present, we can only judge qualitatively the range of quality. There is still a long way to go to calculate the quality accurately.

4.1 The π meson structure

4.1.1 π^0 meson structure

The strong deformation of charges will lead to dimensional space splitting. The splitting follows the following rules: charge $e^{\pm}$ binary splitting constitutes charged $\pi^{\pm}$ meson state, and trisomic splitting constitutes proton p or antiproton.

---

① In the Dragon Boat Festival of 2019, I visited my student Long Wenkai in Yiliang. He proposed that plastic deformation could not exist because of vacuum characteristics. I have thought about this seriously and revised the strong interaction vacuum model. I am very grateful for this.

Chapter 4　Hadron Structure

Lepton intrinsic space is half-space. Lepton field only have x, y, zone-way coordinate 3-dimension. Anti-lepton field have $-x$, $-y$, $-z$ three one-way dimensions, as shown in Fig. 4.1.1(a). The total amount of substance of a basic unit of vacuum is 1, and each dimension substance of a basic unit is 1/3 (Using charge to detect). Consider the basic unit of vacuum P is non-free state in the ground state, without charge. Meson π^0 states can be constructed by basic unit of vacuum P deformation.

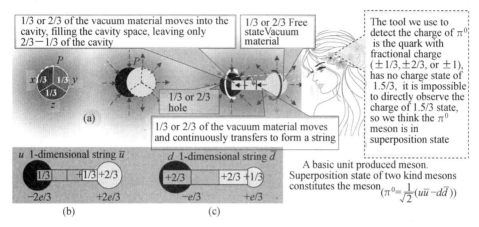

Fig. 4.1.1　The composition of the meson π^0

1. π^0 meson production

Mesons π^0 consists of holes and free vacuum matter, which will inevitably cause the vacuum matter connecting them to move towards the holes, thus forming a channel. Because strain has the smallest deformation on the spin axis, the channel formed by movement of vacuum materials products becomes the spin axis of two quarks, i.e. two quarks coaxial, and the spin direction of two quarks is opposite.

There are three states between a vacuum basic unit substance and a vacuum basic unit hole. In the first case, the relationship between a complete vacuum basic unit material and a vacuum basic unit hole is the relationship between a positive charge and a negative charge, which is known as the electromagnetic force.

Case 2 (Fig. 4.1.1(b)): 1/3 of the material of vacuum cell moves into the cavity of vacuum cell, filling the cavity space, leaving only 2/3 of the cavity, the vacuum material of vacuum cell moves and continuously transfers to form a string. They form $u - \bar{u}$ meson states. Vacuum cell is the basic unit of vacuum.

Case 3 (Fig. 4.1.1(c)): 2/3 of the material of vacuum cell moves into the cavity of vacuum cell, filling the cavity space, leaving only 1/3 of the cavity, the vacuum material of vacuum cell moves and continuously transfers to form a string. They form $d - \bar{d}$ meson states.

2. π^0 meson is a superposition state

Vacuum has 3-dimensional structure, 1/3 is the smallest unit, only exist $e/3$ and no $e/2$ state exists, and therefore it only can be considered in a superposition state that is π^0 meson.

$$\pi^0 = \frac{1}{\sqrt{2}}(u\bar{u} - d\bar{d}) \tag{4.1.1}$$

The meson state will eventually have to return to the complete 3-dimensional structure, i. e. the stable groundstate. Quarks and antiquarks will be annihilated when they encounter. The string pulled ud and $\bar{u}\bar{d}$ together, ud and $\bar{u}\bar{d}$ encounter to annihilation. Consider the conservation of momentum, the annihilation quantum wave to be separated to form two momentums opposite quantum wave. This is very easy to understand

$$\pi^0 \to \gamma + \gamma \tag{4.1.2}$$

3. π^0 meson field function

$$\phi_{\pi^0}(\xi) = \exp\left\{-i\begin{bmatrix} E(\xi) & \Delta\varepsilon_{01} & \Delta\varepsilon_{02} & \Delta\varepsilon_{01} \\ \Delta\varepsilon_{10} & \begin{matrix}\delta(i-1)(\vec{\partial}_u[\varepsilon]_u + \vec{\partial}_{\bar{d}}[\varepsilon]_d + \vec{\partial}_{u-\bar{u}}\varepsilon_{Lu-\bar{u}} \\ -\vec{\partial}_{\bar{u}}[\varepsilon]_{\bar{u}} - \vec{\partial}_{\bar{d}}[\varepsilon]_{\bar{d}} + \vec{\partial}_{d-\bar{d}}\vec{\varepsilon}_{Ld-\bar{d}}) + \Delta\varepsilon_1\end{matrix} & \Delta\varepsilon_{12} & \Delta\varepsilon_{13} \\ \Delta\varepsilon_{20} & \Delta\varepsilon_{21} & \begin{matrix}\delta(i-2)(\vec{\partial}_u[\varepsilon]_u + \vec{\partial}_{\bar{d}}[\varepsilon]_d + \vec{\partial}_{u-\bar{u}}\varepsilon_{Lu-\bar{u}} \\ -\vec{\partial}_{\bar{u}}[\varepsilon]_{\bar{u}} - \vec{\partial}_{\bar{d}}[\varepsilon]_{\bar{d}} + \vec{\partial}_{d-\bar{d}}\vec{\varepsilon}_{Ld-\bar{d}}) + \Delta\varepsilon_2\end{matrix} & \Delta\varepsilon_{23} \\ \Delta\varepsilon_{30} & \Delta\varepsilon_{31} & \Delta\varepsilon_{32} & \begin{matrix}\delta(i-3)(\vec{\partial}_u[\varepsilon]_u + \vec{\partial}_{\bar{d}}[\varepsilon]_d + \vec{\partial}_{u-\bar{u}}\varepsilon_{Lu-\bar{u}} \\ -\vec{\partial}_{\bar{u}}[\varepsilon]_{\bar{u}} - \vec{\partial}_{\bar{d}}[\varepsilon]_{\bar{d}} + \vec{\partial}_{d-\bar{d}}\vec{\varepsilon}_{Ld-\bar{d}}) + \Delta\varepsilon_3\end{matrix} \end{bmatrix} \times \check{\gamma}_5 \begin{pmatrix} \xi_0 & \xi_0 & \xi_0 & \xi_0 \\ \xi_1 & \{(\partial_u+\partial_{\bar{u}}+\partial_d+\partial_{\bar{d}})(\zeta)+\partial_{u-\bar{u}}+\partial_{d-\bar{d}}+1\}\xi_1 & \xi_1 & \xi_1 \\ \xi_2 & \xi_2 & \{(\partial_u+\partial_{\bar{u}}+\partial_d+\partial_{\bar{d}})(\zeta)+\partial_{u-\bar{u}}+\partial_{d-\bar{d}}+1\}\xi_2 & \xi_2 \\ \xi_3 & \xi_3 & \xi_3 & \{(\partial_u+\partial_{\bar{u}}+\partial_d+\partial_{\bar{d}})(\zeta)+\partial_{u-\bar{u}}+\partial_{d-\bar{d}}+1\}\xi_3 \end{pmatrix}\right\}$$

$$\tag{4.1.3}$$

Here, $\Delta\varepsilon_{13}$ is quark background strain in π meson internal space. For $\delta(i-1)$, $i = 1,2,3$ make it possible to have only one existence in 3-dimensional space, which guarantees that π meson have 1-dimensional structure. ∂_i is scale factor selection, satisfying (4.1.4); $[\varepsilon]_q$ is quark strain matrix; $\varepsilon_{Lq-\bar{q}}$ is strain (gluon transport) in quark chord channels. $\Delta\varepsilon_i$ is momentum of π meson. Embedding 1-dimensional space into 3-dimensional space has the above expression.

Chapter 4   Hadron Structure

$$[\varepsilon]_q = -Q_q \begin{bmatrix} \varepsilon_{00} & \varepsilon_{01} & \varepsilon_{02} & \varepsilon_{03} \\ \varepsilon_{10} & \varepsilon_{11} & \varepsilon_{12} & \varepsilon_{13} \\ \varepsilon_{20} & \varepsilon_{21} & \varepsilon_{22} & \varepsilon_{23} \\ \varepsilon_{30} & \varepsilon_{31} & \varepsilon_{32} & \varepsilon_{33} \end{bmatrix}, (\zeta) = \begin{pmatrix} \zeta_0 & \zeta_0 & \zeta_0 & \zeta_0 \\ \zeta_1 & \zeta_1 & \zeta_1 & \zeta_1 \\ \zeta_2 & \zeta_2 & \zeta_2 & \zeta_2 \\ \zeta_3 & \zeta_3 & \zeta_3 & \zeta_3 \end{pmatrix} \begin{pmatrix} \vec{\ni}_i \ni_i = \ni_i, \vec{\ni}_i \ni_j = 0, i \neq j; \\ i,j = u, \bar{u}, d, \bar{d}, u-\bar{u}, d-\bar{d} \end{pmatrix}$$

(4.1.4)

Here, Q_q is quark charge value, $Q_{u,c,t} = 2/3$, $Q_{\bar{u},\bar{c},\bar{t}} = -2/3$, $Q_{\bar{d},\bar{s},\bar{b}} = -1/3$. It should be noted that the positive charge is a negative strain and the negative charge is a normal strain. (ζ) is quark intrinsic space; $\ni_u, \ni_{\bar{u}}, \ni_d, \ni_{\bar{d}}$ are scaling factors, $\vec{\ni}$ is scaling factor base vectors, $|\ni| = 1$, $0 \leq \ni_{u-\bar{u}}, \ni_{d-\bar{d}} \leq 1$ are chord length scaling factors connecting $u, \bar{u}$ and $d, \bar{d}$. The scaling factors have vector characteristics. The vector scaling factor ensures that the described strain corresponds to space one by one, just like a key unlocking a lock.

$E(\xi)$ is the kinetic energy of the π^0 in static intrinsic space ξ, $E(\xi) = 0$.

$$\phi_{\bar{u}}(\xi) = \exp\left\{i\begin{bmatrix} E_{\bar{u}} & 0 & 0 & 0 \\ 0 & \delta(i-1)[\varepsilon]_{\bar{u}} & 0 & 0 \\ 0 & 0 & \delta(i-2)[\varepsilon]_{\bar{u}} & 0 \\ 0 & 0 & 0 & \delta(i-3)[\varepsilon]_{\bar{u}} \end{bmatrix} \cdot \right.$$
$$\left. \overset{\smile}{\gamma_5} \begin{pmatrix} \xi_0 & 0 & 0 & 0 \\ 0 & (\ni_{\bar{u}}(\zeta) + 1)\xi_1 & 0 & 0 \\ 0 & 0 & (\ni_{\bar{u}}(\zeta) + 1)\xi_2 & 0 \\ 0 & 0 & 0 & (\ni_{\bar{u}}(\zeta) + 1)\xi_3 \end{pmatrix} \right\} \quad (4.1.5)$$

The structure of π^0 can be simplified as a string connecting the positive and negative quarks.

$$\phi_{\pi^0}(\xi) = \exp\left\{(-i) \times \begin{bmatrix} E_{u+\bar{u}+L_u+d+\bar{d}+L_d} & 0 & 0 & 0 \\ 0 & \delta(i-1)([\varepsilon_q] + [\varepsilon_{Lq-\bar{q}}] + [\varepsilon_{\bar{q}}])\Delta\varepsilon_1 & 0 & 0 \\ 0 & 0 & \delta(i-2)([\varepsilon_q] + [\varepsilon_{Lq-\bar{q}}] + [\varepsilon_{\bar{q}}]) \Delta\varepsilon_2 & 0 \\ 0 & 0 & 0 & \delta(i-3)([\varepsilon_q] + [\varepsilon_{Lq-\bar{q}}] + [\varepsilon_{\bar{q}}]) + \Delta\varepsilon_3 \end{bmatrix} \times \right.$$

$$\left. \overset{\smile}{\gamma_5} \begin{pmatrix} \xi_0 & 0 & 0 & 0 \\ 0 & [\ni_q(\zeta) + \ni_{q-\bar{q}} + \ni_q(\zeta')]\xi_1 & 0 & 0 \\ 0 & 0 & [\ni_q(\zeta) + \ni_{q-\bar{q}} + \ni_q(\zeta')]\xi_2 & 0 \\ 0 & 0 & 0 & [\ni_q(\zeta) + \ni_{q-\bar{q}} + \ni_q(\zeta')]\xi_3 \end{pmatrix} \right\}$$

(4.1.6)

It is further simplified into a quality string embedded in 3-dimensional space.

$$\phi_{\pi^0}(\xi) = \exp\left\{-i\begin{bmatrix} E_{u+\bar{u}+L_u+d+\bar{d}+L_d} & 0 & 0 & 0 \\ 0 & \varepsilon_1 \mathfrak{z}(i-1) + \Delta\varepsilon_1 & 0 & 0 \\ 0 & 0 & \varepsilon_2 \mathfrak{z}(i-2) + \Delta\varepsilon_2 & 0 \\ 0 & 0 & 0 & \varepsilon_3 \mathfrak{z}(i-3) + \Delta\varepsilon_3 \end{bmatrix} \right. $$
$$\left. \underset{\gamma_5}{\sim} \begin{pmatrix} \xi_0 & 0 & 0 & 0 \\ 0 & \xi_1 & 0 & 0 \\ 0 & 0 & \xi_2 & 0 \\ 0 & 0 & 0 & \xi_3 \end{pmatrix} \right\} \qquad (4.1.7)$$

Since all the strains are confined to the intrinsic space, the total charge of the charged quark is zero, the spin superposition of the quark, antiquark is zero, and only the mass effect is left, so the π^0 meson has no charge and spin in the external observation space, only the mass. That is a real scalar field particle.

For Magnetic line coupling, quarks with opposite charges spin opposite, which leads to the coupling of magnetic lines, forming a closed magnetic line.

4.1.2 $\pi^{\pm}$ meson structure

π^-: 1/3 of the material of vacuum cell moves into the cavity of vacuum cell, filling the cavity space (Fig. 4.1.2(a)), leaving only 2/3 of the cavity, and the 2/3 of the material of vacuum cell moves into another cavity of vacuum cell, filling the cavity space, leaving only 1/3 of the cavity, the medium of vacuum cell moves and continuously transfers to form a string. They form d-$\bar{u}$ meson states, this is π^-.

$$\pi^- = d\bar{u} \qquad (4.1.8)$$

π^+: 2/3 of the basic unit material in free vacuum squeezes the vacuum medium around it. Formation of u quarks with 3-dimensional characteristics and $+2e/3$ charge. Another third of the vacuum medium constitutes an independent dimension, and push the material in front of the dimension forward. The material moving forward constitutes a chord channel. At the end of the string, 1/3 of the free vacuum basic unit material squeezes the surrounding material to form a fractional charge ($e/3$) with 3-dimensional characteristics, i.e. $\bar{d}$ quark (Fig. 4.1.2(b)). They form u-$\bar{d}$ meson states, i.e. π^+.

$$\pi^+ = u\bar{d} \qquad (4.1.9)$$

Spin wave exists in lepton fields, both positive and negative lepton field opposite spin is superposed that spin is zero. Meson 2-dimensional separation state is non-steady state.

π^0 meson is real scalar field (Fig. 4.1.3), K^0 meson is pseudo-scalar field particles, $\pi^{\pm}$ is complex scalar field. The real scalar field can roughly as "source" and "sink" superposition state. Here, "source" can be regarded as a positively charged quark, "sink" can be regarded as the negatively charged quarks.

Chapter 4 Hadron Structure

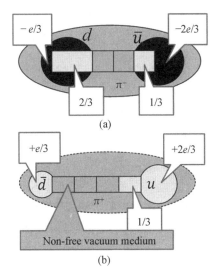

Fig. 4.1.2 $\pi^{\pm}$ meson

(1) The "source" structure: $\varphi_a(\xi) = \exp[\varepsilon\xi]$.

(2) The "sink" structure: $\bar{\varphi}_a(\xi) = \exp[-\varepsilon\xi]$.

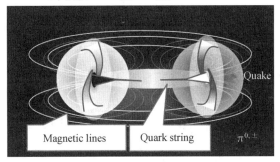

Fig. 4.1.3 π meson structure

"Source" and "sink" superposition get the intrinsic function of scalar particles.

The meson $\varphi_{\text{Meson}}(\xi)$ is the scalar field, the spin is zero. Consider the 4-dimensional space-time structure

$$\phi_{\pi^+}(\xi) = \exp\left\{-i\begin{bmatrix} E(\xi) & 0 & 0 & 0 \\ 0 & \delta(i-1)(\vec{\partial}_{\bar{d}}[\varepsilon]_{\bar{d}} + \vec{\partial}_{u-\bar{d}}\varepsilon_{Lu-\bar{d}} + \vec{\partial}_u[\varepsilon]_u) + \Delta\varepsilon & 0 & 0 \\ 0 & 0 & \delta(i-2)(\vec{\partial}_{\bar{d}}[\varepsilon]_{\bar{d}} + \vec{\partial}_{u-\bar{d}}\varepsilon_{Lu-\bar{d}} + \vec{\partial}_u[\varepsilon]_u) + \Delta\varepsilon & 0 \\ 0 & 0 & 0 & \delta(i-3)(\vec{\partial}_{\bar{d}}[\varepsilon]_{\bar{d}} + \vec{\partial}_{u-\bar{d}}\varepsilon_{Lu-\bar{d}} + \vec{\partial}_u[\varepsilon]_u) + \Delta\varepsilon \end{bmatrix}\right\} \times$$

$$\tilde{\gamma_5}\begin{pmatrix} \xi_0 & 0 & 0 & 0 \\ 0 & \{(\partial_u+\partial_{\bar d})(\zeta)+\partial_{u-\bar d}+1\}\xi_1 & 0 & 0 \\ 0 & 0 & \{(\partial_u+\partial_{\bar d})(\zeta)+\partial_{u-\bar d}+1\}\xi_2 & 0 \\ 0 & 0 & 0 & \{(\partial_u+\partial_{\bar d})(\zeta)+\partial_{u-\bar d}+1\}\xi_3 \end{pmatrix} \quad (4.1.10)$$

Embedding 1-dimensional space into 3-dimensional space has the above expression.

$$[\varepsilon] = \begin{bmatrix} \varepsilon_{00} & \varepsilon_{01} & \varepsilon_{02} & \varepsilon_{03} \\ \varepsilon_{10} & \varepsilon_{11} & \varepsilon_{12} & \varepsilon_{13} \\ \varepsilon_{20} & \varepsilon_{21} & \varepsilon_{22} & \varepsilon_{23} \\ \varepsilon_{30} & \varepsilon_{31} & \varepsilon_{32} & \varepsilon_{33} \end{bmatrix} ; \quad (\zeta) = \begin{pmatrix} \zeta_0 & \zeta_0 & \zeta_0 & \zeta_0 \\ \zeta_1 & \zeta_1 & \zeta_1 & \zeta_1 \\ \zeta_2 & \zeta_2 & \zeta_2 & \zeta_2 \\ \zeta_3 & \zeta_3 & \zeta_3 & \zeta_3 \end{pmatrix} \quad (\partial_i\partial_i=\partial_i,\partial_i\partial_j=0,i\ne j)$$

$$(i,j = u, \bar u, d, \bar d, u-\bar u, d-\bar d)$$

$$[\varepsilon_{\bar Q}](\zeta) = -[\varepsilon_Q](\zeta), \partial(\xi) = \begin{cases} 0 \\ 1, \xi_0 < \xi < \xi_0 + \Delta\xi \end{cases} \quad (4.1.11)$$

4.1.3 The neutral π^0 meson and the strong interaction of nucleon

Starting from the nature of the vacuum, we have established the model of hadrons. Hadron internal quark structure can be simplified, regardless of the intrinsic structure of hadrons, can be researched the strong interaction $\mathscr{L}_1$ by determine the symmetry.

1. Simplification of π^0 meson structure

$$\phi_{\pi^0}(\xi) = \exp\left\{-i\begin{bmatrix} E_{u+\bar u+L_u+d+\bar d+L_d} & 0 & 0 & 0 \\ 0 & \varepsilon_1\partial\delta(i-1) & 0 & 0 \\ 0 & 0 & \varepsilon_2\partial\delta(i-2) & 0 \\ 0 & 0 & 0 & \varepsilon_3\partial\delta(i-3) \end{bmatrix} \tilde{\gamma_5} \begin{pmatrix} \xi_0 & 0 & 0 & 0 \\ 0 & \xi_1 & 0 & 0 \\ 0 & 0 & \xi_2 & 0 \\ 0 & 0 & 0 & \xi_3 \end{pmatrix}\right\}$$

$$(4.1.12)$$

2. Lagrange function density of strong interaction

Hadrons are reduced to point particles. The interaction between π^0 meson and nucleon is strong interaction. That is, Hermite scalar field and spinor field are coupled to form a strong interaction. Lagrange function density of Hermite scalar field-spinor field system is

$$\mathscr{L} = \mathscr{L}_\phi + \mathscr{L}_\psi + \mathscr{L}_I \quad (4.1.13)$$

π^0 meson is a real scalar field.

$$\mathscr{L}_\phi = \frac{1}{2}\partial_\mu\phi\partial^\mu\phi + \frac{1}{2}m^2\phi\phi \quad (4.1.14)$$

Ignoring the intrinsic structure of nucleon, from the external observation space, nuclear is a spinor field.

$$\mathscr{L}_\psi = \bar\psi(X)(\partial_n)\psi(X) = -\bar\psi(i\gamma_\mu\partial_\mu + M)\psi \quad (4.1.15)$$

When there is interaction

$$\mathscr{L}_0 + \mathscr{L}_g = \bar\psi[i\gamma^\mu(\partial_\mu + iG\gamma_5\phi) - M]\psi \quad (4.1.16)$$

Chapter 4 Hadron Structure

Here, nuclear particles treated as a point, m and M are mass of π^0 mesons and nuclear. Determine $\mathscr{L}_I$ according to the following considerations: First, $\mathscr{L}_I$ must have the space inversion invariance (experiments showed that parity conservation in the strong interaction process) and positive Lorentz invariance. Second, the form of $\mathscr{L}_I$ is as much as possible simple. Because parity of π^0 meson is odd, therefore ϕ is a pseudo-scalar and spinor field functions constitute the simplest pseudo-scalar is $\bar{\psi}\gamma_5\psi$, γ_5 is the interaction vertex space, it is desirable as

$$\mathscr{L}_I = iG\,\bar{\psi}\gamma_5\psi\phi \qquad (4.1.17)$$

Wherein the imaginary number i is introduced in order to ensure $\mathscr{L}_I$ has Hermiticity. G is the coupling constant of strong interaction. It is similar as electromagnetic coupling. $\mathscr{L}_I$ is called Yukawa coupling in above formula.

3. Vertex space γ_5 of strong interaction

What is the strong interaction vertex space γ_5? There is an intuitive physics image of strong interaction in the vacuum theory which is interaction of 1-dimensional strings. For a 1-dimension string within the 3-dimensional space only has the properties of axial symmetry. Consider x, y, z 3-dimensions, if only x-dimension involved in interactions, y, z dimension independent with x-dimension in 3-dimensional space, the interaction is x-axis symmetry (Fig. 4.1.4). Thus the strong interaction vertex angle space γ_5 is x-axis symmetry space, it's derived from the string interaction characteristics of the strong interaction.

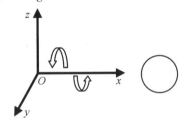

Fig. 4.1.4 Strong interaction vertex space has axial symmetry
(Only the x-dimension involved in the interaction, y, z and x-dimension independent then there is rotational symmetry that is axis symmetry)

4. Coupling constants

Where $\mathscr{L}_I$ in (4.1.17) compared with electromagnetic interaction Lagrangian $\mathscr{L}_I = ie\,\bar{\psi}\gamma_\mu\psi A_\mu$, their difference between the two is the coupling strength vary greatly.

$$\alpha = \frac{e^2}{4\pi} = \frac{1}{137}; \quad \frac{G^2}{4\pi} = 1 - 15 \qquad (4.1.18)$$

Now we already know that the electromagnetic coupling constant is caused by the fiberstructure of the electron and photon. Let's consider why the coupling constant of π meson with nucleon is 1—15, which is determined by quark structure. Every single hadron with 3-dimensions corresponding three chords, each chord is likely to become the channel of interaction. The greatest contribution of each string is 1 (Fig. 4.1.5), we can roughly estimate the strong coupling constant of π meson and nucleon in low-energy. Any two quarks with opposite charges can directly form chord channels. Any two quarks with opposite charges can directly form chord channels. In strong interactions, it is possible for each quark to form chord channels in 3-dimensions.

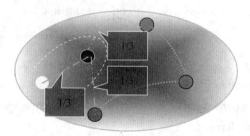

Fig. 4.1.5 π mesons with nucleons strong coupling constant

5. Maximum strong coupling constant

Considering nucleon there are 3 quarks, π meson has 2 quarks, a total of 5 quarks participate in the strong interaction. Each quark has 3-dimensions, and each dimension has a strong interaction channel (Fig. 4.1.6), so the total number of interaction channels is

$$5 \times 3 = 15 \qquad (4.1.19)$$

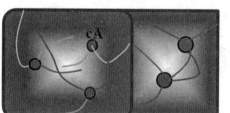

Fig. 4.1.6 Quark chord with chord interaction of the strongest and the weakest

(In low-energy interactions, a string of quark A will be interacting up to with three strings of quark B, so the maximum contribution rate of a string is: $\frac{1}{3} + \frac{1}{3} + \frac{1}{3} = 1$. High energy, the quark are close to each other, A quark is often in a state of minimal strings intersect, only one chord interactions with another chord of B quark, the minimum contribution rate of a string is 1/3)

The coupling constant definition are established in the single particle interaction probability, this definition does not apply to the interaction of multi-particle system. The strong interaction of π meson and nuclear is the multi-quark interaction, interaction coupling constant can be greater than 1, so the number of channels determines the strength of the interaction.

6. The minimum strong coupling constant

Minimal strong coupling constant is π meson into the nuclear intrinsic space (high-energy, short distance), the two quarks in mesons which a channel (Once in the intrinsic space, quark of π mesons seen the quark strings in nuclear interior space is thin and long, a string local chord interaction with another local chord, the minimum contribution of a chord is 1/3 (Fig. 4.1.5). Each quark has 3-dimensions channel, wherein 1-dimension is 1/3) and

Chapter 4 Hadron Structure

nuclear have 3 quarks, 1-dimensional channels simultaneously interact with 3 quarks, totally

$$\frac{1}{3} + \frac{1}{3} + \frac{1}{3} = 1 \quad (4.1.20)$$

The total number of interaction channels 1, the coupling strength is 1 (the dimension channel definition inconsistent that in inside and outside intrinsic space).

7. Strong interaction field equation

A strong interaction is the transfer of 4-momentum between quarks by strings. A nucleon can be roughly regarded as a spinor field ψ, π meson as a scalar field, and the coupling constant as G, then

$$\varphi'(X) = \phi(x)\psi(\xi) = \exp\{-i([\varepsilon^\psi]_\alpha X_\alpha + g[\varepsilon^\phi]_\alpha \cdot \gamma_5 X_\alpha^{1-\gamma_S})\} \quad (4.1.21)$$

$$\varphi'(X) = \exp(-i)\left\{ \left[\begin{pmatrix} \varepsilon_0 & 0 & 0 & 0 \\ 0 & \varepsilon_1 & 0 & 0 \\ 0 & 0 & \varepsilon_2 & 0 \\ 0 & 0 & 0 & \varepsilon_3 \end{pmatrix} \cdot \widetilde{\gamma}_\mu \begin{pmatrix} x^0 & 0 & 0 & 0 \\ 0 & -x^1 & 0 & 0 \\ 0 & 0 & -x^2 & 0 \\ 0 & 0 & 0 & -x^3 \end{pmatrix} \right. \right.$$

$$+ \underbrace{\begin{pmatrix} E_0\varepsilon_{0M} & 0 & 0 & 0 \\ 0 & [\varepsilon_1] & 0 & 0 \\ 0 & 0 & [\varepsilon_2] & 0 \\ 0 & 0 & 0 & [\varepsilon_3] \end{pmatrix}}_{\text{Nucleon intrinsic mass, } [p_0]\text{Nucleon intrinsic strain field}} \cdot$$

$$\widetilde{\gamma}_M \begin{pmatrix} x^0 & 0 & 0 & 0 \\ 0 & -(\partial_\psi[\xi]+1)x^1 & 0 & 0 \\ 0 & 0 & -(\partial_\psi[\xi]+1)x^2 & 0 \\ 0 & 0 & 0 & -(\partial_\psi[\xi]+1)x^3 \end{pmatrix} -$$

∂_ψ is the scale factor of nucleon in observation space; $[\xi]$ is nucleon interior space

$$g\begin{bmatrix} \varepsilon_0 & 0 & 0 & 0 \\ 0 & \Delta\varepsilon_1\delta(i-1) & 0 & 0 \\ 0 & 0 & \Delta\varepsilon_2\delta(i-2) & 0 \\ 0 & 0 & 0 & \Delta\varepsilon_3\delta(i-3) \end{bmatrix} \cdot$$

$$\left. \left. \widetilde{\gamma}_5 \begin{pmatrix} x_0 & 0 & 0 & 0 \\ 0 & -(\partial_\phi+1)(x^1)^{1-\gamma_S} & 0 & 0 \\ 0 & 0 & -(\partial_\phi+1)(x^2)^{1-\gamma_S} & 0 \\ 0 & 0 & 0 & (\partial_\phi+1)(x^3)^{1-\gamma_S} \end{pmatrix} \right] \right\}$$

$$(4.1.22)$$

$$\varphi'(X) = \exp\{(-i)(p_\mu \cdot \gamma_\mu X^\mu + M_0 \cdot \gamma_M \cdot X^M) - ig\gamma_5 \cdot \Delta p_\mu(X^\mu)^{1-\gamma_S}\} \quad (4.1.23)$$

After strong interaction, the momentum obtained by the nucleus is Δp_μ. The probability of obtaining momentum is g, and the total effect is $G = g \cdot \Delta p_\mu$. $1 - \gamma_5$ indicates that the dimension of interaction is 1-dimensional, and the anomalous dimension of interaction is γ_S.

$\mathscr{L}$ are substituting the Euler-Lagrane equation obtained ψ and ϕ the strong interaction (strong interaction vertex γ_5) field equations (in Pauli metric).

$$D_\mu \varphi'(X) = 0 \Rightarrow (\gamma_\mu \partial_\mu - iG\gamma_5 \phi + \gamma_M \partial_M)\psi = 0 \quad (4.1.24)$$

We can get the equation

$$(\gamma_\mu \partial_\mu + M)\psi = iG\gamma_5 \psi \phi \quad (4.1.25)$$

$$\overline{\psi}(\gamma_\mu \overleftarrow{\partial}_\mu - M) = iG\overline{\psi}\gamma_5 \phi \quad (4.1.26)$$

$$(\Box - m^2)\phi = -iG\overline{\psi}\gamma_5\psi \quad (4.1.27)$$

Field equation satisfied by this is the neutral π^0 meson and nucleon strong interaction. The canonical coordinates and canonical conjugate momentum of the field are

$$\phi, \pi = \dot{\phi}, \psi ; \pi_\psi = i\psi^+$$

Hamilton function:

$$H = \int_V d^3x \mathscr{H}(x)$$

$$\mathscr{H}(x) = \pi\dot{\phi} + \pi_\psi \dot{\psi} - \mathscr{L} = \underbrace{\frac{1}{2}[\pi^2 + (\nabla\phi)^2 + m^2\phi^2]}_{\mathscr{H}_\phi} + \underbrace{\overline{\psi}(\gamma \cdot \nabla + M)\psi}_{\mathscr{H}_\psi} + \underbrace{(-iG\overline{\psi}\gamma_5\psi\phi)}_{\mathscr{H}_I = -\mathscr{L}_I}$$

$$(4.1.28)$$

4.2 Proton structure

4.2.1 The center structure of electron

1. Stable structure of e^-

Now look to the formation of stimulate state of the basic unit of vacuum. When a basic unit is excited from A prolapsed to C form a free state of the basic unit which constituting the excited state, as shown in Fig. 4.2.1.

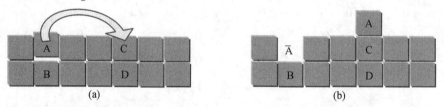

Fig. 4.2.1 Electronic produce

At point A appeared a basic unit hole, forming a lepton field e^-, at point C is more a basic unit, is extruded around the vacuum formation anti-lepton field e^+. The formation of the lepton intrinsic space is a half-space. Anti-lepton intrinsic space only have $-x$, $-y$, $-z$ three coordinates dimension, positive lepton field is opposite its intrinsic space only have x, y, z three directional dimension. The center of e⁻ is a hole, it is no dimension structure form a very stable structure, when have disturbed the center hole are likely to form quasi-steady state structures of e⁻.

2. Quasi-steady state structure of e⁻

What's the structure of center point of $e^{\pm}$? The formation of quasi-steady state structure is defined by a basic unit A into the cavity of lepton center and expansion caused by dimension division, the A quality closely with hole constituting the half-space dimension structure (Fig. 4.2.2 (b)), which is a quasi-steady state lepton structure, can also become steady e⁻ lepton structure. This change is no observable effect.

3. Quasi-steady state structure of e⁺

The free state basic unit of vacuum e⁺ is pressed into vacuum medium, the basic unit of vacuum e⁺ is strongly compressed by external vacuum medium and formed a small ball (Fig. 4.2.2(a)), there is no dimension separation, the strain of non-spherically symmetric prone to deformation to form a quasi-steady state antilepton.

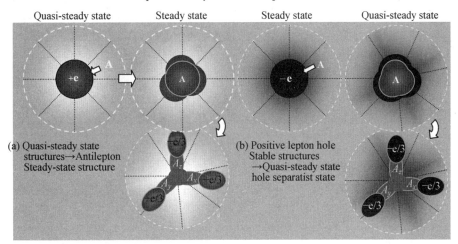

Fig. 4.2.2 The structure of electron center

4. Steady-state structure of e⁺

The superposition state of the e⁺ and A makes it to be compression deformation, A is extruded and appeared very weak dimension split, the structure shown in Fig. 4.2.2 (a). The free matter e⁺ in lepton intrinsic space only has half-space structure, and as same that A only has half-space structure. A is not a free state matter, thus uncharged. The free matter e⁺ is squeezed resulting in e⁺ is divided into three equal parts, each vacuum matter has $e^+/3$ charge, form a stable structure, corresponding to the e⁻ anti-lepton field with the dual structure is quasi-steady state structure(Fig. 4.2.2 (b)).

4.2.2 The formation of the proton

While the 3-dimensional separation of basic unit two fields are will result in the generation of the baryon, in other words, proton is formed by 3D lepton field splitting (Fig. 4.2.3), firstly consider the center of the lepton field, such as the central point of the positron, it is a basic unit in the free state(Can be thought as the anti-basic unit material). The overlapping of free state basic unit and non-free state basic unit result in part of the center point is compressed to deform, in other words, the free states with a non-free state basic unit of vacuum tightly coupled into one but there is a slight dimension separation, thus positron center point of the non-steady-state structure. This is the potential incentives of generation of protons.

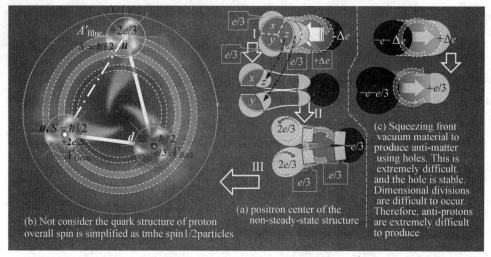

Fig. 4.2.3 Proton structure diagram

When an unsteady antielectron is squeezed (Fig. 4.2.3), the vacuum material squeezed at the front produces free matter $+\Delta e$, while the back produces small voids $-\Delta e$, and further squeezing produces dimensional splitting. The specific process is that $+e/3$ in the original $+e$ charge is combined with x in the vacuum medium to form a $+2e/3 u$ quark. The free matter of $+e/3$ combines with y of vacuum medium to form another $+2e/3$ u quark. When an unsteady antielectron is squeezed (Fig. 4.2.3), the vacuum material squeezed at the front produces free matter $+e$, while the back produces small voids, and further squeezing produces dimensional splitting. The specific process is that $+e/3$ in the original $+e$ charge is combined with x in the vacuum medium to form a $+2e/3u$ quark. The free matter of $+e/3$ combines with y of vacuum medium to form another $+2e/3u$ quark. $z(+e/3)$ of vacuum medium and free vacuum matter($\Delta e = +e/3$) produced by extrusion, they fill the back hole $-e$ to form $-e/3 d$ quarks. The vacuum medium of vacuum cell moves and continuously transfers to form two strings connecting d to u and d to u'. Quark field has $\hbar/2$ spin. When an unsteady antielectron $+e$ eventually forms a separated state (uud) consisting of fractional charges, it is a stable state, which is known as a proton.

$$p = uud \qquad (4.2.1)$$

Since the background field is 3-dimensional, quarks are never separated, and 2-dimensional or 1-dimensional structure fields are not allowed to exist independently, but only

Chapter 4 Hadron Structure

3-dimensional structures are allowed. 3D is the inherent nature of the basic unit of fibrosis vacuum(Fig. 4.2.3 (b)), the proton having a 3-dimensional stable structure. Because quark strings directly connect quarks with opposite charges, there is no direct chord connection between $u-u$. When the spin is coaxial, there exists a magnetic line channel. The strong interaction channel between u and u is $u-d-u$ channel. Quarks pass through the spin axis of d quark, therefore, the most natural structure of protons is to connect $u-d-u$ to form a linear body through a strong action string. When quarks split, the original position of the +e rotation axis is replaced by d quarks. This results in free-state vacuum matter $+2e/3$ at both ends of the external string, which squeezes the surrounding vacuum medium. The center is a hole $-e/3$, stretching the surrounding material. In this way, a convergent wave is formed, which is characterized by proton spin.

Particular attention should be paid to the fact that proton spin waves propagate in the inner space of protons where quarks exist, and quarks and quark strings do not rotate.

The theoretical value of Rand factor g_{pz} in proton spin magnetic moment is 2. On the other hand, the $u-d-u$ quark can be roughly regarded as a cylindrical spin wave with a total charge of $+E$ and its theoretical value of g_{px} is 2 (Fig. 4.2.4(a)). Because the quark spin angular momentum and proton spin angular momentum are orthogonal to each other, for the experimental observation:

$$J_z + J_x = \sqrt{2} J_{p+} \qquad (4.2.2)$$

It can be seen that the theoretical value of proton Rand factor is $g_p = 2$, then

$$\mathscr{g}_P = \sqrt{2} g_P \approx 2.828 \qquad (4.2.3a)$$

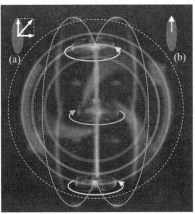

Fig. 4.2.4 The image of handron

It's very close to the experimental value of 2.792 8. If considered further, proton g_{pz} and neutron g_{Nz} are exactly equal, and $g_{Px} = 2$, then

$$\mathscr{g}_P = g_{Nz}\cos(\pi/4) + g_{px}\cos(\pi/4) = 2.797\,6 \qquad (4.2.3b)$$

What is the quark color? d quark is the center of proton. According to Pauli incompatibility principle, two u and u' quarks at both ends of string can not be at the same energy level.

The closer d quark is to u quark, the lower the energy level is, u and u' quarks are in different energy levels. Energy level is proportional to the string tension corresponding length of string, energy level is minimum, the chord length is should be minimized. Two quarks cannot exist at the same energy level and the energy levels are only three states: high, medium and small. The energy level state of three quarks corresponds to the R, G, B three colors state. Quark strings are the channels through which gluons are transmitted. Quarks change their color state (energy level state) by transferring gluons between them (Fig. 4.2.4).

Neutron and all baryon are formed by three quarks, they can be obtained by two basic unit which to exchange the quality of dimensions, and internal structure of baryons same as the internal structure of proton, the difference is the quark mass and charge. The essence of strong interaction is the quarks interacting by strings, quark own string in addition can also induce non excited basic unit of vacuum drawing forms strings to pass strong interaction.

There is strain in the proton intrinsic space, and the strain of the background field of the proton intrinsic space is expressed by $\varepsilon_{\mu\nu(B)}$. There is the quark strings in this context. Here, the strain field of quarks in proton is expressed by $[\varepsilon_{\mu\nu(Q)}]$, and it is a 4-dimensional strain field. $[\varepsilon_{\mu\nu(S)}]$ represents the strain field of the proton string. Since the string has a 1-dimensional structure, it forms a 2-dimensional spatiotemporal strain field. (ζ_q) is the quark space frame; (ξ_s) is the spatial framework of quark strings; (ξ) is the frame of proton intrinsic space. The specific expressions of quark's taste and color are as follows.

$$\varphi_{g-c}(\xi) = \exp\left\{-i\begin{bmatrix} \varepsilon_{00} & 0 & 0 & 0 \\ 0 & g_R\begin{bmatrix}\delta(i-1)\varepsilon_R \\ \delta(i-1)\bar{\varepsilon}_R\end{bmatrix} + g_G\begin{bmatrix}\delta(i-2)\varepsilon_G \\ \delta(i+2)\bar{\varepsilon}_G\end{bmatrix} + g_B\begin{bmatrix}\delta(i-3)\varepsilon_B \\ \delta(i+3)\bar{\varepsilon}_B\end{bmatrix} + \vec{\partial}_1[\varepsilon_{q1}] & 0 & 0 \\ 0 & 0 & g_G\begin{bmatrix}\delta(i-1)\varepsilon_G \\ \delta(i-1)\bar{\varepsilon}_G\end{bmatrix} + g_B\begin{bmatrix}\delta(i-2)\varepsilon_B \\ \delta(i-2)\bar{\varepsilon}_B\end{bmatrix} + g_R\begin{bmatrix}\delta(i-3)\varepsilon_R \\ \delta(i+3)\bar{\varepsilon}_R\end{bmatrix} + \vec{\partial}_2[\varepsilon_{q2}] & 0 \\ 0 & 0 & 0 & g_B\begin{bmatrix}\delta(i-1)\varepsilon_B \\ \delta(i+1)\bar{\varepsilon}_B\end{bmatrix} + g_R\begin{bmatrix}\delta(i-2)\varepsilon_R \\ \delta(i+2)\bar{\varepsilon}_R\end{bmatrix} + g_G\begin{bmatrix}\delta(i-3)\varepsilon_G \\ \delta(i+3)\bar{\varepsilon}_G\end{bmatrix} + \vec{\partial}_3[\varepsilon_{q3}] \end{bmatrix}\right.$$

The Strong interact with the gluon in the intrinsic space of nucleon, and the quark obtains momentum $p(R,G,B)$. The probability is g_i. The momentum absorbed by quarks does not destroy the intrinsic structure of quars. This momentum is represented by the color state of quarks has changed

$$\left.\begin{pmatrix} \xi_0 & 0 & 0 & 0 \\ 0 & \partial_1[\zeta_1] + \xi_1 & 0 & 0 \\ 0 & 0 & \partial_2[\zeta_2] + \xi_2 & 0 \\ 0 & 0 & 0 & \partial_3[\zeta_3] + \xi_3 \end{pmatrix}\right\}$$

Internal space of nucleon $[\zeta_i]$,

internal space of guark q_i,

∂_i is quark scale factor

Chapter 4 Hadron Structure

$$[\varepsilon_P](X) = \begin{bmatrix} \varepsilon_{0(Q)} + \varepsilon_{0(S)} + \varepsilon_{0(B)} & \varepsilon_{01(B)} & \varepsilon_{02(B)} & \varepsilon_{03(B)} \\ \varepsilon_{10(B)} & [\varepsilon_{ij(Q_1)}(\zeta_q)] + [\varepsilon_{kl(S_1)}(\xi_s)] + \varepsilon_{11(B)} & \varepsilon_{12(B)} & \varepsilon_{13(B)} \\ \varepsilon_{20(B)} & \varepsilon_{21(B)} & [\varepsilon_{ij(Q_2)}(\zeta_q)] + [\varepsilon_{kl(S_2)}(\xi_s)] + \varepsilon_{22(B)} & \varepsilon_{23(B)} \\ \varepsilon_{30(B)} & \varepsilon_{31(B)} & \varepsilon_{32(B)} & [\varepsilon_{ij(Q_3)}(\zeta_q)] + [\varepsilon_{kl(S_3)}(\xi_s)] + \varepsilon_{33(B)} \end{bmatrix} \times$$

$$\underbrace{\begin{pmatrix} \xi_0 & \xi_0 & \xi_0 & \xi_0 \\ \xi_1(\partial_u(\zeta_u) + \partial_{Ls} + \partial_{\bar{L}s} + 1)\xi_1 & \xi_1 & \xi_1 & \xi_1 \\ \xi_2 & \xi_2 & (\partial_{u'}(\zeta_{u'}) + \partial_{Ls} + \partial_{\bar{L}s} + 1)\xi_2 & \xi_2 \\ \xi_3 & \xi_3 & \xi_3 & (\partial_d(\zeta_d) + \partial_{Ls} + \partial_{\bar{L}s} + 1)\xi_3 \end{pmatrix}}_{\text{Space-time structure of proton}}$$

(4.2.4)

Here, $\varepsilon_{0(Q)}$ and $\varepsilon_{0(S)}$ represent the influence of quark chords on the temporal dimension of background space.

4.2.3 Proton spin

1. Spin formation

This is equivalent to at O missed a basic unit that similar to appeared a hole effect which same as being formed lepton field effect which produce spin waves it's proton spin. Spin waves in the proton center far away from quark, and thus the proton spin and quark spin almost no direct link, independent from each other. This proton field has no hole, and thus will not form a field of fibrous structure, so there is no charge. Here R_0 is intrinsic space radius of proton. Antiprotons and protons situation exactly the same, we can get the antiprotons structure by the lepton field quasi-steady state structure in Fig. 4.2.2, antiprotons structure is the dual structure of proton.

$$\bar{p} = \bar{u}\bar{u}\bar{d} \quad (4.2.5)$$

Quark with fractional charge is naturally with a fiber structure, spin-wave propagation in a fiber of electric field generates quark magnetic moment. Thus the baryon magnetic moment is caused by the quark spin and baryon spin. Because of this, we can through quark models to estimate the magnetic moments of the baryon.

Recent experiments have shown that the spin of quark (including all valence quarks and sea quark) accounting for 25% of the total spin of proton, which component − 0.10 ± 0.03 is provided by quarks, where the minus sign refers to the s quark contribution opposite to the proton spin direction. The above values are directly measured results, and the two high energy accelerators (CERN and SLAC) results converge. The composition of the proton spin $\hbar/2$ is

$$\omega(\Sigma/2) + \omega(L_q) + \omega(L_G) + \omega(\Delta G) = \frac{\hbar}{2} \qquad (4.2.6)$$

Here, $\omega(\Sigma/2)$ is the percentage of all quark, $\omega(L_q)$ is the contribution of the quark orbital angular momentum, $\omega(L_q)$ is the contribution of the orbital angular momentum of the gluon, $\omega(\Delta G)$ is the contribution of gluon spin. Considering the internal structure of hadron, the rotational strain can be written as

$$[\Omega](\xi) = \begin{bmatrix} 0 & 0 & 0 \\ 0 & 0 & [\Omega_{(B)} + \partial_q \omega_{(q)} + \partial_L \omega_{(L)} + \partial_G \omega_{(G)}]_{12} \\ 0 & [\Omega_{(B)} + \partial_q \omega_{(q)} + \partial_L \omega_{(L)} + \partial_G \omega_{(G)}]_{21} & 0 \\ 0 & [\Omega_{(B)} + \partial_q \omega_{(q)} + \partial_L \omega_{(L)} + \partial_G \omega_{(G)}]_{31} & [\Omega_{(B)} + \partial_q \omega_{(q)} + \partial_L \omega_{(L)} + \partial_G \omega_{(G)}]_{32} \end{bmatrix}$$

$$\underbrace{}_{\text{Rotational strain, spin of proton } \hbar/2}$$

$$\begin{matrix} 0 \\ [\Omega_{(B)} + \partial_q \omega_{(q)} + \partial_L \omega_{(L)} + \partial_G \omega_{(G)}]_{13} \\ [\Omega_{(B)} + \partial_q \omega_{(q)} + \partial_L \omega_{(L)} + \partial_G \omega_{(G)}]_{23} \\ 0 \end{matrix} \cdot$$

$$\begin{pmatrix} 0 & 0 & 0 & 0 \\ 0 & 0 & (1 + \partial_q(\partial) + \partial_L + \partial_G)\xi_1 & (1 + \partial_q(\partial) + \partial_L + \partial_G)\xi_1 \\ 0 & (1 + \partial_q(\partial) + \partial_L + \partial_G)\xi_2 & 0 & (1 + \partial_q(\partial) + \partial_L + \partial_G)\xi_2 \\ 0 & (1 + \partial_q(\partial) + \partial_L + \partial_G)\xi_3 & (1 + \partial_q(\partial) + \partial_L + \partial_G)\xi_3 & 0 \end{pmatrix}$$

$$(4.2.7)$$

This is a system of strain. From the view of large-scale structure, the rotational strain of quark, gluon and chord is limited to a very small range by the scale factor, which can be ignored. Therefore, the large-scale low-energy experiment can only detect the spin of hadron background space, which is $\hbar/2$.

$$S_{\mu\nu} = \omega_{\mu\nu} = \left(\frac{\partial A_\mu(\xi)}{\partial \xi^\nu} - \frac{\partial A_\nu(\xi)}{\partial \xi^\mu} \right) \quad (\mu, \nu = 0, 1, 2, 3) \qquad (4.2.8)$$

Gluon is string from the point of view of the field, and there is no the spin of real sense. Same as the proton spin, the spin waves of hadrons manifested as a whole, the center point of quark and the center point of spin waves is relatively independent, so that the quark spin contribution is small to the overall spin of Hadrons. Taking into account the spin of quark q_z in a dimensional completely overlap with hadrons spin wave on z-axis, the direction of q_x and q_y spin perpendicular to z that without the contribution, therefore q_z maximum contribution to a hadron's spin cannot exceed 1/3, this simple estimation is obvious. Quark, string and the background field relative to the spin wave is stationary, their contribution to the hadrons spin should be zero.

2. Spin crisis

Hadron spin waves (Fig. 4.2.3) exist in the inner space of hadrons. In addition, there

Chapter 4 Hadron Structure

are quarks, quark strings and gluons in the inner space of hadrons, and there is no direct correlation between spin waves and parts of hadrons. We can make an inaccurate analogy. Just as there are wind power piles and trees in the area where tornadoes exist, there is no direct relationship between these wind power piles and trees, but because tornadoes cannot be measured directly, we can judge the strength of tornadoes according to the speed of blades of power pile and the bending degree of trees. Therefore, we believe that the rotation of the blades of the power pile and the bending of the trees constitute a tornado, which is wrong. However, the strength of tornado can be determined by measuring the speed of blades of power pile and the bending degree of trees. Our correct understanding of hadron spin should be that the effect of hadron spin wave on quark and quark string and gluon can be determined experimentally by measuring the part of the hadron, and can be described theoretically by strain.

Strain caused by hadron spin $\varepsilon_{ii} = \underbrace{\varepsilon'_{ii}}_{\text{Background space principal strain of hadrons}} + \underbrace{\varepsilon_{qi}}_{\text{Quark strain}} + \underbrace{L_i}_{\text{String strain}}$. Hadron spins

strain $\omega_{ij} = \frac{1}{2}\left(\frac{\partial u_i}{\partial \xi_j} - \frac{\partial u_j}{\partial \xi_i}\right)$. Spin of the proton configuration $\frac{\hbar}{2} = \frac{\Sigma}{2} + L_q + L_G + \Delta G$, Σ is the percentage which provided by all quarks, L_q is contribution of quark orbital angular momentum, corresponding to $\frac{\Sigma}{2} + L_q = \frac{1}{2}\left(\frac{\partial u_{qi}}{\partial \xi_j} - \frac{\partial u_{qj}}{\partial \xi_i}\right)$. L_G is the contributions of gluon orbital angular momentum. ΔG is gluon spin contribution. Corresponding to

$$L_G + \Delta G = \frac{1}{2}\left(\frac{\partial u'_{ii} + u_{Li}}{\partial \xi_j} - \frac{\partial u'_{jj} + u_{Lj}}{\partial \xi_i}\right) \tag{4.2.9}$$

This explains the so-called "hadron spin crisis".

4.2.4 Non-existence of antimatter

Why is there no antimatter in the universe? This is due to the difference in structure between electronic and anti-electronic center. Antiproton and proton is a dual structure that is particularly worth mentioning the different is a proton with negative electron, it is a free basic unit hole field which stretched around the non-free-state field (can also be understood as a non-free-state field collapses to the hole of center point) form a charged e^- lepton field. The center point is a hole without matter naturally without dimension split, and thus the central point is extremely stable, to constitute very stable lepton field. And lepton e^- is positive material which formed by the free-state hole (Fig. 4.2.5). e^+ is too different from the e^-, the structure of e^+ center point has two kinds of forms (Fig. 4.2.2). This two form centre point exist in the free-state which are

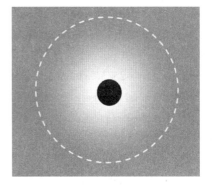

Fig. 4.2.5 The center of e^-
(The hole of a basic unit)

squeezed by ground state, if they are disturbed, it is prone to produce dimension division into a proton. Due to the form of e⁻ central point makes this hole lepton field exceptionally stable, it is difficult to form a dimension separation and the formation of anti-baryons, so this is why we have seen a great deal of material in the form of e⁻ and proton rarely seen antimatter. From here may be obvious to know that the universe has no antimatter world exists.

So far, The spacecraft has been launched to do close observation, the results of all the observations failed to find the existence of antimatter. Estimated on galactic scales, the antimatter content will not exceed one millionth. From the analysis of the existing theory, it should satisfy the symmetry, baryon number and antibaryon number should be exactly equal evolution, from the Big Bang to the present, the proportion of the baryon number of the universe (including positive and negative) and the number of photons should be 10^{-19}. But in fact, the observed ratio is 10^{-10}. Reasonable explanation for this phenomenon is only that the baryon number and antibaryon number is not equal at original. This phenomenon is called symmetry breaking, the extent of breaking degree 10^{-10}, the breaking already existed when positive and negative particles generated in the vacuum.

4.2.5 Neutron structure

When a basic unit vacuum medium is excited, the vacuum material squeezed at the front produces free matter $+\Delta e$, while the back produces small voids $-\Delta e$, and further squeezing produces dimensional splitting. Free state two basic unit vacuum material and two basic unit voids are generated. The two basic units vacuum matter of free state split in dimension under strong impact. 2/3 of the two basic units vacuum matter of free state fill two basic unit holes, forming two d quarks with $-e/3$ charge, and the remaining $+2e/3$ of the basic unit vacuum matter of free state form u quarks.

$$n = udd \qquad (4.2.10)$$

Neutron spin: The structure of neutron is a linear structure of $d-u-d$ series. Negative charged quarks d at both ends and positive charged quarks u at the middle. $d-u-d$ quark spin coaxial, magnetic field coupling, equivalent to a particle with negative charge spinning. $J_{nz} = J_{n-}$, on the other hand, in the 2-dimensional plane S of $d-u-d$ series, there exists spin wave in the 2-dimensional plane S. The two ends of the string are holes $-e/3$, and the center is free vacuum matter $+2e/3$. The spin wave is from inside to outside, just opposite to the proton, but because the total charge is 0, the magnetic moment produced by the neutron spin wave is 0, $J_{nx} = 0$. The theoretical value of g of the negatively charged 1/2 spin particle is -2, so the neutron is

$$g_n = -2 \qquad (4.2.11)$$

It can be roughly judged. Considering the existence of $d-u-d$ tandem in 3-dimensional space, the spin axis of the cylindrical spin wave will change. Therefore, it is reasonable to regard neutron as a small sphere model with negative external charge and positive internal charge.

Chapter 4　Hadron Structure

The antineutron situation is the opposite. Since it is difficult for two stretched holes to split in dimension, antineutrons are difficult to produce.

$$\bar{n} = \bar{u}\bar{d}\bar{d} \tag{4.2.12}$$

The neutron itself is produced in a non-free vacuum medium, so there is no charge. The function $\delta(i - n)$ $(i, n = 1, 2, 3)$ is introduced to separate and distinguish the described quarks and quark strings.

$$[\varepsilon]_n(X) = \underbrace{\begin{bmatrix} \varepsilon_{0R(u)}\delta(i-1) + \varepsilon_{0G(u)} \cdot \\ [\delta(i-2) + \varepsilon_{0B(u)}\delta(i-3)] & 0 & 0 & 0 \\ & \partial_u[\varepsilon_{ij(u)}] + \begin{bmatrix}\delta(i-1)\varepsilon_R \\ \delta(i+1)\bar{\varepsilon}_R\end{bmatrix} + \\ 0 & \begin{bmatrix}\delta(i-2)\varepsilon_G \\ \delta(i+2)\bar{\varepsilon}_G\end{bmatrix} + \begin{bmatrix}\delta(i-3)\varepsilon_B \\ \delta(i+3)\bar{\varepsilon}_B\end{bmatrix} & 0 & 0 \\ 0 & 0 & 0 & 0 \\ 0 & 0 & 0 & 0 \end{bmatrix}}_{u \text{ quark strain field}} \cdot$$

$$\underbrace{\begin{pmatrix} X_{0(u)} & 0 & 0 & 0 \\ 0 & \partial_u[(\zeta_u) + 1]\xi_1 & 0 & 0 \\ 0 & 0 & 0 & 0 \\ 0 & 0 & 0 & 0 \end{pmatrix}}_{\text{Space-time structure of } u \text{ quark strain field}} +$$

$$\underbrace{\begin{bmatrix} \varepsilon_{0R(u)}\delta(i-1) + \varepsilon_{0G(u)} \cdot \\ \delta(i-2) + \varepsilon_{0B(u)}\delta(i-3) & 0 & 0 & 0 \\ 0 & 0 & 0 & 0 \\ & \partial_d[\varepsilon_{ij(d)}] + \begin{bmatrix}\delta(i-1)\varepsilon_R \\ \delta(i+1)\bar{\varepsilon}_R\end{bmatrix} + \\ 0 & \begin{bmatrix}\delta(i-2)\varepsilon_G \\ \delta(i+2)\bar{\varepsilon}_G\end{bmatrix} + \begin{bmatrix}\delta(i-3)\varepsilon_B \\ \delta(i+3)\bar{\varepsilon}_B\end{bmatrix} & 0 & 0 \\ 0 & 0 & 0 & 0 \end{bmatrix}}_{d \text{ quark strain field}} \cdot$$

$$\underbrace{\begin{pmatrix} X_{0(d)} & 0 & 0 & 0 \\ 0 & 0 & 0 & 0 \\ 0 & 0 & \partial_d[(\zeta_d) + 1]\xi_2 & 0 \\ 0 & 0 & 0 & 0 \end{pmatrix}}_{\text{Space-time structure of } d \text{ quark strain field}} +$$

$$\underbrace{\begin{bmatrix} \varepsilon_{0R(u)}\delta(i-1)+\varepsilon_{0G(u)}\cdot \\ \delta(i-2)+\varepsilon_{0B(u)}\delta(i-3) & 0 & 0 & 0 \\ 0 & 0 & 0 & 0 \\ 0 & 0 & 0 & 0 \\ 0 & 0 & 0 & \partial_{d'}[\varepsilon_{ij(d')}]+\begin{bmatrix}\delta(i-1)\varepsilon_R\\ \delta(i+1)\bar\varepsilon_R\end{bmatrix}+ \\ & & & \begin{bmatrix}\delta(i-2)\varepsilon_G\\ \delta(i+2)\bar\varepsilon_G\end{bmatrix}+\begin{bmatrix}\delta(i-3)\varepsilon_B\\ \delta(i+3)\bar\varepsilon_B\end{bmatrix} \end{bmatrix}}_{s\text{ quark strain field}}\cdot$$

$$\underbrace{\begin{pmatrix} X_{0(d)} & 0 & 0 & 0 \\ 0 & 0 & 0 & 0 \\ 0 & 0 & 0 & 0 \\ 0 & 0 & 0 & \partial_d[(\zeta_{d'})+1]\xi_3 \end{pmatrix}}_{\text{Space-time structure of }s\text{ quark strain field}}+$$

$$\underbrace{\begin{bmatrix} [\varepsilon_{kl(S_0)}(\xi_s)] & 0 & 0 & 0 \\ 0 & [\varepsilon_{kl(S_1)}(\xi_s)] & 0 & 0 \\ 0 & 0 & [\varepsilon_{kl(S_2)}(\xi_s)] & 0 \\ 0 & 0 & 0 & [\varepsilon_{kl(S_3)}(\xi_s)] \end{bmatrix}}_{\text{Neutron three string strain field}}\cdot$$

$$\underbrace{\begin{pmatrix} (\partial_{Ls}+1)\xi_0 & 0 & 0 & 0 \\ 0 & (\partial_{Ls}+1)\xi_1 & 0 & 0 \\ 0 & 0 & (\partial_{Ls}+1)\xi_2 & 0 \\ 0 & 0 & 0 & (\partial_{Ls}+1)\xi_3 \end{pmatrix}}_{\text{Space-time structure of neutron string}}+$$

$$\underbrace{\begin{bmatrix} [\varepsilon_{kl(S_0)}(\xi_s)] & 0 & 0 & 0 \\ 0 & [\bar\varepsilon_{kl(S_1)}(\xi_s)] & 0 & 0 \\ 0 & 0 & [\bar\varepsilon_{kl(S_2)}(\xi_s)] & 0 \\ 0 & 0 & 0 & [\bar\varepsilon_{kl(S_3)}(\xi_s)] \end{bmatrix}}_{\text{Three inverse string strain fields of neutron}}\cdot$$

$$\underbrace{\begin{pmatrix} (\partial_{\bar Ls}+1)\xi_0 & 0 & 0 & 0 \\ 0 & (\partial_{\bar Ls}+1)\xi_1 & 0 & 0 \\ 0 & 0 & (\partial_{\bar Ls}+1)\xi_2 & 0 \\ 0 & 0 & 0 & (\partial_{\bar Ls}+1)\xi_3 \end{pmatrix}}_{\text{Space-time structure of neutron string}}+$$

Chapter 4 Hadron Structure

$$\underbrace{\begin{bmatrix} \varepsilon_{0(B)} & 0 & 0 & 0 \\ 0 & \varepsilon_{22(B)} & 0 & 0 \\ 0 & 0 & \varepsilon_{22(B)} & 0 \\ 0 & 0 & 0 & \varepsilon_{33(B)} \end{bmatrix}}_{\substack{\text{Neutron intrinsic background} \\ \text{principal strain}}} \underbrace{\begin{pmatrix} \xi_0 & 0 & 0 & 0 \\ 0 & \xi_1 & 0 & 0 \\ 0 & 0 & \xi_2 & 0 \\ 0 & 0 & 0 & \xi_3 \end{pmatrix}}_{\substack{\text{Space-time structure of} \\ \text{neutron intrinsic background}}} +$$

$$\frac{1}{2}\underbrace{\begin{bmatrix} 0 & 0 & 0 & 0 \\ 0 & 0 & \gamma_{12(B)} & \gamma_{13(B)} \\ 0 & \gamma_{21(B)} & 0 & \gamma_{23(B)} \\ 0 & \gamma_{31(B)} & \gamma_{32(B)} & 0 \end{bmatrix}}_{\text{Spatial-temporal shear strain in nucleon}} \begin{pmatrix} 0 & \xi_0 & \xi_0 & \xi_0 \\ \xi_1 & 0 & \xi_1 & \xi_1 \\ \xi_2 & \xi_2 & 0 & \xi_2 \\ \xi_3 & \xi_3 & \xi_3 & 0 \end{pmatrix} +$$

$$\underbrace{\begin{bmatrix} 0 & \varepsilon_{01(B)} & \varepsilon_{02(B)} & \varepsilon_{03(B)} \\ \varepsilon_{10(B)} & 0 & 0 & 0 \\ \varepsilon_{20(B)} & 0 & 0 & 0 \\ \varepsilon_{30(B)} & 0 & 0 & 0 \end{bmatrix}}_{\text{Nucleon intrinsic 4-momentum flow}} \begin{pmatrix} 0 & \xi_0 & \xi_0 & \xi_0 \\ \xi_1 & 0 & 0 & 0 \\ \xi_2 & 0 & 0 & 0 \\ \xi_3 & 0 & 0 & 0 \end{pmatrix} +$$

$$\frac{1}{2}\underbrace{\begin{bmatrix} 0 & 0 & 0 & 0 \\ 0 & 0 & \omega_{12(B)} & \omega_{13(B)} \\ 0 & \omega_{21(B)} & 0 & \omega_{23(B)} \\ 0 & \omega_{31(B)} & \omega_{32(B)} & 0 \end{bmatrix}}_{\text{Nuclear spin}} \begin{pmatrix} 0 & 0 & 0 & 0 \\ 0 & 0 & \xi_1 & \xi_1 \\ 0 & \xi_2 & 0 & \xi_2 \\ 0 & \xi_3 & \xi_3 & 0 \end{pmatrix} \quad (4.2.13)$$

The field function of neutron is

$$\varphi_n(X) = \exp\{-i[\varepsilon]_n(X)_n\} \quad (4.2.14)$$

4.2.6 The representation of hadrons

Similar to the color representation, three different flavors of quarks u, d, s can be considered as three base of $SU(3)$ groups.

$$\underbrace{\begin{bmatrix} \varepsilon_{0(u)} & 0 & 0 & 0 \\ 0 & [\varepsilon_{ij(u)}(\zeta_u)] & 0 & 0 \\ 0 & 0 & 0 & 0 \\ 0 & 0 & 0 & 0 \end{bmatrix}}_{u \text{ quark strain field}} \Rightarrow u = \begin{pmatrix} 1 \\ 0 \\ 0 \end{pmatrix} \quad \underbrace{\begin{bmatrix} \varepsilon_{0(d)} & 0 & 0 & 0 \\ 0 & 0 & 0 & 0 \\ 0 & 0 & [\varepsilon_{ij(d)}(\zeta_d)] & 0 \\ 0 & 0 & 0 & 0 \end{bmatrix}}_{d \text{ quark strain field}} \Rightarrow d = \begin{pmatrix} 0 \\ 1 \\ 0 \end{pmatrix}$$

$$\underbrace{\begin{bmatrix} \varepsilon_{0(s)} & 0 & 0 & 0 \\ 0 & 0 & 0 & 0 \\ 0 & 0 & 0 & 0 \\ 0 & 0 & 0 & [\varepsilon_{ij(s)}(\zeta_s)] \end{bmatrix}}_{s \text{ Quark strain field}} \Rightarrow s = \begin{pmatrix} 0 \\ 0 \\ 1 \end{pmatrix}$$

Without considering chord structure and background space:

$$\phi_n = \exp(-i) \left\{ \begin{bmatrix} \varepsilon_{0(n)} & 0 & 0 & 0 \\ 0 & \vec{\partial}_u[\varepsilon_{ij(u)}] & 0 & 0 \\ 0 & 0 & \vec{\partial}_d[\varepsilon_{ij(d)}] & 0 \\ 0 & 0 & 0 & \vec{\partial}_d'[\varepsilon_{ij(d)}] \end{bmatrix} \right.$$

$$\left. \underbrace{\begin{bmatrix} X_{0(d)} & 0 & 0 & 0 \\ 0 & \{\vec{\partial}_u[X_{ij(u)}(\zeta_u)]+1\}\xi_1 & 0 & 0 \\ 0 & 0 & \{\vec{\partial}_d[X_{ij(d)}(\zeta_d)]+1\}\xi_2 & 0 \\ 0 & 0 & 0 & \{\vec{\partial}_d[X_{ij(d)}(\zeta_d)]+1\}\xi_3 \end{bmatrix}}_{\text{Quark intrinsic strain field in neutron internal space}} \right\} \Rightarrow$$

$$\phi_n = \begin{pmatrix} u \\ d \\ d \end{pmatrix} \tag{4.2.15}$$

Regardless of the internal structure of Hadron, the hadron simplified strain field is

$$\varphi_n(X) = \exp\{-i[(E_0 + \Delta E)x_0 + (\Delta\varepsilon_{1(n)} \cdot i)x_1 + (\Delta\varepsilon_{2(n)} \cdot i)x_2 + (\Delta\varepsilon_{3(n)} \cdot i)x_3]\}$$
(4.2.16)

From the strain structure of protons, we can see that the internal strain of hadrons is extremely complex. The construction of neutrons with three quarks alone has lost a lot of information and only retained the most important information of three quarks.

For quarks, the background space of quarks is the interior space of neutrons, and there is a strong space-time distortion. Therefore, it is impossible to determine the mass of quarks by experiment alone and accurately. From the space relationship, $x[\xi(\zeta_Q)]$. Quark's internal space (ζ_Q) is embedded into neutron's internal space $[\xi]$. The change of internal space will directly lead to the change of (ζ_Q) space. Neutron's internal space $[\xi]$ is embedded into observation space. The gravitational field of observation space $\{x\}$ will affect neutron's internal space $[\xi]$. In the process of experiment, we can only determine the neutron's characteristics, but also want to confirm it by experiment. Quarks must be involved in the interior space of the neutron, and the disturbance of the interior space of the neutron will inevitably affect the strain characteristics of the quark, which leads to the inability to accurately measure the quality of the quark in the observation space.

$$[\varepsilon]_q(\zeta) = \underbrace{\begin{bmatrix} \varepsilon_{00} & \varepsilon_{01} & \varepsilon_{02} & \varepsilon_{03} \\ \varepsilon_{10} & \varepsilon_{11} & \varepsilon_{12} & \varepsilon_{13} \\ \varepsilon_{20} & \varepsilon_{21} & \varepsilon_{22} & \varepsilon_{23} \\ \varepsilon_{30} & \varepsilon_{31} & \varepsilon_{32} & \varepsilon_{33} \end{bmatrix}}_{=\varepsilon_{qi}} \begin{pmatrix} \zeta_0 & \zeta_0 & \zeta_0 & \zeta_0 \\ \zeta_1 & \zeta_1 & \zeta_1 & \zeta_1 \\ \zeta_2 & \zeta_2 & \zeta_2 & \zeta_2 \\ \zeta_3 & \zeta_3 & \zeta_3 & \zeta_3 \end{pmatrix}$$

(In quark interior space, quark intrinsic strain can be seen)

$$[\varepsilon]_Q(\xi) = \underbrace{\begin{bmatrix} \varepsilon_{00} & 0 & 0 & 0 \\ 0 & \varepsilon_{q1} & 0 & 0 \\ 0 & 0 & \varepsilon_{q2} & 0 \\ 0 & 0 & 0 & \varepsilon_{q3} \end{bmatrix}}_{=\varepsilon_H} \cdot \begin{pmatrix} \partial_0 & 0 & 0 & 0 \\ 0 & \partial_{q1}(\partial) + \xi_1 & 0 & 0 \\ 0 & 0 & \partial_{q2}(\partial) + \xi_2 & 0 \\ 0 & 0 & 0 & \partial_{q3}(\partial) + \xi_3 \end{pmatrix}$$

(In the inner space of hadron, we can see three quarks)

$$[\varepsilon]_{\text{Hadrons}}(x) = \underbrace{\begin{bmatrix} \varepsilon_{00} & 0 & 0 & 0 \\ 0 & \varepsilon_{H1} & 0 & 0 \\ 0 & 0 & \varepsilon_{H2} & 0 \\ 0 & 0 & 0 & \varepsilon_{H3} \end{bmatrix}}_{\varepsilon_{Hi} = \varepsilon_{H0} + \Delta\varepsilon_{Hi} = M_{H0} + \Delta p_i} \cdot \begin{pmatrix} \zeta_0 & 0 & 0 & 0 \\ 0 & \delta(i-1)\partial_H(\xi) + x_1 & 0 & 0 \\ 0 & 0 & \delta(i-1)\partial_H(\xi) + x_2 & 0 \\ 0 & 0 & 0 & \delta(i-1)\partial_H(\xi) + x_3 \end{pmatrix}$$

(Only the momentum distribution of point hadrons can be seen in the observation space)

(4.2.17)

For the observer, to measure the specific parameters of quarks, we must not only enter the inner space of the nucleon, but also the inner space of the quark in order to measure the parameters of the quark. On the other hand, the measured parameters must pass through the inner space of the quark to the inner space of the nucleon, and finally enter the observation space to be verified. In this process, the probe itself will disturb the two interior spaces and distort the interior space-time, so it is theoretically impossible to measure the real physical properties of quarks.

$[\zeta_{(u)}]$, $[\zeta_{(d)}]$ and $[\zeta_{(s)}]$ are the three independent internal spaces of u, d and s, respectively. These three subspaces move rapidly inside the neutron intrinsic space. The neutron intrinsic space is regarded as a local stationary frame of reference, which satisfies the Lorentz transformation. There is a conservation of quark four momentums in the proton.

$$\begin{cases} \varepsilon_{0(u)}^2 - \varepsilon_{1(u)}^2 - \varepsilon_{2(u)}^2 - \varepsilon_{3(u)}^2 = 0 \\ \varepsilon_{0(d)}^2 - \varepsilon_{1(d)}^2 - \varepsilon_{2(d)}^2 - \varepsilon_{3(d)}^2 = 0 \\ \varepsilon_{0(s)}^2 - \varepsilon_{1(s)}^2 - \varepsilon_{2(s)}^2 - \varepsilon_{3(s)}^2 = 0 \\ \varepsilon_0 - \varepsilon_{0(u)} - \varepsilon_{0(d)} - \varepsilon_{0(s)} = 0 \end{cases} \quad (4.2.18)$$

$$E_0 + \Delta E = E\gamma \quad (4.2.19)$$

1. Proton spin coupling

The experimental results consistently show that the quark hardly bears any spin of the proton. This surprising and puzzling result is called the proton spin crisis. From the previous discussion, we know that in the 2-dimensional plane S of u–d–u series, there exists spin wave in the

2-dimensional plane S. There is no direct correlation between proton spin wave and quark spin. In the non-polarized state, the spin axis of the u–d–u quark tandem is orthogonal to the spin axis of the proton spin wave, so the quark spin to the proton spin is 0. But in the polarized state, the spin axis of u–d–u quark tandem body and the rotation axis of proton spin wave are less than $\pi/2$. There is a partial coupling between quark spin J_{sq} and proton spin J_p.

$$J_{\text{coupling}} = J_{sq} \cdot J_p \qquad (4.2.20)$$

J_{coupling} can be understood as the share of quark spin J in proton spin.

2. Proton electromagnetic radius

The experimental value of proton charge radius $\sqrt{\langle r^2 \rangle} = 0.8751(61)$ fm is detected by using electrons as probe particles. If the electron is replaced by a μ particle, the mass effect should be considered. The measured value is $0.84087(39)$ fm. The radius detected at the edge of proton charge radius A and B is different from that detected at the center position C (Fig. 4.2.6). The Fermion radius $R'_p = kh_f/2m_{0p}c$ detected at position A and B (the electromagnetic radius coefficients of k-atypical point particles, protons consisting of three quarks). Both proton and e^-, μ^-, τ^- have central convergent waves. In the central region, the detected particle mass superposes the proton mass, and the detected fermion mass $R''_p = kh_f/2(m_{0p}c+m_{0\mu}c)$, the detected particles are from $a \to c \to b$, each location occupies one third of the space. Using simple weighted average

$$R_{p\mu} = (2R'_{p\mu} + R''_{p\mu})/3 \qquad (4.2.21)$$

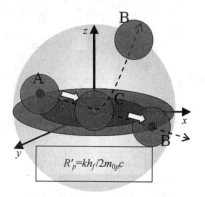

Fig. 4.2.6 Detection of proton electromagnetic radius

We can easily calculate $R_{p\mu}/R_{pe} \approx 0.9644$ and the experimental value is $0.8409/0.8751 = 0.9698$.

Prediction: If τ^- replace electrons, $R_{p\tau}/R_{pe} = 0.61345$ can be predicted by the same estimation method, the detected proton electromagnetic radius will be further reduced to about 0.53683 fm.

3. Meson octet

That would be based on $SU(3)$ group theory can be obtain the two quark system to compose meson eight state (it's same as quark color eight state in mathematics, just different symbols only), the eight state and 10 states of hadrons composed by three quarks (Fig. 4.2.7), the specific composition shown in Table 4.2.1 and Table 4.2.2.

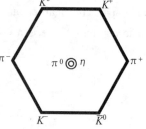

Fig. 4.2.7 Meson octet

Chapter 4 Hadron Structure

Table 4.2.1 $J^P = 0^-$ Quarks composition meson

Quarks	$J^P = 0^-$
$d\bar{u}, (u\bar{u} - d\bar{d})/\sqrt{2}, u\bar{d}$	π^-, π^+, π^0
$d\bar{s}, u\bar{s}$	K^0, K^+
$s\bar{u}, s\bar{d}$	$K^-, \bar{K}^0$
$(u\bar{u} + d\bar{d} + 2s\bar{s})/\sqrt{6}$	$\eta(549)$
$(u\bar{u} + d\bar{d})/\sqrt{2}$	π^0

The flavor and color of the hadron comparison is not strictly symmetrical. The reason is as follows.

Flavor corresponds to the quality state, which has three generations, so there are three generations of quarks. Each generation of quarks corresponds to a background field. For example, the low-strain background field there will be only a low mass state of quark-lepton field, high strain background field there will be a high mass state of quark lepton field. Thus u, d quarks and t, b quarks cannot occur simultaneously in one baryon constitute a hadron, because intrinsic space background field is the same in a same baryon.

Table 4.2.2   Baryon's quark ingredients

Quarks	$J^P = 1/2^+$	$J^P = 3/2^+$	MeV	
udd, uud	n, p		938	Octet state
$(du-ud)/\sqrt{2}$	Λ		1 116	
$dds, (du+ud)s\sqrt{2}, uus$	$\Sigma^-, \Sigma^0, \Sigma^+$		1 189	
$(uu+dd)/\sqrt{2}$	Ξ^-, Ξ^0		1 315	
ddd, udd, uud, uuu		$\Delta^-, \Delta^0, \Delta^+, \Delta^{++}$	1 232	Ten state
dds, uds, uus		$\Sigma^{*-}, \Sigma^{*0}, \Sigma^{*+}$	1 385	
dss, uss		Ξ^{*-}, Ξ^{*0}	1 530	
sss		Ω^-	1 672	

For the same generation of quarks, if the mass difference is too big, require existence the differences background field. Due to the limitations of the background field lead to differences generation quark is not easy to combination, and the same generation quark easier to combination (of course to satisfy the integer charge conditions). Formation of High generation of leptons field need high strain background field. If consider the high strain background field as a local reference system, then this reference system has three states: moving very close to the speed of light; in high strain static zone, such as inside a black hole; or at the area in the moment of the strong impact of high-energy particles. These harsh conditions make the high generation quarks are not easy to occur. On the other hand, the low-energy-generation quark constitute a hadron. Because the strain are directly determines the propagation properties of lepton field, in other words, the lepton field generation cannot be precisely determined its mass. Taking these two reasons, the flavor $SU(3)$ symmetry is inaccurate and cannot constitute a larger group.

4.3 Quark

Quarks and leptons have the same internal structure (Fig. 4.3.1). The difference between them is that the background vacuum is different. Background vacuum of the electron is the observation space, and the background space of the quark is the internal space of the hadron.

$$[\varepsilon]_{\text{Quark}} = \underbrace{\left[\frac{\partial u_\mu}{\partial \zeta_\nu}\right]}_{\text{Quark internal strain}} + \underbrace{\left[\frac{\partial A_\mu}{\partial \xi_\nu}\right]}_{\substack{\text{The fiber field in the} \\ \text{hadronic interior space,} \\ \text{electric flux line strain}}} + \underbrace{[\varepsilon_L]}_{\text{Quark chord}} + \underbrace{\left[\frac{\partial U_\mu}{\partial \xi}\right]}_{\substack{\text{Quark background} \\ \text{space strain}}}$$

$$= \underbrace{\varepsilon_{\mu\mu}(\zeta)}_{\text{4-momentum}} + \underbrace{\frac{1}{2}\gamma_{\mu\nu}(\zeta)}_{\text{Weakly acting strain}} + \underbrace{\frac{1}{2}\omega_{\mu\nu}(\zeta)}_{\text{Spin}} + \underbrace{\Delta E_{\mu\mu}(\xi)}_{\substack{\text{electromagnetic field of} \\ \text{scalar and vector potential}}} +$$

$$\underbrace{\frac{1}{2}\Omega_{\mu\nu}(\xi)}_{\text{Electromagnetic field}} + \underbrace{\varepsilon_L}_{\text{Quark chord}} + \underbrace{\left[\frac{\partial U_\mu}{\partial \xi}\right]}_{\substack{\text{Quark background} \\ \text{space strain}}} \quad (4.3.1)$$

$$\begin{cases}
\varepsilon_r = -H\dfrac{\pi}{\Phi}\sin\dfrac{\pi r}{\Phi} + \delta(\theta + \pi)\dfrac{h_f}{3LS}; \gamma_{r\varphi} = -\dfrac{H}{r}\cos\dfrac{\varphi}{2} \\[2mm]
\varepsilon_\phi = \dfrac{H}{r}\cos\dfrac{\pi r}{\Phi} - \dfrac{1}{2}\dfrac{H}{r}\sin\dfrac{\varphi}{2}; \gamma_{\varphi\theta} = -\dfrac{H}{r}\cot\varphi\sin\theta \\[2mm]
\varepsilon_\theta = \dfrac{H}{r}\cos\dfrac{\pi r}{\Phi} + \dfrac{H}{r}\left(\cot\varphi\cos\dfrac{\varphi}{2} + \dfrac{\cos\theta}{\sin\varphi}\right); \gamma_{\theta r} = -\dfrac{H}{r}\sin\theta
\end{cases} \quad (4.3.2)$$

Rotational strain: $\omega_\varphi = \varepsilon_\varphi(\theta + d\theta) - \varepsilon_\varphi(\theta)$

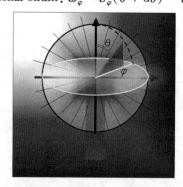

Fig. 4.3.1 Single quark structure

It should be noted that quark string strain ε_L, quark string itself does not change the vacuum structure, strain is 0, but when quark string transfers gluon, the string is observed, so the string is caused by the gluon transfer between quarks. The repeated subscript represents the spherically symmetric strain. The strain structure of quarks is exactly the same as that of electron, the only difference is that quark has a string. The string is the spin axis of quark itself. String is the translational motion of vacuum medium ±n/3 basic unit. The average shape variable of a single string is $h_f/3LS$, that is, the strain of string, and S is the cross sectional area of string.

4.3.1 The intrinsic energy level structure and mass of e, μ, τ

There are three quality states, respectively m_e, m_μ, m_τ. We know that the lepton electric charges are same witch having nothing to do with intrinsic 4-momentum.

1. Intrinsic structure of e

More or missing a basic unit of vacuum in flat vacuum will formation of the dual lepton, considered a hole that generating converged wave to the center of the lepton. Just able to form stable aggregation wave is the lowest energy state. For the same background field that is a constant (Fig. 4.3.2). Thus the electron rest mass is a constant (low-energy case).

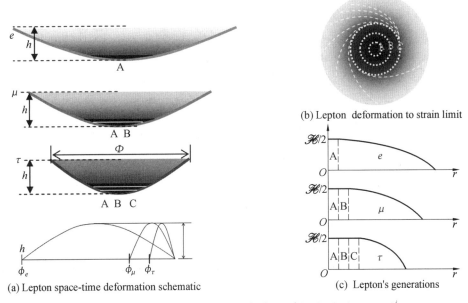

Fig. 4.3.2 Schematic diagram of relationships intrinsic space ϕ^i

The vacuum basic unit A (Fig. 4.3.2) adjacent the center of the lepton, it along the direction r in the deformation limit state, while B and C are basic unit, but did not meet the deformation limit state. This is the lowest energy state of the lepton, so it is the most stable structural state, as long as the center of the lepton vacuum is missing or more of 1 basic unit of the vacuum, the electrons always exist. Electronic energy level is lowest then the volume is largest.

2. Intrinsic structure of μ

For the lepton μ center gains and losses is also a basic unit of the vacuum, but the intrinsic momentum than electrons its deformation rate along r to the center of the lepton is greater.

From the point of view of the vacuum is electrons background field further strain form μ lepton. In addition to the adjacent center of the lepton vacuum A achieve deformation limit, second stage vacuum B also reached the field deformation limit as shown in Fig. 4.3.2(a), lepton center deformation limit zone becomes larger, along r 1-dimensional space-time shape variables such as Fig. 4.3.2(b) τ shows, $R_\mu < R_e, P_\mu > P_e$, P is four momentum. A and B two layers of the basic unit of vacuum of lepton central region has reached strain limit then intrinsic space field strain increased, transmissibility of the lepton μ is greatly reduced that performance the rest mass has increased $m_\mu c = h_f / \Phi_\mu$.

3. Intrinsic structure of τ

τ is similar to μ, the third layer C of the basic unit of the vacuum also reached the deformation limit (Fig. 4.3.2(b)) along r the 1-dimensional space-time deformation as shown in Fig. 4.3.2(c) shown, we can see A, B, C have reached the deformation limit. Here pay special attention to the size of A, B, C compared to R is much smaller, $l_A = l_B = l_C \ll R$. For ease of understanding, the description in Fig. 4.3.2 is exaggerated.

In Fig. 4.3.2(a), B, we can see that the A, B, C are vacuum material, when the A, B, C three layer to achieve deformation limit does not affect the formation of lepton center internal area electric flux line. Determine the total amount of the internal area electric flux line is ultimately determined by the amount of gain or loss of vacuum material in the center of the lepton.

Above, we can understand the fundamental questions such as why e, μ, τ with same charge, their spins are same and quality are difference. Electronics is the lowest level which is the most stabilization stable.

For the quality we know $m_0 c = h_f / \phi$, the quality difference is caused by the diameter of the intrinsic space (Fig. 4.3.2). Now we look at why the intrinsic space diameter is so different. Based on the deformation effect in above model image, we can estimate the diameter of lepton intrinsic space.

We know that the fiber field strain $u(r)$ in the intrinsic space will result in change of the intrinsic space. The space-time compression coefficient is γ in special relativity, the intrinsic space-time compression coefficient of lepton is γ. The function relationship is $\phi' = \gamma \phi$.

$$m'_0 c = h_f / \phi' \qquad (4.3.3)$$

We first consider the wave of convergence. Due to the vacuum is the elastic medium, $u(r)$ is spatial elastic deformation.

The change of static mass comes from the existence of additional vacuum strain. The simplest consideration is radial strain. When the mass of an object approaches the speed of light, the mass increases along the propagation direction and the space is compressed. For leptons and quarks, the mass is compressed along 3-dimensions, and the compression coefficient is γ.

Chapter 4 Hadron Structure

$$[p'_{\mu}](\xi) = \begin{bmatrix} E_0 & 0 & 0 & 0 \\ 0 & m_0c/3 & 0 & 0 \\ 0 & 0 & m_0c/3 & 0 \\ 0 & 0 & 0 & m_0c/3 \end{bmatrix} \begin{pmatrix} \xi_0\gamma & 0 & 0 & 0 \\ 0 & \xi_1\gamma & 0 & 0 \\ 0 & 0 & \xi_2\gamma & 0 \\ 0 & 0 & 0 & \xi_3\gamma \end{pmatrix} \quad (4.3.4)$$

We get the expression of the quality $R_e = 2.4263102367(11) \times 10^{-12}$ m, $\phi_0 = 2R_e$. Three lepton mass relation is estimated according to the e, μ, τ. Vacuum strain that provides an image for our understanding relation of the three generations of lepton mass. Referring to 1-dimensional mass change $M = M_0\gamma^{-1} = M_0 \left(\sqrt{1 - \dfrac{u^2}{c^2}}\right)^{-1}$, we can see that $m_{0i}c = \dfrac{h_f}{\phi_0 \cdot \gamma_i^3}$. The essence of velocity is vacuum 1-dimensional strain. In the process of considering strain superposition, the elasticity is gradually lost $H_0 \cos\theta_i$. The deformation of lepton ΔH_i is superposed with the deformation of previous generation lepton ΔH_{i-1}, then

$$\gamma_i = \sqrt{1 - \frac{\sum_{i=1}^{N} \Delta H_i^2}{H^2}} \quad (4.3.5)$$

Here, $\Delta H_1 = H[\sin(\theta_0 + \theta_1) - \sin\theta_0]$; $\Delta H_0 = H\sin\theta_0$ (background deformation required to exist leptons); $\Delta H_2 = H[\sin(\theta_2 + \theta_0) - \sin(\theta_1 + \theta_0)]$; $\Delta H_3 = H[\sin(\theta_3 + \theta_2 + \theta_0) - \sin(\theta_2 + \theta_0)]$.

$$m_i c = \frac{h_f}{\phi_0 \cdot (\gamma_i)^3} \quad (i = 1,2,3; \text{ Correspondence } e, \mu, \tau) \quad (4.3.6)$$

ΔH_0 is background space vacuum deformation; ΔH_i is vacuum deformation between two generations of leptons, $\Delta H_i = H_0 \cos\theta_i$, θ_i - amplitude; H is vacuum deformation of first generation lepton.

The above formula is only a conjecture of the relationship between lepton mass generations.

4. Unresolved issues

The unsolved problem in the theory of super large unified field based on vacuum strain is the calculation of lepton and quark mass and their relationship. This problem has been plaguing new theories. In terms of theoretical framework, although we know that mass comes from vacuum strain and vacuum rupture, the calculation of the relationship between quality and quality generation should be solved, the new theory cannot accurately describe the vacuum breakdown. On the other hand, the dimension of space is not what we understand as 3-dimensional space in the internal space of basic particles, this may lead to the failure of the description of strain. In the future, we need to explore an important area.

4.3.2 Quark flavor

Similar to leptons, quarks there are corresponding to the three generation. A is field quality center of quark (Fig. 4. 3. 3), its existence in fractional state rather than the basic unit integer (or hole). This non-integer states field quality is not independently exist in 3-dimensional space. And same as lepton generations principle, the spot quality A deformation limit is reached, the field in lowest energy state, the quark field charge is fractional charge, as the first generation of the quark, space variable is $\Delta \xi^1$ (Table 4. 3. 1).

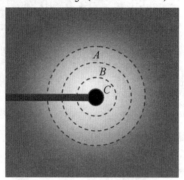

Fig. 4. 3. 3 Generations of quarks

Table 4. 3. 1 **Generation of quarks and lepton**

Generation	Lepton			Quark					
	Symbol	Charge /Q	Mass /MeV	Symbol	Charge /Q	Mass	Symbol	Charge /Q	Mass
I	e	-1	0.511	u	2/3	1.5—4 MeV	d	-1/3	4—8 MeV
II	μ	-1	105.66	c	2/3	1.15—1.35 GeV	s	-1/3	80—130 MeV
III	τ	-1	1,784.1	t	2/3	169—174 GeV	b	-1/3	4.1—4.4 GeV

Base on the first generation of deformation quark, the field of the outer layer B is the vacuum deformation again, reaches the maximum (A, B have reached the deformation limit), the spatial variable is $\Delta \xi^2$, at this time the free state field A has not changed, so the charge value of second-generation quark unchanged. Obviously, the 4-momentum becomes large, rest mass was significantly greater than the first generation of quarks.

The third generation of the quark deformation is based on the second generation of quarks, and A, B, C have reached the deformation limit, charge value is unchanged, but the rest mass is greater than the second generation of quark, spatial deformation is $\Delta \xi^3$.

$$\gamma_i = \cosh \Phi; \quad \Phi = \text{arccosh } \gamma_i \qquad (4.3.7)$$

The following calculation is made according to the quality.

$\Phi_e = \text{arccosh}(0.511/0.511) = 1$;

$\Phi_\mu = \text{arccosh}(105.66/0.511) = \text{arccosh}(206.77) = 6.02$; $\Phi_\mu - \Phi_e = 5.02$

$\Phi_\tau = \text{arccosh}(1,784.1/0.511) = \text{arccosh}(3,491.4) = 8.85$; $\Phi_\tau - \Phi_\mu = 2.8$

Spatial elasticity $(5.02+2.8)/2 = 3.91$;

For quarks, the same charge can be compared with the lepton, so we modified the charge.

$\Phi_d = \text{arccosh}(6/0.511/3) = \text{arccosh}(35.3) = 4.25$; $\Phi_s - \Phi_e = 3.25$

$\Phi_s = \text{arccosh}(105/0.511/3) = \text{arccosh}(616.6) = 7.12$; $\Phi_s - \Phi_d = 2.87$

$\Phi_b = \text{arccosh}(4,150/0.511/3) = \text{arccosh}(24,364) = 10.79$; $\Phi_b - \Phi_s = 3.67$

Spatial elasticity $(2.87+3.67)/2 = 3.27$;

$\Phi_u = \text{arccosh}(2.75/0.511/1.5) = \text{arccosh}(8.07) = 2.79$; $\Phi_u - \Phi_e = 1.79$

$\Phi_c = \text{arccosh}(1,250/0.511/1.5) = \text{arccosh}(3,369) = 8.9$; $\Phi_c - \Phi_u = 6.1$

$\Phi_t = \text{arccosh}(171,500/0.511/1.5) = \text{arccosh}(503,424.6) = 13.82$; $\Phi_t - \Phi_c = 4.9$

Spatial elasticity $(6.1+4.9)/2 = 5.5$.

We can see that the mass properties of quarks and leptons have the same intrinsic properties as that of space compression (Fig. 4.3.4).

$$m_i c = \frac{h_f}{\phi_0 \cdot |Q_i^{-1}| \gamma_i} ; \quad |Q_i| = \frac{1}{3}, \frac{2}{3} \text{ (Charge values of quarks)}$$

$$i = e, \mu, \tau, d, s, b, u, c, t \tag{4.3.8}$$

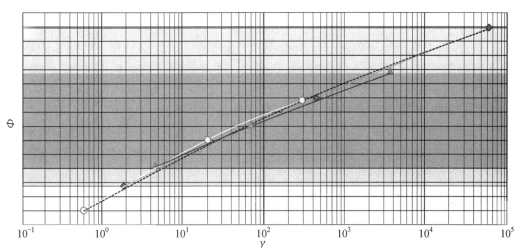

Fig. 4.3.4 Schematic diagram of relationship between space shrinkage and mass

In the above discussion, we must note that there is a background space. The strain in this background vacuum is a tensile strain, which is the same as the negative charge. In this tension strain background space, quarks with negative charges further stretch the background tension space and further harden the space. This makes it more difficult to form three generations of quarks, whose mass is distributed in a smaller range than a flat vacuum.

$$\Phi_{\text{lepton}} = 7.85 ; \quad \Phi_{q(-1/3)} = 6.54.$$
$$\Phi_{\text{lepton}} > \Phi_{q(-1/3)} \tag{4.3.9a}$$

On the other hand, if quarks with positive charges exist in the background space with tensile properties, because the positive charge is a compressive strain, which is contrary to the strain characteristics of the background vacuum, the tensile strain of the background vacuum can be relaxed, showing that quarks have greater elasticity, so the mass of three generations of quarks is distributed in a larger space than that of a flat vacuum. $\Phi_{q(2/3)} = 11.03$.

$$\Phi_{\text{lepton}} < \Phi_{q(2/3)} \qquad (4.3.9b)$$

Because the neutron internal convergence wave is outward, and the proton convergence wave is inward, this will lead to the same quark mass detected by the same probe particle is different. The range of specific measurements is shown in the table below (Table 4.3.2).

Table 4.3.2 Prediction of u and d quark mass relations in protons and neutrons

Mass	Proton P (convergent wave inward, in tensile state)		Neutron n (convergent wave outward, in squeezed state)	
Hadron structure	$u(+2e/3)-d(-e/3)-u(+2e/3)$		$d(-e/3)-u(+2e/3)-d(-e/3)$	
Detection quality of e^-: m'	$m'_d > \bar{m}_d (6 \text{ MeV})$	$m'_u < \bar{m}_u (2.75 \text{ MeV})$	$m'_d < \bar{m}d$	$m'_u > \bar{m}_u$
Detection quality of μ^-: m''	$m''_d > m'_d > \bar{m}_d$	$m''_u < m'_u < \bar{m}_u$	$m''_d < m'_d < \bar{m}_d$	$m''_u > m'_u > \bar{m}_u$

$\bar{m}_d, \bar{m}_u$ are the u, d quark mass takes the median value of the error range.

4.3.3 Quark color

The quark has three "colors" in QCD. They are defined as red, green and blue. Color as a degree of freedom no deeper explanation in QCD, the physical nature of the color degrees of freedom. Now look at a baryon structure (Fig. 4.2.4), in this structure, the essence of color is the level state of level. The relative level only in three states: namely large, medium and small. The energy levels of three quarks cannot be the same at the same time, there are always differences at the same time. So its physical essence is the three dimensions of energy-level state corresponding to R, G, B three colors of quark.

$$\varphi_q(\xi) = \exp\left\{-i\left(\begin{bmatrix} \varepsilon^e_{00} & 0 & 0 & 0 \\ 0 & \varepsilon^e_{11} & 0 & 0 \\ 0 & 0 & \varepsilon^e_{22} & 0 \\ 0 & 0 & 0 & \varepsilon^e_{33} \end{bmatrix} \gamma_\mu \begin{pmatrix} x_0 & 0 & 0 & 0 \\ 0 & \xi_1 & 0 & 0 \\ 0 & 0 & \xi_2 & 0 \\ 0 & 0 & 0 & \xi_3 \end{pmatrix} + \right.\right.$$

$$\left.\begin{bmatrix} \Delta\varepsilon_{0,0} & 0 & 0 & 0 \\ 0 & 0 & 0 & 0 \\ 0 & 0 & 0 & 0 \\ 0 & 0 & 0 & g_B \begin{bmatrix} \delta(i-1)\varepsilon_B \\ \delta(i+1)\bar{\varepsilon}_B \end{bmatrix} + g_R \begin{bmatrix} \delta(i-2)\varepsilon_R \\ \delta(i+2)\bar{\varepsilon}_R \end{bmatrix} + g_G \begin{bmatrix} \delta(i-3)\varepsilon_G \\ \delta(i+3)\bar{\varepsilon}_G \end{bmatrix} \end{bmatrix} \gamma_5 \cdot \right.$$

$$\left.\left.\begin{pmatrix} x_0 & 0 & 0 & 0 \\ 0 & 0 & 0 & 0 \\ 0 & 0 & 0 & 0 \\ 0 & 0 & 0 & \xi_3 \end{pmatrix}\right)\right\} \qquad (4.3.10)$$

For the hadron system, strongly interacting gluons cannot escape from strong subsystems, therefore, the energy conservation of the system is the conservation of color of the system. Color can be expressed as

$$\phi_N(\xi) = \exp\left\{-i\begin{bmatrix}\begin{bmatrix}\varepsilon_\infty & 0 & 0 & 0\\ 0 & g_R\begin{bmatrix}\delta(i-1)\varepsilon_R\\ \delta(i-1)\bar\varepsilon_R\end{bmatrix}+g_G\begin{bmatrix}\delta(i-2)\varepsilon_G\\ \delta(i+2)\bar\varepsilon_G\end{bmatrix}+g_B\begin{bmatrix}\delta(i-3)\varepsilon_B\\ \delta(i+3)\bar\varepsilon_B\end{bmatrix}+\mathfrak{z}_1[\varepsilon_{q1}] & 0 & 0\\ 0 & 0 & g_G\begin{bmatrix}\delta(i-1)\varepsilon_G\\ \delta(i-1)\bar\varepsilon_G\end{bmatrix}+g_B\begin{bmatrix}\delta(i-2)\varepsilon_B\\ \delta(i-2)\bar\varepsilon_B\end{bmatrix}+g_R\begin{bmatrix}\delta(i-3)\varepsilon_R\\ \delta(i+3)\bar\varepsilon_R\end{bmatrix}+\mathfrak{z}_2[\varepsilon_{q2}] & 0\\ 0 & 0 & 0 & g_B\begin{bmatrix}\delta(i-1)\varepsilon_B\\ \delta(i+1)\bar\varepsilon_B\end{bmatrix}+g_R\begin{bmatrix}\delta(i-2)\varepsilon_R\\ \delta(i+2)\bar\varepsilon_R\end{bmatrix}+g_G\begin{bmatrix}\delta(i-3)\varepsilon_G\\ \delta(i+3)\bar\varepsilon_G\end{bmatrix}+\mathfrak{z}_3[\varepsilon_{q3}]\end{bmatrix}\right.$$

The strong interact with the gluon in the intrinsic space of nucleon, and the quark obtains momentum $p(R,G,B)$. The probability is g_i. The momentum absorbed by quarks does not destroy the intrinsic structure of quarks. This momentum is represented by the color state of quarks has changed

$$\left.\begin{pmatrix}\xi_0 & 0 & 0 & 0\\ 0 & \mathfrak{z}_1[\zeta_1]+\xi_1 & 0 & 0\\ 0 & 0 & \mathfrak{z}_2[\zeta_2]+\xi_1 & 0\\ 0 & 0 & 0 & \mathfrak{z}_3[\zeta_3]+\xi_1\end{pmatrix}\right\} \quad (4.3.11)$$

Internal space of nucleon $[\zeta_i]$,

internal space of quark q_i,

$\mathfrak{z}_i$ is quark scale factor

Here, the inner space $[\xi]$ of the nucleon, in which there are three quark systems (q_1, q_2, q_3), each quark can have only one momentum state. The three momentum state are $p_R < p_G < p_B$. $\vec{\mathfrak{z}}_i$ is basis vectors, $\mathfrak{z}_i$ is scale factor $|\mathfrak{z}_i| \ll 1$, $|\vec{\mathfrak{z}}_i\mathfrak{z}_i| = |\mathfrak{z}_i|$, $|\vec{\mathfrak{z}}_i\mathfrak{z}_j| = 0, i \neq j$; $[\xi_i]$ is the intrinsic space of q_i quark, and the three kinetic energy satisfies the conservation of energy.

$$\Delta E_Q^2 = (p_R c)^2 + (p_G c)^2 + (p_B c)^2 \quad (4.3.12)$$

In the process of energy level transition, the quark gains momentum and does not change the internal structure of quarks. The third level is expressed in the existing theory.

$$R = \begin{pmatrix}1\\ 0\\ 0\end{pmatrix}; G = \begin{pmatrix}0\\ 1\\ 0\end{pmatrix}; B = \begin{pmatrix}0\\ 0\\ 1\end{pmatrix} \quad (4.3.13)$$

This is a simple formulation.

You can put the highest level state defined as B state. The lowest level state is defined as R state, and intermediate level is defined as G state. Of course, such regulations are man-made and do not affect the results of physical and physical meanings. A quark is not possible in R, G, B states at the same time, only one state. The strong interaction is the process of quarks momentum exchange between the strings. When any two quarks have strong interactions, such as B quarks and G quark, then G quark transfer energy to R quark by gluon through the link string, the original G quark level becomes low its string is shorter. Gluon state changed the length of the string, so two kinds of quark color swap occurred. With the Feynman diagram shows the results (Fig. 4.3.5).

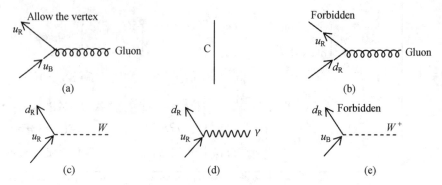

Fig. 4.3.5 Quarks and bosons interacting Feynman diagrams

The interaction between quarks and gluons in both cases does not occur, one is quark change into another quark that is to change the quark flavor; second is resulting in the overall color is not conserved. For example input is R and B, but the output is G and B, this process does not occur, the corresponding physical fact that the input R and B level output level B and G, energy level greater than R and B, so the system energy is out of thin air to increase, energy is not conserved.

Electromagnetic, weak interaction occurs in the lepton field, lepton no dimension division which has nothing to do with 3-dimensional string, and thus do not change color. Quark and strings are inseparable, they are composed by two basic units (free and non-free) topological deformation form 3D string structure, basic properties of 3-dimensional resulting quarks and color cannot exist independently. As shown in Fig. 4.2.3, proton can be written as

$$p \sim \sum \varepsilon_{ijk} u_i u_j d_k \tag{4.3.14}$$

Here, i,j,k take 1, 2, 3, is the state (level) of the color.

Meson with 1-dimensional string structure, there is no color, so it can be expressed as (such as π^+ meson)

$$\pi^+ \sim u\bar{d} \tag{4.3.15}$$

In QCD, the baryon three-color concepts are extended, the π^+ meson expressed as

$$\pi^+ \sim \sum_i u_i \bar{d}_i \tag{4.3.16}$$

i take R,G,B three states, and $R+G+B$ is a colorless state, thus eliminating the possibility of color singlet. Gluons transmit energy is exchange color state only inside hadrons. You can be expressed $SU(3)$ groups based representation by fill in three color state. $SU(3)$ group based representation is a triplet state. In this representation, generator is a 3×3 matrix, write as λ_i ($i=1,2,\cdots,8$), called Gell-Mann matrix, where

$$\lambda_1 = \begin{pmatrix} 0 & 1 & 0 \\ 1 & 0 & 0 \\ 0 & 0 & 0 \end{pmatrix}; \lambda_2 = \begin{pmatrix} 0 & -i & 0 \\ i & 0 & 0 \\ 0 & 0 & 0 \end{pmatrix}; \lambda_3 = \begin{pmatrix} 1 & 0 & 0 \\ 0 & -1 & 0 \\ 0 & 0 & 0 \end{pmatrix}; \lambda_4 = \begin{pmatrix} 0 & 0 & 1 \\ 0 & 0 & 0 \\ 1 & 0 & 0 \end{pmatrix}$$

$$\lambda_5 = \begin{pmatrix} 0 & 0 & -i \\ 0 & 0 & 0 \\ i & 0 & 0 \end{pmatrix} ; \lambda_6 = \begin{pmatrix} 0 & 0 & 0 \\ 0 & 0 & 1 \\ 0 & 1 & 0 \end{pmatrix} ; \lambda_7 = \begin{pmatrix} 0 & 0 & 0 \\ 0 & 0 & -i \\ 0 & i & 0 \end{pmatrix} ; \lambda_8 = \frac{1}{\sqrt{3}} \begin{pmatrix} 1 & 0 & 0 \\ 0 & 1 & 0 \\ 0 & 0 & -2 \end{pmatrix}$$
(4.3.17)

Their common eigenvector is $\begin{pmatrix}1\\0\\0\end{pmatrix}, \begin{pmatrix}0\\1\\0\end{pmatrix}, \begin{pmatrix}0\\0\\1\end{pmatrix}$. The lift and fall of six generators in $SU(3)$ are shown in Fig. 4.3.6. $\lambda_1, \lambda_2, \lambda_3$ corresponds to three Pauli matrices. Gelman matrix has the following commutative relations.

$$\left[\frac{\lambda_i}{2}, \frac{\lambda_j}{2}\right] = i \sum_k f_{ijk} \frac{\lambda_k}{2} \quad (4.3.18)$$

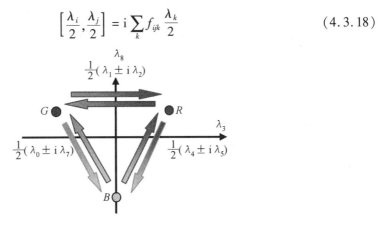

Fig. 4.3.6 Lift and fall operators of $SU(3)$ six generators

Where f_{ijk} are fully anti-symmetric structure constants, non-zero has

$$f_{123} = 1 ; f_{458} = f_{678} = \frac{\sqrt{3}}{2} f_{147} = f_{165} = f_{246} = f_{257} = f_{345} = f_{376} = \frac{1}{2} \quad (4.3.19a)$$

These matrices also satisfy the anti-commutation relations.

$$\left\{\frac{1}{2}\lambda_i, \frac{1}{2}\lambda_j\right\} = \frac{1}{3}\delta_{ij} + d_{ijk}\left(\frac{1}{2}\lambda_k\right) \quad (4.3.19b)$$

Where, d_{ijk} swap any two index are fully symmetrical.

$$d_{118} = d_{228} = d_{338} = -d_{888} = \frac{1}{\sqrt{3}} ; d_{146} = d_{137} = d_{236} = d_{344} = d_{355} = \frac{1}{2} ;$$

$$d_{247} = d_{366} = d_{377} = -\frac{1}{2} ; d_{448} = d_{558} = d_{668} = d_{778} = -\frac{1}{2\sqrt{3}} \quad (4.3.20)$$

Various flavored quark (u, d, s etc.) within baryon there are three colors: R, G, B. Quark is a triplet state of color $SU(3)$ group. It's different with $SU(N)$ flavor symmetry, flavor corresponds to the polarization state, the nature of polarization state corresponding to the strain of field, strain combinations of different quark lepton field combinations baryons is not simple addition and subtraction, but nonlinear. In other words, the lepton field generation cannot be precisely determined its mass, so the $SU(3)$ symmetry of the flavor is not accurate, and the

$SU(3)$ symmetry of color is accurate, there are perfect performance in QCD.

The gluon which transfer color interactions have 8 different color combinations, the distributions are $R\bar{G}, R\bar{B}, G\bar{R}, G\bar{B}, B\bar{R}, B\bar{G}, \frac{1}{\sqrt{2}}(R\bar{R} - G\bar{G}), \frac{1}{\sqrt{6}}(R\bar{R} + G\bar{G} - 2B\bar{B})$. This is the $SU(3)$ octet state, and There is an $SU(3)$ color singlet $\frac{1}{\sqrt{3}}(R\bar{R} + G\bar{G} + B\bar{B})$, it cannot transfer color interactions.

4.3.4 Strong coupling constant α_s

Electromagnetic coupling constant between two quarks related to the amount of the field fiber, coupling strength are $e_1 e_2 \alpha$, where e_i as a charge ($e_i = 2/3$ or $-1/3$), α is the fine structure constant. Similarly, in QCD, the strength of strong coupling that the single gluon exchange between the two color charge is $\frac{1}{2} c_1 c_2 \alpha_s$, where, c_1 and c_2 are associated with the vertex color coefficient, customarily called

$$C_F \equiv \frac{1}{2} |c_1 c_2| \qquad (4.3.21)$$

This is color factor. Let's think about α_s while not to consider C_F.

The strong interaction of three quarks is realized by exchanging gluons through strings to transfer 4-momentum in hadron interior space. The probe particles enter the hadron interior space at high energy. The probability of strong interaction between a test particle and one of the three quarks and the establishment of a channel is 1/3. The probability of the detected particle seeing the test particle and establishing the interaction channel is 1/3. The probability of interaction between two quarks is coupling constants.

$$\frac{1}{3} \times \frac{1}{3} = 0.111 \qquad (4.3.22)$$

We get the minimum possible value of strong interaction coupling constant is 0.11, and the coupling constant of strong interaction in QCD is defined as $\frac{1}{2} c_1 c_2 \alpha_s$, factor 1/2 is the legacy of QCD. Here still complied with previous habits, therefore, the estimates should be multiplied by a factor of 2, namely $\alpha_s = 0.22$. In QCD perturbation theory, when in large Q^2 short-range interactions then α_s is small, experimental data show that $\alpha_s \approx 0.22$ with estimated values issame.

Here special note is: when the coupling constant $\alpha \ll 1$, perturbation calculations of quantum field theory is valid. Because single-channel field fiber (i.e. string) coupling makes the coupling constant $\alpha \ll 1$, and thus in the intrinsic space of hadrons, quark string coupling is prerequisite of perturbation calculations that used in quantum field theory Feynman diagrams. If the probe particle momentum is small that cannot penetrate the hadron, it as a whole, a larger coupling constant the coupling constant α increases (>1), then the perturbation theory failure.

Chapter 4 Hadron Structure

Similar as QCD, color factor $|c_1 c_2|$ can be calculated by color combination of different forms of eight gluons, the results are as follows.

$$-\alpha \rightarrow \begin{cases} -\dfrac{4}{3}\alpha_s, \text{Meson} \\ -\dfrac{2}{3}\alpha_s, \text{Baryon} \end{cases} \quad (4.3.23)$$

Here meson is colorless, introduced color concept that color is three levels. Different level, chord length is different. The coupling strength is not the same, show different color factor.

4.4 Gluon and string

Gluons are waves in Quark strings. Strings tightly connected together to form the quark mesons and baryons, momentum transfer between the quarks by chord volatility. 1-dimensional wave exhibit strong effect change color but does not change the quark flavor. This fluctuation (gluon) also changes the energy level state (color) of quarks, and thus gluon with the color quantum number, with no flavor quantum numbers.

4.4.1 The stress waves in a string

When quark vibratesin a string, it will compress the string, the string section occurs longitudinal displacement slightly, it will be generated adjacent portions of tension or compression. So its adjacent part will produce tension or compression. The tiny longitudinal displacement in the form of longitudinal wave propagation, to describe the longitudinal displacement effects of string by 1-dimensional wave equation. To establish the coordinate x along the axial string (Fig.4.4.1), quark axial displacement is u, the elastic modulus of string is E, density is ρ, cross-sectional area of the string is A, considering the element dx, can be obtain equation of motion was

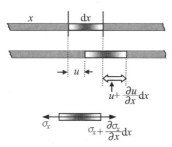

Fig. 4.4.1 String fluctuations

$$AE\left(\dfrac{\partial u}{\partial x} + \dfrac{\partial^2 u}{\partial x^2}dx\right) - AE\dfrac{\partial u}{\partial x} = A\rho dx \dfrac{\partial^2 u}{\partial t^2} \quad (4.4.1)$$

or

$$c^2 \frac{\partial^2 u}{\partial x^2} = \frac{\partial^2 u}{\partial t^2} \qquad (4.4.2)$$

Here, $c^2 = E/\rho$. c is the wave propagation velocity in the string, the wave propagation in deformation field which should be the speed of light. Note that E and ρ are the classical concept does not apply to field theory, here only as a borrowing. The general solution of wave equation (4.4.1) is

$$u = f(x - ct) + g(x + ct) \qquad (4.4.3)$$

This solution is called the D'Alembert solution of the wave equation. Where to determine the functions f and g must satisfy initial boundary conditions. Consider this part $f(x-ct)$. If we take the coordinate axis ξ moves along the positive direction x with speed c, the relationship between the moving coordinate ξ and the fixed coordinate x is $\xi = x - ct$. There by obtaining

$$f(x - ct) = f(\xi) \qquad (4.4.4)$$

This indicates that in the moving coordinate system, the value of the function depends on the coordinates ξ which has nothing to do with the time t, namely the graphics of functions relative to the moving coordinate unchanged. Thus $f(x-ct)$ represents moving wave with speed c which spread to x-negative and wave shape and size maintain invariant. Similarly, $g(x+ct)$ represents a traveling wave which travels at speed c in the x-negative direction. Fig. 4.2.4 shows the quark q_j at point B to the center O compression string, the amount of compression is u, releasing a wave $f(x-ct)$, dropped by B state into G states, and the original G-state absorption of a quark q_i wave $f(x+ct)$, promoted by G-state into B-state, quark q_j and quark q_i interchangeable color state. Special note here is that $f(x-ct)$ is a solitary wave. The formula (4.4.3) seeking t partial derivatives can be obtained up or down quark transition speed.

$$v = \frac{\partial u}{\partial t} = -cf'(x - ct) + cg'(c + ct) = -cf'(\xi) + cg'(\eta) \qquad (4.4.5)$$

Partial differentiating x in (4.4.3), we obtain the strain string

$$\varepsilon_x = \frac{\partial u}{\partial x} = f'(x - ct) + g'(x + ct) \qquad (4.4.6)$$

Wherein, f' and g' are the derivative of each of the independent since the variables. Gluon wave with color quantum numbers used in Fig. 4.4.2 Feynman diagrams shown to express in QCD.

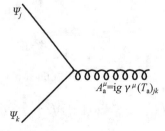

Fig. 4.4.2 Feynman diagrams of gluon wave

The essence of color is the energy level. A quark is converted from a high-energy blue

Chapter 4 Hadron Structure

state to a low energy red state by passing the red blue gluon. The kinetic energy of gluons can be obtained by such a process. Gluons propagate in a string system. The dimension of a string is $1/3$, and its total strain is $h/3$, so there is such a relationship

$$\nu = \frac{|E_n - E_m|}{h/3} \tag{4.4.7}$$

Here, ν is the frequency consisting of three kinds ν_R, ν_G, ν_G, taking into account the gluon in the string system which consisting by three strings

$$\int_{L_1+L_2+L_3} P d\xi = 3 \cdot h/3 \tag{4.4.8}$$

P is the gluon momentum. The gluon is propagated through quark string, so it is 1-dimensional P-wave. The frequencies of longitudinal gluon waves can be divided into three types: low frequency ν_R (corresponding red state), intermediate frequency ν_G (nearest green state) and high frequency ν_B (corresponding blue state). The wavelength of gluon is λ_G and the time of gluon generation is t_G. Then there are the following relations:

$$h/3 = p_G \lambda_G = E_G t_G \tag{4.4.9}$$

Here, E_G is Kinetic energy of gluon, P_G is Gluon momentum.

4.4.2 Interaction between gluon and quark

The quarks transfer the gluons through the string to change the color state of the quarks. Its field function is

$$\varphi_{G-Q}(\xi) = \exp\left\{-i \begin{bmatrix} \varepsilon_{00} & 0 & 0 & 0 \\ 0 & g_R\begin{bmatrix}\delta(i-1)\varepsilon_R\\\delta(i+1)\bar{\varepsilon}_R\end{bmatrix}+g_G\begin{bmatrix}\delta(i-2)\varepsilon_G\\\delta(i+2)\bar{\varepsilon}_G\end{bmatrix}+g_B\begin{bmatrix}\delta(i-3)\varepsilon_B\\\delta(i+3)\bar{\varepsilon}_B\end{bmatrix}+\varepsilon_{L1} & 0 & 0 \\ 0 & 0 & g_G\begin{bmatrix}\delta(i-1)\varepsilon_G\\\delta(i+1)\bar{\varepsilon}_G\end{bmatrix}+g_B\begin{bmatrix}\delta(i-2)\varepsilon_B\\\delta(i+2)\bar{\varepsilon}_B\end{bmatrix}+g_R\begin{bmatrix}\delta(i-3)\varepsilon_R\\\delta(i+3)\bar{\varepsilon}_R\end{bmatrix}+\varepsilon_{L2} & 0 \\ 0 & 0 & 0 & g_B\begin{bmatrix}\delta(i-2)\varepsilon_B\\\delta(i+2)\bar{\varepsilon}_B\end{bmatrix}+g_R\begin{bmatrix}\delta(i-2)\varepsilon_R\\\delta(i+2)\bar{\varepsilon}_R\end{bmatrix}+g_G\begin{bmatrix}\delta(i-3)\varepsilon_G\\\delta(i+3)\bar{\varepsilon}_G\end{bmatrix}+\varepsilon_{L3} \end{bmatrix}\right.$$

_{The strong interact with the gluon in the intrinsic space of nucleon, and the quark obtains momentum $p(R,G,B)$. The probability is g_i. The momentum absorbed by quarks does not destroy the intrinsic structure of quarks. This momentum is represented by the color state of quarks has changed}

$$\left.\begin{pmatrix} \xi_0 & 0 & 0 & 0 \\ 0 & \eth_1[\zeta_1]+\xi_1 & 0 & 0 \\ 0 & 0 & \eth_2[\zeta_2]+\xi_1 & 0 \\ 0 & 0 & 0 & \eth_3[\zeta_3]+\xi_1 \end{pmatrix}\right\} \tag{4.4.10}$$

Internal space of nucleon,

$[\zeta_i]$ is internal space of quark q_i,

$\eth_i$ is quark scale factor

$$\varphi_N(X) \to \varphi'_N(X) = \exp\{-ig_i[A^G_{\mu,\mu}](\xi_\mu)\}\varphi_N(\xi_\mu) \qquad (4.4.11)$$

Here, g_i is Gluon coupling constant. The gluon gauge field has two dimensional space-time structure.

$$[A_G](\xi) = \begin{bmatrix} \varepsilon_{00} & 0 & 0 & 0 \\ 0 & g_R\begin{bmatrix}\delta(i-1)\varepsilon_R \\ \delta(i+1)\bar\varepsilon_R\end{bmatrix} + g_G\begin{bmatrix}\delta(i-2)\varepsilon_G \\ \delta(i+2)\bar\varepsilon_G\end{bmatrix} + g_B\begin{bmatrix}\delta(i-3)\varepsilon_B \\ \delta(i+3)\bar\varepsilon_B\end{bmatrix} & 0 & 0 \\ 0 & 0 & g_G\begin{bmatrix}\delta(i-1)\varepsilon_G \\ \delta(i+1)\bar\varepsilon_G\end{bmatrix} + g_B\begin{bmatrix}\delta(i-2)\varepsilon_B \\ \delta(i+2)\bar\varepsilon_B\end{bmatrix} + g_R\begin{bmatrix}\delta(i-3)\varepsilon_R \\ \delta(i+3)\bar\varepsilon_R\end{bmatrix} & 0 \\ 0 & 0 & 0 & g_B\begin{bmatrix}\delta(i-2)\varepsilon_B \\ \delta(i+2)\bar\varepsilon_B\end{bmatrix} + g_R\begin{bmatrix}\delta(i-2)\varepsilon_R \\ \delta(i+2)\bar\varepsilon_R\end{bmatrix} + g_G\begin{bmatrix}\delta(i-3)\varepsilon_G \\ \delta(i+3)\bar\varepsilon_G\end{bmatrix} \end{bmatrix} [\lambda_{ij}] \cdot$$

$$\begin{pmatrix} \xi_0 & 0 & 0 & 0 \\ 0 & \partial_1\zeta_1 + \xi_1 & 0 & 0 \\ 0 & 0 & \partial_2\zeta_2 + \xi_1 & 0 \\ 0 & 0 & 0 & \partial_3[\zeta_3] + \xi_1 \end{pmatrix} \qquad (4.4.12)$$

The path of $[\lambda_{ij}]$ gluon transition constitutes the gluon space characteristic. Because the wavelength of gluon is very short and has strong particle characteristics, $\partial_1\zeta_1 = \lambda_G$, λ_G is the gluon wavelength. Strain structure of string is

$$[\varepsilon_L] = \begin{bmatrix} \varepsilon_{L0} & 0 & 0 & 0 \\ 0 & \varepsilon_{L1} & 0 & 0 \\ 0 & 0 & \varepsilon_{L2} & 0 \\ 0 & 0 & 0 & \varepsilon_{L3} \end{bmatrix} \begin{pmatrix} \xi_0 & 0 & 0 & 0 \\ 0 & \xi_1 & 0 & 0 \\ 0 & 0 & \xi_1 & 0 \\ 0 & 0 & 0 & \xi_1 \end{pmatrix} \qquad (4.4.13)$$

4.4.3 $SU_c(3)$ structure of quantum chromodynamics

Hadron is a system of three quarks. The field function of hadron can be written as

$$\psi = \begin{pmatrix}\psi_1\\\psi_2\\\psi_3\end{pmatrix}, \quad \psi' = A_3\begin{pmatrix}\psi_1\\\psi_2\\\psi_3\end{pmatrix} = \begin{pmatrix}a_{11}&a_{12}&a_{13}\\a_{21}&a_{22}&a_{23}\\a_{31}&a_{32}&a_{33}\end{pmatrix}\begin{pmatrix}\psi_1\\\psi_2\\\psi_3\end{pmatrix}; \quad A_3^+A_3 = A_3A_3^+ = I, \quad \det A_3 = 1$$

Each quark has three color states. Considering a certain flavor quark q_f, the quark is a spinor field with 4 degrees of freedom.

$$q_f = \begin{pmatrix}q_{1f}\\q_{2f}\\q_{3f}\end{pmatrix}, \quad m_1 = m_2 = m_3 = m$$

A quark field is represented by a Dirac spin with two internal indices. The two internal standards are taste index f and color index c.

Chapter 4 Hadron Structure

$$q \equiv (q_{\alpha cf}), \begin{cases} f = 1,\cdots,n(\text{Flavor}) \\ c = 1,2,3(\text{Colour}) \\ \alpha = 1,2,3,4(\text{Dirac spinor}) \end{cases} \tag{4.4.14}$$

A global $SU(3)$ transformation is

$$q'(x) = \exp[i\alpha_k \frac{\lambda_k}{2}] q(x)$$

Among them, $\lambda'_k S$ is the eight generators of the group (3×3 matrix). For infinitesimal transformation, Dirac index and taste index are omitted, then

$$q'_c(x) = (\delta_{cc'} + i\alpha_k) \left(\frac{\lambda_k}{2}\right)_{cc'} q_{c'}(x) \tag{4.4.15}$$

λ_k see (4.4.13).

Hadrons have no color. That level (color mode is level state) of quarks are confined in hadrons intrinsic space, this system is energy conservation. Energy levels swap meet $SU_c(3)$ symmetry, $SU_c(3)$ is the color symmetry as a local symmetry. Unitary symmetry is flavor symmetry, flavor changes lead to the change of hadrons. Unitary symmetry is the overall symmetry. From the overall gauge symmetry extended to local symmetry, it implies that exists a gauge field. Quark has $SU_c(3)$ local symmetry, it means that $SU_c(3)$ is a local gauge group. The group element is

$$u(\theta) = \exp\left[-i\alpha^k \frac{\lambda^k}{2}\right] \tag{4.4.16}$$

Where, $\theta^k(x)$ is a group parameters which associated with temporal and spatial, $\lambda^k/2 (\alpha = 1, 2, \cdots, 8)$ is a generator of 3-dimensional representation. $\theta^k(x)$ is moving along $\lambda^k/2$. $\theta^\alpha(x)$ is the movement amount. λ^k is a Gell-Mann matrices satisfy the commutation relations (4.3.18).

The coordinate base of λ^k is 8 channels of strings, each string produces a moving wave, this wave is gluon fields which use to transmitted strongly interacting, there are eight gluon fields, their quantum is the gluon.

It is assumed that the Lagrangian function of quarks is invariant under the local color $SU(3)$ group: $q'(x) = \exp[ig\Lambda_k(x) \frac{\lambda_k}{2}] q(x)$. Among them, $\Lambda_k(x)$ is the 8 functions that determine the transformation. In the case of infinitesimal:

$$q'(x) = [I + ig\Lambda_k(x) \frac{\lambda_k}{2}] q(x) \tag{4.4.17}$$

Covariant derivatives replace ordinary derivatives

$$D_\mu q(x) = [\partial_\mu + ig\Lambda_{\mu,k}(x) \frac{\lambda_k}{2}] q(x)$$

Among them, $\Lambda_{\mu,k}(x)$ is eight gauge fields. These derivatives are transformed like the color spinor of (4.4.15).

$$D'_\mu q'(x) = \exp[ig\Lambda_k(x) \frac{\lambda_k}{2}] D_\mu q(x)$$

· 215 ·

This requires that the transformation law of the color gauge field be determined as

$$D'_\mu q'(x) = \exp[ig\Lambda_k(x)\frac{\lambda_k}{2}]D_\mu q(x)$$

Here, $k=1,2,\cdots,8$ in the range of 8, and $U = \exp[ig\Lambda_k(x)\frac{\lambda_k}{2}]$. In order to obtain Lagrangian functions of gauge fields, we define tensors:

$$F_k^{\mu\nu} = \partial^\nu A_k^\mu - \partial^\mu A_k^\nu + gf_{knl}A_n^\mu A_l^\nu \;;\; F_{\mu\nu}^\alpha = \partial_\mu A_\nu^\alpha - \partial_\nu A_\mu^\alpha + gf^{\alpha\beta\gamma}A_\mu^\beta A_\nu^\gamma$$

Gluon field is the gauge field. The color gauge field, or gluon field, it is a group of eight massless vector fields interacting with themselves. The gluon field interacts with quarks through the degree of color freedom. In other words, eight generators corresponds to 8 gauge field $A_\mu^\alpha(x)$ and $A_\mu^\alpha(x)$ can be called the gauge potential, $F_{\mu\nu}^\alpha(x)$ called the canonical field strength. Lagrange functions can be obtained

$$\mathscr{L} = -\frac{1}{4}F_k^{\mu\nu}F_{\mu\nu,k} + \bar{q}(i\gamma^\mu D_\mu - m)q$$

If it contains all the flavors of quarks $f=1,2,\cdots,n$, then

$$\mathscr{L} = -\frac{1}{4}F_{\mu\nu}^\alpha F^{\mu\nu,\alpha} + \sum_{f=1}^n \bar{q}_{nf}[i\gamma^\mu(D_\mu)_{ab} - m\delta_{ij}]q_s^j$$

Ignore Dirac spinor marking. The infinitesimal transformation law of $F_k^{\mu\nu}$ is

$$F'^\alpha_{\mu\nu} = F^\alpha_{\mu\nu} + gf^{\alpha\beta\gamma}A_\mu^\beta A_\nu^\gamma$$

This is the first order transformation law of the color vector g. This system equation ignores the Fadeev-Popov phase and is in the following form

$$D_{v,kl}F_l^{\mu\nu} = gj_k^\mu \qquad (4.4.18)$$

$$D_{kl}^v F_l^{\alpha\beta} + D_{kl}^\beta F_l^{\nu\alpha} + D_{kl}^\alpha F_l^{\beta\nu} = 0$$

$$i\gamma^\alpha D_\alpha q - mq = 0 \qquad (4.4.19)$$

Among them, $j_k^\mu = \sum_f \bar{q}_f \gamma^\mu \frac{\lambda_k}{2} q_f$ is the material flow.

The action which used to describe quarks and gluons and their interaction is

$$S = \int d^4x [\bar{q}_\gamma^i(i\partial\delta_{ij} + g\frac{\lambda_{ij}^\alpha}{2}i^\alpha - m_i^\gamma \delta_{ij}^{rs})q_s^j - \frac{1}{4}F_{\mu\nu}^\alpha F^{\alpha\mu\nu}]$$

4.4.4 Strings and quark sea

1. Strings

Gluons are composed by strings and strings fluctuations. Gluon strings in addition to passing waves, it also directly passed force, its presence three vertices of field effect. Because of the action of three vertices that makes QCD theory has the nature of asymptotic freedom. Feynman diagram is as follows (Fig. 4.4.3), the three gluon vertex coupling relation is

$$-gf_{abc}[g_{ab}(p_1-p_2)_\gamma + g_{\beta\gamma}(p_2-p_3)_\alpha + g_{\gamma\alpha}(p_3-p_1)_\beta]$$

Due to there are only three strings in hadron, so there is no self-interaction of gluon

4-piont, only when four quarks together to exist four gluon interactions.

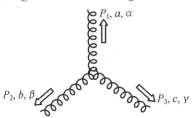

Fig. 4.4.3　Feynman diagrams of self-interaction by three Gluons

2. Dual copy effect-quark sea

According neutrinos (and other leptons e,μ) and nuclear inelastic scattering experimental data, we have more specific understanding on nuclear intrinsic space. Structure function has approximate nothing to do with scaling that indicate there are like-point particles inside the proton which are called partons.

Protons composition by the various "like-point" partons ($i = u, d, \cdots$, corresponding charge is e_i, of course have gluons). Now we introduce the momentum distribution of partons, namely is the share of probability x of partons i after collision which carries the proton momentum p.

$$f_1(x) = \frac{dp_i}{dx} \tag{4.4.20}$$

All share to add up is 1, i.e.

$$\sum_{i'} \int dx x f_i'(x) = 1 \tag{4.4.21}$$

Here, i' is the sub to all partons. Where

$$x = \frac{1}{\omega} = \frac{Q^2}{2M\nu} \tag{4.4.22}$$

Among them, $Q^2 = -q^2$ is 4-momentum transfer. $\nu = E - E'$, E and E' respectively are the energy of probe particles which in the initial and final states; M is the proton mass. Nucleon structure function $F_2(x)$ is the product that the partons probability $f(x)$ with momentum xp in nuclear internal and x, i.e.

$$F_2(x) = xf(x) \tag{4.4.23}$$

Large Q^2, obtained proton inelastic structure function by electron scattering, the corresponding formula is

$$\frac{1}{x}F_2^{ep}(x) = \left(\frac{2}{3}\right)^2 [u^p(x) + \bar{u}^p(x)] + \left(\frac{1}{3}\right)^2 [d^p(x) + \bar{d}^p(x)] + \left(\frac{1}{3}\right)^2 [s^p(x) + \bar{s}^p(x)] \tag{4.4.24}$$

Obtained neutron inelastic structure function by electron scattering, the corresponding formula is

$$\frac{1}{x}F_2^{en}(x) = \left(\frac{2}{3}\right)^2 [u^n(x) + \bar{u}^n(x)] + \left(\frac{1}{3}\right)^2 [d^n(x) + \bar{d}^n(x)] + \left(\frac{1}{3}\right)^2 [s^n(x) + \bar{s}^n(x)]$$

(4.4.25)

There are total of 6 unknown quark structure function.

$$u^p(x) = d^n(x) \equiv u(x)$$
$$d^p(x) = u^n(x) \equiv d(x) \qquad (4.4.26)$$
$$s^p(x) = s^n(x) \equiv s(x)$$

Proton has three valence quarks $u_v\, u_v\, d_v$, and quark sea of quark-antiquark pairs

$$u_s(x) = \bar{u}_s(x) = d_s(x) = \bar{d}_s(x) = s_s(x) = \bar{s}_s(x) = S(x)$$
$$u(x) = u_v(x) + u_s(x)$$
$$d(x) = d_v(x) + d_s(x) \qquad (4.4.27)$$

Where $S(x)$ is the sea quark distributions of all public of quark flavor. Gives the contributions of all partons are summed must be used proton quantum numbers: Charge 1, baryon number 1, singular number 0, which requires

$$\begin{cases} \int_0^1 [u(x) - \bar{u}(x)] dx = 2 \\ \int_0^1 [d(x) - \bar{d}(x)] dx = 1 \\ \int_0^1 [s(x) - \bar{s}(x)] dx = 0 \end{cases} \qquad (4.4.28)$$

Substituting equations (4.4.25), (4.4.22) and (4.4.23) together, can be get

$$\begin{cases} \dfrac{1}{x}F_2^{ep} = \dfrac{1}{9}[4u_v + d_v] + \dfrac{4}{3}S \\ \dfrac{1}{x}F_2^{en} = \dfrac{1}{9}[u_v + 4d_v] + \dfrac{4}{3}S \end{cases} \qquad (4.4.29)$$

Where 4/3 is the summation results of six kinds sea quark distributions e_i^2. Since quark-antiquark pairs generated by the gluon in the sea, when we expect at small x, $S(x)$ has similar bremsstrahlung spectrum (Fig. 4.4.4), so when $\nu \to 0, x \to 0$, the number of sea quarks in logarithmic growth, for a given Q^2, If the virtual photon-proton total cross section is constant, there is $f_i(x) \xrightarrow[x \to 0]{} 1/x$, the number of partons logarithmic growth at small x. It is noteworthy that in a small momentum ($x \approx 0$), Three valence quarks will be obscured by newly formed sea $S(x)$ of multiple low-momentum $q\bar{q}$ pairs. According to equation (4.4.27), there should be

$$\frac{F_2^{en}(x)}{F_2^{ep}(x)} \xrightarrow[x \to 0]{} 1 \qquad (4.4.30)$$

This conclusion is confirmed by experimental data. On the other hand, a large momentum ($x \approx 1$), the high speed valence quark mainly:

$$\frac{F_2^{en}(x)}{F_2^{ep}(x)} \xrightarrow[x \to 1]{} \frac{u_v + 4d_v}{4u_v + d_v} \qquad (4.4.31)$$

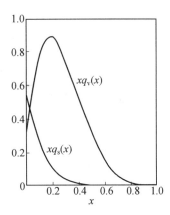

Fig. 4.4.4 Momentum distribution of proton intrinsic space

For protons, when is large x clearly has $u_v \gg d_v$, formula (4.4.29) ratio approaches 1/4.

Integrating using the experimental data of F_2^{ep} and F_2^{en} to obtain the following results:

$$\int dx F_2^{ep}(x) = \frac{4}{9}\varepsilon_u + \frac{1}{9}\varepsilon_d = 0.18$$

$$\int dx F_2^{en}(x) = \frac{1}{9}\varepsilon_u + \frac{4}{9}\varepsilon_d = 0.12 \qquad (4.4.32)$$

$$\varepsilon_g \approx 1 - \varepsilon_u - \varepsilon_d$$

Whereby get $\varepsilon_u = 0.36, \varepsilon_d = 0.18, \varepsilon_g = 0.46$. Therefore, gluons carry the momentum about 50%. It is particularly important to note that this conclusion that when we enter 1 momentum perturbations within hadrons, then $2u$ quarks account for 0.36, d quarks account for 0.18, and string systems account for 0.46. In other words, the strings are easily poked. Toggle strings, strings can produce waves, and the gluon in the string is the vibration wave. The gluons within the string occupy nearly half of the momentum.

According to experiments, the structure function has approximate scaling independence, $F_2(x, q^2) \approx F_2(x)$, indicating that the nucleon internal containing the Dirac particles like-point.

This component has a spin quantum number 1/2, $2xF_1(x) \approx F_2(x)$. You can also prove that the group element has a fractional charge. And these are consistent with quark model. The study of quantum number on the parton can equate them with quarks. About 10% anti-quark in nuclear, which mainly distributed in the small range x. The experimental results are given in more detail, the anti-quark in nuclear mainly in the small x ($x<0.3$) range. As shown in Fig. 4.4.4. Pay particular attention that about 10% anti-quark exists only in being d protons and other proton does not exist anti-quark. Here Anti-quark is the excitation effects when the probe particles into the nuclear intrinsic space. In other words, the seeing by probe particles is not real. Probe will produce excitation, excitation will produces dual copy effects, and non-free state antiparticle exists in particle sea, which is what we can detect antiparticles particles sea (Fig. 4.4.5).

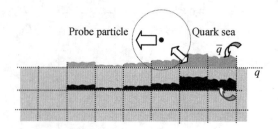

Fig. 4.4.5　Probing particle perturbations lead to fluctuations in quantum field

Quark momentum is only half of the nucleon momentum and the remaining half occupied by the gluons, gluons transmit the strong interaction between quarks and changing the color of quarks. Most of the shares of the proton momentum have been carried by a neutral partons, rather than quarks. The neutral parton is the gluon in QCD, they do not interact with photons. From the proton structure, gluon string with rest mass, satisfies the equation (4.3.3), and the wave on string no rest mass, but it there is momentum. A fluctuation strings is gluon, the total momentum is half of the nucleon momentum. Probe particles acting on the strings, the result is to detect neutral partons.

The total deformation of the proton field consists of three parts: ①quark field strain; ②strain of dimension division of basic unit of vacuum constituting the chord; ③the strain of Intrinsic space area to the center of the proton. This three-part field strain of the intrinsic space constitutes a proton field strain.

Here, we are interested in the sea of particle. All elementary particles were disturbed to exciting corresponding particles sea, so vacuum of disturbed there will be elementary particles sea, vacuum structure becomes very complex.

The quantum field spread itself a disturbance in quiet vacuum. Detecting particles through the hadron intrinsic space that the space is disturbed (Fig. 4.4.5), making internal system of hadron vibrates, quark and chord are vibration to form quark sea, so detecting particles "to see" the anti-quarks inside the proton. Achieve quantum wave launch conditions can produce a large number of photons. When the probe lepton into the intrinsic space of nuclear, if the probe particle 4-momentum p is larger and probe lepton size is smaller and probing particle itself perturbation is smaller to nucleon intrinsic field, it can detect the more fine structure of quark sea. On the contrary, it detected the quark sea structure more rough.

Strong disturbance is extremely strong that resulting formation of meson states, including virtual meson state and real meson state (i.e. lot of K, π meson). The existence of meson states makes strong coupling constant becomes large, becoming a complex function. If the disturbance is strong enough to break the string that can produce lepton pairs and quark pairs.

Chapter 5   Gravitational Field

Einstein's concept of space-time bending was replaced by the concept of vacuum hardening. In other words, space-time hardening produces space-time bending effect. Gravitation comes from the 2-order effect of the intrinsic vacuum strain of elementary particles. Vacuum will not cause any contradiction between relativity and quantum field theory, and will make Einstein's relativity more harmonious.

The exact description of a quantum field should be green strain (1.1.1) at some point P in space-time. It consists of two kinds of strain tensors.

$$E_{\mu\nu} = \frac{1}{2}\left(\frac{\partial u_\mu}{\partial \xi_\nu} + \frac{\partial u_\nu}{\partial \xi_\mu}\right) + \frac{1}{2}\frac{\partial u_m}{\partial \xi_\mu}\frac{\partial u_m}{\partial \xi_\nu} = [\varepsilon_{\mu\nu}]\big|_p + \frac{1}{2}[\mathscr{G}_{\mu\nu}]\big|_p \quad (p \text{ is in the inner space})$$

(5.0.1a)

Here, $[\mathscr{G}_{\mu\nu}]\big|_p$ is second order strain of intrinsic space at point p.

The quantum field intrinsic strain $\varepsilon_{\mu\nu}$ is expressed as energy momentum $T_{\mu\nu}$ in the observation space-time. We know that the space-time inside a quantum field is broken. The space-time outside the center area are smooth space-time. Therefore, the quantum field is a tiny broken point in smooth space-time. This leads to the quantum field theory being a local theory. Broken space-time is completely different from smooth space-time. It is shown that general relativity and quantum field theory are incompatible. This incompatibility is an objective existence of physics.

The 2-order effect of the intrinsic strain $\varepsilon_{\mu\nu}$ of the quantum field is $\mathscr{G}_{\mu\nu}$, i.e. there exists a small 2-order strain $\mathscr{G}_{\mu\nu}$ of the background vacuum, $\varepsilon_{\mu\nu}$ and $\mathscr{G}_{\mu\nu}$ are the strains at the same point in space-time. $\mathscr{G}_{\mu\nu} = \kappa g_{\mu\nu}$. The 2-order vacuum strain $\mathscr{G}_{\mu\nu}$ corresponds to the metric $g_{\mu\nu}$ of the observation space-time of bending. The space-time interval of internal space at point p satisfies $c^2 d\tau^2 = g_{\mu\nu} dx^\mu dx^\nu$. $g_{\mu\nu}$ characterizes the curvature of intrinsic space-time of quantum field at point p. The curvature of space-time extends from the interior the observed space-time, $p(\xi)$ to $p'(x)$, then the metric from $g_{\mu\nu}(x)$ to $R[g_{\mu\nu}(x)]$. It is shown that the observed space-time curves and the curvature $R[g_{\mu\nu}(x)]$ is a combination of gauges related to the 4 momentum distribution, characterizes the vacuum 2-order strain state of the external space-time point p'.

$$E_{\mu\nu} = [\varepsilon_{\mu\nu}(\xi)]\big|_p + \frac{1}{2}[\mathscr{G}_{\mu\nu}(\xi)]\big|_p + \frac{1}{2}R[\mathscr{G}'_{\mu\nu}(x)]\big|_{p'} \quad (5.0.1b)$$

Here, $[\varepsilon_{\mu\nu}(\xi)]\big|_p$ is the strain of point p in the interior space. $\varepsilon_{\mu\nu}(\xi)$ is energy-momentum tensor $T_{\mu\nu}(x)$ in outer observation space-time. $[\mathscr{G}_{\mu\nu}(\xi)]\big|_p$ is the 2-order strain of point p in the interior space. The inner space of the point-like particles considered above is very small. If the radius of intrinsic space is r_0, then $r < r_0 < R$. r is the radius within the inner space and R is the radius outside the inner space. If the inner space is relatively large and the intrinsic strain is asymmetric, the 2-order strain effect $r[\mathscr{G}_{\mu\nu}(x)]$ in the inner space should also be considered. $r[g_{\mu\nu}(x)]\mid_{p''}$ is a combination of gauges related to the intrinsic 4-momentum distribution, characterizes the vacuum 2-order strain state of the inner space-time point p''.

$$\Phi(\xi,x) = \exp\left\{-i([\varepsilon_{\mu\nu}](\xi)\mid_p + \frac{1}{2}[\mathscr{G}_{\mu\nu}](\xi)\mid_p + \frac{1}{2}r[\mathscr{G}_{\mu\nu}](\xi)\mid_{p''} + \frac{1}{2}R[\mathscr{G}_{\mu\nu}](x)\mid_{p'})\right\} \quad (5.0.1c)$$

In observation space, $\frac{1}{2}[\mathscr{G}_{\mu\nu}](\xi)\mid_p + \frac{1}{2}r[\mathscr{G}_{\mu\nu}](\xi)\mid_{p''}$ has no observability to be omitted, and is rephrased as

$$\Phi(\xi,x) = \exp\left\{-i([\varepsilon_{\mu\nu}](\xi)\mid_p + \frac{1}{2}R[\mathscr{G}_{\mu\nu}](x)\mid_p)\right\} \quad (5.0.1d)$$

This is the real scalar particle field function containing the gravitational field. The above formula can also be written as

$$\Phi(\xi,x) = \exp\left\{-i\left[[\varepsilon_{\mu\nu}](\xi)\mid_p + \frac{1}{2}R[\mathscr{G}_{\mu\nu}](x)\mid_{p'}\right]\right\} = \underbrace{\Phi_q(\xi)\mid_p}_{\text{Quantum field function}} \cdot \underbrace{\Phi_G(x)\mid_{p'}}_{\text{Gravitational field function}} \quad (5.0.2)$$

Here, $R[\mathscr{G}_{\mu\nu}]$ is the superposition of the 2-order strain at point p' in outer space-time.

The generalized coordinate system $(X_0, X_1, \cdots, X_\mu, \cdots, X_N)$ is introduced. The external temporal and spatial coordinates $X_\mu \to X^\nu$; $X_\mu = g_{\mu\nu}X^\nu$. $g_{\mu\nu}$ of 4-dimensional space-time corresponds to $g_{\mu\nu}$. $X_\mu = g_{\mu\nu}X^\nu$; $\mu,\nu = 0,1,2,3$. That is $x_\mu = g_{\mu\nu}x^\nu$.
Here

$$\begin{pmatrix} x_0 & 0 & 0 & 0 \\ 0 & 0 & 0 & 0 \\ 0 & 0 & 0 & 0 \\ 0 & 0 & 0 & 0 \end{pmatrix} = X_0; \quad \begin{pmatrix} 0 & 0 & 0 & 0 \\ 0 & x_1 & 0 & 0 \\ 0 & 0 & 0 & 0 \\ 0 & 0 & 0 & 0 \end{pmatrix} = X_1$$

$$\begin{pmatrix} 0 & 0 & 0 & 0 \\ 0 & 0 & 0 & 0 \\ 0 & 0 & x_2 & 0 \\ 0 & 0 & 0 & 0 \end{pmatrix} = X_2; \quad \begin{pmatrix} 0 & 0 & 0 & 0 \\ 0 & 0 & 0 & 0 \\ 0 & 0 & 0 & 0 \\ 0 & 0 & 0 & x_3 \end{pmatrix} = X_3$$

Considering a planet, it can be regarded as a particle from the macro cosmological point of view, which oppresses the space-time around the point. It can be regarded as a cosmic scalar particle and described by the cosmic scalar particle field function. If the vacuum strain is considered at the same point p', $[\varepsilon_{\mu\nu}](\xi)\mid_p = [T_{\mu\nu}](X)\mid_{p'}$. In other words, the energy-momentum tensor of mass charge is independent of the spatial position. The influence of the 2-order strain of mass charge on its background space-time intensity coefficient is κ. See Formula (5.3.2) for details. $\kappa R(\mathscr{G}_{\mu\nu})\mid_{p'} = R(g_{\mu\nu})\mid_{p'}$ is available at point p'.

$$\Phi(X) = \exp\left\{-i([T_{\mu\nu}](X)\mid_p + \frac{1}{2}\kappa^{-1}R[g_{\mu\nu}](X)\mid_{p'})\right\} \quad (5.0.3)$$

For example, considering only energy, momentum and mass (real scalar field particles),

Chapter 5 Gravitational Field

the real scalar quantum field containing gravity is represented of observation space-time.

Gravitational field:

$$\phi = g_{00}/2 + c \Leftrightarrow \frac{1}{2}\kappa^{-1} R[g_{\mu\nu}](X) \tag{5.0.4}$$

In Newtonian mechanics:

$$\nabla^2 \varphi = 4\pi G\rho \Leftrightarrow [T_{\mu\nu}] \tag{5.0.5}$$

The 4-momentum of matter mass is bending its own background space-time. The strain inside the mass charge is strictly proportional to its 2-order strain. The following relations exist

$$\kappa T_{\mu\nu} = R(g_{\alpha\beta}) \tag{5.0.6}$$

In the theory of vacuum superunification, only $R(g_{\mu\nu})$ is known to be a combination of many space-time metric, but its expression can not be obtained directly. Einstein found this relationship $R_{\mu\nu} - \frac{1}{2}g_{\mu\nu}R = \frac{8\pi G}{c^4}T_{\mu\nu}$, which is Einstein gravitational field equation according to the corresponding relationship to know $\kappa = \frac{8\pi G}{c^4}$, $R(g_{\mu\nu}) = R_{\mu\nu} - \frac{1}{2}g_{\mu\nu}R \equiv G_{\mu\nu}$. $G_{\mu\nu}$ is Einstein tensor. Set: $X|_p = X_T$, $X|_{p'} = X_R$. You can write (5.0.2) as $\Phi(X) = \exp\{-i([T_{\mu\nu}](X_T) + \frac{1}{2}\kappa^{-1}[G_{\mu\nu}](X_R))\}$. According to the conservation of vacuum strain, we can know $\hat{\partial}_\mu \Phi(X) = 0$.

Set

$$\hat{\partial}_G = i\frac{\partial}{\partial X_T}; \ \hat{\Gamma}_G = -i\frac{1}{\kappa}\frac{\partial}{\partial X_R} \tag{5.0.7}$$

$$\Rightarrow (\hat{\partial}_G + \hat{\Gamma}_G)\Phi(X) = ([T_{\mu\nu}] - \frac{1}{\kappa}[R_{\mu\nu}])\Phi(X) = 0 \tag{5.0.8}$$

The eigenvalue equation is

$$\kappa [T_{\mu\nu}]_p - [G_{\mu\nu}]_{p'} = 0 \tag{5.0.9a}$$

The gravitational field equation can be further written as

$$\hat{D}_G \Phi(X) = 0 \tag{5.0.9b}$$

Consider a Cosmic scalar particle

$$\Phi(X) = \exp\{-i(\kappa[T_{\mu\nu}](X_T) + [G_{\mu\nu}](X_R))\} = \Phi(X_T)\Phi(X_R) \tag{5.0.10}$$

There is a gravitational interaction with another cosmic scalar particle at same points. This cosmic scalar particle field function is

$$\Phi_G(X) = \exp\{-i(\kappa[T_{\mu\nu}](X_T) + [G_{\mu\nu}](X_R))\} \tag{5.0.11}$$

Another gravitational field function is

$$\Phi'_G(X) = \exp\{-i(\kappa[T'_{\mu\nu}](X_T) + R'[G_{\mu\nu}](X_R))\} \tag{5.0.12}$$

Cosmic particles with gravitational interactions

$$\Phi(X) = \Phi_G(X)\Phi'_G(X) = \exp\{-i(\kappa[T_{\mu\nu} + T'_{\mu\nu}](X_T) + ([G_{\mu\nu}] + [G'_{\mu\nu}])(X_R))\} \tag{5.0.13}$$

In the area of gravity interaction, gravity superposition is linear because there is no negative gravity. Vacuum strain is further intensified and vacuum hardening in the gravitational region is intensified. The expression is the enhancement of space-time curvature.

5.1 The vacuum principle of special relativity

All of vacuum strain is gravitational source. A static cosmic level scalar particle field function is $\Phi(X) = \exp\{-i([T_{\mu\nu}](X_T) + kR[g_{\mu\nu}](X_R))\} \Rightarrow$ From static reference frame to arbitrary motion reference frame (Set $k^{-1} = \kappa$), then

$$\Phi(X') = \exp\{-i(\underbrace{[T_{\mu\nu}]L_{\alpha\beta}(X)}_{\text{Special theory of relativity}} + \underbrace{k[G_{\mu\nu}](X')}_{\text{General relativity}})\} \quad (5.1.1)$$

The field function can be written as

$$\Phi(X) = \Phi_T(X') \cdot \Phi_G(X') \; ; \; X' = L_{\alpha\beta}(X)$$

$$\Phi_T(X') = \exp\{-i([T'_{\mu\nu}](X'))\} \; ; \; \hat{\partial}_T\Phi_T(X') = T'_{\mu\nu}\Phi_T(X') \quad (5.1.2)$$

The intrinsic first-order vacuum strain is energy-momentum tensor in the observation space. As the observation space changes from static reference frame to inertial motion reference frame, the measurement value of $T_{\mu\nu}$ will change. The propagation efficiency of the first-order strain of quantum field constitutes the inertial mass, and the second-order strain constitutes the gravitational mass. The theory describing this change is special relativity.

Cosmological scale particle space-time structure field function is

$$\Phi_G(X') = \exp\{-ik[G_{\mu\nu}](X')\} \; ; \; \hat{\partial}_G\Phi_G(X') = kG_{\mu\nu}\exp\{-i(k[G_{\mu\nu}](X'))\}$$

(5.1.3)

Φ_G is the field function of the space-time of the curved background. The eigen value $R'(g_{\mu\nu})$ is Einstein tensor $G_{\mu\nu}$ in motion inertial reference system, which describes the degree of curvature of background space-time and corresponds to general relativity.

1. Special theory of relativity

Any material group can be regarded as a cosmological scalar particle, large to a galaxy, small to a macroscopic object. Every star in the universe can be regarded as a dust in the galaxy. For a macroscopic object, it is made up of atoms and molecules. Because the macroscopic material block does not consider the charge, it can be roughly regarded as the superposition state of N scalar particles. The scalar particles are dispersed in a volume V, and are composed of atoms.

In a inertial reference frame (x^0, x^1, x^2, x^3), macroscopic real scalar field functionin static state is $\Phi_T(x) = \exp\{-i[T_{\mu\nu}](x)\}$, further written as

$$\Phi_T(x) = \exp\left\{-i\begin{bmatrix} E_0 & 0 & 0 & 0 \\ 0 & -p_1 & 0 & 0 \\ 0 & 0 & -p_2 & 0 \\ 0 & 0 & 0 & -p_3 \end{bmatrix}\begin{pmatrix} x^0 & 0 & 0 & 0 \\ 0 & x^1 & 0 & 0 \\ 0 & 0 & x^2 & 0 \\ 0 & 0 & 0 & x^3 \end{pmatrix}\right\}$$

Chapter 5   Gravitational Field

$$P_0 = mc^2 \text{ ; } P_1 = P_2 = P_2 = mc \tag{5.1.4}$$

When the motion state changes, the real scalar particles into the high-speed inertial system (X'^0, X'^1, X'^2, X'^2).

$$\Phi_T(X') = \exp\left\{-i\begin{bmatrix} E_0 & 0 & 0 & 0 \\ 0 & -p_1 & 0 & 0 \\ 0 & 0 & -p_2 & 0 \\ 0 & 0 & 0 & -p_3 \end{bmatrix}\begin{pmatrix} x'^0 & 0 & 0 & 0 \\ 0 & x'^1 & 0 & 0 \\ 0 & 0 & x'^2 & 0 \\ 0 & 0 & 0 & x'^3 \end{pmatrix}\right\}$$

$$= \exp\left\{-i\begin{bmatrix} E_0 & 0 & 0 & 0 \\ 0 & -p_1 & 0 & 0 \\ 0 & 0 & -p_2 & 0 \\ 0 & 0 & 0 & -p_3 \end{bmatrix}L_\mu^\nu\begin{pmatrix} x^0 & 0 & 0 & 0 \\ 0 & x^1 & 0 & 0 \\ 0 & 0 & x^2 & 0 \\ 0 & 0 & 0 & x^3 \end{pmatrix}\right\} \tag{5.1.5}$$

Consider moving along the x^1 direction (Fig. 5.1.1)

$$\begin{pmatrix} x'^0 & 0 & 0 & 0 \\ 0 & x'^1 & 0 & 0 \\ 0 & 0 & x'^2 & 0 \\ 0 & 0 & 0 & x'^3 \end{pmatrix} = \begin{bmatrix} \gamma & -\beta\gamma & 0 & 0 \\ -\beta\gamma & \gamma & 0 & 0 \\ 0 & 0 & 1 & 0 \\ 0 & 0 & 0 & \gamma \end{bmatrix}\begin{pmatrix} x^0 & 0 & 0 & 0 \\ 0 & x^1 & 0 & 0 \\ 0 & 0 & x^2 & 0 \\ 0 & 0 & 0 & x^3 \end{pmatrix} \tag{5.1.6}$$

$$\beta = \frac{v}{c}; \gamma = \left(\sqrt{1 - \frac{v^2}{c^2}}\right)^{-1} \tag{5.1.7}$$

$$i\partial_\mu \Phi_T(X') = P_\mu \Phi_T(X') \tag{5.1.8}$$

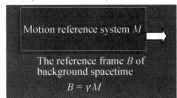

Fig. 5.1.1   The relationship between two reference frames

The eigenvalues of P_μ can be obtained by using 4-momentum operators, where $P_0 = m_0 c^2 + m_0 v^2/2 + \cdots$, $m = m_0\gamma$. The real scalar field is placed in the observation space, and the static reference frame is transformed into the moving inertial reference frame to obtain the results of the special relativity theory. Since the motion of all physical particles is essentially the propagation of quantum fields, the propagation of matter in different space-time satisfies the Lorentz transformation, so that we can re-understand the special relativity from a deeper perspective.

2. Special relativity space-time

Establish two basic principles of special relativity.

(1) The principle of the same speed of light. The speed of light c is the same in all inertial frames. Light propagation speed with the same value in Free-space, It is nothing to do with the velocity of inertial frames, and nothing to do with the speed of light source.

(2) Special relativity principle. The laws of nature are the same in all inertial frames. That is the inertia reference frame completely equivalent.

Lorenz transformation. Popular speaking is the invariance of light speed result that we can happen clock slow and ruler shrinkage effect when we movement close to light speed. For example, motion along x direction, Lorenz transformation is

$$x' = (x - ut)/\sqrt{1 - u^2/c^2}\,; y' = y; z' = z; t' = (t - \frac{u}{c}x)/\sqrt{1 - u^2/c^2}$$

Lorenz transformation shows time and space associate through the factor u/c. The effects of high-speed movement include

①Length contraction: $l = l_0/\sqrt{1 - u^2/c^2}$;

②Time delay: $\tau = \tau_0/\sqrt{1 - u^2/c^2}$;

③Relative mass: $M = M_0/\sqrt{1 - u^2/c^2}$;

④Doppler effect: $\nu = \nu_0(1 - \frac{u}{c}\cos\theta)/\sqrt{1 - u^2/c^2}$.

(3) Space-time interval: $ds^2 = (cdt)^2 - dx^2 - dy^2 - dz^2$. Interval has nothing to do with the frame of reference. Four dimensional space-time is divided into three areas:

①$ds^2 = 0$, Events on light cone face;

②$ds^2 > 0$, Events are within light cone;

③$ds^2 < 0$.

Events are outside light cone. The events in same area are correlation, and no correlation between different areas.

5.1.2 The original concept of time

1. Time

Time is the expression of the result of the degree of freedom of motion of matter, that is, all basic particles exist in the form of waves, such as photons in the form of electromagnetic waves, and the rest leptons in the form of 3-dimensional spherical spin waves. In other words, matter exists in the form of motion. In order to express the effect of motion, there is a comparison between them, which is defined as the time dimension. There is no geometric effect.

2. Clock

This is a device that moves in a circle at the same uniform angular velocity, and is used to compare the motion effects of objects with those of other objects called clocks. Obviously this is an artificial rule. The advantage of uniform angular velocity is that the rotation angle has nothing to do with the size of the circumference, so we can make the clock arbitrary size. In a certain area of space, using a recognized clock motion to compare with other motion(Fig. 5.1.2) effects is the relative time. The amount of clock

Fig. 5.1.2 Clock motion

Chapter 5 Gravitational Field

motion is called the proper time.

3. The link between time and space

Light is a natural ruler. We choose photons as the measurement standard of time and space. The motion of all substances is compared with that of photons. The motion of photons is the fastest in the real world. This advantage makes all measurements have the same symbols. The speed of photon propagation slows down, the clock slows down correspondingly, the path of photon propagation bends, and the space bends. Photon is the best ruler for space-time measurement. There is a conversion factor c, $r = ct$ between space and time, which forms a light cone.

4. Time one-way

Matter exists in the form of motion, as small as elementary particles and as large as cosmic galaxies. For a region, there are many moving objects, which constitute a motion system. The motion effect of the motion system constitutes a time dimension of the region. If the time in the region is reversed, the motion system in the region must satisfy the following conditions:

(1) All moving bodies must return along the original path.

(2) Its return must be simultaneous.

(3) The relationship between systems must remain unchanged. Even in very small areas, such harsh conditions are difficult to achieve. Because it is impossible to have a mirror force field at the same time, all substances of the whole system (at least one region) return to the original path. It is an impossible form of motion, so there is no reverse time. The unidirectionality of the arrow of time is determined by the non-repeatability of the system motion.

5. The speed of clock time

The speed of the clock pointer is determined by the background field. When the background vacuum becomes "hard", it will slow down the propagation of the material system (Fig. 5.1.3), slow the clock in the region, and vice versa. If all the matter in the region enters the black hole and the propagation rate of the matter in the background vacuum of the black hole approaches zero, then the time in the region will be frozen. There is no force of background field that can "Please Return by the Way You Came" all the elementary particles that make up matter in this region, and black holes are no exception.

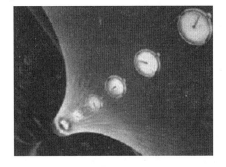

Fig. 5.1.3   Time is frozen on black hole surface

6. Time reversal

In many very rigorous scientific discussions, there is a space-time structure called "wormhole", which can realize time reversal (Fig. 5.1.4). If it is realized, a large-scale motion system or at least one city must be put into the wormhole, and then the wormhole must make every basic particle form the whole city to return accurately in its original way. Regardless of whether it violates the law of cause and effect, wormholes need not only tremendous energy, but more importantly, wormholes must have a complex means, and all moving matter will return to its original path. This is clearly impossible.

5.1.2 The interval of time and space

Fig. 5.1.4 Constitute time tunnel by wormhole

$\Phi_R(X) = \exp\{-iR[g_{\mu\nu}](X_\alpha)\}$ can be understood as the space-time fluctuation of cosmic scalar particles. For a simple local space-time $R[g_{\mu\nu}] = g_{\mu\mu}$, the 2-order strain of the background vacuum satisfies the conservation of strain $g_{00} - g_{11} - g_{22} - g_{33} = 0$, it is on lightcone. For a particle with mass, it cannot propagate at the speed of light in the optical cone (Fig. 5.1.5).

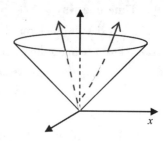

Fig. 5.1.5 Space-time world line

Light is the only legitimate ruler for measuring space and time. Since all substances are excited states of the vacuum background, the background space is curved and the optical ruler is curved. Measuring covariance is inevitable. If we are in a free-falling state, the instantaneous inertial reference system can be regarded as a physical "flat" coordinate system without external force. Therefore, to observe other reference systems from this reference, other reference systems are "bent". The inertial frame of reference we are in is x, y, z, t.

$$ct = x^0, x = x^1, y = x^2, z = x^3$$

These 4-dimensional coordinates can be written as x^μ, $\mu = 0, 1, 2, 3$. The space-time interval is

$$(dx^0)^2 - (dx^1)^2 - (dx^2)^2 - (dx^3)^2 = ds^2 \qquad (5.1.9)$$
$$-ds^2 = g_{\mu\nu} dx^\mu dx^\nu \qquad (5.1.10)$$

Here, $g_{\mu\nu}$ is the metric tensor. The expression (5.1.10) remains unchanged under Lorentz transformations, called Lorentz metric.

The superiority of the background space-time. Our material world is an excited state in the vacuum background, and the waves of various forms of material motion propagate in the vacuum. Therefore, vacuum itself constitutes a superior and most natural static reference system. In this reference system, the strain of background vacuum is 0, which has the natural advantage of isotropy and is a real sense of flat space-time.

5.1.3 Energy momentum tensor

Consider the simplest dust particles. The velocity of dust particles relative to the background space-time is v. The density of dust particles is n particles per unit volume, and the density of dust particles is $\rho \equiv nm$. The particle density $n_0 \equiv n\sqrt{1-v^2} = \gamma^{-1}n$ measured in a co-moving reference frame, i. e. $\rho_0 \equiv n_0 m$, the mass density measured in a local stationary reference frame of particles, is called the intrinsic mass density, and the energy-momentum tensor is

$$T_{\mu\nu} = T_{\nu\mu} = \rho_0 u^\mu u^\nu = \gamma\rho u^\mu u^\nu$$

Consider the simplest case

$$T_{\mu\nu} = \begin{bmatrix} \rho_0 c^2 + \frac{1}{2}\rho_0 v_x^2 & 0 & 0 & 0 \\ 0 & \gamma\rho_0 v_x^2 & 0 & 0 \\ 0 & 0 & \rho_0 v_y^2 & 0 \\ 0 & 0 & 0 & \rho_0 v_z^2 \end{bmatrix} ; \quad T^{\mu\nu} = n_0 m u^\mu u^\nu \quad (5.1.11)$$

The definition of $T^{\mu\nu}$ can be summarized as follows:
$T^{00} = [\text{Energy density}]$
$T^{0k} = T^{k0} = [\text{Momentum density}] = [\text{Energy flow density}]$
$T^{lk} = T^{kl} = [K \text{ momentum flux density in } L \text{ direction}]$

5.2 The spatial characteristics of gravitational field

General relativistic framework:
(1) $ds^2 = g_{\mu\nu} dx^\mu dx^\nu$. Line element, When $\mu = \nu$, is the space-time interval.
(2) $\nabla_\mu \phi^\nu = \partial_\mu \phi^\nu - \Gamma_{\mu\lambda}{}^\nu \phi^\lambda$. Covariant derivative, gravity, and interaction.
(3) $\frac{d^2 x^\mu}{d^2 \tau} + \Gamma^\mu_{\alpha\beta} dx^\alpha dx^\beta = 0$. Particle geodesic equation.
(4) $R_{\mu\nu} - \frac{1}{2} g_{\mu\nu} R = \frac{8\pi G}{c^4} T_{\mu\nu}$. Einstein gravitational field equation.

5.2.1 Measurement covariant and covariant field

1. The concept of flat vacuum

When the vacuum without any deformation, the basic unit of the vacuum strain

$$\varepsilon_{ij} \equiv 0 \qquad (5.2.1)$$

Countless and infinite vacuum elements with zero strain "bond" together to form a complete flat vacuum state of space-time Ω_0, without observable effect (Fig. 5.2.1). The non-uniform hardening vacuum in the macroscopic region is formed by the bonding of the basic vacuum unit of the whole non-uniform strain, which makes the propagation path of light bend (Fig. 5.2.2). If a coordinate system X is set up in a flat space-time, any slight bending will lead to 2-order strain $g_{\mu\nu} \neq 0$.

Here, the "bending" is not a sense of the whole geometry curved of vacuum. The space-time curvature effects caused by vacuum is "hardening", which shows as the no uniform of the propagation speed of light. Nonuniform results in the propagation path consisting of space-time curvature. Because any substance is constituted by the quantum field, bending effects are also exist in any substance, quantum field propagation in this bending background space. The bending of the space-time is the bending of background space of quantum field.

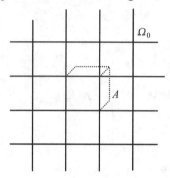 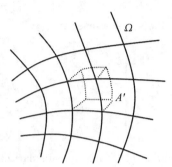

Fig. 5.2.1 Flat vacuum   Fig. 5.2.2 Curved space

The whole vacuum of non-uniform hardening is space-time bending, which can be described by Riemann geometry.

2. The conditions need to be meet by gravity field

The strain compatibility equation ensures that the vacuum does not break down and remains 3-dimensional. The strain compatibility equation gives the boundary conditions of vacuum strain.

$$\varepsilon_{ij,kl} + \varepsilon_{kl,ij} - \varepsilon_{ik,jl} - \varepsilon_{jl,ik} = 0 \tag{5.2.2}$$

The necessary and sufficient condition is that there is a continuous single-valued displacement field. The nature of (5.2.3b) meets the deformed continuous conditions, which is the strain compatibility equation. The strain compatibility equation determines the difference between the gravitational field and the quantum field. It shows that when the strain compatibility equation is not satisfied, the vacuum will break down and lead to vacuum fibrosis, which is the most important characteristic of the quantum field.

For quantum fields, the intrinsic vacuum of quantum fields is fragmented and does not satisfy strain compatibility equations.

$$\varepsilon_{ij,kl}(\xi) + \varepsilon_{kl,ij}(\xi) - \varepsilon_{ik,jl}(\xi) - \varepsilon_{jl,ik}(\xi) \neq 0 \tag{5.2.3a}$$

For the external observable space-time of quantum field, vacuum is continuous, satisfying strain compatibility equations.

$$\varepsilon_{ij,kl}(x) + \varepsilon_{kl,ij}(x) - \varepsilon_{ik,jl}(x) - \varepsilon_{jl,ik}(x) = 0 \tag{5.2.3b}$$

Vacuum satisfies strain compatibility equation, which shows spatiotemporal continuity. Einstein's theory of relativity is based on continuous space-time. This equation is important criterion of judging gravitational field or quantum field.

5.2.2 The principle of measurement covariantion

The two fundamental assumptions of special relativity are summarized as follows:

Chapter 5 Gravitational Field

(1) The laws of nature are the same in all inertial frames.

(2) The speed of light c is the same in all inertial frames.

The inertia frames in special relativity had been popularized to arbitrary reference system in general relativity. All reference system is equal right. And it's called as the first basic assumption the principle of general covariant.

Gauss had ever considered the following question. Imaging there is a two dimension curved surface. On the surface, there live the two dimensions animals are called as "book insect" who have intellect. Can they measure the curve of their space? Gauss found they could make it. At first, we accord a regular to mark the points on the curved surface with a freewill way. Getting any two clusters curves (x_1 = constant and x_2 = constant), we make them a coordinate system (Fig. 5.2.3). Directly measuring the length between point a and b, we get g_{22}. Like that, measuring length of ac and cd, we get g_{12} and g_{11}. Gauss got a set of equations and the equations can show curvature with metric and it's differential. Curvature is an intrinsic attribution. Any curvature on a given point has the same value in each coordinate. Gravitation theory of Einstein has established one connect between the distribution of energy and curvature of space. Here, when ruler goes from one point to another, the length of ruler does not change, the ruler and curved surface are independent, no influence each other.

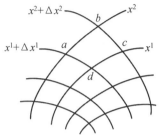

Fig. 5.2.3 Dimensions curve of Manifold

Such as Gauss's "book insect", Riemann had put forward a kind of physics picture. A kind of two dimensions life is called as "book insect" which live in the paper. When we put it on a wrinkles paper, they will still deduce that their world is completely smooth, because their body will be also wrinkled, these books insects never notice their world was twisted. Riemannian pointed out that the book insect attempt to exercise in crepe paper. They will feel a mysterious invisible "force" to stop their motion along a straight line. Here the "force" implies a planar which embedding curved 3D spaces, 2-dimensional surface itself constitute the constraint surface, the 2-dimensional book insect in non-free state felt the "force". This Force comes from background space. When a piece of paper is embedded in the 2-dimensional book insect in a free falling state, then the movement of the bookworm on crepe paper has no sense, this effect is called "measurement covariance". If the 2-dimensional book insect is not embedded in any high-dimensional space, then there is no constraint problem.

Now, consider another thing, the 4-dimensions space-time that we live in. Because all materials made of Fermions are in the excitation state of the vacuum of ground state, ruler itself also is made of Fermions, which is in the excitation state of the field of ground state. So ruler and human self are in the excitation state of field. In other words, our intellect and measure tool "embedding" in the vacuum. It leads to measurement covariance. Now, we establish the concept of "measurement covariantion".

In order to understand the perceptual covariance effect, we use Fig. 5.2.3 as a sketch of the flat ground state field to obtain the simplest rectangular coordinates. We are immersed in the ground state field with the measuring tools. Can we feel it when the shape of the whole vacuum changes? Now the original coordinate system changes from x, y, z, t to x', y', z', t'. Can we measure the coordinate changes? Is the Gauss method feasible? We were surprised at the results.

Imagine that there is an angle square $L(L_x, L_y)$ (Fig. 5.2.4(a)). Consider moving at v-uniform speed along the y-direction, ruler contracts from static length L_y to L'_y. Time expands from t to t', fitting following:

$$L'_y = L_y \sqrt{1 - (v/c)^2} \ ; \quad t' = t/\sqrt{1 - (c/v)^2} \qquad (5.2.4)$$

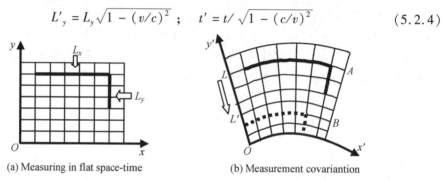

(a) Measuring in flat space-time (b) Measurement covariantion

Fig. 5.2.4 Measure of space-time

Different velocities correspond to different forms of space-time contraction, and each of them corresponds to a frame of reference. For uniform motion in inertial reference frame, the invariance of physical law is Einstein's weak equivalence principle. Because the form of background vacuum non-uniform hardening is arbitrary, the space-time inertial reference frame can be extended to arbitrary reference frame here. It corresponds to the strong equivalence principle.

When we move with a ruler, we get the same results at different positions. In addition, when the space-time bends, the light ruler bends accordingly, and there is no change in the measurement of the light ruler. Human beings themselves are vacuum quantum excited states, and perception and measurement are changing with the change of background vacuum, so we can't perceive the change of space in the end. We define this effect as measuring covariance. The reference frame with covariance effect is inertial reference frame. Thus it can be seen, the ruler, curved surface and intellect which Gauss talked about are independent each other. Under these conditions, we can perceive the state of 2-dimensional surface and establish corresponding geometry. The covariant perception makes flat space and curved space lose their standard of judgment. Because of existence of the covariant perception, we cannot perceive the curved space. Everything we have observed in curved space is the same as in flat space. There is no change when physics laws F changes from a flat field to a transformed field. Thus

Chapter 5 Gravitational Field

$$F \equiv F', \partial F/\partial F' \equiv 1 \qquad (5.2.5)$$

It shows that the laws of physics should have the same form in any reference system, and there is no difference between "good" reference system and "bad" reference system. This is measure covariance principle. For example, there is a gravitational field in a curved space. When objects fall freely, we lose the ability to judge space. So we believe that this space is flat and straight. Thus

$$R_{\mu\nu} = 0 \qquad (5.2.6)$$

That is the measurement covariant effect. The measurement covariance shows strong equivalence principle. We have a deeper understanding of Einstein's strong equivalence principle from the perspective of vacuum.

5.2.2 The relativity between reference systems

"Vacuum is not empty" may raise the question that matter in vacuum can form a natural frame of reference, which means that there exists an absolute frame of reference. Will it destroy the relativity between the frames of reference? In fact, the existence of measurement covariance makes the relationship between reference systems satisfy the relativity theory.

Now, we consider two reference systems S and S', and establish a "straighten" coordinate department $x^\mu(x^0, x^1, x^2, x^3)$ in reference system S. Here, we specially want to notice is due to the existence of measurement covariantion effect, all Fermions and Boson in reference system bend along with the bending of background vacuum, the space-time whether bend which is depend on Light-ruler. thus, we know nothing to our space-time in self-reference system that is curved. Flat space and curved space has lost the standard that differentiated. Spacetime does not exist absolutely flat, "flat" have no meaning. For another reference system S' also can establish "flat" coordinate department $x'^\mu(x'^0, x'^1, x'^2, x'^3)$. We use a light ruler to "calibrate" the coordinate system, also think that coordinate x'^μ is "straighten". From reference system S surveys reference system S', think that the coordinate department in reference system is bent, otherwise, from reference system S surveys reference system S', think that the coordinate department x'^μ in reference system S' is bent. Existence is as follows relation between two coordinate systems

$$dx'^\mu = \frac{\partial x'^\mu}{\partial x^\nu} dx^\nu \quad (\mu, \nu = 0, 1, 2, 3) \qquad (5.2.7)$$

Imagine the measure tool dx^ν in reference system S such as straightedge, laser source and clock take in reference system S' to "calibration" coordinate department of S' whether flat, after the measure tool in reference system S enters in reference system S', these tools are also bent with bend reference system S', make $dx^\nu = dx'^\mu$. Thus measure result that in coordinate department x'^μ is still flat which measured by these outside tools, when two reference systems are mutually to split and independence of each other. We measure for space-time in big scope, instead of measure in region space-time, we can get rid of measurement covariantion effect, and can measure space-time bend. So from reference system S can survey reference system S' is bend. And reference system S' can survey also think reference system S is bend. Deserve to be noticed is measurement covariantion effect occur in locally the space-time. Here "bend" is relative, which was decided by the location of reference system. The location of reference system difference, the result of surveying is also different, reference system are same right, reference system do not exist "good" or "bad". This is just relativity principle.

5.2.3 Principle of invariance of light speed

Particles are the excitation states of the vacuums. γ is the photon in the frame of reference S(Fig. 5.2.5). i and j are fermions with different masses and velocities. The speed of i is v_i, and j is v_j. In a gravitation field, when a frame of reference S which is in a free falls state down to B from A along direction y, its speed is v. S into to S' from low speed state to high speed state. S is compressed to S', the speed of photon γ' is c', the speed of i' and j' are respectively v'_i and v'_j. The transformation of space-time makes

$$c : v_i : v_j = c' : v'_i : v'_j \qquad (5.2.8)$$

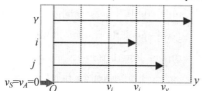
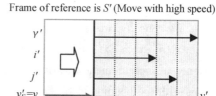

Fig. 5.2.5 The principle of constant velocity of lights

The ratio relationship maintain invariable. The measurement covariance occurs obviously. In S', the space-time is compressed along direction y with kinetic photon as standard size and standard clock. The standard size and standard clock are compressed together, thus the results of measure do not change. From the view of i', the speed of γ' is still c. The speed of j' does not change.

For two reference systems of uniform motion, it is assumed that two identical systems are in low-speed S and high-speed S, respectively. The Fermion strain increases synchronously from low-speed state to high-speed state. This effect is equivalent to background vacuum hardening. The propagation effect satisfies $c : v_i : v_j = c' : v'_i : v'_j$, that is to say, the speed of light remains unchanged. Therefore, the essence of the principle of invariance of light speed comes from measurement covariance. Let's take an example. Suppose we are falling into a black hole, and the background vacuum hardening is intensifying, so the propagation speed of photons will decrease, and the corresponding Fermion propagation speed will also decrease. Since all elementary particles propagate in the form of waves, the relative proportions between the propagation rates of elementary particles remain unchanged. For example, spacecraft, clocks and scales are made up of fermions, which synchronously strain, so the measurement results are

Chapter 5 Gravitational Field

invariable, which will obviously take place in the measurement of covariance, as shown by the speed of light invariance.

Vacuum vacancy will lead to measurement covariance effect, which is the basis for the existence of hypotheses of relativity.

5.2.4 Tidal forces

1. The formation of tidal forces

We discussed the measurement of covariance. If the spacecraft were in a free-falling state, the astronauts thought they were in a gravity-free environment ($g=0$). Can astronauts discover by an experiment that they fall into a gravitational field? The answer is yes. Astronauts can judge by the tidal force produced by gravitational field. Astronauts placed a large droplet in the center of their spacecraft, and the surface tension of the liquid made it spherical (non-gravitational). In free fall, they will find that it is not entirely spherical, just like an egg. The tip points to the earth. This is because the existence of gravitational field makes the gravitational force near the end of the droplet larger, while the gravitational force at the other end is weaker. This is tidal forces. We can calculate tidal forces. Consider a reference point $(0,0,z)$ of free-fall, the gravitational acceleration is

$$-\frac{GM}{(r_0 + z)^2} \tag{5.2.9}$$

Here, r_0 is the distance from the center of the Earth to the coordinate origin. M is the mass of the Earth. As the origin has an acceleration $-GM/r_0^2$, acceleration of the particle relative to the origin in the case that z is very small.

$$-\frac{GM}{(r_0 + z)^2} + \frac{GM}{r_0^2} \approx 2z\frac{GM}{r_0^3} \tag{5.2.10}$$

$$f_z = 2z\frac{Gmm}{r_0^3} \tag{5.2.11}$$

This is the tidalforces. It is proportional to the distance from the particle to the origin and is repulsive. For a particle at $(0, y, 0)$, tidal forces point to the direction on $-y$ and is equal to $f_y = -yGmm/r_0^3$, this is a tide restoring force towards the origin. For a particle located at $(0, y, 0)$, tidal forces point to the direction on $-x$ and is equal to $f_x = -xGmm/r_0^3$.

2. Tidal forces destructs measurement covariance

The existence of tidal force makes astronauts know the gravity field, which destroys the principle of measuring covariance. For very strong gravitational fields, such as the surface of a black hole, strong tidal forces are enough to tear apart spacecraft and astronauts. Assuming that the area of the observation space is q and the area of non-uniform hardening in the background space is Q, if $q \ll Q$, the measurement covariance is strictly established. The ratio of the two regions determines the damage degree of the covariance measurement.

5.3 The relationship between gravitational field and quantum field

At present, the contradiction between relativity and quantum mechanics is considered as follows. We can draw the following conclusions

(1) Quantum field is a non-commutative field, which satisfies the non-commutative relation $[x^i, p_i] \equiv i\hbar$. The non-commutative relation constitutes the condition of field quantization. Time and space are discontinuous.

(2) The metric of quantum field is a fixed metric $\eta_{\mu\nu}$.

1. Spatiotemporal structure

Relativity theory holds that time and space depend on the distribution of matter, and metric $g_{\mu\nu}$ is a variable. But quantum mechanics holds that time and space are stationary, flat, and unaffected by matter, $g_{\mu\nu} = \eta_{\mu\nu}$.

Quantum fields are first-order strains. The gravitational field is the 2-order strain. The existence of matter can also cause space-time bending inside and outside quantum fields, because the space-time inside the quantum field has no observable effect in the observational space, compared with the first-order strain of quantum field, the 2-order strain is a small amount, it degenerates to $\eta_{\mu\nu}$. Quantum field is observable for the 2-order strain effect caused by background space-time, which can be described by the metric tensor $g_{\mu\nu}$. The distribution of substances composed of elementary particles determines that the background space-time metric $g_{\mu\nu}$ is a variable.

2. Vacuum fluctuation

The micro-space-time view of general relativity is continuous and smooth, but quantum mechanics holds that there are quantum fluctuations in vacuum, spatiotemporal discontinuity.

Gravitational field is the 2-order effect of vacuum strain of quantum field. Although the vacuum inside the quantum field is fragmented, it does not fluctuate without measurements. But when the probe particle enters the micro-region for detection, it will disturb the quantum field. This misleads us to believe that there is a vacuum fluctuation in the quantum field. The gravitational field is macroscopic. The disturbance of the probe particle to the gravitational field is too weak to be observed.

5.3.1 Gravitational properties of mass

The vacuum is breakdown to form the quantum field. The intrinsic space-time of quantum field is fragmented, the quantum field has a fibrous structure, which not only produces electromagnetic field effect, but also produces gravitational effect. In other words, gravity originates from the relative deformation of the background vacuum caused by the concave and convex surface of the mass-loaded fiber structure (Fig. 5.3.1).

The quantum field pierces the smooth space-time like a small nail. The process is that the

Chapter 5 Gravitational Field

tip of a small nail compresses the smooth surface of space-time, and the surface of space-time is punctured after reaching the endurance limit. The edge of the smooth space-time surface is able to withstand the deformation limit. Without considering the temporal and spatial effects of other quantum fields, the total deformation of a smooth surface is

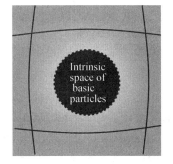

Fig. 5.3.1 Curved background space of particle

$$\int_{-\infty}^{\infty} \mathscr{G}_{\mu\nu} dV = h_G \quad (5.3.1)$$

From the point of view of internal space-time of Fermion quantum field, the total deformation of background space-time is $\int_{-\infty}^{\infty} \varepsilon_{\mu\nu} dV = h_f$. Here, we need to explain is that this deformation of surrounding particle internal field caused a wide range of tiny overall deformation vacuum field is defined as the gravitational field, impact size on the surrounding vacuum field is defined as the gravitational mass.

From the above, we can see that the gravitational mass is only an observable effect after the vacuum field is deformed. Gravity mass and inertia mass are the results of Fermion field strain observation from different angles. Different manifestations have the same essence, so gravity mass and inertia mass are strictly equal (Fig. 5.3.2).

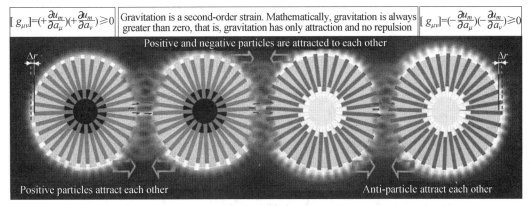

Fig. 5.3.2 Gravitation

1. Gravitational mass

The gravitational mass is the deformation of space-time caused by presence mass to generate a force field, this force field to another mass produces a force. And meet $F = m_G g$. Here g is accelerated speed in gravitational field. This force is called gravity.

From the above analysis, we can see that the electrostatic force produced by the charge is very similar to the gravity produced by the mass. Both forces are derived from the fiber properties of the quantum field. Therefore, there exists the same expression for the interaction force between the charges $F_e = kQ_1 Q_2 / r^2$ and the interaction force between matters $F_G = GM_1 M_2 / r^2$.

This explains why the two forces are so similar.

2. Inertial mass

An object in the state of rest or uniform motion, this objects to maintain the original state, and did not show mass effect. When applying an external force on the object, the object try to maintain the original state of transmission will to be generates a resistance, the force to meet $F = m_I a$, m_I is called the inertial mass.

The gravitational mass and the inertia mass are strictly equivalent, which has been confirmed by the experiment. The mass concept is a classic concept, and gravitational mass or inertia mass are not touched the essence of mass, Einstein's general relativity let us to have a more profound understanding to gravitation. Why the gravitational mass would be equal to the inertial mass, existing theories do not give a reasonable explanation.

5.3.2 Calculation of gravitational coupling strength

Quantum fields have fiber properties, the small protuberances with fiber structure on the surface of mass charge (Fig. 5.3.3). These small protuberances suppress the background vacuum and create gravity. When there is no small protuberance, consider that the mass charge is a sphere and the surface area of the sphere is A. When there are small protuberances, the outer space is pushed outward by small projections. The area of space expansion is ΔS (i.e. the second-order fiber offset), which is caused by the existence of fiber field. Therefore, the area is directly represents the strength of gravitational field. The top area of the small protuberance is the cross-section area of the fibers. The real gravitational effect is the top area of the small protuberance. Because the basic vacuum unit used to make up the fibers is exactly the same as the basic vacuum field unit next to it. This characteristic causes the fibrosis to protruding the top area $s = \Delta S/2$. We set the height of the small projection to be Δr, $\Delta r = [\mathscr{G}_{\mu\nu}] = \dfrac{\partial u_m}{\partial a_\mu} \dfrac{\partial u_m}{\partial a_\nu}$, This is the second-order strain $\varepsilon_{\mu\nu}$ of quantum fibers. Set the radius of mass charge sphere is R.

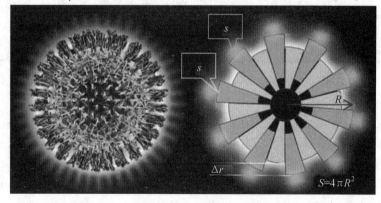

Fig. 5.3.3 Mass charge surface structure and background space

The surface area and volume of the mass charge are changed by the fibrous structure.

Chapter 5 Gravitational Field

Based on these changes, we can get the following equation.

$$\begin{cases} 4\pi(R+\Delta r)^2 - 4\pi R^2 = 2s \\ 2s \cdot \Delta r = \frac{4}{3}\pi(R+\Delta r)^3 - \frac{4}{3}\pi R^2 \end{cases} \quad (5.3.2)$$

Simultaneous solution is available.

$10s\pi \cdot \Delta r^2 = s^2 + 4\pi^2 \cdot \Delta r^4$, then $s^2 - 5s \cdot 2\pi\Delta r^2 + (2\pi\Delta r^2)^2 = 0$

Simultaneous solution is available.

$10s\pi \cdot \Delta r^2 = s^2 + 4\pi^2 \cdot \Delta r^4$, then $s^2 - 5s \cdot 2\pi\Delta r^2 + (2\pi\Delta r^2)^2 = 0$

$$s = [5 \cdot 2\pi\Delta r^2 \pm \sqrt{(5 \cdot 2\pi\Delta r^2)^2 - 4(2\pi\Delta r^2)^2}]/2$$

$$s_1 = (5 - \sqrt{21})\pi \cdot \Delta r^2 = (5-\sqrt{21})\pi\left(\frac{\partial u_m}{\partial a_\mu}\frac{\partial u_m}{\partial a_\nu}\right)^2 \text{ (Corresponding anti strain)}$$

$$s_2 = (5 + \sqrt{21})\pi \cdot \Delta r^2 = (5+\sqrt{21})\pi\left(\frac{\partial u_m}{\partial a_\mu}\frac{\partial u_m}{\partial a_\nu}\right)^2 \text{ (Corresponding normal strain)}$$

For substances, both positive and negative substances are equal, so the actual gravitational effect is the average of the two coefficients. The area of the top of the small protuberance determines the strength of the gravitational field $g_G = (s_1+s_2)/2$. The weak interaction strain determines the strength of weak interaction $g_W = k\varepsilon_{\mu\nu}$(See(7.07)). g_w is weak interaction constant.

$$g_G = \frac{s_2 + s_2}{2} = 5\pi\left(\frac{\partial u_m}{\partial a_\mu}\frac{\partial u_m}{\partial a_\nu}\right)^2 = 5\pi[(g_W \cdot g_W)(g_W \cdot g_W)] = 1.57(k_W)^4 \times 10^{-39}$$

(5.3.3)

According to $g_W = k_W \times 10^{-5} \approx 0.71 \times 10^{-5}$. Set $k' = 1.57(k_W)^4$, you can get $g_G = k' \times 10^{-39} \approx 0.4 \times 10^{-39}$. The conclusion is consistent with the experiment. Therefore, as long as we determine the weak coupling constant, the gravitational constant is also determined.

General relativity is used for large-scale astronomy. In that distance, Einstein's theory shows that there is no material that means that space is flat.

We think that such a stable flat space will always remain to any of the distance scale, but quantum mechanics completely changed this idea. All things could not escape quantum fluctuations provides by the uncertainty principle, and the gravitational field is no exception. Quantum mechanics shows that although the average gravitational field is zero, in fact, the gravitational field fluctuates due to the existence of quantum fluctuations. The uncertainty principle of quantum mechanics tells us that the smaller the area you observe, the larger the vacuum fluctuation (Fig. 5.3.4).

Above makes General relativity lose efficacy when it comes into the quantum field failed. There exists an insurmountable gap between general relativity and quantum field.

The unified vacuum theory holds that a vacuum without any disturbance is flat. As long as the intrinsic space of quantum field is not entered, the space-time of any scale is smooth and continuous. The prerequisite is that the vacuum is not disturbed and the vacuum does not break. Cognition of vacuum fluctuations is measured. Our experimental observation tool is the

Fig. 5.3.4 The vacuum fluctuation
(Amplified region of space, reveal its characteristic of ultra-micro Amplified spatial region, closer look will find that the presence of quantum fluctuations)

elementary particle. The smaller the space observed, the shorter the corresponding distance observed. The tool itself is a disturbance source, which causes the calm quantum field to produce quantum fluctuations. The shorter distance measurement, the greater the impact on the measured space. It's similar to watching a calm lake by helicopter (Fig. 5.3.5). Helicopter propeller would cause disturbance to the lake face. The greater the distance observed, the more calm lake face, contrary closer observation, the lake face is the more rolling waves, and the waves rolling in the observation area. Similarly, when the observation distance is close to a certain extent, we will see the fluctuation of the virtual particle sea caused by our own observation disturbance. The quantum fluctuation caused by close observation is no longer the intrinsic fluctuation of gravitational field. Because the fluctuation of gravitational field will lead to the destruction of vacuum (background space-time) structure, which does not conform to the strain compatibility equation, the "gravitational field" observed at close range is no longer gravitational field, but quantum field. When the macroscopic gravitational field is cut into quantum field by means of observation tools, the gravitational field no longer exists. Quantization itself destroys the gravitational field.

As long as we do not enter the inner space-time of the quantum field, the curved and straight space-time on any scale are smooth and continuous, andthere is no quantum fluctuation. Therefore, in very small space, vacuum satisfies the strain coordination function, and space-time has the characteristics of continuity and non-rupture. In this sense, the object of

Chapter 5   Gravitational Field

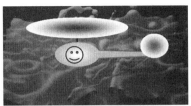

(a) Close observation, leading to the quantum fluctuations.

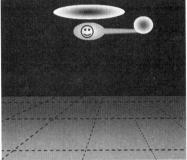

(b) Remote observation, no quantum fluctuations.

Fig. 5.3.5   The quantum fluctuation

(Space is flat, amplified region of space, means that close observation. The observation itself causes the quantum field spatiotemporal fluctuations)

study of relativity and quantum field is different.

Einstein's gravitational field equation is applicable to the second-order strain of background vacuum caused by the intrinsic vacuum strain of quantum field $\varphi(X) = \exp i\{[\varepsilon]_m g_{mn} X^n\}$. Vacuum deformation is continuous and must satisfy the strain field equation. The existence of quantum field results in vacuum hardening of quantum field background. It shows that the background space-time bends (Fig. 5.3.6). The result is the existence of gravitational field.

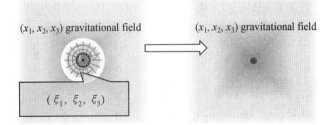

Fig. 5.3.6   Point like particles constitutes the source of the gravitational field

5.3.4   Mass charge constitutes gravitational source in observation space

The mass charge of quantum field in gravitational field is defined as gravitational charge. This very weak 2-order strain $[\mathscr{G}_{\mu\nu}]$ hardens the background vacuum and bends the background space. The strain is spherically symmetric. The generalized quantum field function containing gravity can be written as

$$\varphi(X) = \exp(-i)\{[\mathscr{G}_{\mu\nu}](X_R)\}\exp(-i)\{[\varepsilon_{\mu\nu}](X_T)\} = \Phi_G(X)\varphi_G(X)$$

The intrinsic of gravitational charge is obtained. We only consider the gravitational effect produced by intrinsic strain, so we define the spherically symmetric irrotational field function of gravitational charge.

$$\Phi_G(X) = \exp(-i)\{[\mathscr{G}_{\mu\mu}](X_R)\} \quad \text{(Gravitational charge)} \tag{5.3.4a}$$

$$\phi_M(X) = \exp(-i)\{[\varepsilon_{\mu\mu}](X_T)\} \quad \text{(Mass charge)} \tag{5.3.4b}$$

In static state, consider flat outer space. The wave function of gravitational charge is

$$\Phi_G(X) = \exp\left\{(-i)\begin{bmatrix}\mathscr{G}_{00} & 0 & 0 & 0\\ 0 & \mathscr{G}_{11} & 0 & 0\\ 0 & 0 & \mathscr{G}_{22} & 0\\ 0 & 0 & 0 & \mathscr{G}_{33}\end{bmatrix}\eta^{\mu}_{\nu}\begin{pmatrix}x_0 & 0 & 0 & 0\\ 0 & x_1 & 0 & 0\\ 0 & 0 & x_2 & 0\\ 0 & 0 & 0 & x_3\end{pmatrix}\right\}$$

$$= \exp i\left\{k\begin{bmatrix}g_{00} & 0 & 0 & 0\\ 0 & g_{11} & 0 & 0\\ 0 & 0 & g_{22} & 0\\ 0 & 0 & 0 & g_{33}\end{bmatrix}\begin{pmatrix}x^0 & 0 & 0 & 0\\ 0 & -x^1 & 0 & 0\\ 0 & 0 & -x^2 & 0\\ 0 & 0 & 0 & -x^3\end{pmatrix}\right\}$$

Here, $[g_{\mu\mu}]$ is Schwarzschild metric. Gravitational charges are curved in space and time.

$$\phi_M(X) = \exp\left\{(-i)\left[\begin{bmatrix}\varepsilon_{00} & 0 & 0 & 0\\ 0 & 0 & 0 & 0\\ 0 & 0 & 0 & 0\\ 0 & 0 & 0 & 0\end{bmatrix}\begin{pmatrix}x^0 & 0 & 0 & 0\\ 0 & 0 & 0 & 0\\ 0 & 0 & 0 & 0\\ 0 & 0 & 0 & 0\end{pmatrix} - \begin{bmatrix}0 & 0 & 0 & 0\\ 0 & \varepsilon_m & 0 & 0\\ 0 & 0 & \varepsilon_m & 0\\ 0 & 0 & 0 & \varepsilon_m\end{bmatrix}\begin{pmatrix}0 & 0 & 0 & 0\\ 0 & x^1 & 0 & 0\\ 0 & 0 & x^2 & 0\\ 0 & 0 & 0 & x^3\end{pmatrix}\right]\right\}$$

After Lorenz transform, from static state to moving state $\varphi_M(X) = \exp(-i)\{[\varepsilon_{\mu\mu}]g_{\mu\nu}L(X)\}$.

Static system $x^1 = x^2 = x^3 = 0$, $\begin{bmatrix}\varepsilon_0 & 0 & 0 & 0\\ 0 & 0 & 0 & 0\\ 0 & 0 & 0 & 0\\ 0 & 0 & 0 & 0\end{bmatrix} = E_0$; $\begin{bmatrix}0 & 0 & 0 & 0\\ 0 & \varepsilon_m & 0 & 0\\ 0 & 0 & \varepsilon_m & 0\\ 0 & 0 & 0 & \varepsilon_m\end{bmatrix} = m_0$ then

$$\phi_M(X) = \exp\left\{(-i)\left[\begin{bmatrix}c\varepsilon'_0 & 0 & 0 & 0\\ 0 & 0 & 0 & 0\\ 0 & 0 & 0 & 0\\ 0 & 0 & 0 & 0\end{bmatrix}\begin{pmatrix}\dfrac{x'^0}{\sqrt{1-u^2/c^2}} & 0 & 0 & 0\\ 0 & 0 & 0 & 0\\ 0 & 0 & 0 & 0\\ 0 & 0 & 0 & 0\end{pmatrix} - \begin{bmatrix}0 & 0 & 0 & 0\\ 0 & \varepsilon'_m & 0 & 0\\ 0 & 0 & \varepsilon'_m & 0\\ 0 & 0 & 0 & \varepsilon'_m\end{bmatrix}\right]\right\}.$$

$$\underbrace{}_{= m_0}$$

Chapter 5 Gravitational Field

$$\left. \begin{pmatrix} 0 & 0 & 0 & 0 \\ 0 & \dfrac{u_x x'^1}{\sqrt{1-u^2/c^2}} & 0 & 0 \\ 0 & 0 & \dfrac{u_{x2} x'^2}{\sqrt{1-u^2/c^2}} & 0 \\ 0 & 0 & 0 & \dfrac{u_{x3} x'^3}{\sqrt{1-u^2/c^2}} \end{pmatrix} \right\} \right\} \quad (5.3.5)$$

Mass charge eigenvalue equation $\hat{K}_T \phi_M(X) = T_{\mu\nu} \phi_M(X)$. Eigenvalue is

$$T_{00} = \frac{mc^2}{\sqrt{1-u^2/c^2}} \; ; \; T_{ii} = \frac{mu_i}{\sqrt{1-u_i^2/c^2}} \quad (5.3.6)$$

$$\phi_M(X) = \exp\left\{(-\mathrm{i}) \left[\begin{pmatrix} \dfrac{-m_0 c}{\sqrt{1-u^2/c^2}} & 0 & 0 & 0 \\ 0 & \dfrac{-m_0 u_1}{\sqrt{1-u_1^2/c^2}} & 0 & 0 \\ 0 & 0 & \dfrac{-m_0 u_2}{\sqrt{1-u_2^2/c^2}} & 0 \\ 0 & 0 & 0 & \dfrac{-m_0 u_3}{\sqrt{1-u_3^2/c^2}} \end{pmatrix} \begin{pmatrix} x'^0 & 0 & 0 & 0 \\ 0 & x'^1 & 0 & 0 \\ 0 & 0 & x'^2 & 0 \\ 0 & 0 & 0 & x'^3 \end{pmatrix} \right] \right\}$$

$$= \exp(-\mathrm{i} T_{\mu\nu} X_G) \quad (5.3.7)$$

This is the expression of the quantum field eigenvalue equation of special relativity. Because of $\dfrac{1}{\sqrt{1-u^2/c^2}} \approx 1 + \dfrac{1}{2}\dfrac{u^2}{c^2} + \dfrac{3}{8}\left(\dfrac{v}{c}\right)^4 + \cdots$, then $\phi_G(X) = \exp[(-\mathrm{i})(T_{\mu\nu}^0 + \Delta T_{\mu\nu}) X_G]$, $T_{\mu\nu}^0$ is intrinsic strain of mass charge, when $u_i \neq 0$.

$$\phi_M(X) = \exp\left\{ (-\mathrm{i}) \left[\begin{pmatrix} m_0 c^2 & 0 & 0 & 0 \\ 0 & -m_0 c & 0 & 0 \\ 0 & 0 & -m_0 c & 0 \\ 0 & 0 & 0 & -m_0 c \end{pmatrix} + \right. \right.$$

$$\left. \left. \frac{1}{2} \underbrace{\begin{pmatrix} m_0 u^2 & 0 & 0 & 0 \\ 0 & -m_0 u_1^2 & 0 & 0 \\ 0 & 0 & -m_0 u_2^2 & 0 \\ 0 & 0 & 0 & -m_0 u_3^2 \end{pmatrix}}_{u^2 = u_1^2 + u_2^2 + u_3^2} \begin{pmatrix} x'^0 & 0 & 0 & 0 \\ 0 & x'^1 & 0 & 0 \\ 0 & 0 & x'^2 & 0 \\ 0 & 0 & 0 & x'^3 \end{pmatrix} \right] \right\} \quad (5.3.8)$$

$$\phi_M(X) = \exp[(-\mathrm{i})(T_{\mu\nu}^0 + \Delta T_{\mu\nu}) X_G] \quad (5.3.9)$$

Special relativity is the space-time description of the eigenvalue of mass charge field function in motion state.

5.3.5 Cosmological real scalar particle background space-time

Cosmological real scalar particle field functions including gravity

$$\phi(X) = \exp\{(-i)[[T_{\mu\nu}](\xi_\alpha) + k[G_{\mu\nu}](X_\alpha)]\} \quad (5.3.10)$$

Cosmological real scalar particles can be regarded as a mass charge simply. A black hole is a real scalar particle in universe. For a real cosmic scalar particle in a flat vacuum, there is no vacuum strain outside. The gravitational field of the scalar particle is caused by its own mass and has nothing to do with the outside world. Gravitation is the second-order strain of the intrinsic vacuum of spherically symmetric real scalar particles. Considering that a point P outside the intrinsic space has no rotation, the intrinsic four momentum of matter can be seen in the observation space-time $[T_{\mu\mu}](\xi_\alpha) = [T_{\mu\mu}](X_\alpha)$, which satisfies $[T_{\mu\mu}](X)|_p - k[G_{\mu\mu}](X)|_p = 0$, that is $\partial_\alpha \varphi(X) = 0$. This is the wave equation formulation of Schwarzschild's solution.

Without considering the internal structure of the cosmic scalar particle, the influence of the particle on the background space-time satisfies the Schwarzschild solution of Einstein's gravitational field equation. The rest mass is M_0 which can be regarded as a spheres of mass charge oppression on the background space-time (Fig. 5.3.7), such effects is spherically symmetric. Consider a point P on the spherical shell, the radius of the spherical shell is r, static condition means using static coordinate system, $g_{\mu\nu}$ has nothing to do with time x^0, have the same, $g_{0m} = 0$. Take spherical polar coordinates, $x^0 = t, x^1 = r, x^2 = \theta, x^3 = \Phi$, ds^2 meet ball symmetric, the most common form is

$$ds^2 = Udt^2 - Vdr^2 - Wr^2(d\theta^2 + \sin^2\theta d\Phi^2)$$

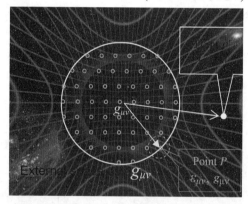

Fig. 5.3.7 The mass of the celestial body is evenly distributed in a spherical shell

U, V, w are just the functions of r, so we can use any $f(r)$ instead of r, do not affecting the symmetry of the ball. So we can simplify the problem, set $W=1$, ds^2 can be written as

$$ds^2 = e^{N(r)}dt^2 - e^{L(r)}dr^2 - r^2 d\theta^2 - r^2 \sin^2\theta d\Phi^2 \quad (5.3.11)$$

We can get the value of $g_{\mu\nu}$ from the formula, that is

$$g^{00} = e^N;\ g^{11} = -e^L;\ g^{22} = -r^{-2};\ g^{33} = -r^{-2}\sin^{-2}\theta \quad (5.3.12)$$

And

Chapter 5 Gravitational Field

$$g_{\mu\nu} = 0 \quad (\text{when } \mu \neq \nu)$$

The apostrophe is differential offunction, The following Christoffel symbols are not equal to zero.

$$\Gamma^1_{00} = \frac{1}{2} N' e^{N-L}, \quad \Gamma^0_{10} = \Gamma^0_{01} = \frac{1}{2} N'; \quad \Gamma^1_{11} = \frac{1}{2} L', \quad \Gamma^2_{12} = \Gamma^3_{13} = r^{-1}$$

$$\Gamma^1_{22} = -re^{-L}, \quad \Gamma^3_{23} = \Gamma^3_{32} = \cot\theta; \quad \Gamma^1_{33} = -r\sin^2\theta e^{-2\lambda}, \quad \Gamma^2_{33} = -\sin\theta\cos\theta$$

The Ricci tensor and curvature invariants can be calculatedby the Christoffel symbols, only the diagonal component R^ν_μ is not zero, Einstein equation in vacuum $R^\nu_\mu - \frac{1}{2}\delta^\nu_\mu R = 0$, the result is

$$R^0_0 - \frac{1}{2} R = - e^{-L} \left(\frac{L'}{r} - \frac{1}{r^2} \right) - \frac{1}{r^2} = 0, \tag{5.3.13}$$

Can be written as

$$e^{-L}(-rL' + 1) = 1 \tag{5.3.14}$$

$$R_1^{\ 1} - \frac{1}{2} R = e^{-L} \left(\frac{N'}{r} + \frac{1}{r^2} \right) - \frac{1}{r^2} = 0 \tag{5.3.15}$$

$$R_2^{\ 2} - \frac{1}{2} R = e^{-L} \left(\frac{N''}{2} - \frac{L'N'}{4} + \frac{N'^2}{4} + \frac{N' - L'}{2r} \right) = 0 \tag{5.3.16}$$

$$R_{33} = R_{22}\sin^2\theta \tag{5.3.17}$$

$R_{\mu\nu}$'s other components are zero. Consider (5.3.16) $e^{-L}(-rL' + 1) = 1$, That is $(re^{-L})' = 1$, then $re^{-L} = r - C$.

$$e^L = \frac{1}{1 - C/r} \tag{5.3.18}$$

Here C is constant. (5.3.18) minus (5.3.17), we can see $L' = -N'$. When $r \to \infty$, space must be close to flat space, λ and ν tend to zero. Thus $L+N=0$, $N(r)$'s solution is

$$e^{N(r)} = 1 - C/r \tag{5.3.19}$$

C is the integral constant, take $C = 2GM$

$$g_{00} = e^N = 1 - \frac{2GM}{r} \tag{5.3.20}$$

g_{00} is Potential. Complete solution

$$ds^2 = \left(1 - \frac{2GM}{r}\right) dt^2 - \left(1 - \frac{2GM}{r}\right)^{-1} dr^2 - r^2 d\theta^2 - r^2\sin^2\theta d\Phi^2 \tag{5.3.21}$$

This is the Schwarzschild solution of the mass charge, here c^2 is a very large number, R and is intrinsic space radius of mass charge. When $r \geqslant R$ it's in the mass charge background space, 4-dimensional smooth space-time. $r \leqslant R$ internal space of cosmological real scalar particle.

$$\begin{bmatrix} \rho_{M_0} c^2 & 0 & 0 & 0 \\ 0 & -\rho_{M_0} c & 0 & 0 \\ 0 & 0 & -\rho_{M_0} c & 0 \\ 0 & 0 & 0 & -\rho_{M_0} c \end{bmatrix} \begin{pmatrix} \tau & 0 & 0 & 0 \\ 0 & \xi^1 & 0 & 0 \\ 0 & 0 & \xi^2 & 0 \\ 0 & 0 & 0 & \xi^3 \end{pmatrix} \Rightarrow \text{Observation space}$$

$$\begin{bmatrix} \rho_{M_0}c^2 & 0 \\ 0 & -\rho_{M_0}c \end{bmatrix} \begin{pmatrix} t & 0 \\ 0 & r \end{pmatrix} \tag{5.3.22}$$

Here, $(\xi^1)^2+(\xi^2)^2+(\xi^3)^2=r^2$. (5.3.22) satisfaction (5.0.5). This is a function of a real scalar particle black hole in the universe.

$$R = 2GM_0/c^2 \quad \text{(Intrinsic space radius)} \tag{5.3.23}$$

5.3.6 Gravitational range of masscharge

Fermion with static mass m_0 is the mass charge, which deforms the background vacuum, and the total deformation is $h_G = h_f$. The smooth backgroundspace-time hardening is caused by fiber strain of Fermion intrinsic space. Fermion's fiber structure can simplify the problem. The influence coefficient of fiber strain on the background space-time is Gravitational constant G', meet

$$GE_G \cdot T = G'P_G \cdot R = G'mc^2 \cdot T = G'mc \cdot R = h_f \tag{5.3.24}$$

This is the quantization of the gravitational field. Here, E_G is the particle energy of gravitational field; P_G is the intrinsic gravitational field momentum of mass charge.

We consider the 4-dimensional space-time volume same as the total deformation of the mass charge that is h_f, see definition of charge, we can get

$$\nabla \cdot G = G\frac{m_0}{h_f} \tag{5.3.25}$$

1. The range of mass charge of gravity

The influence coefficient of different mass charge G' are not the same. Here, we can use the relationship between gravity and electric field strength to determine the coefficient, $G' = G/k_e$, intrinsic gravitational field momentum of mass charge is $G'm_0c$. Energy of gravitational field of mass charge $G'm_0c^2$. According to the quantization of the gravitational field $RG'm_0c = h_f$, Gravity range is limited, the radius to meet

$$R = \frac{h_f}{G'm_0c} \tag{5.3.26}$$

The vacuum curvature is minimal, the bending radius of deformation vacuum is great, and in other words, gravity is long range force.

A large number of particle mass gravity super position which has been observed gravity. Here, we look for the gravitational radius of electrons and protons. Consider

(1) Planck constant: $h = 6.626\ 068\ 96(33) \times 10^{-34}$ J·s (or $h = 4.135\ 667\ 43(35) \times 10^{-15}$ eV·s).

(2) Electronic quality $= 9.1 \times 10^{-31}$ kg, or 511 keV (1 eV $= 1.602 \times 10^{-19}$ J).

(3) Electric and gravitational strength ratio: $G'^{-1} = ke^2/Gm^2$.

(4) The speed of light in vacuum: 299 792 458 m/s.

2. Electronic gravitational radius

The electrostatic force between electrons is $f = ke^2/r^2$. Here, $k = 9 \times 10^9$, $e = 1.6 \times 10^{-19}$,

Chapter 5 Gravitational Field

$G=6.67\times10^{-11}$. The gravitation between electrons: $f'=Gm^2/r^2$; $f/f'=ke^2/Gm^2$. We can get the numerical: $f/f'=4.17\times10^{42}$, and can be calculated electronic gravitational radius by the formula $R=h_f/G'm_0c$.

$$R_{Ge}=4.17\times10^{42}\times\frac{2\times6.626\,068\,96\times10^{-34}\,\text{kg}\cdot\text{m}^2\cdot\text{s}^{-1}}{9.109\times10^{-31}\,\text{kg}\times3\times10^9\,\text{m/s}}=2.024\times10^{30}\,\text{m}$$

3. The gravitational radius of the proton

Proton mass $1.672\,623\,1\times10^{-27}$ kg into $G'^{-1}=ke/Gm^2$, can be calculated $G'^{-1}=1.235\times10^{36}$.

$$R_{Gp}=1.235\times10^{36}\times\frac{2\times6.626\,068\,96\times10^{-34}\,\text{kg}\cdot\text{m}^2\cdot\text{s}^{-1}}{1.672\,6\times10^{-31}\,\text{kg}\times2.998\times10^9\,\text{m/s}}=2.688\times10^{24}\,\text{m}$$

The radius of Milky Way is about 5—6 million light-years (1 light-year = 9,454,254,955,488,000 m), be equal to $5.5\times10^4\times9.454\,254\,955\,488\times10^{15}$ m $=5.2\times10^{20}$ m. The radius of universe is 137×10^8 light-years, equal to $137\times10^8\times9.454\,254\,955\,488\times10^{15}$ m $=1.295\times10^{26}$ m.

Therefore, we know the gravitational radius of electrons than the universe radius, the gravitational radius of the proton radius smaller than the universe, but more than the Milky Way.

5.4 the curved space-time description

5.4.1 Vector translation

Levi-Civita introduced the concept of generalized of "translation". Setting there is a vector A and its starting point is at M on 2-dimensional surfaces embedded in 3-dimensional Euclidean spaces. If you make A' which in M translation infinitesimal distance into an infinitesimal neighborhood M', according to the definition of Levi-Civita: first make A into M' by conventional translation, then projection in the surface of tangent plane M' obtain A'. It's called "translation" that vector A at point M to A', or $A'//A$. This translation is called Levie-Civita translation (Fig.5.4.1).

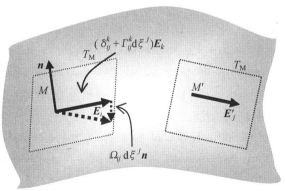

Fig.5.4.1 Levi-Civita translation

Set 2-dimensional Riemann space R_2 coordinate systems are $\{\xi^i\}$ ($i=1,2$) in 3-dimensional Euclidean space E_3, curve C in surface R_2, and the parameter equation is $\xi^i = \xi^i(t)$, with the change from t to $t+\mathrm{d}t$, the point in the curve change correspondingly from $M(\xi^i)$ to $M'(\xi^i + \mathrm{d}\xi^i)$. At this time, the point M' is in the infinitely small neighborhood region of the point M.

Set in the E_3, the points R_2 in M and M' tangent space (plane), are denoted by T_M, T'_M and then recorded the diameter as $r = r(\xi^i)$. Apparently, tangent frame of each point on the curve C while in their respective tangent space, but not unrelated to each other. Set the tangent frame vector of point M as

$$E_i = \frac{\partial r}{\partial \xi^i}\bigg|_M = \frac{\partial x^\alpha}{\partial \xi^i}\bigg|_M e_\alpha$$

Here, e_α ($\alpha = 1,2,3$) is the basis vectors of a fixed frame (rectilinear coordinate system) which selects in E_3, $x^\alpha = x^\alpha(\xi^i)$ is coordinate frame of r in $\{0 - e_\alpha\}$, So the tangent frame vector of point M' is

$$E_i = \frac{\partial r}{\partial \xi^i}\bigg|_{M'} = \frac{\partial x^\alpha}{\partial \xi^i}\bigg|_{M'} e_\alpha = \left(\frac{\partial x^\alpha}{\partial \xi^i} + \frac{\partial^2 x^\alpha}{\partial \xi^i \partial \xi^j}\mathrm{d}\xi^j\right)\bigg|_M e_\alpha$$

$$= E_i + \frac{\partial^2 r}{\partial \xi^i \partial \xi^j}\bigg|_M \mathrm{d}\xi^j \tag{5.4.1}$$

$\frac{\partial^2 r}{\partial \xi^i \partial \xi^j}\big|_M$ is quantity of the point M, usually in E_3, not in T_M. Let through the point M make a vector non T_M, so the vector $\frac{\partial^2 r}{\partial \xi^i \partial \xi^j}$ can be expressed as

$$\frac{\partial^2 r}{\partial \xi^i \partial \xi^j} = \Gamma_{ij}^k E_k + \Omega_{ij} n \tag{5.4.2}$$

Amongit, Γ_{ij}^k is the component of vector $\frac{\partial^2 r}{\partial \xi^i \partial \xi^j}$ of the projection vector in T_M about tangent frame $\{E'_i\}$ of M point, Ω_{ij} is normal component of vector. The (5.4.2) into (5.4.1) we can obtain:

$$E'_i = E_i + \Gamma_{ij}^k \mathrm{d}\xi^j E_k + \Omega_{ij}\mathrm{d}\xi^j n = (\delta_i^k + \Gamma_{ij}^k \mathrm{d}\xi^j) E_k + \Omega_{ij}\mathrm{d}\xi^j n$$

From the point of E_3 view (Fig. 5.4.1), the above equation shows that tangent frame vector M' on E'_i translate (parallel move) to the point M, which is the sum of the vector $(\delta_i^k + \Gamma_{ij}^k \mathrm{d}\xi^j) E_k$ and the vector $\Omega_{ij}\mathrm{d}\xi^j n$ at normal. $n \perp T_M$, so the vector $E_i - \Omega_{ij}\mathrm{d}\xi^j n$ is the projection vector of E'_i on T_M, called as $(\delta_i^k + \Gamma_{ij}^k \mathrm{d}\xi^j) E_k$.

According to the Levi-Civita definition of translation, the vector above-mentioned is the result that the vector E'_i translation from the point M' by the Levi–Civita translation. Then here is $(\delta_i^k + \Gamma_{ij}^k \mathrm{d}\xi^j) E_k \parallel E'_i$.

Now, setting there is a vector A' in M' that is translated to M by the Levi-Civita translation. We find the relationship between before and after coordinate's translation. If $A' = A'^i E_i$,

Chapter 5 Gravitational Field

then
$$A'^{i}(\delta_i^k + \Gamma_{ij}^k d\xi^j)E_k \mathbin{/\mkern-6mu/} A'^{i}E'_i$$

Marker the vector as $A = A^i E_i$, and A' is translation to the point M. That is
$$A^i E_i = A'^{i}(\delta_i^k + \Gamma_{ij}^k d\xi^j)E_k = (A'^{k} + \Gamma_{ij}^k A'^{i} d\xi^j)E_k$$

Change the dummy index on the right hand side, and compare the coefficient of E_i, then
$$A^i = A'^{i} + \Gamma_{jk}^i A'^{j} d\xi^k \tag{5.4.3}$$
$$\delta A^i = A'^{i} - A^i = -\Gamma_{jk}^i A'^{j} d\xi^k = -\Gamma_{jk}^i (A^j + \delta A^j) d\xi^k \tag{5.4.4}$$

Leave out the second-order infinitesimal, here is
$$\delta A^i = A'^{i} - A^i = -\Gamma_{jk}^i A^j d\xi^k \tag{5.4.5}$$
$$A'^{i} = A^i - \Gamma_{jk}^i A^j d\xi^k \tag{5.4.6}$$

This is the relationship between before and after coordinatetranslation. It Can be shown, to translation the tensor A_i on R_2, have
$$\delta A_i = \Gamma_{jk}^i A_j d\xi^k \tag{5.4.7}$$

And
$$A'_i = A_i + \Gamma_{ik}^j A_j d\xi^k \tag{5.4.8}$$

Euclidean space has a uniform linear scale. If the vector components are the same at two different points, we think they are the same, or parallel, or each of them produced by another translation. This shift is called as ordinary translation. However, ordinary translation is meaningless in the Riemannian space. That also cannot be considered they are equal when vector on different points which have the same component, because different points have different tangent frame.

5.4.2 Covariant differential

$$A^\mu_{;\sigma} = \frac{\partial A^\mu}{\partial x^\sigma} + \Gamma_{\sigma\alpha}^\mu A^\alpha \tag{5.4.9}$$

$$D_\sigma = \frac{\partial}{\partial x^\sigma} + \delta_\mu^\alpha \Gamma_{\sigma\alpha}^\mu \tag{5.4.10}$$

$$A_{\mu;\sigma} = \frac{\partial A_\mu}{\partial x^\sigma} - \Gamma_{\mu\sigma}^\alpha A_\alpha \,;\, A_{\mu\nu;\sigma} = \frac{\partial A_{\mu\nu}}{\partial x^\sigma} - \Gamma_{\mu\sigma}^\alpha A_{\alpha\nu} - \Gamma_{\nu\sigma}^\alpha A_{\mu\nu} \tag{5.4.11}$$

The relationship of Γ_{jk}^i and g_{jk}:
$$\Gamma_{\mu\alpha}^\gamma = \frac{1}{2} g^{\gamma\nu}\left[\frac{\partial g_{\nu\mu}}{\partial x^\alpha} + \frac{\partial g_{\nu\alpha}}{\partial x^\mu} - \frac{\partial g_{\mu\alpha}}{\partial x^\nu}\right] \tag{5.4.12}$$

That's the second Christoffel symbols.
$$\Gamma_{\nu,\mu\alpha} = \frac{1}{2}\left[\frac{\partial g_{\nu\mu}}{\partial x^\alpha} + \frac{\partial g_{\nu\alpha}}{\partial x^\mu} - \frac{\partial g_{\mu\alpha}}{\partial x^\nu}\right] \tag{5.4.13}$$

That's first Christoffel symbol.

5.4.3 Curvature tensor

The curved space is not the bend of geometric meaning, and the essence of the vacuum which exist the vacuum strain. We know that any form of strain in a quantum field will cause a small 2-order strain in the background vacuum, which is characterized by the curvature of the background space-time. Background vacuum strain causes vacuum "hardening". In the unified theory of vacuum, the essence of curved space-time is the non-uniform hardening of vacuum, which is represented by the curved background space-time. In an inhomogeneous hardening vacuum, the inhomogeneity of the propagation velocity of photons (and other forms of quantum fields) makes the propagation path curved, which is similar to the refraction caused by light entering a convex lens. The inhomogeneous strain of background vacuum is understood as curved space-time.

In the curved space (Fig. 5.4.2), along the closed line translation, generally do not give the original vector. Covariant derivative DA_j is equal to the ordinary differential dA_j subtracts translation variation δA_j. Accumulation around the closed line in a circle is zero. The accumulation of translate variation δA_j is equal to the accumulation of covariant differential DA_j. Therefore, the accumulation around closed circuit in a circle characteristics the degree of bended space. Now let us consider a vector component changes when it displaces in parallel along the infinitesimal closed route, and the closed route is composed of four curves them amongtwo parameter family of curves.

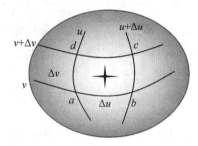

Fig. 5.4.2 The image of the curvature tensor

$$x^\alpha = f^\alpha(u,v) \tag{5.4.14}$$

The route is considered in Fig. 5.4.2 surrounded the four side of the quadrilateral, the change of n^μ in the closed path as the follow formula.

$$\delta n^\mu = -\oint \Gamma^\mu_{\alpha\beta} n^\alpha dx^\beta \tag{5.4.15}$$

Taking into account the Algebraic sum of contribution which is generated by the sides of the "parallelogram" and retaining once item of Δu and Δv, the result is as following equation.

$$\delta n^\mu = \frac{\partial}{\partial v}(\Gamma^\mu_{\alpha\beta} n^\alpha) \Delta v \frac{\partial x^\beta}{\partial u}\Delta u - \frac{\partial}{\partial u}(\Gamma^\mu_{\alpha\gamma} n^\alpha) \Delta u \frac{\partial x^\gamma}{\partial v}\Delta v \tag{5.4.16}$$

To complete the operation, and use n^μ to indicate changing formula which under parallel displacement, the result can be written as

$$\delta n^\mu = -\left[\frac{\partial \Gamma^\mu_{\alpha\gamma}}{\partial x^\beta} - \frac{\partial \Gamma^\mu_{\alpha\beta}}{\partial x^\gamma} + \Gamma^\mu_{\sigma\beta}\Gamma^\sigma_{\alpha\gamma} - \Gamma^\mu_{\sigma\gamma}\Gamma^\sigma_{\alpha\beta}\right] \times n^\alpha \frac{\partial x^\gamma}{\partial v}\frac{\partial x^\beta}{\partial u}\Delta u \Delta v \tag{5.4.17}$$

$$\lim_{\substack{\Delta u \to 0 \\ \Delta v \to 0}} \frac{\delta n^\mu}{\Delta v \Delta u} = -\left[\frac{\partial \Gamma^\mu_{\alpha\gamma}}{\partial x^\beta} - \frac{\partial \Gamma^\mu_{\alpha\beta}}{\partial x^\gamma} + \Gamma^\mu_{\sigma\beta}\Gamma^\sigma_{\alpha\gamma} - \Gamma^\mu_{\sigma\gamma}\Gamma^\sigma_{\alpha\beta}\right] n^\alpha \frac{\partial x^\gamma}{\partial v}\frac{\partial x^\beta}{\partial u} \tag{5.4.18}$$

The left-side of the equation (5.4.18) is a vector, because u and v is parameter, $\partial x^\beta/\partial u$ and $\partial x^\gamma/\partial v$ is also the vectors. Therefore, according to the following equation

Chapter 5 Gravitational Field

$$R^{\mu}_{\alpha\beta\gamma} = \frac{\partial \Gamma^{\mu}_{\alpha\gamma}}{\partial x^{\beta}} - \frac{\partial \Gamma^{\mu}_{\alpha\beta}}{\partial x^{\gamma}} + \Gamma^{\mu}_{\sigma\beta}\Gamma^{\sigma}_{\alpha\gamma} - \Gamma^{\mu}_{\alpha\gamma}\Gamma^{\sigma}_{\alpha\beta} \qquad (5.4.19)$$

Defining $R^{\mu}_{\alpha\beta\gamma}$ as the tensor, called Raman-Christoffel tensor or referred to as the curvature tensor. For an infinitesimal closed routes

$$\delta n^{\mu} = - R^{\mu}_{\alpha\beta\gamma} n^{\alpha} \frac{\partial x^{\beta}}{\partial u} \frac{\partial x^{\gamma}}{\partial v} du dv \qquad (5.4.20a)$$

And because $\delta(n_{\mu}n^{\mu})$ is equal to zero under parallel displacement, has

$$\delta n_{\mu} = R^{\alpha}_{\mu\beta\gamma} n_{\alpha} \frac{\partial x^{\beta}}{\partial u} \frac{\partial x^{\gamma}}{\partial v} du dv \qquad (5.4.20b)$$

Through (5.4.20), for any finite closed routes (along the curve integral to u and v), it guaranteed any vector's invariant under the parallel displacement when $R^{\mu}_{\alpha\beta\gamma}$ is equal to zero. Lorenz metric in a given region, because of $g_{\mu\nu}$ is constant, curvature tensor to be zero. Carrying on coordinate transformation to the space, due to $R^{\mu}_{\alpha\beta\gamma}$ tensor property, its component remains zero after transformation. Thus, the necessary condition of the space flat is $R^{\mu}_{\alpha\beta\gamma}$ equal to zero. On the other hand, if R^{μ} are zero anywhere, any vector all component is zero under parallel displacement. All first-order derivative of the metric will be zero everywhere.

As the metric of the starting point for the Lorenz metric, apparently, any metric are the Lorenz metric anywhere. The results show that: the necessary and sufficient condition of the-vacuum of space for flat is that $R^{\mu}_{\alpha\beta\gamma}$ all components are zero everywhere. The expression (5.4.19) tells us, $R^{\mu}_{\alpha\beta\gamma}$ is antsymmetric to β and γ. Tensor $R_{\delta\alpha\beta\gamma}$ is given by

$$\begin{aligned} R_{\delta\alpha\beta\gamma} &= g_{\mu\delta} R^{\mu}_{\alpha\beta\gamma} \\ &= \frac{1}{2}\left(\frac{\partial^2 g_{\delta\gamma}}{\partial x^{\alpha} \partial x^{\beta}} + \frac{\partial^2 g_{\alpha\beta}}{\partial x^{\delta} \partial x^{\gamma}} - \frac{\partial^2 g_{\delta\beta}}{\partial x^{\alpha} \partial x^{\gamma}} - \frac{\partial^2 g_{\alpha\gamma}}{\partial x^{\delta} \partial x^{\beta}} \right) + g_{\mu\nu}(\Gamma^{\mu}_{\alpha\beta}\Gamma^{\nu}_{\delta\gamma} - \Gamma^{\mu}_{\alpha\gamma}\Gamma^{\nu}_{\delta\beta}) \end{aligned} \qquad (5.4.21)$$

By the formula (5.4.21) can be seen

$$R_{\alpha\delta\beta\gamma} = - R_{\delta\alpha\beta\gamma} = - R_{\alpha\delta\gamma\beta} = R_{\beta\gamma\alpha\delta} \qquad (5.4.22)$$

By (5.4.21) and (5.4.22) can be verified

$$R^{\alpha}_{\beta\gamma\delta} + R^{\alpha}_{\delta\beta\gamma} + R^{\alpha}_{\gamma\delta\beta} = 0 \qquad (5.4.23)$$

$$R_{\alpha\beta\gamma\delta} + R_{\alpha\delta\beta\gamma} + R_{\alpha\gamma\delta\beta} = 0 \qquad (5.4.24)$$

5.4.4 Geodesic

We consider this question "each line elementis results of the first element parallel displacement", what kind of curve equation is defined by this one requirement? Any curve equation is a single parameter point family $x^{\alpha} = f^{\alpha}(s)$. The tangent vector is dx^{α}/ds, and new tangent vector can be attained by parallel displacement as following:

$$T'^{\rho} = \frac{dx^{\rho}}{ds} - \Gamma^{\rho}_{\alpha\beta} \frac{dx^{\alpha}}{ds} dx^{\beta} \qquad (5.4.25)$$

New tangent vector T'^{ρ} can be given by the following equation

$$T'^{\rho} = \frac{dx^{\rho}}{ds} + \frac{d}{ds}\left(\frac{dx^{\rho}}{ds} \right) ds \qquad (5.4.26)$$

Order (5.4.25) be equal to (5.7.26), then curve equation is

$$\frac{d^2 x^\rho}{ds^2} + \Gamma^\rho_{\alpha\beta} \frac{dx^\alpha}{ds} \frac{dx^\beta}{ds} = 0 \qquad (5.4.27)$$

Each line element of the curve is the first element parallel displacement results. (5.4.27) represents extreme length curve called a geodesic equation. The equation describes the path of light propagation, that is, geodesic line.

5.5 Gravitational field equation

5.5.1 Einstein gravitational field equation

1. Establishing Einstein field equations

The basic idea of establishing Einstein's field equation is that the distribution and motion of matter determine the space-time structure. This principle is called Einstein's principle. This principle is that local field deformation affects the space-time of background field in vacuum.

Einstein's principle concludes that a good basic quantity used to describe the motion of matter should be proportional to the basic good quantity used to describe the structure of space-time. Both basic good quantities aresymmetrical. We consider good basic quantity now.

(1) Basic quantity of describe the space-time structure should be energy-momentum tensor T^{ik}.

(2) Describe the space-time structure of basic tensor is metric $g_{\mu\nu}$ and curvature tensor R^{ik} and R. Note Γ^i_{kl} is also an amount of description space-time structure, but not tensor, so it can't correspond tensor T^{ik}. The gravitation and acceleration in curved space is described by Riemann metric. A non-uniform gravitational field, in each small area, with a proper acceleration system is equivalent. So it can be describe gravitational field by Riemann metric. The line element square of deformation vacuum space is given by $-ds^2 = g_{\mu\nu} dx^\mu dx^\nu$, the metric tensor of gravitational field equation with $g_{\mu\nu}$, this is a natural result of the 2-order strain ofvacuum. In general, space-time curvature tensor

$$R^{ik}_{;k} \neq 0 \qquad (5.5.1)$$

So it cannot be directly used $R_{ik} \propto T_{ik}$ to establish the field equations. R_{ik} had to be improved. Find a tensor $G^{ik}_{;k} = 0$, and it's closer to R^{ik}, this tensor does exist.

Because $G^{ik}_{;k} = 0$, and G^{ik} is closer to R^{ik}, should be able to export by the curvature tensor covariant derivative equation. This tensor is discussed earlier in the Einstein tensor, i.e.

$$(R^\nu_\gamma - \frac{1}{2}\delta^\nu_\gamma R)_{;\nu} = G^\nu_{\gamma;\nu} = 0 \qquad (5.5.2)$$

You can calculate $T^{ik}_{;k} = 0$. According to Einstein's principle, we can see that the G^{ik} and T^{ik} is proportional

$$G^{ik} = \kappa T^{ik}$$

Chapter 5 Gravitational Field

Here, κ is a constant of proportionality; G^{ij} is Einstein tensor. Field equations will take the following form:

$$R_\mu^\nu - \frac{1}{2}\delta_\mu^\nu R - \lambda \delta_\mu^\nu = KT_\mu^\nu \quad (5.5.3)$$

Here, λ and κ is a constant. In the weak field, gravitational gauge φ describe by Poisson equation

$$\nabla^2 \varphi = 4\pi G \rho_M \quad (5.5.4)$$

Where G is the gravitational constant, ρ_M is the mass per unit volume. At least the (5.5.4) is modified into Lorenz invariant, it is written as

$$\Box \phi = 4\pi G \rho_M \quad (5.5.5)$$

Equation (5.5.5) is similar to the 4-dimensional potential in the electrodynamics equations. (5.5.3) constant κ chosen as $8\pi G/c^4$, G is the gravitational constant $G=6.667\times10^{-8} \mathrm{cm}^3 \cdot \mathrm{g}^{-2} \cdot \mathrm{s}^{-2}$. Then the field equation becomes

$$R_{\mu\nu} - \frac{1}{2}g_{\mu\nu}R = \frac{8\pi G}{c^4}T_{\mu\nu} \quad (5.5.6)$$

This is the well-known Einstein equation of gravitational field.

2. Vacuum physics image of the field equations

The concept of "space-time bending" based on vacuum is inconsistent with that of Einstein, where "space-time bending" is vacuum non-uniform hardening (Fig. 5.5.1). Vacuum hardening reduces the ability of wave propagation, slows down light and refracts it. For example, when light enters a high density medium from a low density medium, changes in the speed of propagation cause the light to bend. Large-scale non-uniform hardening in vacuum is characterized by curved space-time, curved propagation path of light and existence of gravitational field. We define space-time as follows, because the large-scale inhomogeneous field strain in the background vacuum, the propagation rates of photons and fermions are inhomogeneous, which results in the continuous refraction effect that bends light. This effect is called space-time bending.

The left side of Fig. 5.5.1 is the vacuum of bending corresponding to the observed effect of the right mass.

The essence of space-time bending is the refraction of light propagation which caused by vacuum hardening. Therefore, the total angle of light bending through a single gravitational field is less than or equal to 90°.

$$\alpha = \int_0^\infty d\alpha \leqslant \pi/2 \quad (5.5.7)$$

Here, $0\text{---}\infty$ measuring range, that is, the distance of light passing through the gravitational field, α is the total angle of light deflection.

This conclusion is different from general relativity. According to the basic properties of vacuum, we know:

The Vacuum Super-unified Theory

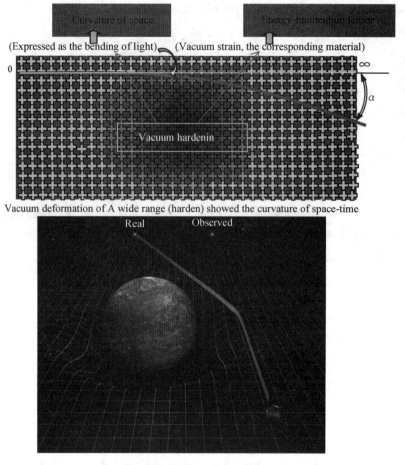

Fig. 5.5.1 Curved space

(1) Vacuum will break up, and space-time cannot be bent into closed space.

(2) Vacuum hardening characteristics do not allow light propagation paths to form closed curves.

(3) Both positive and negative substances cause the background vacuum hardening, and the bending of space-time are positive curvature.

5.5.2 The gravitational field equation of vacuum strain

Every quantum field is aspace-time breakdown point. Matter composed of quantum field is a collection of space-time breakdown points. The 1 – order strain $\varepsilon_{\mu\nu}$ (4-momentum $T_{\mu\nu}$) of quantum field determines the 2-order strain $g_{\mu\nu}/2 = \varepsilon'_{\mu\nu}$ of quantum field. The 2-order strain $g_{\mu\nu}$ of the quantum field is the strain of the background vacuum of the quantum field. The strain of the background vacuum of the quantum field determines the hardening degree of the background vacuum of the quantum field, which shows the curvature of the space-time of the background of the quantum field. The 1-order strain $\varepsilon_{\mu\nu}$ of quantum field characterizes the degree of vacuum breakdown. Any strain applied to the quantum field will aggravate the degree of vacuum break-

Chapter 5 Gravitational Field

down. In other words, the motion and distribution of matter can be characterized by the degree of vacuum rupture. Therefore, it is concluded that the 4-momentum tensor $T_{\mu\nu}$ of matter determines the breakdown tensor $L_{\mu\nu}$ of background vacuum and the curvature tensor $R_{\mu\nu}$ of background space-time. The geometric effect of space-time breakup inside matter on background space-time is the gravitational effect of quantum field. The expression of this relationship is the quantum gravitational equation. According to this principle, we can construct the gravitational field equation. $T_{\mu\nu} = KL_{\mu\nu} = K'R_{\mu\nu}$.

The gravitational field strain can be reduced to $\varphi_G(X) = \exp\{(-i)[[\varepsilon_G]_\mu g_{\mu\nu} X(\xi)^\nu + [\varepsilon'_G]_k g_{kl} X(x)^l]\}$ from the point of view of quantum field strain. It consists of two parts: internal strain $\varepsilon_{\mu\nu}$ and background vacuum 2-order strain $\varepsilon'_{\mu\nu}$. It is noteworthy that the quantum field propagates in a vacuum and does not twist against the background vacuum, so the background vacuum is inflexible, in other words, the gravitational field is inflexible. $\varepsilon_{ij} = 0, i \neq j$. This makes the strain of the gravitational field the most simple. From the view of quantum field structure, space-time beyond the intrinsic space of a quantum field are continuous and smooth. The intrinsic strain of the quantum field determines its spatiotemporal strain state $T_{\alpha\beta}$. Thus, there exists the following relationship:

Strain incompatible tensor $L_{\alpha\beta} \propto T_{\alpha\beta}$ (The 4-momentum shownoutside the intrinsic space of the quantum field), that is the gravitational field equation

$$\hat{L}_{\alpha\beta} \phi_G(\xi, x) = K\hat{T}_{\alpha\beta} \phi_G(\xi, x)$$

The logical relationship with relativity is: Internal space-time breakdown in quantum field $L_{\alpha\beta} \rightarrow$ strain $T_{\alpha\beta}$ in observation space of quantum field $\rightarrow$ 4-dimensional space-time bending $R_{\alpha\beta}$.

1. Fracture degree of internal space-time of quantum field

The strain compatibility equation is introduced to describe vacuum fracture. The strain compatibility equation is that the deformation does not satisfy the relation (5.5.8), then the deformed element cannot be reconstituted into a continuum, during which there will be cracks or fractures. In order to keep the deformed object in continuum, the strain component must satisfy the equation (5.5.8). For the local space-time of quantum field, the local space coordinate frame $(\xi_0, \xi_1, \xi_2, \xi_3)$. $\varepsilon_{\mu\nu}$ is the strain of quantum field. If the vacuum is continuous, then the strain compatibility equation is satisfied.

$$\frac{\partial^2 \varepsilon_{\mu\nu}}{\partial \xi_\rho \partial \xi_\sigma} + \frac{\partial^2 \varepsilon_{\rho\sigma}}{\partial \xi_\mu \partial \xi_\nu} - \frac{\partial^2 \varepsilon_{\sigma\nu}}{\partial \xi_\rho \partial \xi_\mu} - \frac{\partial^2 \varepsilon_{\mu\rho}}{\partial \xi_\nu \xi_\sigma} = 0 \qquad (5.5.8)$$

2. Equivalence of spatiotemporal bending and vacuum fracture

The uncoordinated strain tensor is $\underbrace{\frac{\partial^2 \varepsilon_{\mu\nu}}{\partial \xi_\rho \partial \xi_\sigma}}_{(1)} + \underbrace{\frac{\partial^2 \varepsilon_{\rho\sigma}}{\partial \xi_\mu \partial \xi_\nu}}_{(2)} - \underbrace{\frac{\partial^2 \varepsilon_{\sigma\nu}}{\partial \xi_\rho \partial \xi_\mu}}_{(3)} - \underbrace{\frac{\partial^2 \varepsilon_{\mu\rho}}{\partial \xi_\nu \partial \xi_\sigma}}_{(4)}$. The expression is built in Cartesian coordinates, to enter the general relativity is the bending space, the ordinary derivative must change the covariant derivative, we calculate the covariant derivative

$$L_{\mu\rho\sigma\nu} = \underbrace{\varepsilon_{\mu\nu;\rho;\sigma}}_{(1)} + \underbrace{\varepsilon_{\rho\sigma;\mu;\nu}}_{(2)} - \underbrace{\varepsilon_{\sigma\nu;\rho;\mu}}_{(3)} - \underbrace{\varepsilon_{\mu\rho;\nu;\sigma}}_{(4)} \qquad (5.5.9)$$

Direct utilization formula $\nabla_k \nabla_m A^i = \dfrac{\partial^2 A^i}{\partial u^k \partial u^m} + \dfrac{\partial \Gamma^i_{pm}}{\partial u^k} A^p - \Gamma^i_{qm} \Gamma^q_{pk} A^p$, available

$$D_\rho = \frac{\partial}{\partial x_\rho} - \delta^\alpha_\mu \Gamma^\alpha_{\mu\rho} - \delta^\alpha_\nu \Gamma^\alpha_{\mu\rho}$$

$$\varepsilon_{\mu\nu;\rho\sigma}^{(1)} = D_\rho \left(\frac{\partial \varepsilon_{\mu\nu}}{\partial \xi_\sigma} - \Gamma^\alpha_{\mu\sigma} \varepsilon_{\alpha\nu} - \Gamma^\alpha_{\nu\sigma} \varepsilon_{\mu\alpha} \right)$$

$$= \frac{\partial^2 \varepsilon_{\mu\nu}}{\partial x_\rho \partial x_\sigma} - (\delta_\mu^{\ \alpha} \Gamma^\alpha_{\mu\rho} + \delta^\alpha_\nu \Gamma^\alpha_{\mu\rho}) \frac{\partial \varepsilon_{\mu\nu}}{\partial x_\sigma} - D_\rho (\Gamma^\alpha_{\mu\sigma} \varepsilon_{\alpha\nu} + \Gamma^\alpha_{\nu\sigma} \varepsilon_{\mu\alpha})$$

$$\varepsilon_{\rho\sigma;\mu\nu}^{(2)} = \frac{\partial^2 \varepsilon_{\rho\sigma}}{\partial x_\mu \partial x_\nu} - (\delta_\rho^{\ \alpha} \Gamma^\alpha_{\rho\mu} + \delta^\alpha_\sigma \Gamma^\alpha_{\rho\mu}) \frac{\partial \varepsilon_{\rho\sigma}}{\partial x_\nu} - D_\mu (\Gamma^\alpha_{\rho\nu} \varepsilon_{\alpha\sigma} + \Gamma^\alpha_{\sigma\nu} \varepsilon_{\rho\alpha})$$

$$- \varepsilon_{\sigma\nu;\rho\mu}^{(3)} = -\frac{\partial^2 \varepsilon_{\sigma\nu}}{\partial x_\rho \partial x_\mu} + (\delta_\sigma^{\ \alpha} \Gamma^\alpha_{\sigma\rho} + \delta^\alpha_\nu \Gamma^\alpha_{\sigma\rho}) \frac{\partial \varepsilon_{\sigma\nu}}{\partial x_\mu} + D_\rho (\Gamma^\alpha_{\sigma\mu} \varepsilon_{\alpha\nu} + \Gamma^\alpha_{\nu\mu} \varepsilon_{\sigma\alpha})$$

$$- \varepsilon_{\mu\rho;\nu\sigma}^{(4)} = -\frac{\partial^2 \varepsilon_{\mu\rho}}{\partial x_\nu \partial x_\sigma} + (\delta_\mu^{\ \alpha} \Gamma^\alpha_{\mu\nu} + \delta^\alpha_\rho \Gamma^\alpha_{\mu\nu}) \frac{\partial \varepsilon_{\mu\rho}}{\partial x_\sigma} + D_\nu (\Gamma^\alpha_{\mu\sigma} \varepsilon_{\alpha\rho} + \Gamma^\alpha_{\rho\sigma} \varepsilon_{\mu\alpha})$$

Merge these four items, pay attention to the exchange of dummy index, and the consideration of the contact coefficient can be got

$$(1) + (2) + (3) + (4) = \frac{\partial^2 \varepsilon_{\mu\nu}}{\partial x_\rho \partial x_\sigma} + \frac{\partial^2 \varepsilon_{\rho\sigma}}{\partial x_\mu \partial x_\nu} - \frac{\partial^2 \varepsilon_{\sigma\nu}}{\partial x_\rho \partial x_\mu} - \frac{\partial^2 \varepsilon_{\mu\rho}}{\partial x_\nu \partial x_\sigma} + \underbrace{\Gamma^\alpha_{\rho\sigma} D_\nu (\varepsilon_{\mu\alpha})}_{(a)} - \underbrace{\Gamma^\alpha_{\rho\nu} D_\mu (\varepsilon_{\alpha\sigma})}_{(b)}$$

$$(a) + (b) = \Gamma^\alpha_{\rho\sigma} \left(\frac{\partial \varepsilon_{\mu\alpha}}{\partial x_\nu} - \Gamma^\beta_{\mu\nu} \varepsilon_{\beta\alpha} - \Gamma^\beta_{\alpha\nu} \varepsilon_{\mu\beta} \right) - \Gamma^\alpha_{\rho\nu} \left(\frac{\partial \varepsilon_{\alpha\sigma}}{\partial x_\mu} - \Gamma^\beta_{\alpha\mu} \varepsilon_{\beta\sigma} - \Gamma^\beta_{\sigma\mu} \varepsilon_{\alpha\beta} \right)$$

$$= \Gamma^\alpha_{\rho\nu} \Gamma^\beta_{\alpha\mu} \varepsilon_{\beta\sigma} + \Gamma^\alpha_{\rho\nu} \Gamma^\beta_{\sigma\mu} \varepsilon_{\alpha\beta} - \Gamma^\alpha_{\rho\sigma} \Gamma^\beta_{\mu\nu} \varepsilon_{\beta\alpha} - \Gamma^\alpha_{\rho\sigma} \Gamma^\beta_{\alpha\nu} \varepsilon_{\mu\beta}$$

$$= \Gamma^\alpha_{\rho\nu} \Gamma^\beta_{\alpha\mu} \varepsilon_{\beta\sigma} - \Gamma^\alpha_{\rho\sigma} \Gamma^\beta_{\alpha\nu} \varepsilon_{\mu\beta} \quad \text{(Change } \sigma, \mu \text{ into } \alpha, \text{ pay attention to the symmetry)}$$

$$= (\Gamma^\alpha_{\rho\nu} \Gamma^\beta_{\alpha\mu} - \Gamma^\alpha_{\rho\sigma} \Gamma^\beta_{\alpha\nu}) \varepsilon_{\alpha\beta}$$

$$L_{\mu\rho\sigma\nu} = \frac{\partial^2 \varepsilon_{\mu\nu}}{\partial \xi_\rho \partial \xi_\sigma} + \frac{\partial^2 \varepsilon_{\rho\sigma}}{\partial \xi_\mu \partial \xi_\nu} - \frac{\partial^2 \varepsilon_{\sigma\nu}}{\partial \xi_\rho \partial \xi_\mu} - \frac{\partial^2 \varepsilon_{\mu\rho}}{\partial \xi_\nu \partial \xi_\sigma} + (\Gamma^\alpha_{\rho\nu} \Gamma^\beta_{\alpha\mu} - \Gamma^\alpha_{\rho\sigma} \Gamma^\beta_{\alpha\nu}) \varepsilon_{\alpha\beta}$$

(5.5.10)

Compared with (5.4.21) and (5.5.10), the Riemann curvature tensor $R_{\delta\alpha\beta\gamma}$ and the vacuum fracture tensor $L_{\mu\rho\sigma\nu}$ have exactly the same mathematical form, which shows that the metric $g_{\alpha\beta}$ and $\varepsilon_{\mu\nu}$ have the same mathematical properties. $\mathscr{G}_{\alpha\beta}$ is the 2-order strain of the internal vacuum of the quantum field, corresponding to the metric tensor $g_{\mu\nu}$ of the background space-time. Einstein proposed that physical space is not an abstract space, but restricted by matter (energy). That is to say, physical space has a geometry defined by metric tensor $g_{\alpha\beta}$, which is itself controlled by the distribution of $T_{\mu\nu}$ in the universe. It can be said that this geometry already contains the properties of material distribution, and space is bent in the sense of differential (or affine) geometry. The metric exists in a smooth manifold, it is a 2-order tensor used to measure distance and angle in metric space. This tensor describes the properties of space-time manifolds. The line element square of deformation vacuum field space is given by $\mathrm{d}s^2 = g_{\mu\nu} \mathrm{d}x^\mu \mathrm{d}x^\nu$.

3. Gravitational field equation of quantum space-time fragmentation

When $L_{\mu\rho\sigma\nu}=0$, the space-time is in a smooth state, which ensures that space-time does not fracture. The smooth and continuous space-time of large-scale vacuum field is relativistic space-time, and the curvature of space-time corresponds to gravitational field.

$L_{\mu\rho\sigma\nu} \neq 0$, the fragmentation of smooth space-time means the emergence of quantum fields. Such space-time is no longer relativistic space-time. Time and space are fragmented, and cracks are enclosed in the inner space of quantum fields (Fig. 5.4.2). The strain incompatibility $L_{\mu\rho\sigma\nu}$ represents the degree of intrinsic spatial-temporal fragmentation. The degree of space-time fragmentation determines the space-time curvature of the quantum field background $kL_{\mu\rho\sigma\nu} = R_{\delta\alpha\beta\gamma}$. The strain compatibility equation constitutes the watershed between quantum field space-time and relativistic space-time.

Bianchi identities exist for space-time curvature tensors $R^{\mu}_{\delta\beta\gamma;\nu} + R^{\mu}_{\delta\nu\beta;\gamma} + R^{\mu}_{\delta\gamma\nu;\beta} = 0$. Similarly, by (5.5.10) Bianchi identities also exist for space-time curvature tensors $L^{\alpha}_{\mu\rho\sigma;\nu} + L^{\alpha}_{\mu\nu\rho;\sigma} + L^{\alpha}_{\mu\sigma\nu;\rho} = 0$. So there is also rich tensor $L_{\mu\nu}$. For a single particle quantum field, its internal space-time breakdown is $L^i_{\mu\nu}$. Then the average space-time rupture degree of macroscopic material consisting of n particles is $L_{\mu\nu} = \frac{1}{n}\sum_{i=1}^{n} L^i_{\mu\nu}$, $kL_{\mu\nu} = R_{\mu\nu}$. If the 4-momentum of particle is $t^i_{\mu\nu}$, from the point of view of energy conservation, the 4-momentum of quantum field is independent of the reference system. In other words, the total 4-momentum of the quantum field is equivalent to the 4-momentum observed outside. So the 4-momentum of macroscopic matter consisting of n particles is $T_{\mu\nu} = \sum_{i=1}^{n} t^i_{\mu\nu}$. Considering that there is only one elementary particle in a large space-time, the gravitational field equation of the effect of fragmentation on space-time of single particle is expressed as follows.

$$k\left(L^i_{\mu\nu} - \frac{1}{2}g_{\mu\nu}L^i\right) = \frac{8\pi G}{c^4}t^i_{\mu\nu} \tag{5.5.11}$$

Momentum tensor $t^i_{\mu\nu}$ of quantum field after accumulation is ultimately macroscopically expressed as $T_{\mu\nu}$. Considering Einstein's gravitational field equation, then

$$\frac{k}{n}\sum_{i=1}^{n}(L^i_{\mu\nu} - \frac{1}{2}g_{\mu\nu}L^i) = R_{\mu\nu} - \frac{1}{2}g_{\mu\nu}R = \frac{8\pi G}{c^4}\sum_{i=1}^{n}t^i_{\mu\nu} = \frac{8\pi G}{c^4}T_{\mu\nu} \tag{5.5.12}$$

The formula shows that the degree of rupture determines the curvature of the background space. This is the gravitational field equation of space-time fragmentation of quantum field.

4. Coefficient determination

$kL_{\mu\nu} = R_{\mu\nu}$ can be considered that it is caused by the observation scale factor. $L_{\mu\nu}$ is the degree of micro-scale space-time rupture, $R_{\mu\nu}$ is the macro-scale space-time bending, and the micro-scale coordinate system is ξ_μ, and the macro-scale coordinate system is x_μ. The simplest way is to enlarge the micro-scale to the macro-scale $L_{\mu\nu}$ is equivalent to $R_{\mu\nu}$, so k can be understood as an enlargement coefficient $k = x^\mu/\xi_\mu$. On the other hand, according to $[\xi_\mu, p_\mu] \equiv \hbar$,

$P_\mu = \sum_{i=1}^{n} p_\mu^i$. Gravitational field equation of quantum space-time fragmentation can be obtained

$$\frac{x^\mu}{\xi_\mu}\left(L_{\mu\nu} - \frac{1}{2}g_{\mu\nu}L\right) = R_{\mu\nu} - \frac{1}{2}gR = \frac{8\pi G}{c^4}T_{\mu\nu}; [\xi_\mu, P_\mu] = \hbar \qquad (5.5.13)$$

Here, P_μ is four momentum of macroscopic object, $T_{\mu\nu}$ is the energy-momentum tensor of the described macroscopic object. Imaginally speaking, macroscopic objects can be regarded as a "quantum field" with huge mass. For macroscopic objects in the universe, P_μ is very large, making $\xi_\mu \to 0$. Fine cracks only exist in the space-time distortion cell of basic particles, which will not damage relativistic space-time, but will bend relativistic space-time.

Vacuum fragmentation exists in the inner space-time of the quantum field composed of basic particles, but the micro-characteristics of the inner space-time fragmentation determine the macro-characteristics of its background space-time curvature. We have a deeper understanding of Einstein's gravitational field equation. The space-time bending is caused by the space-time breakdown in quantum interior.

5.6 Gravitational waves

On February 11, 2016, LIGO scientific cooperation and virgo cooperative team detected gravitational waves for the first time.

The gravitational strain is $\mathscr{G}_{\mu\nu} = \frac{A_m}{\partial_\mu}\frac{A_m}{\partial_\nu}$, gravitational wave satisfies equation $\mathscr{G}_{\mu\nu} = 0$. The weak field approximation is $h_{\mu\nu}$, it satisfying the vacuum 2-order strain conditions $\Box h_{\mu\nu} = 0$, $h_{\mu,\nu}^\nu - \frac{1}{2}h_{\nu,\mu}^\nu = 0$. It can be concluded that the gravitational wave is a transverse wave, and the graviton has no static mass and propagates at the speed of light. Longitudinal waves are intrinsic characteristics of gravitational waves, with spherical symmetry and unobservable.

5.6.1 The weak field approximation

The Einstein field equation $R_{ik} - \frac{1}{2}g_{ik}R = \frac{8\pi G}{c^4}T_{ik}$ gives the weak-field solution. Assume that the metric is close to the Minkowski metric η_{ik}, this solution is based on the assumption

$$g_{ik} = \eta_{ik} + h_{ik} \qquad (5.6.1)$$

$|h_{ik}| \ll 1$, η_{ik} is first-order quantity.

$$\Box h_{\mu\nu} = 0 \qquad (5.6.2)$$

Consider the standard wave equation, $h_{\mu\nu}$ represent the gravitational field. The gravitational field can be in the form of wave spreading around, the propagation speed is the speed of light c, and this is a gravitational wave.

5.6.2 Gravitational wave

1. Transversal wave characteristics of gravitational wave

$$h_{\mu\nu}(x) = e_{\mu\nu}\exp(ik_\lambda x^\lambda) + e^*_{\mu\nu}\exp(-ik_\lambda x^\lambda) \tag{5.6.3}$$

$$k_\mu k^\mu = 0 \tag{5.6.4}$$

Meet (5.6.2), if

$$k_\mu e^\mu_\nu = \frac{1}{2}k_\nu e^\mu_\mu \tag{5.6.5}$$

Matrix has symmetry $e_{\mu\nu} = e_{\nu\mu}$, called the polarization tensor.

Coordinate transformation is $x^\mu \to x^\mu + \varepsilon^\mu(x)$, we can change the metric $\eta_{\mu\nu} + h_{\mu\nu}$ to new metric $\eta_{\mu\nu} + h'_{\mu\nu}$. Suppose we choose

$$\varepsilon^\mu(x) = i\varepsilon^\mu \exp(ik_\lambda x^\lambda) - i\varepsilon^{\mu*}\exp(-ik_\lambda x^\lambda) \tag{5.6.6}$$

By (5.6.3), get

$$h'_{\mu\nu}(x) = e'_{\mu\nu}\exp(ik_\lambda x^\lambda) + e'^*_{\mu\nu}\exp(-ik_\lambda x^\lambda) \tag{5.6.7}$$

Here

$$e'_{\mu\nu} = e_{\mu\nu} + k_\mu \varepsilon_\nu + k_\nu \varepsilon_\mu \tag{5.6.8}$$

This wave is still to satisfy the concorde coordinate conditions (5.6.4). Only $6-4=2$ have physical meaning of $e_{\mu\nu}$. In order to clarify the difference between the different components of polarization tensor $e_{\mu\nu}$, we can consider a wave propagating along the direction $+z$, so that the wave vector

$$k^1 = k^2 = 0\ ;\ k^3 = k^0 = k > 0 \tag{5.6.9}$$

By (5.6.4) we have

$$e_{01} + e_{31} = e_{02} + e_{32} = 0$$

$$e_{03} + e_{33} = -e_{00} - e_{03} = \frac{1}{2}(e_{11} + e_{22} + e_{33} - e_{00})$$

Given from the two above

$$e_{01} = -e_{31}\ ;\ e_{02} = -e_{32}\ ;\ e_{03} = -\frac{1}{2}(e_{33} + e_{00})\ ;\ e_{22} = -e_{11} \tag{5.6.10}$$

To perform the coordinate system transformation by (5.6.7) and (5.6.6) defined, these 6 independent components of $e_{\mu\nu}$ change in accordance with the equation (5.6.8).

$$e'_{11} = e_{11}\ ;\ e'_{12} = e_{12}\ ;\ e'_{13} = e_{13} + k\varepsilon_1$$
$$e'_{23} = e_{23} + k\varepsilon_2\ ;\ e'_{33} = e_{33} + 2k\varepsilon_3\ ;\ e'_{00} = e_{00} - 2k\varepsilon_0$$

The only e_{11}, e_{22} have absolute physical significance. Implement the following transformation

$$\varepsilon_1 = -\frac{\varepsilon_{13}}{k}\ ;\ \varepsilon_2 = -\frac{\varepsilon_{23}}{k}\ ;\ \varepsilon_3 = -\frac{\varepsilon_{33}}{2k}\ ;\ \varepsilon_0 = \frac{\varepsilon_{00}}{2k}$$

To be eliminated e'_{11}, e'_{22} and $e'_{22} = -e'_{11}$, all the components are 0. It shows that gravitational wave is transverse wave, there is no amplitude along the propagation direction.

2. Spin angular momentum of gravitational wave

R rotates an angle θ around the z-axis, which could further given the differences between the different components of polarization tensor $e_{\mu\nu}$. The rotation is follows Lorentz transformation

$$R = \begin{pmatrix} 1 & 0 & 0 & 0 \\ 0 & \cos\theta & -\sin\theta & 0 \\ 0 & \sin\theta & \cos\theta & 0 \\ 0 & 0 & 0 & 1 \end{pmatrix} \qquad (5.6.11)$$

Here, the components $R_1^1 = \cos\theta$, $R_1^2 = \sin\theta$, $R_2^1 = -\sin\theta$, $R_2^2 = \cos\theta$, $R_3^3 = R_0^0 = 1$ and the remainder $R_\mu^\nu = 0$. Rotation remains k_μ unchanged, that is $R_\mu^\nu k_\nu = k_\mu$, which makes

$$e_{\mu\nu} = R_\mu^\rho R_\nu^\sigma e_{\rho\sigma} \qquad (5.6.12)$$

Coordinate implement the tworotations $R_\mu^\rho R_\nu^\sigma$ which in 3-dimensional space, and electromagnetic field rotation only in the 2-dimensional space which is not the same, can be seen from here that graviton has a 3-dimensional space structure. Use relations (5.6.10), we get

$$e'_{\pm} = \exp(\pm 2i\theta) e_{\pm} \qquad (5.6.13)$$
$$f_{\pm} = \exp(\pm i\theta) f_{\pm} \qquad (5.6.14)$$
$$e'_{33} = e_{33}, \quad e'_{00} = e_{00} \qquad (5.6.15)$$

Here

$$e'_{\pm} = e_{11} \mp ie_{12} = -e_{22} \mp ie_{12} \qquad (5.6.16)$$
$$f'_{\pm} \equiv e_{31} \mp ie_{32} = -e_{01} \mp ie_{02} \qquad (5.6.17)$$

In general, any planewave ψ rotation any angle θ which around the propagation direction

$$\psi' = e^{ih\theta}\psi \qquad (5.6.18)$$

Above helicity amount is h. (5.6.13) shows that the gravitational plane wave can be broken down as part $e_{\pm}$ by the helix amount $h = \pm 2$. Electromagnetic waves can be decomposed into two parts by helicity are ± 1 and 0, and only helicity ± 1 have physical meaning. At that time, we say that the spin of the electromagnetic wave is 1, and now we can say that the spin of the gravitational waves is 2.

3. Plane wave

The linear field equations in vacuum

$$\partial_\lambda \partial^\lambda \varphi^{\mu\nu} = 0 \qquad (5.6.19)$$

The gauge condition is

$$\partial_\mu \varphi^{\mu\nu} = 0 \qquad (5.6.20)$$

Consider simple plane wave solutions

$$\varphi^{\mu\nu} = \varepsilon^{\mu\nu} \cos k_a x^a \qquad (5.6.21)$$

Here, $\varepsilon^{\mu\nu}$ is called the polarization tensor, k_a is a constant tensor, is the wave vector. Meet

$$k^a k_a = 0 \text{ (Conservation of space-time strain)} \qquad (5.6.22)$$
$$\varepsilon^{\mu\nu} k_\mu = 0 \qquad (5.6.23)$$

Wave frequency is

$$\omega = k^0 \qquad (5.6.24)$$
$$k^0 = (k_x^2 + k_y^2 + k_z^2)^{1/2} = |k|; \quad \omega/|k| = 1 \qquad (5.6.25)$$

Chapter 5 Gravitational Field

That wave speed is the speed of light. Set up waves propagate along the z-axis, and

$$k^a = (\omega, \ 0, \ 0, \ \omega) \tag{5.6.26}$$

$$\varphi^{\mu\nu} = \varepsilon^{\mu\nu}\cos(\omega t - \omega z) \tag{5.6.27}$$

The linearly independent solution of (5.6.23) is

$$\varepsilon^{\mu}_{(1)}\varepsilon^{\nu}_{(1)} - \varepsilon^{\mu}_{(2)}\varepsilon^{\nu}_{(2)} \tag{5.6.28}$$

$$\varepsilon_{(1)}{}^{\mu}\varepsilon_{(2)}{}^{\nu} + \varepsilon_{(1)}{}^{\nu}\varepsilon_{(2)}{}^{\mu} \tag{5.6.29}$$

$$\begin{cases} \varepsilon_{(1)}{}^{\mu}k^{\nu} + \varepsilon_{(1)}{}^{\nu}k^{\mu} \\ \varepsilon_{(2)}{}^{\mu}k^{\nu} + \varepsilon_{(2)}{}^{\nu}k^{\mu} \\ k^{\mu}k^{\nu} \\ \varepsilon_{(1)}{}^{\mu}\varepsilon_{(1)}{}^{\nu} + \varepsilon_{(2)}{}^{\mu}\varepsilon_{(2)}{}^{\nu} \end{cases} \tag{5.6.30}$$

Do not carry energy and momentum, not a physical wave.

Which the vector $\varepsilon_{(1)}{}^{\mu}$ and $\varepsilon_{(2)}{}^{\mu}$ are defined as follows:

$$\varepsilon_{(1)}{}^{\mu} = (0, \ 1, \ 0, \ 0), \tag{5.6.31}$$

$$\varepsilon_{(2)}{}^{\mu} = (0, \ 0, \ 1, \ 0) \tag{5.6.32}$$

Gravitational field same as the photon, they are transverse, not longitudinal wave. Does not existing longitudinal wave that because of exists the measurementcovariantion to make longitudinal wave no observation.

4. Gravitational longitudinal wave

Exist the gravitational longitudinal waves cannot be avoidedin vacuum, any strain reaches the limit will be in the form of waves, a strain limit corresponds to a wave to produce the compression strain. This strain is the longitudinal momentum $p_{//}$. It is necessary to join the longitudinal wave vector

$$\varepsilon_{(3)}{}^{\mu} = (1, \ 0, \ 0, \ -1) \tag{5.6.33}$$

$$\phi_{//}{}^{\mu} = \varepsilon_{(3)}{}^{\mu}A_{//} \exp(k_0\xi^0 - k_3\xi^3) \tag{5.6.34}$$

Amplitude of longitudinalwave $A_{//}$ is the deformation which along the propagation direction of graviton. Photon intrinsic space along the vertical deformation of the vacuum is h. Graviton strain rate is very tiny, the deformation is continuous and smooth, along the ξ^3 direction, the direction of deformation also causes deformation of the $\xi^1\xi^2$-plane. This deformation is 3-dimensional spherical strain, deformation is $h_G = h_f$, which distinct photonic fiber structure field. The intrinsic wave function of Graviton longitudinal wave is

$$\phi_{//}{}^{\mu}(\xi) = \varepsilon_{(3)}{}^{\mu}_G \exp(k_0\xi^0 - k_3\xi^3) \tag{5.6.35}$$

There is no amplitude along the propagation direction and the problem is not difficult to understand. Quantum wave has the following characteristics: it is the solution of the longitudinal wave whose period is π, the longitudinal wave exists in theintrinsic space and is closed in the intrinsic space. Gravity is also a longitudinal solution. Gravity energy exists in intrinsic longitudinal waves. If there is no gravitational longitudinal wave, the gravitational wave will not propagate forward. For observation, P-wave has no observation effect, and of course there is no wave expression.

5. Gravitational transverse wave

The polarization tensor(5.6.26) and (5.6.27) is called horizontal. Specific form is

$$\varepsilon_{\oplus}{}^{\mu\nu} = \varepsilon_{(1)}{}^{\mu}\varepsilon_{(1)}{}^{\nu} - \varepsilon_{(2)}{}^{\mu}\varepsilon_{(2)}{}^{\nu} = \begin{pmatrix} 0 & 0 & 0 & 0 \\ 0 & 1 & 0 & 0 \\ 0 & 0 & -1 & 0 \\ 0 & 0 & 0 & 0 \end{pmatrix} \quad (5.6.36)$$

$$\varepsilon_{\otimes}{}^{\mu\nu} = \varepsilon_{(1)}{}^{\mu}\varepsilon_{(2)}{}^{\nu} + \varepsilon_{(1)}{}^{\nu}\varepsilon_{(2)}{}^{\mu} = \begin{pmatrix} 0 & 0 & 0 & 0 \\ 0 & 0 & 1 & 0 \\ 0 & 1 & 0 & 0 \\ 0 & 0 & 0 & 0 \end{pmatrix} \quad (5.6.37)$$

This corresponds to the unique polarization of the gravitational waves of physics.

6. Energy and momentum of gravitational waves

Wave solutions of (5.6.21) into the energy-momentum tensor (one approximation) $t_{(1)}^{\mu\nu}$

$$t_{(1)}^{\mu\nu} = \frac{1}{4}\left[2\phi^{\alpha\beta,\mu}\phi_{\alpha\beta}^{,\nu} - \phi^{,\mu}\phi^{,\nu} - \eta^{\mu\nu}\left(\phi^{\alpha\beta,\sigma}\phi_{\alpha\beta,\sigma} - \frac{1}{2}\phi_{,\sigma}\phi^{,\sigma}\right)\right]$$

Got

$$t_{(1)}^{\mu\nu} = \frac{1}{2}\left(\varepsilon^{\alpha\beta}\varepsilon_{\alpha\beta} - \frac{1}{2}\varepsilon_{\alpha}^{\alpha}\varepsilon_{\beta}^{\beta}\right)k^{\mu}k^{\nu}\sin^{2}k_{\sigma}x^{\sigma} \quad (5.6.38)$$

Transverse wave of amplitude $A_{\oplus}$ and $A_{\otimes}$

$$\phi_{\oplus}^{\mu\nu} = A_{\oplus}\varepsilon_{\oplus}^{\mu\nu}\cos k_{\sigma}x^{\sigma} \quad \text{Into a complex wave form} \rightarrow A_{\oplus}\varepsilon_{\oplus}{}^{\mu\nu}\exp(ik_{\sigma}x^{\sigma})$$
$$(5.6.39)$$

$$\phi_{\otimes}^{\mu\nu} = A_{\otimes}\varepsilon_{\otimes}^{\mu\nu}\cos k_{\sigma}x^{\sigma} \quad \text{Into a complex wave form} \rightarrow A_{\otimes}\varepsilon_{\otimes}{}^{\mu\nu}\cos k_{\sigma}x^{\sigma} \quad (5.6.40)$$

Wave energy-momentum tensor

$$t_{(1)}{}^{\mu\nu} = (A_{\oplus})^{2}k^{\mu}k^{\nu}\sin^{2}k_{\sigma}x^{\sigma} \quad (5.6.41)$$
$$t_{(1)}{}^{\mu\nu} = (A_{\otimes})^{2}k^{\mu}k^{\nu}\sin^{2}k_{\sigma}x^{\sigma} \quad (5.6.42)$$

$t_{(1)}^{03}$ Component is given to z-direction of energy flow. The direction of the average energy flow $\sin^{2}k_{\sigma}x^{\sigma}$ instead of by $1/2$, get

$$t_{(1)}{}^{\mu\nu} = \frac{1}{2c}(A_{\oplus})^{2}\omega^{2} \quad (5.6.43)$$

$$t_{(1)}{}^{\mu\nu} = \frac{1}{2c}(A_{\otimes})^{2}\omega^{2} \quad (5.6.44)$$

Gravitational waves and electromagnetic waves carry energy and momentum. Definition the circular polarized wave

$$(\varepsilon_{\oplus}{}^{\mu\nu} - i\varepsilon_{\otimes}{}^{\mu\nu})\exp(ik_{\sigma}x^{\sigma}) \quad \text{(Positive spiral)} \quad (5.6.45)$$
$$(\varepsilon_{\oplus}{}^{\mu\nu} + i\varepsilon_{\otimes}{}^{\mu\nu})\exp(ik_{\sigma}x^{\sigma}) \quad \text{(Negative spiral)} \quad (5.6.46)$$

These circular polarized waves carry angular momentum. Angular momentum is proportional to the energy of wave

$$[\text{Wave angular momentum}] = \frac{2}{\omega}[\text{Wave energy}] \quad (5.6.47)$$

Chapter 5 Gravitational Field

The quantum mechanics explanation is that the graviton has a spin $2\hbar$. Angular momentum/energy of each quantum is

$$\frac{2\hbar}{\hbar\omega} = \frac{2}{\omega} \tag{5.6.48}$$

A zero quality particles have to make it spin along the direction of movement (positivehelicity) or opposite (negative helicity), any other directions of spin are forbidden.

7. Tidal forces line

Consider the action on a particle by gravitational wave. The equations of motion of the particles is

$$\frac{du_\mu}{d\tau} = -\kappa\left(h_{\mu\alpha,\beta} - \frac{1}{2}h_{\alpha\beta,\mu}\right)u^\alpha u^\beta \tag{5.6.49a}$$

For arest particle, $u_0 = 1$ and $u_k = 0$ thus have

$$\frac{du_\mu}{dt} = -\kappa\left(h_{\mu 0,0} - \frac{1}{2}h_{00,\mu}\right)u^\alpha u^\beta \tag{5.6.49b}$$

The $h_{0\mu} = 0$ initial rest particle is to remain stationary. However, considering the two particles are placed in the x-axis, one at $x = x_0$, and the other at $x = -x_0$, receives a gravitational wave of the type (5.6.39). In accordance with (5.6.51), the distance between the two particle is $2x_0$, but with the meter stick to measure the physical distance, not only depends on Δx, but also depends on the space-time metric

$$g_{\mu\nu} = \eta_{\mu\nu} + \kappa h_{\mu\nu} \tag{5.6.50}$$

The distance between the two particles by measured

$$\Delta l^2 = -g_{11}\Delta x^2 = (1 - \kappa h_{11})(2x_0)^2 = (1 - \kappa A_\oplus \cos\omega t)(2x_0)^2$$

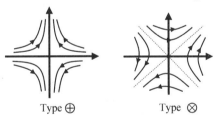

Type $\oplus$ Type $\otimes$

Fig. 5.6.1 Tidal lines of gravitational waves

The linear approximation the amplitude $A_\oplus$ is small, thus

$$\Delta l \approx \left(1 - \frac{\kappa}{2}A_\oplus \cos\omega t\right)2x_0 \tag{5.6.51}$$

Fory-axis, similar available

$$\Delta l \approx \left(1 + \frac{\kappa}{2}A_\oplus \cos\omega t\right)2y_0 \tag{5.6.52}$$

Each particle's acceleration (Fig. 5.6.1) which toward the particles is

$$\frac{d^2}{dt^2}\frac{\Delta l}{2} = -\frac{\kappa}{2}A_\oplus x_0 \frac{d^2}{dt^2}\cos\omega t = \frac{\kappa}{2}A_\oplus \omega^2 x_0 \cos\omega t \tag{5.6.53}$$

Therefore, the particles at x-axis and y-axis its quality is m, the tidal forces:

$$f_x = \left(m \frac{\kappa}{2} A_\oplus \omega^2 \cos \omega t \right) x_0 \; ; f_y = -\left(m \frac{\kappa}{2} A_\oplus \omega^2 \cos \omega t \right) y_0 \qquad (5.6.54)$$

5.6.3 The intrinsic structure of graviton

The source of graviton emission comes from Fermi particles with rest mass. Gravitons are emitted when a particle vibrates or accelerates.

1. The non-proliferation of graviton

Consider a Fermion with a stationary mass and spin of 1/2 (Fig. 5.6.2). When the particle accelerates, the center of the particle compresses and stretches the background vacuum. The center point of the inner space reaches the strain limit and the condition of wave generation is reached, and the graviton is emitted. In the process of propagation, only along the longitudinal axis of propagation can the strain limit be reached, which is similar to photons, and the gravitational force does not diffuse.

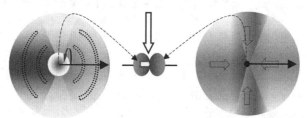

Spin $\hbar/2$. The fermion with rest mass emission graviton

Fig. 5.6.2 Structure of graviton

(The center of the fermions to squeeze in front of the basic unit
of the adjacent field, the basic unit is the center of graviton)

2. Graviton 3-dimensional structure

The mass charge has 3-dimensional spherical symmetry and the graviton has 3-dimensional axisymmetry. The graviton is generated by the mass charge oscillation, the photon is generated by the charge oscillation, the charge is the first-order strain wave of the fiber field, and the force is the second-order strain wave of the fiber field. The center point of graviton moves forward, so that the surface of $\xi^1 \xi^2$ stretches in the $-\xi^3$ region to form a negative hemisphere. Similarly, the surface of $\xi^1 \xi^2$ in the $+\xi^3$ region is compressed to form a positive hemisphere. Because the graviton is a second-order strain wave, the hemisphere at both ends of the graviton will generate gravitation, there is no anti-graviton, which leads to the graviton transmission is always gravitational, not repulsive force. The graviton has only a single gravitational property, so the space of the graviton is the product of two omnidirectional spaces, i. e. the second-order strain. The space characteristic of the graviton is that the spin is 2.

3. Total deformation of graviton space-time

The structure of the graviton is similar to that of the photon, but the surface of the graviton has a continuous smooth upward and downward fluctuation structure. There are eelectric flux

Chapter 5 Gravitational Field

lines on the photon surface. The cross-section of each electric flux line is the cross-section of the basic unit of the field. According to the homogeneity of the basic field elements, it can be seen that it takes two times of fibers to cover the surface of the mass charge to be a continuous smooth ups and downs undulating structure, so we can get that the total space-time deformation of graviton charge is $2h$, $h_g = 2h$ (Fig. 5.6.3).

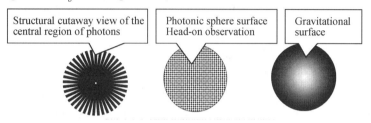

Fig. 5.6.3 Gravitational surface

Graviton deformation with 3-dimensional structure, in the section of the lepton we know thatspace-time of fermions has 3-dimensional spherical symmetric deformation and total deformation is h_C. The left hemisphere space-time deformation is $h_g/2$, There is no negative term for the second order strain, so the right hemisphere space-time deformation is $+h_g/2$ (Fig. 5.6.4). Graviton total space-time deformation is

$$h_g = e\mathcal{H} \tag{5.6.55}$$

Should be noted the limit deformation that time and space along ξ^3-dimension of is H.

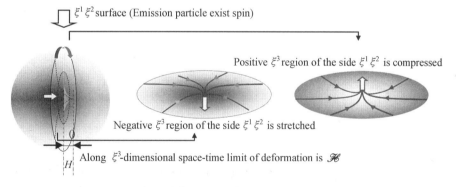

Fig. 5.6.4 Schematic of the structure of graviton

4. Strain structure of gravitational waves

Gravitation is a 2-order strain. Both hemispheres of the graviton hemisphere exhibit gravitation and have no dual structure, the complex representation is wrong and the imaginary number i should be removed. The strain field function of gravitational waves can be written as

$$h(\xi) = \begin{pmatrix} e_0 \\ e_1 \\ e_2 \\ e_3 \end{pmatrix} \exp\left\{ \begin{bmatrix} \mathscr{G}_{00} & 0 & 0 & 0 \\ 0 & \mathscr{G}_{11} & 0 & 0 \\ 0 & 0 & \mathscr{G}_{22} & 0 \\ 0 & 0 & 0 & \mathscr{G}_{33} \end{bmatrix} \begin{pmatrix} x_0 & 0 & 0 & 0 \\ 0 & x_1 & 0 & 0 \\ 0 & 0 & x_1 & 0 \\ 0 & 0 & 0 & \xi_3 \end{pmatrix} \right\} \tag{5.6.56}$$

(1) Along the direction of movement is squeezed then existence strain that makes the

Graviton movement along ξ^3 direction. The strain the momentum of ξ^3 direction is the longitudinal momentum p_3 of the gravitational wave. $p_3 = \mathscr{g}_{33}$ Is momentum of longitudinal graviton, $h_g = p \cdot \frac{4}{3}\pi R^3$. R is radius of graviton. Radial strain is spherically symmetric for observers, and is not observable.

(2) Another dimension strain in the surface of $\xi^1 \xi^2$, makes the center point of graviton around ξ^3-axis screw rotation.

$$h(\xi) = \begin{pmatrix} e_0 \\ e_1 \\ e_2 \end{pmatrix} \exp\left\{ \begin{bmatrix} E & 0 & 0 \\ 0 & p_1 & 0 \\ 0 & 0 & p_2 \end{bmatrix} \begin{pmatrix} x_0 & 0 & 0 \\ 0 & x_1 & 0 \\ 0 & 0 & x_2 \end{pmatrix} \right\} \quad (5.6.57)$$

And similar to the photon spin, the graviton's momentum determines the size of the radius of the spiral. Meet

$$h_G = p'_\perp \cdot \frac{4}{3}\pi R_\perp^3 \quad (5.6.58)$$

$$p'_\perp \cdot \frac{4}{3}\pi R_\perp^3 = \frac{4}{3}\pi \left(\frac{\mathscr{H}}{2}\right)^3 \Rightarrow \sqrt[3]{p'_\perp} \cdot e \cdot R_\perp = e\frac{\mathscr{H}}{2} \Rightarrow p_\perp \cdot R_\perp = h_f \quad (5.6.59)$$

$p'_\perp$ has a 3-dimensional character, no measurable, must consider the 1-dimensional effect. Set $p_\perp \equiv \sqrt[3]{p'_\perp} \cdot e$ graviton moving along the direction of the momentum. $R_\perp$ is radius of spin graviton. The spin momentum of graviton can be merged into the center of graviton so that the graviton spin angular momentum is

$$2\pi R_S \times p_\perp = h_g \Rightarrow R_S \times p_\perp = 2\hbar \quad (5.6.60)$$

Graviton spin angular momentum is $2\hbar$. Relationship between Compression wave momentums-with shear wave momentum satisfies the orthogonal relation.

$$p^2 = p_\parallel^2 + p_\perp^2; \quad E = cp \quad (5.6.61)$$

The strain of Graviton is too small, the tidal forces effect cannot be observed. Graviton space-time deformation meet

$$h_G = \int_0^\Omega p(x)\,d\Omega = \int_0^{\Omega_0} p^0(x)\,d\Omega^0 \quad (5.6.62)$$

$p(x)$ is graviton momentum, $p^0(x)$ is graviton energy. Ω is 3-dimensional deformation scope of space, it is a huge sphere, volume of ballis $\Omega = 4\pi R^3/3$. The formula can be simplified as $p_\mu R_\mu = h_f$. R_0 is the time required to form the graviton. By (5.6.55) we know $p \to 0$ and $R \to \infty$, graviton is huge, a graviton size depends on the single fermions emission graviton that can give momentum size. Countless graviton superposition can increase the strength of the gravitational wave, but cannot change the frequency of gravitational waves. It's same as photon.

Chapter 6　Dark Matter and Dark Energy in the Universe

1. Dark matter

The discovery of dark matter was derived from the measurement of the galaxy rotation curve of Andromeda (M31) which adjacent to our Milky Way galaxy, in 1970 by Vera Rubin. Measure the rotation speed of the matter at the different radius of the galaxy (Fig. 6.0.1). The abscissa is the distance from the center of the galaxy, and the ordinate is the velocity. If the mass distribution of galaxies is known, the relationship between velocity and distance can be easily obtained by Kepler's Law (Redline). The actual measurement results (white lines) show that the rotation speed in the periphery of the galaxy is much faster than that of the calculated ones. This shows that the galaxy has a lot of quality is invisible, because of $V \propto \sqrt{GM/r}$. Since then, astronomers have measured by gravitational lensing. The results of the measurements show that there is dark matter, and the dark matter is distributed in a mass. The interaction between dark matter and normal matter is very weak. According to the possibility of moving speed of dark matter, dark matter particles are classified as hot dark matter (Hot Dark, Matter, HDM, particle moving close to the speed of light), cold dark matter (Cold Dark Matter, CDM, particle velocity is far less than the speed of light) and warm dark matter (Warm Dark Matter, WDM, middle speed). Astronomers now think that the main component of the dark matter in the universe should be cold dark matter. But the cold dark matter model also has some problems. For example, the numbers of companion galaxy of Milky Way we have known are more than the numbers of the galaxy which by the prophecy of cold dark matter. And we know that the total mass ratio of dark matter in the universe is much more than normal material. So it is clear that the event of gravitational interaction on the cosmic scale is dominated by dark matter, such as the formation and combination of galaxies. In general, there is a lot of knowledge about the nature of dark matter.

Dark matter should actually be called invisible material. It cannot be observed by any electromagnetic means. Dark matter occupies 1/4 of the total mass of the universe. The only possible evidence we know that is the rotation speed of the galaxy edge is faster than that of Newton's gravity, because the powerful centrifugal force did not tear it out. Until now scientists do not know what the dark matter is. The image of dark matter has been captured by gravitational lensing. Scientists have synthesized a "cosmic network" image of dark matter filaments that according to the weak gravitational lensing images of more than 23,000 galaxy pairs from 4 billion 500 million light years away.

The effect of dark matter is to bridge the galaxies together. Bright galaxies are shown as

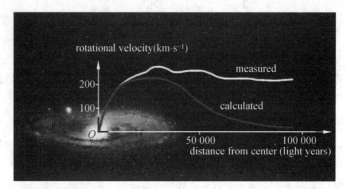

Fig. 6.0.1 The relationship between speed and distance of a galaxy

white areas and dark matter filaments that connect the galaxy pairs are red in this hypothetical diagram. The dark matter signal is the strongest in the galaxy pair with a scale less than 4,000 light years. Dark matter has a net structure in the large scale of the universe, if we can stand outside the universe and watch it, the universe is like a loofah. Clusters of galaxies and galaxiesare connected by dark matter together to form the filamentous structure of the loofah. The cavities in the loofah are the huge hollow structures in the universe. To observe the whole universe, the form of dark matter is similar to the loofah, which is roughly the same structure in each area. This is called the isotropy of the universe in large scale space. According to this theory, dark matter is the scaffold of the universe. It is the most basic structure of the universe. All galaxies and clusters of galaxies are woven on the net structure of dark matter which strongly supports the theory of vacuum field.

2. Dark energy

According to Harbert's Law, the farther the universe is far away from us, the faster the movement is (and there is a larger red shift), because the universe is expanding. Hubble's Law is

$$V_f = H_c \times D \qquad (6.0.1)$$

Here, V_f is velocity(Far Away), km/s; H_c is hubble's Constant, km/(s. Mpc); D is the distance relative to the earth, Mpc is megapars.

Because of the interaction of matter in the universe, the speed of this expansion willchange, we will find that Harbert's law is no longer a simple linear relationship. And it turns into the Figure below. The transverse axis is a red shift (directly related to speed), and the longitudinal axis is the equivalent luminance (directly related to the distance).

Considering that all the matter in the universe has a gravitational contribution, the universe should slow down. However, the data points measured by La Supernova do not support such a universe. The measured data tend to be an accelerated expansion of the universe. That is to say, the dominant thing in the universe is not the gravitational force between matter, but a kind of inexplicable outward pressure. The source of this pressure is called the dark energy. At present, what astronomy can do is to introduce the Einstein constant, but use it to make a universe

Chapter 6 Dark Matter and Dark Energy in the Universe

that can accelerate expansion, and then determine how much it is now.

The most accurate measurements by the Planck satellite is show that 4.9% of the normal matter is in the universe, 26.8% of the dark matter and 68.3% of the dark energy (Characterizing by energy density). It can be found that the universe is dominated by dark energy, followed by cold and dark matter, and the rest is the normal substance that we can see.

Dark energy is being eaten by the dark energy of the universe.

At present, a new study by scientists shows that the dark matter in the "scaffolding" of the universe is gradually disappearing and is being swallowed by dark energy. The latest research is published in "Physical Review Letters". Cosmologists at University of Portsmouth in UK pointed out that the latest astronomical survey data show that dark energy is increasing and interacting with dark matter, which continuously engulf dark matter.

If dark energy increases gradually and at the same time. We will end up in a huge of astrospace and there will be nothing in the universe. Dark matter provides a framework for the growth of the cosmic structure. Our galaxy formation is based on the dark matter structure. The latest research shows that dark matter is evaporating and slowing the growth of the cosmic structure. How does this phenomenon be explained?

The vacuum unified field theory will give a succinct answer, Fermi matter waves propagate outward, and gradually pass through the vacuum crack area of the universe. The background space of the bending of the universe is separated from the observation space of the Milky Way, which eventually leads to such a result. We will discuss these issues in detail in this section.

3. The evolution of the universe

All visible substances are made up of equal amount of substance and antimatter in the theory of vacuum uniform field. The densest neutron stars in a celestial body are made up of neutrons and neutrons are made up of substance and antimatter. A neutron star is the stellar evolution to the end, after the gravitational collapse of a supernova, only a few can become a neutron star. When the star loses the support of the heat radiation pressure, the outer material will fall quickly to the core by the gravity traction, it may lead to the transformation of the kinetic energy of the shell into a heat energy outburst to produce a supernova explosion. The inner district of the star is compressed into white dwarfs, neutron stars and even black holes that depending on the mass of the stars. The star is compressed violently in the process of the white dwarf being compressed into a neutron star. The electrons in the white dwarf matter are compressed into the proton and converted into neutrons, with a diameter of about more than ten kilometers. The weight of substance of 1 cm^3 can reach to one billion tons and the speed of rotation is very fast. Because the magnetic axis and the spin axis does not coincide, the radio waves and other kinds of radiation produced by the rotation of the magnetic field may be transmitted to the earth in the mode of pulse radiation, they are also called pulsars.

The mass of a typical neutron star is between 1.35 and 2.1 times the mass of the sun, and the radius is between 10 and 20 kilometers (the larger the mass is, the greater the radius is).

Therefore, the neutron star density is 8×10^{13} grams to 2×10^{15} grams per cubic centimeter, the density is approximately equal to the nuclear density. Dense star quality if less than 1.44 times the mass of the sun, it may become a white dwarf, but the quality is greater than the Oppenheimer limit (1.5—3.0 times the mass of the sun) neutron star will continue to gravitational collapse occurs, inevitably will produce a black hole.

Because the neutron star preserves most of the angular momentum of the parent star and the radius is just the tiny amount of the parent star. The reduction of radius resulted in rotation speed is rapid increase, producing very high rotation rate. The rotation period is from 1/700 to 30 seconds. The high density of the neutron star also makes it have a strong surface gravity, which is 2×10^{11} to 3×10^{12} times the strength of the earth. From the evolution of the universe we can know that the supernova explosion is a normal state. The problem is that the black hole will explode? In the theory of vacuum unification, any substance is made up of equal amount of substance and antimatter. The material of black hole is no exception means that as long as the pressure is large enough, the black hole will explode. If to make pressure around the black hole large enough, it will be a giant black hole, the total mass of the black hole will not be lower than the total mass of the visible matter in our universe. We call it the cosmic black hole. The explosion of the cosmic black hole will give birth to a new universe.

6.1　The universe come into being and extinction

The universe is infinitely large in the eyes of our human beings, and in fact the universe is limited. The physical significance of the Big Bang is similar to that of a supernova explosion. There are many cosmic black holes in the supermoms, the mass of the cosmic black hole exceeds a limit that will have a Big Bang, all of which spread out after the big bang, and the universe eventually dies. These diffused substances confluence with other matter of the cosmic explosion to form a new material system and re-enter a new evolutionary process.

The wonder of the vacuum is that any strain can cause a vacuum hardening to produce gravitational force. According to the gravitational properties, we can build such a physical image:The cosmic dust accumulates continuously by the gravitational effect of dark matter. The filamentous structure of dark matter produces a gravitational pipe effect, and the small dust becomes large dust and forms a star. A large mass of stars will inhale the stars around them, getting bigger and bigger, and eventually becoming supermassive stars. At the end of its life, the center of super mass stars will leave a huge black hole after the supernova eruption. It's magnitude larger of the mass of "medium mass black hole" than the mass of the ordinary black hole of the former stellar class. The number of middle — mass black holes is huge, and they fall into the center of the new Galaxy along the gravitational pipeline which constructed by dark matter, where they gather and merge into a black hole in the center of a new galaxy. A lot of material into the center of the new galaxy after combining two galaxies, the center of the black hole

Chapter 6   Dark Matter and Dark Energy in the Universe

is becoming more and bigger, finally formed a quasar level of supermassive black holes. According to the calculation simulation, it takes about 1 billion years to collide from two galaxies to the final formation of a quasar. The adjacent supermassive black holes of quasar level will continue to merge each other. Finally, a cosmic class of black holes is formed. Such black holes have laid the foundation for the creation of the universe. The black hole will engulf the other surrounding material, and the black hole will grow up. When the black hole grows up, the black hole constantly engulf the surrounding material. In the end, all the substances, including the surrounding black holes, will converge to the big black holes. These big black holes gather together to form a supermassive black hole. Eventually, all the material in a spherical space is swallowed up. It likes a super vacuum cleaner that takes all the cosmic dust inside the shell (radius is R) into the super black hole until there is no any visible matter. When the total mass of the super black hole exceeds the total visible substance of the existing universe (that is, 5% of the total matter in our universe, excluding dark matter and dark energy), the matter of the black hole is finally crushed by strong gravity. A new universe began to explode. The cosmic substances are thrown out to meet the material of another universe, converging to form a new black hole, the black hole exploded again, so it repeated. Every universe is a brilliant firework (Fig. 6.1.1).

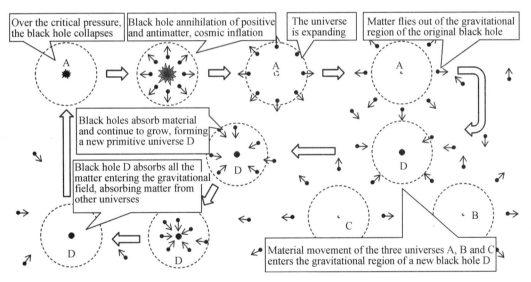

Fig. 6.1.1   The schematic diagram of the cosmic cycle mechanism

The black hole was originally regarded as a star of death and was considered the final destination of the evolution of the star. In fact, a black hole has a rich connotation. It has notonly mechanical properties but also quantum and thermal properties. A black hole is not a dead star, it has a rich life. A black hole is not the final destination of the evolution of a celestial body, but a middle stage of the evolution of the celestial body. We don't discuss black holes in detail here. We only consider that black holes are made up of matter and antimatter. The uncharged black hole is made up of Planck fermions. The number of holes in the center of a charged black hole is equal to that of the external positive charge Free State matter. This structure is extremely unstable. Whatever the structure, black holes are composed of matter and antimatter. When the external pressure is large enough, they will explode. Here, we are most concerned about the curvature of the background space-time of the black hole, because this part involves dark energy.

6.1.1 A physical image of dark matter dark energy

Dark matter and dark energy have a very clear physical image in vacuum unified field theory. Because the vacuum is a hard brittle material with isotropic and weak elasticity, then the big explosion in the cosmic black hole will cause a vacuum to break up. Similar to rock blasting, the visible matter is produced by the annihilation of the matter and antimatter and the vacuum powder fragmentation in the shattered area. It is shown to produce fermion and boson which constitute the visible matter. The visible material produced in the bursting cavity and the edge accounts for 5% of the total energy of the big bang. Before the big bang, the sharp bending of the cosmic black hole causes the background vacuum to break. In the big bang, the vacuum produces a new crack outside the smash district. The broken vacuum absorbs energy and causes the vacuum to be deformed and hardened. The cracks in the vacuum are dark matter. The old crack (before the big bang) and the new crack (the big bang) accounted for 24% of the total energy of the universe. On the other hand, the vacuum inhomogeneous in the universe is energy, because the spherical converging waves are produced during the formation of a black hole (Black holes collapse inward), this makes the black hole angular momentum. In addition, the Big Bang will cause a wide range of fluctuations in the vacuum, this fluctuation causes the inhomogeneity of the vacuum, space-time bend are dark energy. In general, dark energy contains the following parts:

(1) Cosmic black hole causes the bending energy of the background space.

(2) The energy of the vortex converging wave around the center of the cosmic black hole.

(3) The excessive curvature of black hole background space-time results in the breakdown of background vacuum, which produces dark matter energy.

(4) The energy of Big Bang caused a wide range gravitational wave.

All of these dark energy accounts for 70% of the total energy of the universe.

In the theory of rock blasting, the volume wave, especially the longitudinal wave, can cause the compression and tensile deformation of the rock. The longitudinal wave has a great contribution to the crushing of rock. At the periphery of the smash district, the shear wave stretches the rock, which makes the fracture of the rock fracture, which is mainly the contribution of the shear wave. In the existing blasting theory, we know that if the energy of the source radiation is 100, the energy ratio of P-wave is 7% and S-wave is 26%, while the surface wave is 67%. It is coincidental that this corresponds to the proportion of the visible matter, dark matter and dark energy to the total energy of the universe in the great explosions of the universe. This shows that the vacuum matter is indeed similar to the rock. Dark matter is a vacuum crack. New dark matter and dark energy were produced with the Big Bang. As a result, we need to have a holistic grasp of the creation and extinction of the universe (Fig. 6.1.2).

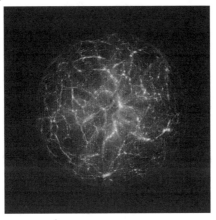

Fig. 6.1.2 Big Bang caused the cosmic vacuum to bread and vacuum cracks constitute dark matter

6.1.2 Black hole classification and image

Why does a black hole explode? It is necessary to discuss the most basic characteristics of a black hole first. Black holes are divided into four categories (Table 6.1.1). See the following table

Table 6.1.1 black hole classification

Charge	Rotation	Type	Existence possibility
No	No	(a) Schwarzschild	Secondary class II, existence
No	Yes	(b) Kerr	Highest class I, existence
Yes	No	(c) Reissner-Nordstrom(R-N)	The lowest class IV, no existence
Yes	Yes	(d) Kerr-Newman	Class III, positive charge, unstable

Negative charge black hole is a negative structure with a black hole in reference Fig. 6.1.3. The center of the black hole is a free vacuum sphere. These free substances lose the background space for the movement of the fibers of the electric flux line. There is no positive charge electric flux line, the central sphere has no charge. The ball without the electric flux line, but will squeeze around the vacuum, causes the surrounding vacuum to break up. The electrons form the spherical shell. The outside of the shell is a background vacuum, the black hole has a sandwich structure. The inward 1/2 electric flux line of the negatively charged spherical shell is terminated in the central free sphere, the electric flux line of the outward 1/2 extends to infinity. The performance is that the black hole is negatively charged, so only half of the negative charge of a black hole is observable (Fig. 6.1.3).

The stability of a negative black hole: From the above structure, it is impossible to have non revolving negative charged black hole. Because a black hole has a negative charge, this means that there is an outward expansion of free vacuum matter in the center of the black hole. The shell of a black hole is cavitation particle, that is, negative electrons. The external vacuum is inwardly compressed, it will inevitably lead to the annihilation of the encounters of the positive and negative substances, therefore, a black hole with a negative charge is extremely unstable, and once it is produced it will disappear immediately. If the high speed spin of a black hole has very high angular momentum, the centrifugal force produced by this angular momentum is enough to separate the electromagnetic force between the positive and the negative. But the sphere of the central free matter will still expand outwards, and the result is still annihilated. If the vacuum free matter is rotating at a high speed, no fragmentation, no expansion, then it is possible to have such a black hole. But this possibility does not exist.

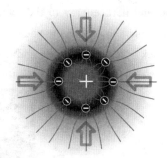

Fig. 6.1.3   Negative charge black hole

(1) A positive charged black hole. All negative charges in the sphere of a positive charge black hole are merged in the central district, it turns into a big hole. Because there is no structure in the hole, the conditions for the movement of a negative charged fiber cannot be produced. There is no condition for the displacement of negatively charged fiber and have no negative charge in the hollow sphere. The effect of this cavity is not to form the electric flux line, but to stretch the surrounding vacuum to cause breakage. The empty hole surface covered with ionized substances, namely positive charge. The total amount of these positive charges is equal to the total amount of cavity, and the matter and anti-matter is equal. The positive charge inward 1/2 electric flux line terminate in the negative sphere cavity; the outward 1/2 electric flux line has been extended; it is shown that a black hole has a positive charge. In the same way, only half the positive charge of the black hole can be observed.

(2) Stability of positive charged black hole. Relative to a negatively charged black hole, the center of a positive charge black hole is a void without any substance, and this structure is relatively stable. Due to the inward contraction of the external vacuum of a non-rotating positive charge black hole, the free material at the edge of the hole is squeezed into the cavity, It will lead to the annihilation of the matter and anti-matter, Therefore, the black hole with positive charge is also unstable, and once it is produced it will disappear immediately. If the high speed spin of a black hole has very high angular momentum, the centrifugal force produced by this angular momentum is sufficient to separate the electromagnetic force between the positive and negative substances to counteract the inward pressure around the cavity. Since there is no structure in the cavity, the vacuum material around the cavity is pulled apart, the central cavity will not expand outward. This makes the high speed rotating black hole with positive charge has the

possibility of existence. The precondition is that the angular momentum is large enough to be able to separate the static electricity between the matter and anti-matter. It can be easily estimated by Newtonian mechanics(Fig. 6. 1. 4).

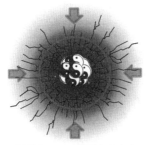

(a) Schwarzschild black hole

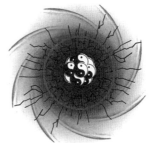

(b) Kerr black hole

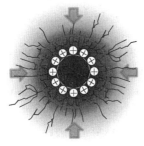

(c) Reissner-Nordstrom black hole

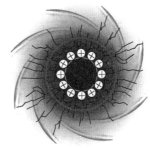

(d) Kerr -Newman black hole

Fig. 6.1.4 The schematic diagram of the structure classifieation of black holes

$$F_{\text{centrifugal force}} = M_{\text{Black hole}}\omega^2 r \ ; \ F_{\text{gravitation}} = GM_{\text{Black hole}}m_{e+}/r^2 \ ; \ F_{\text{Static electric power}} = KQ^2/r^2 \tag{6.1.1}$$

We can get:

$$M\omega^2 = KQ^2/r + GM_{\text{Black hole}}m_{e+}/r . \tag{6.1.2}$$

This is the condition of the existence of a charged black hole.

Here, M is black hole mass; Q is the amount of charge of the observed black hole; r is the radius of the sphere of the black hole; ω is black hole angular velocity; G is universal gravitational constant; m_{e+} is the mass of a positive charge material.

The high speed rotating black hole is the only potential live black hole in the universe.

6.1.3 Schwarzschild black hole

The material of the black hole is composed of Planck fermions and has 1/2 spin without a charge. The mass density of the black hole is Planck density. The possibility of existence is lower than the Kerr black hole.

Schwarzschild gave a rigorous solution of Einstein equation. This is a solution to the external vacuum of a static, spherically symmetric star. The non-zero metric components are

$$\begin{cases} g_{00} = -\left(1 - \dfrac{2GM}{c^2 r}\right) \; ; g_{11} = \left(1 - \dfrac{2GM}{c^2 r}\right)^{-1} \\ g_{22} = r^2 \; ; g_{33} = r^2 \sin^2\theta \end{cases} \qquad (6.1.3)$$

It is written in matrix form

$$(g_{\mu\nu}) = \begin{pmatrix} -\left(1 - \dfrac{2GM}{c^2 r}\right) & 0 & 0 & 0 \\ 0 & \left(1 - \dfrac{2GM}{c^2 r}\right)^{-1} & 0 & 0 \\ 0 & 0 & r^2 & 0 \\ 0 & 0 & 0 & r^2 \sin^2\theta \end{pmatrix}$$

Expressed by line element:

$$ds^2 = -\left(1 - \dfrac{2GM}{c^2 r}\right) c^2 dt^2 + \left(1 - \dfrac{2GM}{c^2 r}\right)^{-1} dr^2 + r^2 d\theta^2 + r^2 \sin^2\theta d\varphi^2 \qquad (6.1.4)$$

The singularity causes the intrinsic singularity at $r=0$. There is no singularity in the theory of vacuum uniform field. The space-time excessively bending can cause vacuum to break up. The essence of space-time bending is vacuum hardening. Because there is no singularity, then have no the difficult for the singularity of the theory of relativity. The general relativity describes the ideal space-time bending, and corresponds to the ideal elastic medium in the vacuum field.

For the infinite red shift surface equal to event horizon surface is

$$r = r_g = \dfrac{2GM}{c^2} \qquad (6.1.5)$$

That is the minimum radius that the black hole can bend the light. Because the time and space can only bend the total amount less than $\pi/2$, it needs to is re understood. There is an event horizon surface in Schwarzschild space-time. It is just the singular surface of the gravitational radius at r_g, which coincides with the infinite red shift surface. In particular, the new understanding is to consider the three points a, b, and c they are near the black hole. At point b emits a beam of light that is tangent to the singular surface of the black hole and the angle of the beam inward bending is equal to $\pi/2$. If point a shoot a beam of light within the singular surface of the black hole, $r_g > r_a$, this beam of light can only inward bending $\theta_a = \pi/2$, the light goes into the black hole (Fig. 6.1.5). There is a fundamental difference between the new theory and the present theory. There is no case where a beam of light is closed to propagate in the event horizon of a black hole, the event horizon surface of a black hole cannot make up a closed space-time. Outside the event horizon surface of the black hole $r_g < r_c$, the

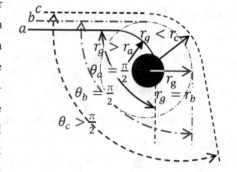

Fig. 6.1.5 The curved angle of light passes through a black hole

point c emits a beam of light, the angle of the beam inwardly bending will be greater than $\pi/2$, $\theta_c > \pi/2$, Light doesn't go into the black hole, it only changes the direction of propagation and it will go through the odd surface of a black hole. In accordance with the original theory, the event horizon surface is defined as the boundary of the black hole. The space-time district of $r < r_g$ is called the interior of a black hole; the space-time district of $r > r_g$ is called the exterior of a black hole. For cosmic black holes, the outer space-time bending of cosmic black hole forms dark energy.

6.1.4 A charged black hole

The interior of R–N black hole is a spherical cavitation. The external cavity is free vacuum material with positive charges. Under the action of strong gravity, the matter and anti-matter of the charged non-revolving black hole is annihilated directly. This kind of black hole does not exist in the universe.

(1) Metric:
$$ds^2 = -\left(1 - \frac{2M}{r} + \frac{Q^2}{r^2}\right) dt^2 + \left(1 - \frac{2M}{r} + \frac{Q^2}{r^2}\right)^{-1} dr^2 + r^2 d\theta^2 + r^2 \sin^2\theta d\varphi^2 \quad (6.1.6)$$

(2) Singular surface: $r_+ = M + \sqrt{M^2 - Q^2}$; $r_- = M - \sqrt{M^2 - Q^2}$.

(3) Infinite red shift surface: From the red shift formula $\nu = \nu_0 \sqrt{-g_{00}}$. We know, if $g_{00} = 0$, there will be an infinite red shift.

(4) The event horizon surface

$r = r_+$ is the boundary of the black hole. But unlike the Schwarzschild black hole is R–N black hole has internal event horizon and external event horizon and two infinite red shift surfaces. These two horizons coincide with the two infinite red shifts, respectively.

6.1.5 Kerr–Newman black hole

The structure of Kerr–Newman black hole is extremely unstable. If a black hole is rotating at a high speed, it can provide strong centrifugal force to separate positive and negative charges. But in the super gravitational field, positive and negative material will annihilate, does not exist in the universe.

Kerr–Newman solution describes the external gravitational field of a rotating charged star, that is, the bending of the outer space-time of the star, the linear element is

$$ds^2 = -\left(1 - \frac{2Mr - Q^2}{\rho^2}\right) dt^2 + \frac{\rho^2}{\Delta} dr^2 + \rho^2 d\theta^2 +$$
$$\left[(r^2 + a^2)\sin^2\theta + \frac{(2Mr - Q^2) a^2 \sin^4\theta}{\rho^2}\right] d\varphi^2 - \frac{2(2Mr - Q^2) a \sin^2\theta}{\rho^2} dt d\varphi$$
$$(6.1.7)$$

Here, $\rho^2 = r^2 + a^2 \cos^2\theta$; $\Delta = r^2 - 2Mr + a^2 + Q^2$.

The event horizon surface: $r_\pm = M \pm \sqrt{M^2 - a^2 - Q^2}$.

Two event horizon surface: $r^s_\pm = M \pm \sqrt{M^2 - a^2\cos^2\theta - Q^2}$.

Under the action of strong gravity, matter will collapse, it will aggravate the collapse of matter and make both positive and antimatter extinguish, so the black hole without charge.

6.1.6 A rotating black hole

It is a rotating Schwarzschild black hole. The centrifugal force generated by rotation slows down the collapse caused by gravity and is relatively morestable. It exists in the universe.

$$ds^2 = -\left(1 - \frac{2Mr}{\rho^2}\right)dt^2 + \frac{\rho^2}{\Delta}dr^2 + \rho^2 d\theta^2 +$$
$$\left[(r^2 + a^2)\sin^2\theta + \frac{2Mra^2\sin^4\theta}{\rho^2}\right]d\varphi^2 - \frac{4Mra\sin^2\theta}{\rho^2}dtd\varphi \qquad (6.1.8)$$

Here, $\rho^2 \equiv r^2 + a^2\cos^2\theta$; $\Delta \equiv r^2 - 2Mr + a^2$.

Infinite red shift surface: $r^s_\pm = M \pm \sqrt{M^2 - a^2\cos^2\theta}$.

The event horizon surface: $r_\pm = M \pm \sqrt{M^2 - a^2}$.

This is the event horizon of Kerr's space-time. We see that the event horizon face of Kerr space-time and the infinite red shift face each have two, and the event horizon face and the infinite red shift face do not coincide. The boundary of a black hole is defined by the event horizon face instead of the infinite red shift face. We call the Kerr black hole is the part which is surrounded by the outer of the r_+ event horizon face.

Because of the strain limit of vacuum, the vacuum is broken in the event horizon of black hole, the space-time rupture occur, the geometric effect of infinite bending only exists in mathematical analysis, and the original theory will fail, no further discussion is made here.

6.1.7 Vacuum substance

The vacuum substance is assumed to be isotropic elastic material. For isotropic materials, the following relationship exists between the three elastic constants of E, G and μ.

$$G = \frac{E}{2(1 + \mu)} \qquad (6.1.9)$$

The modulus of elasticity (E) can be regarded as an indicator of the difficulty of producing elastic deformation of a material. Definition: $L/EA = \Delta L \leq h$. Among them, F is force, L is length, E is the modulus of elasticity, A is the cross-sectional area, ΔL is the length variation, namely deformation. The modulus of elasticity E refers to the stress required by the material to produce a unit elastic deformation under external action. It is an indicator of material resistance to elastic deformation.

Shear modulus (G) is material constant. In the stage of elastic deformation, the material is subjected to shear stress, the ratio of shear stress to shear strain in the limit range of elastic deformation ratio. It is also called the shear modulus or the stiffness modulus. It characterizes the ability of the material to resist the shear strain.

Chapter 6 Dark Matter and Dark Energy in the Universe

The Poisson ratio μ is the absolute value of the ratio of the transverse strain to the corresponding longitudinal strain caused by a uniformly distributed longitudinal stress within the ratio limit of the material. It is an elastic constant that represents the transverse deformation of the material. It's also called the transverse deformation coefficient. It only shows the mutual influence between the deformations of different directions, which are not related to the degree of the deformation.

Vacuum as a medium, we can know the velocity of longitudinal wave (P-wave is the longitudinal wave of gravitational wave. No observability) and transverse wave (S-wave, that is, gravitational shear wave) by reference to elastic mechanics.

$$V_p = \sqrt{\frac{\lambda + 2G}{\rho}} \;;\; V_S = \sqrt{\frac{G}{\rho}} \qquad (6.1.10)$$

Here, $\lambda = \mu E/(1+\mu)(1-2\mu)$ is Lame constant; ρ is vacuum density. In vacuum, spherically symmetric waves are not observable, so the longitudinal waves of the Big Bang can not be observed, only the shear waves can be observed.

This involves the problem of vacuum density in the subsequent analysis, the vacuum density is a topic that we can't go around, and so what is the vacuum density? The traditional definition is the amount of material contained in the unit volume, and the essence of matter is energy. Therefore, the definition of vacuum density is the energy contained in the vacuum volume per unit volume. For the ground state vacuum, the vacuum ground state energy is not 0, in addition to the dark energy, dark matter and so on, so we can define the vacuum density. The vacuum density is the energy contained in a unit volume vacuum.

When the vacuum is hardened, the vacuum energy increases, which leads to decrease of the propagation velocity: $V \propto 1/\sqrt{\rho}$.

6.1.8 The field function of the black hole

The Big Bang is the sudden collapse of the positive and antimatter inside the black hole after it is compressed to its limit, which causes the annihilation of the positive and antimatter. This process is similar to the electron-positron collision, which forms the basic particle soup. Under extreme high pressure, the positive charge is unstable and squeezed into protons. In this process, the vacuum strain is conserved.

Considering the whole black hole as a whole, it is an aggregate of positive and negative matter. The black hole will form a huge non-charged spin sphere. The internal field function of the black hole can be expressed as follows.

$$\begin{aligned}\Phi_{BH}(\xi) &= \exp\{(-i)[|\,[\varepsilon_{\alpha\beta}]\,|\,(\xi) + \frac{1}{2}[\mathscr{G}_{\mu\nu}](x)]\}\exp\{(-i)[|\,[-\varepsilon_{\alpha\beta}]\,|\,(\xi,x) + \frac{1}{2}[\mathscr{G}_{\mu\nu}](x)]\} \\ &= \exp(-i)\{[\mathscr{G}_{\mu\nu}](x) + \end{aligned}$$

$$2\begin{bmatrix}|\varepsilon_0| & 0 & 0 & 0\\ 0 & 0 & 0 & 0\\ 0 & 0 & 0 & 0\\ 0 & 0 & 0 & 0\end{bmatrix}\begin{pmatrix}x_0 & 0 & 0 & 0\\ 0 & 0 & 0 & 0\\ 0 & 0 & 0 & 0\\ 0 & 0 & 0 & 0\end{pmatrix}+$$
(1) Rest Energy of Electron E_0

$$2\begin{bmatrix}0 & 0 & 0 & 0\\ 0 & |\varepsilon_{11}| & 0 & 0\\ 0 & 0 & |\varepsilon_{22}| & 0\\ 0 & 0 & 0 & |\varepsilon_{33}|\end{bmatrix}\begin{pmatrix}0 & 0 & 0 & 0\\ 0 & x_1 & 0 & 0\\ 0 & 0 & x_2 & 0\\ 0 & 0 & 0 & x_3\end{pmatrix}+$$
(2) Fiber radial strain, Mass is M_{0R}

$$\begin{bmatrix}0 & 0 & 0 & 0\\ 0 & 0 & \Omega_{12} & 0\\ 0 & \Omega_{21} & 0 & 0\\ 0 & 0 & 0 & 0\end{bmatrix}\begin{pmatrix}0 & 0 & 0 & 0\\ 0 & 0 & x_1 & 0\\ 0 & x_2 & 0 & 0\\ 0 & 0 & 0 & 0\end{pmatrix}+$$
(3) Spin strain, Spin angular momentum ω_{12}

$$[|\gamma_{\mu\nu}|+|-\gamma_{\mu\nu}|](\xi)+[|\omega_{\mu\nu}|+|-\omega_{\mu\nu}|](\xi)] \quad (6.1.11)$$
(4) Shear strain   (5) Rotational strain

Here, $2|\varepsilon_{\mu\nu}|=|\varepsilon_{\mu\nu}|+|-\varepsilon_{\mu\nu}|$. The above equation, the (1), (2) is the intrinsic energy momentum, with the observability in the black hole outside; (3) is the rotational strain of the black hole, this one can also be observed from the external space, performance for the rotating black hole. The field function of the second-order strain leading to the hardening of the outer space-time of the black hole is expressed as

$$(g_{\mu\nu})=\begin{pmatrix}-\left(1-\dfrac{2Mr}{\rho^2}\right) & 0 & 0 & 0\\ 0 & \dfrac{\rho^2}{\Delta} & 0 & 0\\ 0 & 0 & \rho^2 & -\dfrac{Mra\sin^2\theta}{\rho^2}\\ 0 & 0 & \dfrac{Mra\sin^2\theta}{\rho^2} & (r^2+a^2)\sin^2\theta+\dfrac{2Mra^2\sin^4\theta}{\rho^2}\end{pmatrix}$$
$$(6.1.12)$$

The (4), (5), (6) is enclosed in the inner space of the black hole and cannot be observed in the outer space-time. The Einstein field equations only describe is the observability of physical items. The relationship between them satisfies Einstein's gravitational field equation in observation space-time.

$$[\varepsilon_{\alpha\beta}]=\kappa[g_{\mu\nu}] \quad (6.1.13)$$

According to the principle that the corresponding same position term in the matrix is equal, we can directly write the following equation.

Chapter 6　Dark Matter and Dark Energy in the Universe

$$K\begin{bmatrix} \mathscr{g}_{00} & 0 & 0 & 0 \\ 0 & \mathscr{g}_{33} & 0 & 0 \\ 0 & 0 & \mathscr{g}_{33} & 0 \\ 0 & 0 & 0 & \mathscr{g}_{33} \end{bmatrix} = \begin{pmatrix} -\left(1 - \frac{2Mr}{\rho^2}\right) & 0 & 0 & 0 \\ 0 & \frac{\rho^2}{\Delta} & 0 & 0 \\ 0 & 0 & \rho^2 & 0 \\ 0 & 0 & 0 & (r^2 + a^2)\sin^2\theta + \frac{2Mra^2\sin^4\theta}{\rho^2} \end{pmatrix}$$

(6.1.14)

$$\kappa\Omega_{12} = -\kappa\Omega_{21} = \frac{Mra\sin^2\theta}{\rho^2} \tag{6.1.15}$$

This is the gravitational field equation of a rotating black hole.

6.1.9　The Big Bang and the conservation of vacuum strain

Strain conservation includes all the conserved quantities in physics, such as four momentum conservation, angular momentum conservation, and space-time symmetry and so on. The strain of vacuum field is equal before and after the Big Bang. According to this conservation principle, we can write the following eigenvalue equation:

$$\frac{\partial}{i\partial \xi_\mu}\Phi_{BH}(\xi) = N_{e^-}\frac{\partial}{i\partial_\mu}\varphi_{e^-}(\xi) + N_p\frac{\partial}{i\partial_\mu}\varphi_p(\xi) + N_n\frac{\partial}{i\partial_\mu}\varphi_n(\xi) +$$
$$N_\gamma\frac{\partial}{i\partial_\mu}\varphi_\gamma(\xi) + N_\nu\frac{\partial}{i\partial_\mu}\varphi_\nu(\xi) + \cdots \tag{6.1.16}$$

$N_{e^-} = N_p$ is the number of electrons is equal to the number of protons. $|\Phi_{BH}(x)| = |(N_+ + N_-)| \cdot \exp(\pm i \cdot 0)| \equiv 0$.

Experiments predicted that the collision of two high-energy positrons would inevitably produce protons. Electrons and electrons collide to produce antiprotons, but the probability is much lower than that of protons.

$$N(e^+ + e^+ \to p + p) \gg N(e^- + e^- \to \bar{p} + \bar{p}) \tag{6.1.17}$$

The above formula can be abbreviated as

$$\frac{\partial}{i\partial \xi_\mu}\Phi_{BH}(\xi) = \sum_i^n N_i \frac{\partial}{i\partial_\mu}\varphi_i(\xi); i = e^-, p, n, \gamma, \nu_\nu, e^+, \mu^-, \mu^-, \tau^-, \tau^-, \bar{p}, \pi^+, \pi^-, \cdots$$

(6.1.18)

The types of particles produced by the Big Bang include all the basic types of particles we know. All elementary particles are equal in both positive and negative matter. From this point of view, black holes cannot be charged.

6.2　Spherical stress waves that lead to vacuum fragmentation under the initial impact load of the explosion

The Big Bang is prone to misunderstanding, not explosion as we understand it. The

· 281 ·

process is that the black hole particles inside the external strong gravitational black hole crush, instantly collapse inward, making the positive and negative particles annihilate, and the vacuum material after annihilation constitutes the basic particle soup, and then begin to thermal expansion.

For the big bang, the inner space of the cosmic black hole is not compressed to Planck density, the matter and anti-matter inside the huge black hole are collapsing under the strong pressure of the outside world, and the hole-material surface will produce a negative pressure, an inward convergent wave will be produced for the outer space (Fig. 6.2.1). Because the matter has the hole characteristics, the antimatter is a free matter. After the compression reached Planck density, the matter and anti-matter was annihilated. The big bang starts from a smaller radius, forming positive pressure.

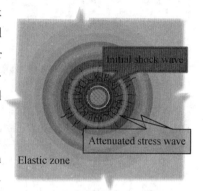

Fig. 6.2.1 A schematic diagram of the range of explosive stress wave

In a moment of black hole collapse, the material of the surface of the black hole is moved inward, and the external pressure is equal to the negative pressure of the surface of the black hole, which can be regarded as a constant value.

Referring to the super position method. In this paper, the stress wave properties of a rectangular pulse loaded with a spherical explosion are studied. The mathematical expressions of the solution of the wave equation are derived and the mechanism of the radial and tangential cracks is analyzed. The total mass of the black hole is M_B, and the total energy of the cosmic explosion is $E_D = M_B c^2$.

Before the big bang, all the visible matter in the universe was absorbed by a black hole. The whole universe is extremely clean. The universe is extremely curved in space-time, forming a huge and stable converging vortex wave. The energy of space-time bending and vortex waves is the energy that exists before the big bang. The energy of space-time bending and vortex wave is the dark energy E_{Dark0}. The violent bending of space-time of a cosmic black hole causes a vacuum to break around the black hole; this constitutes the original dark matter $E_{Dark\ matter0}$. First of all, consider the initial explosion (Fig. 6.2.2). The initial pressure of the wall surface of a thermally expansive black hole is P_b.

$$P_b < P_D \quad (6.2.1)$$

In the upper expression, P_b is the initial pressure of the wall surface of a thermally expansive black hole; P_D is the detonation pressure of the annihilation of matter and anti-matter. The impedance ratio of the explosive force to the vacuum R

$$R = \frac{\rho_e D}{\rho_r C_p} \quad (6.2.2)$$

ρ_e is the energy density (Planck density) in thermally expansive black hole before explosion; ρ_r

Chapter 6 Dark Matter and Dark Energy in the Universe

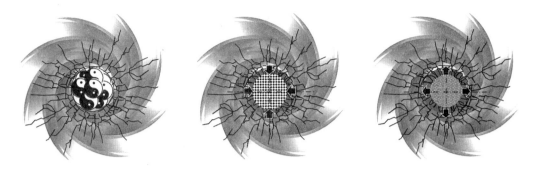

Fig. 6.2.2 Sechematic diagram of the relationship between the big bang and the background space of the cosmic black hole

is vacuum energy density; D is explosion speed of the big bang; C_p is the velocity of the longitudinal wave in the vacuum is the speed of light. If the explosion, the impact load on the vacuum exceeds the critical stress on the vacuum impact deformation curve, and the shock wave is aroused in the vacuum (the critical stress is approximately equal to the volume compression modulus of the vacuum), and the shock waves are aroused in the vacuum. According to the shock wave theory, the initial pressure on the wall of the thermally expansive black hole can be pressed.

$$P_b = f(R) P_D \qquad (6.2.3)$$

As an approximate description, the curve of pressure in the thermally expansive black hole varies with time (Fig. 6.2.3). Due to the rapid release of explosive energy, unloading wave plays an important role in studying the formation mechanism of tangential fissures. The process of producing stress field in a vacuum is a dynamic process, which must be studied from the wave equation.

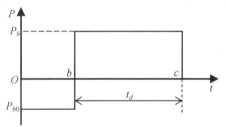

Fig. 6.2.3 Pressure-time curve in blasting cavity

Under the action of pressure $P(t)$ in the cavity, the spherical stress wave propagating outward in the elastic medium causes the medium to produce spherically symmetric expansion, and the motion equation is as follows.

$$\frac{\partial^2 u}{\partial r^2} + \frac{2}{r} \frac{\partial u}{\partial r} - \frac{2u}{r^2} = \frac{1}{C_p^2} \frac{\partial^2 u}{\partial t^2} \qquad (6.2.4)$$

$$C_p = \sqrt{\frac{\lambda + 2G}{\rho_e}} \qquad (6.2.5)$$

Here, u is radial displacement; r is radial coordinate; t is time coordinate; G is lame constant. The relationship between displacement and stress follows Hooke's law.

$$\begin{cases} \sigma_r = \lambda\left(\dfrac{\partial u}{\partial r} + \dfrac{2u}{r}\right) + 2G\dfrac{\partial u}{\partial r} \\ \sigma_\theta = \lambda\left(\dfrac{\partial u}{\partial r} + \dfrac{2u}{r}\right) + 2G\dfrac{\partial u}{\partial r} \end{cases} \qquad (6.2.6)$$

Initial conditions:

$$\begin{cases} u(r,0) = 0 \\ \left.\dfrac{\partial u}{\partial t}\right|_{t=0} = 0 \end{cases} \qquad (6.2.7)$$

If the wall load of the thermally expansive black hole is simplified as the shown in Fig. 6.3.4, the boundary condition can be expressed as:

$$p(t) = \begin{cases} -p_{b0} & 0 < t < t_0 \\ p_b & t_0 < t < t_d \\ 0 & t > t_d \end{cases} \qquad (6.2.8)$$

Here, t_d is pressure action time of thermally expansive black hole wall surface; $u(r,t)$ is radial displacement of medium particle (m); r is the distance between the particle of the vacuum medium and the center of the explosion (m); $\sigma_r, \sigma_\theta, \sigma_\varphi$ The radial stress and the circumferential stress caused by the unloading of spherical waves in the medium (Mpa); λ, G is lame constant; C_p is longitudinal wave velocity of elastic medium (m/s), It's the speed of light $c = 300\,000$ km/s, ρ is vacuum energy density. t_0 is the time of vacuum collapse; t_d is explosion load time. $T = t_0 + t_d$, Time of the big bang. μ is poisson's ratio, the Poisson's ratio of the isotropic material with constant volume in small deformation is 0.5.

Method for solving the problem by Laplace transform of spherical wave elastic medium, Luther Seville function solutions based on the application of superposition method is given for rectangular pulse loads. It can be seen that under load solution of rectangular pulse is equal to the superposition of three separate solutions of Seville Hyde function. The superposition principle is shown in Fig. 6.2.4. In the load (a), (b) and (c), the displacement fields of solutions are as follows.

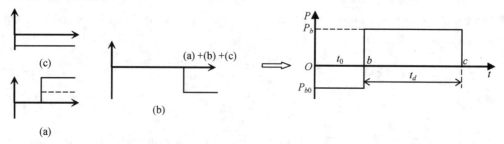

Fig. 6.2.4 The principle diagram of pressure superposition in the wall of an explosion cavity

$$\text{(a)} \quad u_{(a)} = -\dfrac{p_{b0} a^2}{4Gr}\left\{\dfrac{a}{r} + e^{-D\omega t_0}\left[D\left(2 - \dfrac{a}{r}\right)\sin\omega t_0 - \dfrac{a}{r}\cos\omega t_0\right]\right\} \quad (t_0 > 0)$$

$$(6.2.9)$$

(b) $u_{(b)} = H(t_b) \dfrac{p_b a^2}{4Gr} \left\{ \dfrac{a}{r} + e^{-D\omega t_0} \left[D\left(2 - \dfrac{a}{r}\right) \sin \omega t_0 - \dfrac{a}{r} \cos \omega t_0 \right] \right\}$ $(t > t_0)$

(c) $u_{(c)} = - H(t_c) \dfrac{p_b a^2}{4Gr} \left\{ \dfrac{a}{r} + e^{-D\omega t_0} \left[D\left(2 - \dfrac{a}{r}\right) \sin \omega t_0 - \dfrac{a}{r} \cos \omega t_0 \right] \right\}$

$(t > t_0 + t_d)$

Here, a is sphere cavity radius; D is constant, $D = \sqrt{\lambda/\lambda + G}$; ω is constant, $\omega = \omega_0 \sqrt{1 - K^2}$, $\omega^0 = \dfrac{2}{a}\sqrt{\dfrac{G}{\rho_r}}$, $K = \sqrt{\dfrac{G}{\lambda + 2G}}$; t_0 is time, $t_0 = t - t_a$, $t_a = \dfrac{r-a}{C_p}$.

$$u_2(t_1) = -u_1(t_0) \quad (t_1 > 0) \tag{6.2.10}$$

Above: $t_1 = t - t_d = t - t_a - t_d$, The stress field is as follows.

$$\sigma_{rb} = -\dfrac{a^3 p_b}{r^3} + \dfrac{a p_b}{r} e^{-D\omega t_0} \left[\left(1 - \dfrac{2a}{r} + \dfrac{a^2}{r^2}\right) D\sin \omega t_0 - \left(1 - \dfrac{a^2}{r^2}\right) \cos \omega t_0 \right] \quad (t_0 > 0)$$

$$\tag{6.2.11}$$

$$\sigma_{rb}(t_1) = -\sigma_{ra}(t_0) \quad (t_1 > 0) \tag{6.2.12}$$

$$\sigma_{\theta b} = -\dfrac{a^3 p_b}{2r^3} + \dfrac{a p_b}{r} e^{-D\omega t_0} \left[\left(K^2 - \dfrac{a}{r} - \dfrac{a^2}{2r^2}\right) D\sin \omega t_0 - \left(K^2 + \dfrac{a^2}{2r^2}\right) \cos \omega t_0 \right] \quad (t_0 > 0)$$

$$\tag{6.2.13}$$

$$\sigma_{\theta b}(t_1) = -\sigma_{\theta a}(t_0) \quad (t_1 > 0) \tag{6.2.14}$$

The velocity field is

$$v_b = \dfrac{p_b}{\rho_r \omega r} e^{-D\omega t_0} \left[\left(-2K^2 + \dfrac{a}{r}\right) \sin \omega t_0 + \dfrac{2\sqrt{G(\lambda + G)}}{\lambda + 2G} \cos \omega t_0 \right] \quad (t_0 > 0)$$

$$\tag{6.2.15}$$

$$v_b(t_1) = -v_a(t_0) \quad (t_1 > 0) \tag{6.2.16}$$

In the formula, the subscript indicates the displacement, stress and velocity field when the load(a) acts separately. The subscript b indicates the displacement, stress and velocity field when the load (b) acts separately. Known in the Hyde Seville function, the problem of spherical wave in elastic medium can be solved by Laplace transform and so on. Under the action of the rectangular pulse load, the solution after the superposition of two loads should be as follows.

Displacement field:

$$u = u_{(a)} + u_{(b)} + u_{(c)} = -\dfrac{p_{b0} a^2}{4Gr} \left\{ \dfrac{a}{r} + A \right\} + H(t_b) \dfrac{p_b a^2}{4Gr} \left\{ \dfrac{a}{r} + A \right\} - H(t_c) \dfrac{p_b a^2}{4Gr} \left\{ \dfrac{a}{r} + A \right\}$$

$$= \left(-p_{b0} + H(t_b) p_b - H(t_c) p_b\right) \dfrac{a^2}{4Gr} \left\{ \dfrac{a}{r} + e^{-D\omega t_0} \left[D\left(2 - \dfrac{a}{r}\right) \sin \omega t_0 - \dfrac{a}{r} \cos \omega t_0 \right] \right\}$$

$$\tag{6.2.17}$$

Here, $H(t)$ is unit step function, that is: $H(t_1) = \begin{cases} 1, & t_1 > 0 \\ 0, & t < 0 \end{cases}$.

Stress field:

$$\sigma_{r1} = \sigma_{ra} + \sigma_{rb} + \sigma_{rc} = (-p_{b0} + H(t_b)p_b - H(t_c)p_b) \times$$
$$\left\{-\frac{a^3}{r^3} + \frac{a}{r}e^{-D\omega t_0}\left[\left(1 - \frac{2a}{r} + \frac{a^2}{r^2}\right)D\sin\omega t_0 - \left(1 - \frac{a^2}{r^2}\right)\cos\omega t_0\right]\right\} \quad (t_0 > 0)$$
(6.2.18)

$$\sigma_{\theta 1} = \sigma_{\theta 1} + \sigma_{\theta 2} = (-p_{b0} + H(t_b)p_b - H(t_c)p_b) \times$$
$$\left\{-\frac{a^3}{2r^3} + \frac{a}{r}e^{-D\omega t_0}\left[\left(K^2 - \frac{a}{r} - \frac{a^2}{2r^2}\right)D\sin\omega t_0 - \left(K^2 + \frac{a^2}{2r^2}\right)\cos\omega t_0\right]\right\} \quad (t_0 > 0)$$
(6.2.19)

$$v = v_1 + v_2 = (-p_{b0} + H(t_b)p_b - H(t_c)p_b) \times$$
$$\left\{\frac{e^{-D\omega t_0}}{\rho_r \omega r}\left[\left(-2K^2 + \frac{a}{r}\right)\sin\omega t_0 + \frac{2\sqrt{G(\lambda + G)}}{\lambda + 2G}\cos\omega t_0\right]\right\} \quad (t_0 > 0) \quad (6.2.20)$$

Through the above analysis, we know

(1) Under the action of the rectangular pulse pressure, the radial maximum tensile stress generated by the unloading can be obtained by (6.2.18).

$$\sigma_{r\max} = \frac{ap_b}{r}\left(1 - \frac{a^2}{r^2}\right) \quad (6.2.21)$$

(2) The maximum radial velocity (pointing Center) produced when unloading is

$$v_{r\max} = \frac{2p_b}{\rho_r \omega r}\frac{\sqrt{G(\lambda + G)}}{\lambda + 2G} \quad (6.2.22)$$

(3) The faster the unloading speed, the greater the unloading wave intensity, that is, the greater the radial tensile stress is.

(4) The action time t_d of the wall pressure of the thermally expansive black hole has a great influence on the radial tensile stress at the unloading time. The increase of t_d can make the radial compressive stress tend to the steady state value, then suddenly unload, aggravate the tangential fissure.

(5) Because of the discontinuity of the solution, the maximum radial pressure is attenuated according to r^{-1}.

(6) The stress waves produced by the rectangular pulse load have two stress discontinuitiessurface, that is, the arrival time and the unloading time of the stress wave. the time difference between two stress discontinuities surface is a constant at any distance, equal to t_d when the wall pressure is acted on the thermally expansive black hole.

(7) The formation of cracks around the thermally expansive black hole. At first, the thermally expansive black hole collapse, Pressure unloading in the burst cavity, the unloading wave and radial tensile stress are produced. Under the action of radial tensile stress, the annular fissure is produced. And then it exploded. The radial fissure is formed under the action of tangential tensile stress. After the explosion shock stress disappears, the unloading wave is produced. Then the radial tensile stress is produced. Under the action of radial tensile stress, the annular

fissure is produced.

6.3 The mechanism of the production of dark matter caused by vacuum rupture

6.3.1 Stress intensity factor

When a 2-dimensional vacuum is broken, the normal stress σ_n and shear stress τ which act on crack plane, they are

$$\sigma_n = \frac{1}{2}(\sigma_r + \sigma_\theta) + \frac{1}{2}(\sigma_r - \sigma_\theta)\cos(2\beta) \qquad (6.3.1)$$

$$\tau = \frac{1}{2}(\sigma_r - \sigma_\theta)\sin(2\beta) \qquad (6.3.2)$$

1. Stress intensity factors under pressure and shear conditions

When the crack is pressed in a bi-directional state, the crack is pressed and closed, the friction is produced between the crack surfaces, and produce relative sliding under the action of the effective shear stress τ_{eff}, it is Ⅱ type crack. The end stress intensity factor $K_{Ⅱ}$ is

$$K_{Ⅱ} = \tau_{eff}\sqrt{\pi a} = (\tau - \mu\sigma_n)\sqrt{\pi a} \qquad (6.3.3)$$

In the formula, μ is the sliding friction coefficient of the crack surface (Fig. 6.3.1). For vacuum, the interaction between two-way compression cracks will produce low frequency radiation, which means that the dark matter around the black hole will radiate low frequency electromagnetic waves when the space-time changes around the black hole.

When τ_{eff} reaches a critical value, the crack begins to expand and the tip of the initial crack will produce a wing crack. Wing crack growth is bring Ⅱ type strength factor convert to Ⅰ type intensity factor

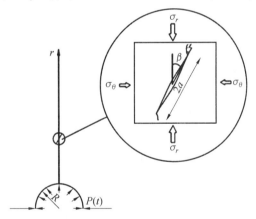

Fig. 6.3.1 Schematic diagram of analysis model

$$K_Ⅰ(\theta) = \frac{3}{2}\sin\theta\cos\frac{\theta}{2}K_{Ⅱ} \qquad (6.3.4)$$

In the formula, θ is the angle of the wing crack growth direction. It is known from the maximum tensile stress criterion. The initial growth direction of the wing crack is $\theta_0 = \pm 70.5°$, when the crack is cracked under the condition of pressure shear stress, the stress intensity factor of the wing crack: $K_Ⅰ = \frac{2}{\sqrt{3}}K_{Ⅱ}$.

2. Stress intensity factors under tension shear conditions

When the bi-directional surface of the crack is drawn, or when one side is pressed and one side is pulled, under the action of shear stress, shear slide will occur in the fracture surface. At the same time, it is influenced by the normal tensile stress. The fracture surface also produces normal tension. In tension shear stress state, composite crack is jointly controlled by the shear stress τ and normal tensile stress σ_n. Using polar coordinates to solve known composite cracks, the stress field at the end of the crack can be obtained. The σ_θ in the crack of the extended wing can be expressed as

$$\sigma_\theta = \frac{1}{\sqrt{2\pi r}}\cos\frac{\theta}{2}\left(\sigma_n\sqrt{\pi a}\cos^2\frac{\theta}{2} - \frac{3\tau\sqrt{\pi a}}{2}\sin\theta\right) \qquad (6.3.5)$$

The stress intensity factor at the end of the crack is defined as

$$K_I = \lim_{r \to 0}\sqrt{2\pi r}\,\sigma_\theta \qquad (6.3.6)$$

So there is

$$K_I(\theta) = \cos\frac{\theta}{2}\left(\sigma_n\sqrt{\pi a}\cos^2\frac{\theta}{2} - \frac{3\tau\sqrt{\pi a}}{2}\sin\theta\right) \qquad (6.3.7)$$

When the stress intensity factor reaches a certain critical value, the crack begins to expand. Computing partial derivative for θ in the formula (6.3.5), and stipulate that it is 0, which can be obtained

$$2\tau\tan^2\frac{\theta}{2} - \sigma_n\tan\frac{\theta}{2} - \tau = 0 \qquad (6.3.8)$$

The angle θ_0 of the wing crack can be obtained by the formula (6.3.8). Substituting the cracking angle θ_0 into the K_I expression, the stress intensity factor of a wing crack under the condition of tensile shear stress

$$K_I = \cos\frac{\theta_0}{2}\left(\sigma_n\sqrt{\pi a}\cos^2\frac{\theta_0}{2} - \frac{3\tau\sqrt{\pi a}}{2}\sin\theta_0\right) \qquad (6.3.9)$$

6.3.2 Cracking mechanism of vacuum district driven by stress wave in vacuum explosion

The radial stress rises to the peak of compressive stress in a very short time. With the removal of the explosive load, the radial stress in the vacuum of the chamber wall is changed from pressure stress to tensile stress. Due to the impact of the explosive load, the wall of the thermally expansive black hole will stimulate the ring pressure stress in the initial stage of the explosion load.

In addition, due to the strong radial compression produced by the explosive load in the cavity wall vacuum, the larger circumferential tensile stress will be derived. In space, the peak value of radial stress and circumferential stress attenuates rapidly with the increase of the distance from the center of the thermally expansive black hole. In time, by comparing the radial stress and the circumferential stress at different times of the same distance, it can be found that the stress state in the vacuum medium will be in pressure shear or tensile shear stress state

Chapter 6   Dark Matter and Dark Energy in the Universe

successively, which can be roughly divided into 3 stages. That is, I: the pressure and shear stress state of the combination of radial stress and circumferential pressure stress; II: the tension shear stress state of the combination of radial stress and circumferential tensile stress. III: the tension shear stress state of the combination of radial tensile stress and circumferential tensile stress. As shown in Fig. 6.3.2. The explosive gas intrudes into the cracks, which results in the distribution of visible materials along the cracks in the shell. The thickness of the shell is equal to the radius of the explosion zone.

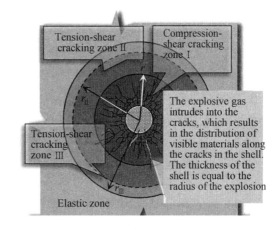

Fig. 6.3.2   Schematic diagram of vacuum cracking range

1. Critical condition of wing crack growth under pressure shear conditions

For the stress state of the pressure shear, the stress intensity factor K_I is related to the stress state in the crack, the length $2a$ of the crack and the friction factor μ of the crack surface, and so on. The radial stress r and the circumferential stress θ produced by the explosive load are all related to the explosion load $P(t)$. To order

$$\begin{cases} \sigma_r = f(p_{(t)}) \\ \sigma_\theta = k(p_{(t)}) \end{cases} \quad (6.3.10)$$

Among them: $k = \sigma_\theta / \sigma_r$. Take the formula (6.3.10) into the (6.3.4), the expression of K_I can be obtained:

$$K_I = \sqrt{\frac{\pi a}{3}} f(p_{(t)}) \{(1-k)[\sin(2\beta) - \mu\cos(2\beta)] - \mu(1+k)\} \quad (6.3.11)$$

Toorder: $K_r = K_I / \sqrt{\frac{\pi a}{3}} f(p_{(t)})$, is there

$$K_r = (1-k)[\sin(2\beta) - \mu\cos(2\beta)] - \mu(1+k) \quad (6.3.12)$$

The K_r can react the influence of the circumferential stress and radial stress ratio K, the crack surface friction factor μ and the crack inclination angle β to the stress intensity factor K_I. The larger K_r, the larger K_I is.

2. Critical conditions for the wing crack growth under the condition of pulling shear

Under the condition of tension shear, the formula (6.3.10) can be replaced (6.3.9) in the same way, and the expression of K_I is obtained.

$$K_I = \frac{\sqrt{\pi a}}{2} f(p_{(t)}) \left\{ \cos^3 \frac{\theta_0}{2} [(1+k) + (1-k)\cos(2\beta)] - 3\cos^3 \frac{\theta_0}{2} \sin \frac{\theta_0}{2}(1-k)\sin(2\beta) \right\}$$

$$(6.3.13)$$

Set
$$K_r = \frac{K_I}{\sqrt{\pi a} f(p_{(t)})/2} \quad (6.3.14)$$

$$K_r = \cos^3\frac{\theta_0}{2}[(1+k)+(1-k)\cos(2\beta)] - 3\cos^3\frac{\theta_0}{2}\sin\frac{\theta_0}{2}(1-k)\sin(2\beta) \quad (6.3.15)$$

By analyzing the variation of explosive stress wave with time and space and its influence on vacuum cracking process, the following conclusions are drawn:

(1) Under the action of explosive load, according to the time sequence, the stress state in vacuum can be divided into three stages: the compression-shear stress state of the combination of radial compressive stress and circumferential compressive stress, the tension-shear stress state of the combination of radial compressive stress and circumferential tension stress, and the tension-shear stress state of the combination of radial tension stress and circumferential tension stress.

(2) In the near zone of the blasting chamber, the vacuum cracking is mainly controlled by the stage I of the explosive stress wave, i.e., the compressive shear stress state; when the explosive stress wave is far from the blasting chamber, the tension-shear stress state triggered by the stage II of the explosive stress wave plays a major role in the rock cracking; with the further increase of the distance, the cracking of the blasting chamber is mainly controlled by the tension-shear stress state triggered by the stage III of the explosive stress wave.

6.4 The destruction scope of Big Bang to the vacuum

To consider the big bang, we must first consider the composition of the black hole. A black hole is made up of matter and anti-matter. All these substances reach the strain limit in the central district of the black hole. When the strain limit is reached, time and space are frozen. Before the big bang, the starting point of the universe was a supermassive black hole. The black hole is bigger, vacuum collapsing inward strength will be greater. If balance of time and space freezing is broken under the strong pressure of the vacuum, and it collapses inward, matter and anti-matter is violent explosions after meeting—this constitutes the big bang of the universe. For ordinary black hole space-time once frozen and the process of inword collapse in a very slow state.

Because vacuum is a brittle elastic material, it has similar physical properties to brittle materials. We can compare and use for reference the characteristics of rock explosion to study the vacuum explosion.

When the matter and anti-matter annihilated to produce a big explosion, a strong stress wave would be produced. The blasting energy is transmitted to the vacuum by the shock wave. The shock wave induced by the vacuum in the nearest area of the blasting point is made to compress the vacuum in radial direction and form a crushing district. The annihilation of matter and anti-matter within the thermally expansive black hole are disintegrated into the vacuum basic unit powder, to form the visible material we are familiar with and can be regarded as a gas, that

Chapter 6 Dark Matter and Dark Energy in the Universe

is the plasma gas made up of fermions and bosons, the pulverization of the vacuum is understood as a vacuum gasification, at the same time it also causes the displacement of the vacuum particle and the expansion of the thermally expansive black hole. The thermally expansive black hole constitutes an absolute vacuum cavity without any material. When the shock wave works on the vacuum, the energy rapidly attenuates, and the shock wave at the edge of the crushing district decays into the elastic stress wave (hereinafter referred to as the stress wave). The propagation of stress waves produces radial fissure in the vacuum tangential stretch. In the process of expansion, the shock wave energy continues to be consumed, and the strength of the stress wave is reduced. So that it can only cause the elastic deformation of the vacuum and the vibration of the particle outside the fractured district. It is particularly important that following the shock wave, the explosive gas expands further, expands the cavity further, and rushing into the fissure, resulting in the "gas wedge effect", which leads to the extension of the fracture and makes the crack gap present vaporization material. On the other hand, after the explosion, the vacuum bursting cavity loses the external energy output, and the external vacuum pressure is very strong, resulting in a retraction. It will form an inward cosmic spin wave. The existence of the cosmic spin wave makes the universe matter have a rotating wave background.

Under the action of blast wave and stress wave, the damage of vacuum is a rather complicated dynamic process. Similarity between setting vacuum blasting and rock blasting. Big Bang produced extremely high temperatures and high pressures at the moment. A shock wave is formed between the wall of the thermally expansive black hole, and it acts instantaneously on the outer vacuum. The shock wave of the superlight velocity exerts a very strong impact compression effect on the cavity wall vacuum. The vacuum around the thermally expansive black hole was crushed and crushed to form a crushing district. The gas produced in the thermally expansive black hole is the plasma gas made up of fermions and bosons. The evolution of this part is described by Thermo cosmic big explosion theory known to us, which is no longer described here. Although the range is very small, it consumes most of the energy of the shock wave, which reduces the shock wave on the interface of the smash district into a compressive stress wave. Under the action of the stress wave, the initial crack is formed in the vacuum. Then the detonation gas (the gas is the plasma gas made up of fermions and bosons) expansion, extrusion, wedge effect contributed to the extension and expansion of crack, Only when the stress wave and the detonation gas are attenuated to a certain degree, the crack growth can be stopped. At the same time, there will be differences in the propagation of cracks. In this way, with the formation, expansion, penetration, crisscrossing, interlacing, internal and external fissures of the radial fissure, circumferential crack and shear fracture, the vacuum is fragmented into different sizes. Near the crushing district, the vacuum block is fine and sticky, away from the smash district, the bulk increases. This broken vacuum state constitutes a substance with a linear viscoelastic property. The nearer the detonating cavity, the stronger the viscosity, the farther from the detonating cavity, the more it tends to the ideal elastic material. The radial fissure area in the vacuum is the main form of the vacuum explosion damage. The extended range of the fractured district is of great significance to the subdivision of dark matter and the evolution of the subsequent universe.

6.4.1 The smash district and fractured district in vacuum

The smash district formed by the joint action of annihilation of matter and anti-matter and shock wave.

The physical mechanism of the Big Bang is a strong explosion at the moment of the annihilation of the matter and anti-matter. The powdered vacuum in the thermally expansive black hole and the lining of the smash district forms gas, and then the detonation gas that is the plasma gas made up of fermions and bosons. The focus of our attention is on the dark matter made up of vacuum rupture in the big bang. To using the common formulas in blasting mechanics to calculate the radius of the fracture district, knowable

$$R_T = r_b(\lambda P_d/S_T)^{1/\alpha} \tag{6.4.1}$$

In the formula, P_d is the initial impact pressure of the shock wave on the vacuum of the wall of the thermally expansive black hole; r_b is the radius of the burst cavity; S_T is the tensile strength of vacuum; λ is a coefficient; $\lambda = \mu/(1-\mu)$; μ is vacuum Poisson ratio. For the ideal elastic medium, take 0.5, In fact, our real material world is not completely ideal. According to the relationship between the physical world and the vacuum, It is preliminarily estimated that the vacuum in the elastic medium is close to 0.5; α is the attenuation index of the stress wave. The effect of shock wave and the existence of the smash district are ignored in the derivation which is calculated directly by the stress wave formed in the vacuum after the explosion.

Outside the crushing district formed by the explosion shock wave, it is the scope of the action of the stress wave, and it follows the attenuation law of the stress wave. Therefore, first, the radius of the crushing district under the impact of the shock wave is calculated. The following relationship exists between the velocity of the wave propagation wave in the vacuum and the velocity of the vacuum particle on the wave front.

$$D_c = a + bu \tag{6.4.2}$$

In the formula, D_c is the internal shock wave velocity in the vacuum. u is the velocity of the particle motion of the vacuum. a, b are determined by the constant of the experiment. At the interface of the smash district, the shock wave attenuates to the stress wave. The wave velocity D_c of the stress wave is the elastic longitudinal wave velocity in the vacuum. That is $D_c = C_p$. The peak pressure P_r here is obtained by the conservation of the shock wave momentum in the vacuum.

$$P_r = \rho_m C_p u_r \tag{6.4.3}$$

In the formula, P_r is the peak pressure on the wave front of the shock wave; ρ_m is vacuum density; C_p is the elastic longitudinal wave velocity in the vacuum. u_r is the velocity of the movement of the vacuum particles on the interface of the smash district.

Chapter 6 Dark Matter and Dark Energy in the Universe

According to (6.4.2), we have to: $u_r = (C_p - a)/b$. Take it into (6.4.3):

$$P_r = \frac{\rho_m C_p (C_p - a)}{b} \tag{6.4.4}$$

On the other hand, the attenuation relationship of the peak pressure of the shock wave with the distance can be approximately expressed as

$$P_c = \frac{P_d}{\bar{r}^3} \tag{6.4.5}$$

In the form: $\bar{r} = R/R_a$; R is the distance from the center of the detonating cavity; R_a is the radius of the smash district. There is no gap when a black hole explodes. The P_c calculation is as follows.

$$P_c = \frac{\rho_0 D^2}{4} \times \frac{2\rho_m C_P}{\rho_m C_P + \rho_0 D} \tag{6.4.6}$$

In the formula, D is the shock wave velocity; ρ_m is the initial density of vacuum; ρ_0 is the black hole density; C_p is the elastic longitudinal wave velocity in the vacuum. On the fractured district interface:

$$P_r = P_d \left(\frac{r_b}{R_0}\right)^3 \tag{6.4.7}$$

In the formula, R_0 is the radius of the smash district. The radius of the smash district can be obtained by (6.4.4) and (6.4.7).

$$R_0 = r_b \left[\frac{bP_d}{\rho_m C_p (C_p - a)}\right]^{\frac{1}{3}} \tag{6.4.8}$$

The symbols in the formula are same as that in the front. So as long as we get two constants a and b by Big bang cosmic observation, the peak pressure and the radius of the smash district can be obtained by the (6.4.4) and (6.4.8).

6.4.2 Stress in fractured district and failure district under the action of stress wave

1. Stress in fractured district

The large amount of energy consumed by the shock wave attenuates rapidly and becomes the stress wave outside the smash district. The pressure wave continues to propagate along the radial direction in the vacuum, causing radial compression and tangential tension, and the gravitational wave is the stress wave. Due to the poor tensile capacity of the vacuum, the crack will occur in the radial direction when the tensile strain exceeds the failure strain. Although the peak pressure of stress wave is far lower than the value of the shock wave, exceed the vacuum compression strength in a certain range, this makes the vacuum to be crushed.

We know, the relationship between radial stress wave stress and distance attenuation is

$$\sigma_r = \frac{P_d}{\bar{r}\alpha} \tag{6.4.9}$$

In the formula, $\bar{r}$ is the distance of ratio; P_d is the initial radial peak stress of the stress wave;

α is the attenuation index. The tensile stress produced in the direction of tangential direction:

$$\sigma_\theta = \frac{b\sigma_r}{\bar{r}\alpha} \tag{6.4.10}$$

In the formula, $b = \mu/(1 - \mu)$, μ – Poisson's ratio.

In the propagation process of the stress wave in the fracture district, the relation between the tangential stress and the radial stress around the thermally expansive black hole, that is, the condition of the cracking in the vacuum.

$$\sigma_\theta = (1 - 2b^2)\sigma_r \tag{6.4.11}$$

In the formula, b is the ratio of the vacuum transverse wave to the longitudinal wave velocity, and $b = C_S/C_p$. When the tangential stress σ_θ in the vacuum exceeds the tensile strength of the vacuum, the vacuum appears cracking, that is

$$\sigma_\theta \geq K_T S_T \tag{6.4.12}$$

In the formula, S_T is the tensile strength of vacuum; K_T is the increasing coefficient of the tensile strength of the vacuum. In the state of dynamic load, $K_T > 1$.

2. Crush district radius under stress wave

In order to distinguish the crushing district under the impact compression of shock wave, the crush range caused by stress wave is called the crush district. The relationship between the peak pressure of the stress wave and the attenuation of the distance is known. That is

$$\sigma_{rmax} = \frac{P_r}{\bar{r}^\alpha} \tag{6.4.13}$$

In the formula, P_r is the peak stress of the stress wave on the interface of the smash district. It can be obtained by (6.4.4); $\bar{r}$: is the comparison distance $\bar{r} = R/R_0$, R_0: the radius of the smash district; R: distance from the center of the detonating cavity; α: the attenuation index of stress wave, $\alpha = 2 - \mu(/1 - \mu)$. The peak pressure σ_{rmax} of the stress wave in the formula which can be replaced by the vacuum compressive strength S_c, we can obtain the calculation formula of the radius of the crush district.

$$R_c = R_0 \left(\frac{P_r}{S_c}\right)^\alpha = R_0 \left[\frac{\rho_m C_p (C_p - \alpha)}{bS_c}\right]^{\frac{1}{\alpha}} \tag{6.4.14}$$

In the formula, the constant a and b is determined by the test.

3. Fracture district under the action of stress wave

The main cause of the radial fissure in the vacuum is caused by the tangential tensile stress. The relation between the known tangential peak tensions with the change of distance is

$$\sigma_{\theta max} = \frac{\lambda P r}{\bar{r}^\alpha} \tag{6.4.15}$$

In the formula, λ is a coefficient $\lambda = \mu/(1 - \mu)$. The vacuum tensile strength S_T is used to replace the $\sigma_{\theta max}$ in the formula. The radial crack can be obtained.

$$R_c T = R_0 \left(\frac{\lambda P_r}{S_c}\right)^{\frac{1}{\alpha}} = R_0 \left[\frac{\lambda \rho_m C_p (C_p - \alpha)}{b S_T}\right]^{\frac{1}{\alpha}} \tag{6.4.16}$$

4. Growth length of explosion crack by the action of stress wave and damage

Due to the strong bending of space-time around the black hole before the big bang, there must be a large number of primary cracks in the vacuum around the black hole. The length and orientation of the cracks are randomly distributed in space. That is, a certain degree of damage has already existed when the vacuum is not subjected to the external load. When such a vacuum is subjected to an external load, the primary crack is interacted with the explosive stress wave. The growth of many new cracks and primary cracks in the vacuum, this will certainly affect the crack growth under the action of detonating gas. This is similar to the case of rock. As early as 1971, it was noticed by Kutter and Fairhurst. After considering the damage effect of the stress wave, the crack length is L'

$$L' = L/(1-D)^2 \qquad (6.4.17)$$

In the formula, D is the damage value of the crack tip; L is the length of the final crack under the action of the stress wave when the damage is not considered; L' is the crack growth length after taking into account the effect of internal vacuum injury. This formula shows that the primary crack of vacuum and the damage of the explosive stress wave are combined action on vacuum, it leads to the final length of the crack will be longer than that of the previous theory. This is obviously the contribution of the stress wave. In view of the above, the crack caused by the Big Bang is longer than that predicted by the theory.

5. The relationship between crack and detonating gas

The total energy of the detonating gas is less than or equal to 5% of the total energy of the universe. In the process of the big bang, the detonating gas is injected into the crack in the comminuted district. Due to the random occurrence of a crack containing a detonating gas, after the crack is closed; the detonating gas is cooled to a hydrogen atom, which forms a material wave and diffuses outward. On the other hand, the crack itself is a dark matter, which has a gravitational effect. Gravitation makes the distribution of visible matter the following characteristics:

After the big bang, near the center, the denser the crack is; the farther away from the center, the thinner it is. At the moment of the big bang, the detonating gas was injected into the crack. The distribution of visible matter in a spherical shell which thickness is ΔR (Fig.6.4.1), all the gases in the cracks (fermions) get momentum and get the same initial velocity to spread out in the big bang. That is when the detonating gases in all the cracks expand outward having the same initial velocity. Our cosmological theory holds that all substances have a same initial velocity at the same time from the center point of the big bang, and there must be an exact Harbert constant. The two are different. But here, all of the visible material distribution in a spherical body, namely the R_0 radius of the smash district, simultaneous start, this leads to the inevitable existence of an error range for the Harbert's constant. The latest WMAP data give Harbert's constant $H_0 = 71$ km/(s·Mpc). In various practical measurement verification work, the distance determination for brightest Cluster Galaxies and type Ia supernovae (SNIa) is still strictly obeyed up to 150 h^{-1}·Mpc Hubble relation with an error of $\Delta H_0/H$

is 0.07. Therefore, the determination of the error range directly determines the radius of the sphere in the early comminuted area of the big bang. In the process of the evolution of the universe; as the universe expands, the sphere evolves into a spherical shell. The shell thickness is between ΔR, $\Delta R > R_0$, R_0 is the radial of the smash district under the impact compression of shock wave. We can according to the (radial) universe observation Harbert law: $V_f = H_c \times D$ to determine the distribution of the thickness of the visible matter $\Delta R = \Delta D$.

$$\Delta D = V_f / H_{c\,min} - V'_f / H_{c\,max} \tag{6.4.18}$$

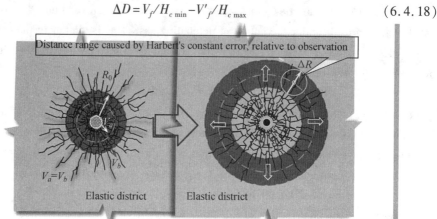

Fig. 6.4.1　Visible matter is distributed in the vacuum fissures

V_f: The velocity of motion of the visible matter we have observed along the radial direction of the universe. Choosing to the same speed of leaving the observer can simplify the problem. For experimental observations, the minimum Harbert constant $H_{c\,min}$ means that the initial position of the observed body is at the outer edge of the R_0; The maximum Harbert constant $H_{c\,max}$ means that the initial position of the observed body is in the center of the R_0; and the current H_c is a statistical average (Fig. 6.4.2). In this sense, the error of the Harbert constant reflects the radius of the cosmic explosion smash district.

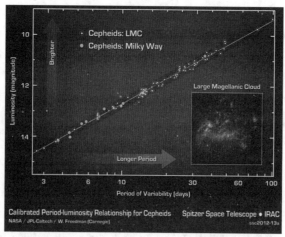

Fig. 6.4.2　Harbert constant measurement data are distributed in an error zone

ΔR's determination: It is known by (6.4.8) that, the radius of the smash district $R_0 =$

Chapter 6 Dark Matter and Dark Energy in the Universe

ΔD, the following equation can be obtained.

$$\Delta R = V_f \left(\frac{1}{H_{c\min}} - \frac{1}{H_{c\max}} \right) = r_b \left[\frac{bP_d}{\rho_m C_p (C_p - a)} \right]^{\frac{1}{3}} \quad (6.4.19)$$

The spherical shell thickness of the visible matter distribution can be obtained by the upper equation. The inevitable error of the Harbert constant will constitute the most important evidence of the new theory.

(1) Visible matter evolution at the center point.

In the process of the big bang, the part of the explosive gas entering the crack is relatively small compared to the number of detonating gases in the thermally expansive black hole. Most of the detonating gases exist in the thermally expansive black hole, so the thermally expansive black hole can be regarded as a point. After the big bang, the material waves propagate from the center of the explosion. In this way, there will be an exact Harbert constant for cosmic observations. The radius of the thermally expansive black hole is the radius of the cosmic black hole.

(2) Visible matter evolution at central point and smash district.

At the beginning of the birth of the universe, the visible matter is accompanied by the existence of a crack. This part of the visible matter is a small part; Most of the visible matter begins to spread out from the center. And the crack itself has the gravitational force, it is shown as a visible substance around the crack. This makes the distribution of the visible matter in the universe consistent with the distribution of the crack, has the shape of the vegetable sponge, that is, the structure of the fiber.

(3) The late stage of the evolution of the universe.

Relative to the matter, the crack is in a relatively stationary state. As a result, the visible material in the spherical shell is expanding outward. In the process of outward diffusion, it is bound to exceed the crack zone. The result of the observation is that the cracks are becoming thinner and thinner. At the same time, far away from the center of the universe, the background is becoming more and more flat, we are speeding away from the center of the universe. This image is in good agreement with the experimental observation. When the expansion of the universe is over the radius of the super crack region, we are going into the area without dark matter. $D \geqslant R_c$, that is

$$\frac{V_f}{H_{c\min}} \geqslant R_0 \left(\frac{\lambda P_r}{S_c} \right)^{1/\alpha} \quad (6.4.20)$$

6.5 Energy distribution of dark energy stress wave in a vacuum

6.5.1 Shock wave energy distribution

The radial displacement of the vacuum particles on the wall of the thermally expansive black hole is caused by the strong compression of the shock wave, and the thermally expansive black hole expands and the diameter increases. When the shock wave propagates to the edge of the smash district, the process of shock compression is over, and the detonation process caused by the shock wave terminates. In this process, the conservation of mass, it can be expressed as follows.

$$(r^2 - r_0^2)\rho_m = \int_{r_1}^{r} 2\rho r dr \tag{6.5.1}$$

In the equation: r is shock wave radius; r_1 is the radius of the detonating cavity corresponding to the r; ρ is the vacuum density on the shock wave front.

The vacuum density is very small which after the shock wave front, it can be treated by equal density, ρ_r is the vacuum density at the wall of the thermally expansive black hole which was compressed. Replacing ρ_0 in the formula with ρ_r, and to integrate the above formula, so it can be available:

$$(r^2 - r_0^2)\rho_m = (r^2 - r_1^2)\rho_r \tag{6.5.2}$$

The expansion law of the thermally expansive black hole under the impact of the shock wave is obtained.

$$r_1 = [r^2 - (r^2 - r_0^2) \cdot \rho_m/\rho_r]^{1/2} \tag{6.5.3}$$

At the edge of the smash district, that is $r = R_C$. The process of detonation ends, the final radius of the thermally expansive black hole is

$$R_0 = [R_C^2 - (R_C^2 - r_0^2) \cdot \rho_m/\rho_r]^{1/2} \tag{6.5.4}$$

The ρ_r can be obtained by the following methods:

At the wall of the thermally expansive black hole, after the shock wave, the vacuum state parameters meet the following two equations.

$$\rho_0 D = \rho_r(D_0 - V_0) \tag{6.5.5}$$

$$D_0 = a + bV_0 \tag{6.5.6}$$

Here, D_0 is shock wave velocity at the wall of a burst cavity; a, b are experimental constant.

Therefore

$$\rho_r = \frac{a + bV_0}{a + (b-1)V_0}\rho_m \tag{6.5.7}$$

The energy of shock wave is mainly consumed in expanding the thermally expansive black hole, producing cracks and causing the elastic deformation of the vacuum. Before the big bang, the total mass of a black hole in the universe is transformed into energy, so the total energy of the big bang cosmic black hole is

$$E_0 = \frac{4\pi}{3}r_0^3\rho_0 c^2 \tag{6.5.8}$$

In the above formula: c is speed of light, ρ_0 is black hole mass density.

6.5.2 The energy E_1 consumed by the formation of an thermally expansive black hole

The energy consumption of the shock wave is equal to the work done by the shock wave to the vacuum. Shock wave work donein the process of bursting cavity expansion is

$$W_1 = \int_{r_0}^{R_0} 4\pi r^2 P_r \, dr \qquad (6.5.9)$$

Take attenuation law of peak pressure of shock wave in vacuum $P_r = P_m(r_0/r)^3$. Substituting the above equation and integrating it.

$$E_1 = W_1 = 4\pi r_0^3 P_m (\ln R_0 - \ln r_0) \qquad (6.5.10)$$

Therefore, The ratio of the consumption energy of formation bursting cavity to the total energy of Big Bang.

$$\eta_1 = \frac{E_1}{E_0} = \frac{3P_m}{\rho_0 c^2} (\ln R_0 - \ln r_0) \qquad (6.5.11)$$

6.5.3 The energy E_2 consumed by the generation of radial fissure

The peak stress of the stress wave outside thethermally expansive black hole is
Radial direction:

$$\sigma_r = P_m \left(\frac{r_0}{r}\right)^\alpha \qquad (6.5.12)$$

Tangential direction:

$$\sigma_\theta = \lambda P_m \left(\frac{r_0}{r}\right)^\alpha \qquad (6.5.13)$$

Set up the tangential stress at the radialr is σ_θ, the corresponding crack length in the area is r, if $\sigma_\theta > [\sigma_t]$, the fissures will expand, δ_θ is used to indicate the expansion of radial fissure caused by δ_θ. The length of the fissure extends from r to $r + \delta_r$, work done by the tangential stress σ_θ is gradually vanished, that is the work done to vacuum, at the same time, the tangential displacement of the crack surface is u_θ. So in this process σ_θ is (two facets of fissures)

$$\delta_W = 2\int_0^{\delta_a} \frac{1}{2} \sigma_\theta u_\theta \, dr \qquad (6.5.14)$$

According to the principle of fracture mechanics, the tangential stress and tangential displacement at the end of the fracture can be respectively expressed as

$$\sigma_\theta = \frac{K_2}{\sqrt{2\pi r}} \qquad (6.5.15)$$

$$u_\theta = \frac{K_1}{E_m} \sqrt{\frac{\delta_\pi - r}{2\pi}} (1 + \nu)(k + 1) \qquad (6.5.16)$$

Here, E_m is elastic modulus of vacuum; the coefficient k is related to the stress and strain state of the vacuum. Considering the background of the Big Bang is a converging wave, so in a larger range, the explosion has an axisymmetric character. The cracks in the vacuum rupture are mainly distributed in a disk range. The axisymmetric explosion can use plane strain problem solution, take $k = (3 - 4\nu)$; K_2 and K_2 respectively correspond to the stress intensity factors at the end of the fissure and the crack length are r and $r + \delta_\pi$, respectively.

When the length of the crack is infinitely short, Because δ_π is an infinitesimal quantity, $\delta_\pi = dr$, You can think of $K_2 = K_1$. Therefore, take (6.5.15), (6.5.16) into (6.5.14) and integral

$$\delta_W = \frac{K_1^2 \delta_\pi}{E_m}(1 - \nu^2) \tag{6.5.17}$$

Therefore, in the process of crack extension, considering the t time crack extension from the original r_0 range to R_t, the work done by the tangential stress to the n crack is

$$W_2 = n\int_{r_0}^{R_t} d\delta_W = \int_{r_0}^{R_t} \frac{nK_2^2}{E_m}(1 - \nu^2) dr \tag{6.5.18}$$

(6.5.13) and (6.5.15) into the upper, the integral is

$$W_2 = \frac{\pi \lambda^2 n(1 - \nu^2) P_m^2 r_0^2}{E_m(1 - \alpha)}\left[\left(\frac{R_t}{r_0}\right)^{2(1-\alpha)} - 1\right] \tag{6.5.19}$$

The n in the above is the number of radial fissures. It is impossible for the radial fissure to develop evenly. For a vacuum, this number can be obtained by astronomical observations. So the ratio of the energy consumed by the fissures to the total energy of the explosion is

$$\eta_2 = \frac{E_2}{E_0} = \frac{W_2}{E_0} = \frac{3\lambda^2 n(1 - \nu^2) P_m^2}{4 r_0 \rho_B c^2 E_m(1 - \alpha)}\left[\left(\frac{R_t}{r_0}\right)^{2(1-\alpha)} - 1\right] \tag{6.5.20}$$

E_2 is the new dark energy produced after the big bang.

6.5.4 Energy E that causes the consumption of elastic deformation E_3

Outside the fissure area, the stress wave can only cause the elastic deformation of the vacuum. The elastic deformation energy of the vacuum in unit volume is

$$\delta_E = \frac{1}{2}(\sigma_r \varepsilon_r + \sigma_\theta \varepsilon_\theta) = \frac{1}{2E_m}(\sigma_r^2 + \sigma_\theta^2) = \frac{\sigma_r^2}{2E_m}(1 + \lambda^2) \tag{6.5.21}$$

Here, σ_r, σ_θ are radial strain and tangential strain. The energy consumed by the vacuum elastic deformation is

$$E_3 = \int_{r_0}^{R_E} 2\pi r \delta_E dr = \frac{\pi(1 + \lambda^2)}{E_m}\int_{r_0}^{R_E} r\sigma_r^2 dr \tag{6.5.22}$$

The upper limit of integral R_E is the range of the vacuum elastic deformation. (6.5.21) into (6.5.22) and integrate:

$$E_3 = \frac{\pi(1 + \lambda^2) P_m^2 r_0^2}{2E_m(\alpha - 1)}\left[(1 - \frac{R_t}{r_0})^{2(\alpha-1)}\right] \tag{6.5.23}$$

$R_E \to \infty$, the limit value of E_3 can be obtained.

$$E_3 = \frac{\pi(1 + \lambda^2) r_0^2 P_m^2}{2E_m(\alpha - 1)} \tag{6.5.24}$$

So the ratio of the energy consumed by vacuum elastic deformation to the total energy of the explosion is

$$\eta = \frac{E_3}{E_0} = \frac{3(1 + \lambda^2) P_m^2}{8E_m(\alpha - 1) r_0 \rho_B c^2} \tag{6.5.25}$$

The stress wave caused by a vacuum explosion is transmitted from the inside to the outside. The stress wave before the explosion is that the converging wave is transmitted from the outside to the inside. Two kinds of elastic strain waves are superimposed, resulting in time and space hardening. The spatial and temporal curvature of the background is superimposed. The outward stress wave caused by the explosion propagates outward and attenuates continuously. At last, the convergent wave is left over in the space-time around the explosion cavity.

6.6 Dark energy of the universe

Dark energy comes from the difference between our observations space-time and our background space-time. When we use a curved space-time to measure another curved space-time, the relative difference of the space-time is display as the dark energy.

6.6.1 The relativity of curvature

Set the curve C to be smooth, the arc length of curve C from M_0 to M is the Δs, the deflection angle of tangent is the $\Delta \alpha$, we call $K_C = \lim_{\Delta s \to 0} \Delta \alpha / \Delta s$ the curvature of the curve C at point M. Consider the smooth curve L, the arc length of the curve L from M_0 to M' is also Δs, the deflection angle of tangent is $\Delta \beta$, the curvature $K_L = \lim_{\Delta s \to 0} \Delta \beta / \Delta s$ at point M'. Here, $\Delta \beta < \Delta \alpha$.

Now, let's consider such a problem, that is, the curvature has the relativity. This problem is discussed in the previous chapter of the gravitational field theory. Again, there is no absolute meaning in what we think of the "so-called" flat space. If we take the curve L as a reference, That is, we have decided that the curve L is a straight line, Then look at the curve C from the point of view of the L. The arc length of curve from M_0 to M is Δs, the deflection angle of tangent is $\Delta \alpha - \Delta \beta$, then the curvature at point M relative to the L curve C is

$$\Delta K = \lim_{\Delta s \to 0} \frac{\Delta \beta - \Delta \alpha}{\Delta s} = K_L - K_C \qquad (6.6.1)$$

Definition: ΔK is the relative curvature, $\Delta K > 0$. Relative curvature is positive. Attraction: $\Delta K < 0$, relative curvature is negative, rejection.

We get a simple conclusion. Use one of the curved spaces C to examine another curved space L, the superposition curvature is directly added or subtract. This conclusion is extended to the general relativity theory. Look at another curved space-time with one of the curved space-time, when the bending directions of two superimposed curved spaces are the same, the space-time with smaller curvature are used to examine the space-time with larger curvature, the larger curvature directly reduces the smaller curvature, and then the relative curvature can be obtained. A space-time with larger curvature is used to investigate the space-time with smaller curvature. The relative curvature can be obtained by the smaller curvature which is directly subtracted from the larger curvature, and it is a negative curvature. If the curvatures of the two curved space-time are the same, one space-time is used to examine another space-time, and the relative curvature is 0. If the direction of the bending of two curved spaces is opposite, the relative curvature is added to the two space-time curvature.

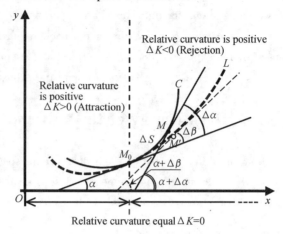

Fig. 6.6.1 The relativity of curvature

6.6.2 Dark energy observation

In 1998, two groups of astronomers independently observed supernovae far away. They measure the distance and red shift of the supernova. And they are attempted to prove that the expansion rate of the universe is slowing down due to the attraction of matter. However, the observations show that the universe is accelerating. The brightness of supernovas exceeds the sum of billions of solar brightness. A special type of supernova is a supernova of type IA. They have the same intrinsic luminosity. We measure the brightness of supernovae. It can be used to judge the rate of the expansion of the universe. Using Harbert space telescope and other ground observation instruments, we observed more supernovae (more than 300). The expansion of the universe is likely to undergo a process of first deceleration and post acceleration. The major change took place about 7 billion years ago. The accelerating process of cosmic expansion proves that there exist negative gravity in the universe, it acts as repulsion, and the energy of negative gravity is called "dark energy".

Experimental observations support that our universe does exist in a curved background space-time, and the closer to the center of the universe, the larger the curvature; the farther away from the center, the smaller the curvature. The curvature of space-time background is equal to that space-time of our Milky way at 7 billion years ago. This area constitutes the inflection point of the expansion of the universe, which first slows down and then accelerates.

6.6.3 Dark energy of negative curvature

In the background of the big bang, the whole universe has a curved space-time A with a small curvature. Our observational space-time is the superposition of the curved space-time of the Milky Way galaxy and the cosmic background. This overly curved space-time is recognized

Chapter 6   Dark Matter and Dark Energy in the Universe

as "flat" time and space. Based on this superposition state curved space-time to study the impact of quality on the background time and space, Einstein established the original gravitational field equation $G_{\mu\nu} = KT_{\mu\nu}$, In this equation, The establishment of the gravitational field equation is based on the space-time which is the overlap by the space-time of solar system in the Milky way and the space-time of cosmic background. The bending of space-time is overestimated. With the expansion of the universe, we are getting farther and farther from the center of the universe, the curvature of the cosmic background space becomes smaller and smaller. This leads to the gradual separation of the cosmic background space from the curved space-time of its own observation. The curvature of the space-time curvature of the superposition state will be smaller. We have been identified as the existence of negative curvature in the universe. It is necessary to introduce a modified amount of space-time bending. The modified Einstein field equation is

$$G_{\mu\nu} - \Lambda_{\mu\nu} = KT_{\mu\nu} \qquad (6.6.2)$$

$g_{\mu\nu}^B$ is the metric of the bending space-time of our observation, That is, the metric of Milky Way background bending space-time; $g_{\mu\nu}^B$ is the metric of cosmic background bending space-time, definition:

$$\Lambda_{\mu\nu} = \rho_\Lambda \Delta g_{\mu\nu} \qquad (6.6.3)$$

$$\Delta g_{\mu\nu} = g_{\mu\nu}^B - g_{\mu\nu}^M \qquad (6.6.4)$$

$\Lambda_{\mu\nu}$ is the relative curvature of the background space-time and the observed time and space.

In order to describe the space-time bending, there must be a "flat" space-time reference system which is identified as "straight", that is our background space. Unfortunately, this "flat" background is curved. This curved background space $B_{\mu\nu}$ is the same as the bending direction of the curved space-time $G_{\mu\nu}$. For our observer: (1) if $B_{\mu\nu} > G_{\mu\nu}$: The background space is more "curved" than the observational space, there is gravity in the background space-time relative to the observer; (2) $B_{\mu\nu} = G_{\mu\nu}$, Relative to the observer's background space-time is "flat" space-time, no force exists; (3) $B_{\mu\nu} < G_{\mu\nu}$, The space-time relative to the observer is "negative curvature", there is repulsive forces between the observer and background space-time. Or the curvature is overestimated; Bending space-time can produce a peculiar effect: When there is a relative motion between the observational space-time and the nonflat background space-time, there is a kind of repulsion force in the background space-time. This is the so-called dark energy.

The existence of a cosmic black hole causes the cosmic background to bend of space-time. Relative to space-time in the universe, our observational space-time with greater curvature; But we firmly believe this greater curvature of the space-time is "straight", so the space-time bending of the cosmic background has the characteristic of negative curvature. In order to analyze the space-time of the cosmic background, we first consider the effect of cosmic black holes space-time.

The flat properties of the universe. The sum of the interior angles of a triangle is equal to

180°, this is we determine the geometric conditions of space-time flat. As far as cosmic observations are concerned, we know that our universe is "straight". For our cosmos space, if the matter in each cubic meter is six hydrogen atoms, the curvature of the universe is positive curvature. If the space of each cubic meter contains four hydrogen atoms, the curvature of the universe is negative; And the observations show that our universe is exactly five hydrogen atoms per cubic meter, and the curvature of the universe is just as good as zero. In this way, we will create a misconception that it is a coincidence. And the reality is that the background space-time for our physics and mathematical cognition is our present cosmic background space-time. this background space-time is a curved space-time formed by the superposition of the galaxy and the background space-time of the whole universe. Such a background is the space-time for which human beings cannot escape. Such a space is recognized by us as an ideal "flat space", so it is not a coincidence, but a cognizance or a rule. This is similar to that water is colorless and tasteless, because water is the main component of the body of our life. We think the light of the sun is white, because we live in the light of the sun. If the aliens came to the earth, the water has a taste, and the sun's light is also colored. Similarly, for a vacuum, the "flat" space is 5 hydrogen atoms per cubic meter. The absolute straight space is zero for each cubic meter.

So we get the Default Principle of Ground State. For human beings, a kind of information that is associated with human beings, who cannot be eliminated for a long time, will eventually be identified by human consciousness as the information base state. Other similar information is considered unusual, and a new cognition is obtained by comparing and analyzing the information of ground state. The flatness of the universe is a kind of ground state information.

6.7 The species of dark energy in the universe

6.7.1 Einstein's cosmological equation and cosmological constant

For a homogeneous and isotropic universe, Robertson–Walker metric is the basis of general relativity discussing cosmology:

$$ds^2 = c^2 dt^2 - \alpha^2(t) \left[\frac{dr^2}{1 - kr^2} + r^2(d\theta^2 + \sin^2\theta d\varphi^2) \right] \qquad (6.7.1)$$

Here, $x^\mu x^\mu = x^i x^i + (x^4)^2 = 1/\kappa$, ($\mu = 1,2,3,4$; $i = 1,2,3$); $x^1 = r\sin\theta\cos\phi$, $x^2 = r\sin\theta\sin\phi$, $x^3 = r\cos\theta$; $d\sigma^2 = dx^\mu dx^\mu = dx^i dx^i + (dx^4)^2$ And $r^2 = [r(x^4)]^2 = x^i x^i = 1/k - (x^4)^2$. $\alpha(t)$ is scale factor of the universe. It characterizes the large scale changes in the universe. And a special coordinate are formed by (r, θ, φ), It's like a flexible grid. The position of the large mass element at the radial grids of the coordinate which is determined by the number of grids to the center. The coordinates (r, θ, φ) with mass element movement and continuous growth, the expansion ratio is determined by the scale factor $\alpha(t)$. k is a symbol factor.

Chapter 6 Dark Matter and Dark Energy in the Universe

$$k = \begin{cases} +1 & \text{Positive curvature space} \\ 0 & \text{Flat space} \\ -1 & \text{Negative curvature space} \end{cases} \quad (6.7.2)$$

Einstein's cosmological equation is

$$G_{\mu\nu} = 8\pi G T_{\mu\nu}.$$

Here, the speed of light is equal to 1. The metric $g_{\mu\nu}$ characterizations the spatial and temporal structure of the background. G is gravitational constant; $T_{\mu\nu}$ is energy momentum tensor, It represents the substance of the universe. It includes particles, radiation, field, dark matter and dark energy. Under the $R-W$ gauge, $g^B_{\mu\nu}$ can be obtained by (2).

$$g^B_{\mu\nu} = \begin{bmatrix} 1 & 0 & 0 & 0 \\ 0 & -\dfrac{\alpha^2(t)}{1-k(t)r^2} & 0 & 0 \\ 0 & 0 & -\alpha^2(t)r^2 & 0 \\ 0 & 0 & 0 & -\alpha^2(t)r^2\sin^2\theta \end{bmatrix} \quad (6.7.3)$$

The matrix $T_{\mu\nu}$ is diagonal:

$$T_{\mu\nu} = \begin{bmatrix} \rho & 0 & 0 & 0 \\ 0 & \dfrac{\alpha^2(t)}{1-k(t)r^2}p & 0 & 0 \\ 0 & 0 & \alpha^2(t)r^2 p & 0 \\ 0 & 0 & 0 & \alpha^2(t)r^2\sin^2\theta p \end{bmatrix} \quad (6.7.4)$$

Here, ρ is the mean mass density of the universe, the p distribution along the diagonal line is the radiation pressure. Einstein's general relativity set up the cosmos model and found that the universe is not expanding or shrinking, but he believes the universe is everlasting. So he modifies the field equation, and adds a cosmological constant Λ. Considering that the energy momentum tensor of vacuum should be Lorentz unchanged. That is, the observers in different sports departments see the same vacuum. So the energy momentum tensor of the vacuum should be

$$T^\Lambda_{\mu\nu} = \rho_\Lambda \Delta g_{\mu\nu} \quad (6.7.5)$$

Here, ρ_Λ is a constant proportional to the cosmological constant Λ. In the concept of dark energy, ρ_Λ is a slowly changing function. That is to say, the property of vacuum energy momentum tensor is related to observer speed. The equation of the field after adding the cosmological constant is

$$G_{\mu\nu} - \rho_\Lambda \Delta g_{\mu\nu} = 8\pi G T_{\mu\nu} \quad (6.7.6)$$

The Milky Way forms a local non-expanding space-time.

$$g^M_{\mu\nu} = \begin{bmatrix} 1 & 0 & 0 & 0 \\ 0 & -\dfrac{\alpha^2}{1-kr^2} & 0 & 0 \\ 0 & 0 & -\alpha^2 r^2 & 0 \\ 0 & 0 & 0 & -\alpha^2 r^2 \sin^2\theta \end{bmatrix} \quad (6.7.7)$$

6.7.2 Calculation of scale factor

1. The space-time of the cosmic background

Generalized cosmic dust. Assuming that the vacuum cracked dark matter is evenly distributed in the large-scale structure of the universe, we can also regard the cosmic dark matter and visible matter as a more generalized cosmic matter, and further regard this matter as the generalized cosmic dust.

The dark matter formed by vacuum cracks is similar to ordinary matter. During the formation of black holes in the universe, the vacuum around them is compressed, and the vacuum around them breaks down. In the Big Bang, new cracks are added. All the generalized cosmic materials synchronously gain an outward momentum of expansion, and the cracks expand outward. In other words, dark matter and visible matter expand outward synchronously. In particular, the density of matter we are discussing already contains dark matter.

In terms of large-scale structure, the universe can be regarded as composed of dust particles. In the process of cosmic expansion, the distance between dust particles is getting farther and farther.

Considering a dust particle i, the external gravitational potential of dust particles is

$$U_i = -\frac{G\rho_i}{r_i} \quad (6.7.8)$$

Because the action range of gravity can be regarded as infinite, so the forces overlap with each other. For the whole universe, the gravitational forces of cosmic dust are overlapping, and the cosmic dust is isotropic and uniformly distributed in space. Therefore, we can take the cosmic center as the coordinate origin, and find out that the gravitational potential in a certain space-time is

$$U = -\int_{R_M}^{R} \frac{G\rho}{r - r_0} dr = -G\rho(t)[\ln(R - r_0) - \ln(R_M(t) - r_0)] \quad (6.7.9)$$

R is the gravitational radius of the proton: $R_{Gp} = 2.2688 \times 10^{24}$ m. This radius determines the size of the universe. r_0 is radius of central dust particles. $R = Nr_0$, R is the range of gravitational action is taken as the radius of the universe. $D = nr_0$. $R_M(t)$ is the distance from the earth to the center of the universe, unit: Mpc. Because the universe is expanding, the distance between the Milky Way and the center of the universe is a variable, which satisfies Hubble's law. Then

$$R_G/R_c = t_G/t_c \quad (6.7.10)$$

t_G is the formation time of the Milky Way; t_c is universe formation time, $R_G = R_c t_G/t_c$. V_f is velocity(Far Away), km/s; H_c is Hubble's Constant, km/(s · Mpc).

If the current time is t_0, the time discussed is t, and the interval between t_0 and t is not too large, $\alpha(t)$ is expanded by Taylor series as follows.

Chapter 6 Dark Matter and Dark Energy in the Universe

$$\alpha(t) = \alpha(t_0) + \frac{d\alpha(t)}{dt}(t-t_0) + \frac{1}{2!}\frac{d^2\alpha(t_0)}{dt^2}(t-t_0)^2 + \cdots$$

$$= \alpha(t_0)\left[1 + \frac{\dot{\alpha}(t_0)}{\alpha(t_0)}(t-t_0) + \frac{1}{2}\frac{\ddot{\alpha}(t_0)}{\alpha(t_0)}(t-t_0)^2 + \cdots\right] \quad (6.7.11)$$

The second coefficient of the formula is the Hubble constant:

$$H_c \equiv \dot{\alpha}(t_0)/\alpha(t_0) \quad (6.7.12)$$

The motion of matter in the universe must be under the action of force. At present, only gravity is considered. That is to say, the retrograde motion of celestial bodies is caused by the gravitational action of matter in the universe. Every ρ can be regarded as a quality source. Every dust in the universe is a source of mass. Ignoring the effect of pressure, gravity can be expressed by Poisson's equation.

$$\nabla \cdot F_i = -4\pi G \rho_i \quad (6.7.13)$$

External forces acting on p_G-point in the middle of two equal-mass $\rho_1\rho_2$ particles is zero. Can be written as $\nabla \cdot F_{pG} = -4\pi G(\rho_1 - \rho_2) = 0$, if $\rho_1 \neq \rho_2$, it can be reconsidered that $\rho_1-\rho_2$ constitutes a new quality source (Fig. 6.7.1), there will be a force $F_{pG} \neq 0$ outside p_G. For our galaxy, our location is 4.6 billion light-years, and the universe formed 13.7 billion light-years, so there is material all around the galaxy.

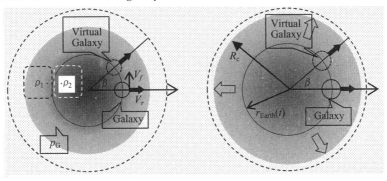

Fig. 6.7.1 Sketch map of cosmic expansion and gravitational field distribution

During the expansion of the universe, the Hubble relation can be written in a more general form as $V=H(t)r$, that is $dr/dt=H(t)r$. Definition scale factor α is

$$\alpha(t) \equiv \frac{r(r)}{r(t_0)} \quad (6.7.14)$$

$r(t)$ is comoving coordinates of celestial bodies; $r(t_0)$ is comoving coordinates of at present time ($t=t_0$, Fig. 6.7.2). $dr/dt=H(t)r$ can be written as

$$\frac{1}{\alpha}\frac{d\alpha}{dr} = H(t) \quad (6.7.15)$$

Under the assumption of homogeneity and isotropy, if the total mass in the universe does not change and the effect of pressure can be neglected, then the density of matter ρ in the universe will inevitably decrease with the decrease of $\alpha(t)^{-3}$, and in general, $\alpha(t_0)=1$.

$$\rho(t)\alpha(t)^3 = \rho(t_0)\alpha(t_0)^3 \equiv \rho_0 \quad (6.7.16)$$

$\rho(t_0) = \rho_0$ is the density at $t = t_0$. Consider $\nabla \cdot F = -4\pi G\rho$, $\nabla \cdot r = 3$, in addition

$$F = \frac{dV}{dt} = \frac{d}{dt}[H(t)r] = r\left[\frac{dH(t)}{dt} + H^2(t)\right] \quad (6.7.17)$$

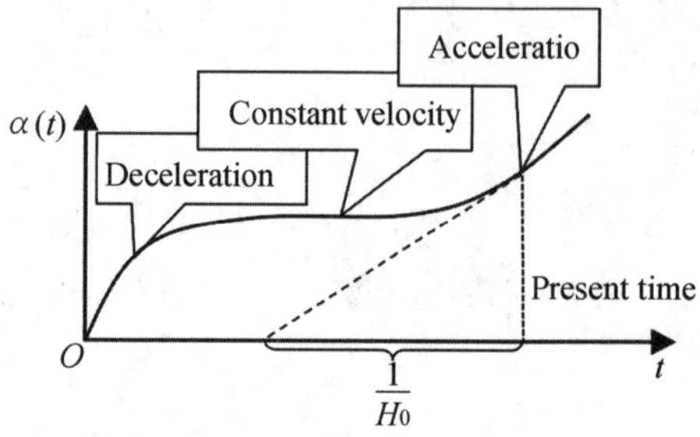

We can get

$$\frac{1}{R}\frac{d^2\alpha(t)}{dt^2} + \frac{4\pi G}{3}\rho = 0 \quad (6.7.18)$$

Fig. 6.7.2 The actual expansion process in which dark energy exists in the universe

The above formula shows that as long as $\rho \neq 0$, it is impossible to be in a static state, and acceleration is inevitable. Multiply the above formula by $2\alpha d\alpha/dt$ and replace it with $\rho(t)\alpha(t)^3 = \rho(t_0)\alpha(t_0)^3 \equiv \rho_0$, can get $\frac{d\alpha}{dt}\left[\left(\frac{d\alpha}{dt}\right)^2 - \frac{8\pi G}{3}\rho_0\frac{1}{\alpha}\right] = 0$, then $\left(\frac{d\alpha}{dt}\right)^2 - \frac{8\pi G}{3}\rho_0\frac{1}{\alpha} = \text{const}$. Introduces curvature K, set const $= -Kc^2$ and ρ_0 are replaced by $\rho\alpha^3$.

$$\left(\frac{d\alpha}{dt}\right)^2 - \frac{8\pi G}{3}\rho\alpha^2 = -Kc^2 \quad (6.7.19)$$

In relativity, K is the curvature of space. The dust of the universe is regarded as an ideal fluid, and the acceleration of the expansion of the universe is

$$\frac{d^2\alpha(t)}{dt^2} = -\frac{4\pi G\alpha}{3}\rho \quad (6.7.20)$$

In fact, the perfect fluid is not in line with the actual situation. On the one hand, the superposition effect of cosmic gravity is not taken into account. On the other hand, the agglomeration effect of the galaxy during the later expansion period of the universe is not taken into account. Therefore, we can directly obtain the acceleration term of the galaxy through the superposition effect of gravity and the agglomeration effect of the galaxy.

$$\frac{d^2\alpha}{dt^2} = -\frac{4\pi G\alpha}{3}(\rho_c - \rho_G) \quad (6.7.21)$$

We need to consider three influence factor:

(1) It has nothing to do with radius, which obviously does not take into account the gravitational superposition of dust particles. The superposition effect of gravitational potential

$$U(r) = \ln(R - r_0) - \ln[R - r_{\text{Earth}}(t)] \quad (6.7.22)$$

Considering that the Milky Way is regarded as a dust $\rho(t)$ expanding from r to r_0 in the course of expansion, it has to overcome the superimposed gravitational potential energy.

$$a\rho(t)(r - r_0) = \rho(t)[\ln(R - r_0) - \ln(r_{\text{Earth}}(t) - r_0)] \quad (6.7.23)$$

It can be seen that in the process of expansion of the universe, to overcome the gravitational potential energy superimposed by dust will lead to a negative acceleration—a. With the constant expansion of the universe, the deceleration a gradually decreases, and finally tends to zero.

Chapter 6 Dark Matter and Dark Energy in the Universe

(2) As the universe expands, the density becomes smaller and smaller, so the change of density should be considered.

$$M_{\text{Universe}} / \frac{4}{3}\pi R_U(t)^3 = \rho_{\text{Universe}} \tag{6.7.24}$$

After considering the above factors, in the same radius location as the Milky Way, choose a region of the same size as the Milky Way, but the density of matter is the average density of the universe for comparison, then you can get

$$\rho(t) = \rho \cdot + \rho[\ln(R - r_0) - \ln(R - r_{\text{Earth}}(t))]\} \tag{6.7.25}$$

(3) Kinetic energy of outward expansion in the Big Bang.

After the Big Bang, all cosmic matter obtained a momentum along the r direction synchronously, showing that matter moved outward along the r-direction with an initial velocity of V_r.

$$\Delta E = \frac{1}{2}M_U V_r^2 \Rightarrow V_r = \sqrt{\frac{2\Delta E}{M_U}} \tag{6.7.26}$$

Considering the expansion of the sphere surface and the radius of the sphere surface is r, then $V_r \Delta t = \Delta r$, $V_f \cdot \Delta t = \Delta r \beta$, it's known that $V_f = V_r \beta$. Consider Hubble's law $V_f = H_c D$, then $V_r = H_c D \beta$, and Hubble constant $H_c = V_r / D\beta$.

$$\alpha(t) = \left[H_c - \frac{G\rho(t)[\ln(R - r_0) - \ln(r_{\text{Earth}}(t) - r_0)]}{r_{\text{Earth}}(t) - r_0} + \frac{4\pi GR}{3}(\rho_G - \rho_U) \right](t_{\text{Earth}} - t_0) \tag{6.7.27}$$

Consider only the Milky Way.

Since the galaxy will not expand, the expansion pressure is zero, and the mass density of the galaxy will not change, the following expressions can be obtained directly.

$$k_{\text{Galaxy}} = 1/(4\sqrt{2}\,G\dot{M}_{\text{Galaxy}})^2 \ ; \ M_{\text{Galaxy}} = \rho_{\text{Galaxy}} \cdot \frac{3}{4}\pi R_{\text{Galaxy}}^3 \tag{6.7.28}$$

Ultimately, we can get the space-time interval of the universe.

$$ds^2 = c^2 d(t_{\text{Earth}} - t_0)^2 - \left[H_c - \frac{G\rho(t)[\ln(R - r_0) - \ln(r_{\text{Earth}}(t) - r_0)]}{r_{\text{Earth}}(t) - r_0} + \frac{4\pi G\alpha}{3}(\rho_G - \rho_U) \right]^2 (t_{\text{Earth}} - t_0)^2 \times$$

$$\left[\frac{dr^2}{1 - kr^2} + r^2(d\theta^2 + \sin^2\theta d\varphi^2) \right] \tag{6.7.29}$$

r_{Earth} is radius of the earth to the center of the universe; ρ_U is average density of cosmic matter; ρ_G is average density of matter in the galaxy; t_0 is start measurement time; t_{Earth} is the time now measured on earth.

From the above formula, we can see that the Big Bang material expands outward after it gains kinetic energy. At first, to overcome the superposition potential of gravity, deceleration will occur. With the expansion of the universe, the deceleration will become smaller and smaller. Finally, it will enter into a constant speed of expansion. After constant expansion, cosmic dust gradually forms galaxy clusters. The universe expands continuously during the formation of galaxy clusters. In the process of expansion, the density of galaxy clusters that have been formed does not change, but the average density of the background space will change, which results in the space-time curvature of the background space less than that of the galaxy clusters, which leads to the accelerated expansion of the universe. The conclusion is that our universe has dark matter but no dark energy.

2. The contribution of dark matter of black hole background space

In modern observational cosmology, it is believed that there exists a huge black hole in the center of the universe and a black hole in the center of the Milky way. In this case, we must consider the effect of dark matter on the expansion of the universe.

This is a vacuum solution to the outside of a static, spherically symmetric star. For the big bang, the cracks produced are distributed in the spherical space. It is similar to a ball formed by a flocculent fiber, so it can be viewed as a ball simply. The gravitational effect produced by a sphere has spherical symmetry. The gravitational effect of dark matter and visible matter is the same, therefore, space-time bending effect of dark matter in the universe which can be described by the Schwarzschild solution.

The metric component of the cosmic background space-time caused by dark matter is not zero.

$$(g_{\mu\nu}^{B2}) = \begin{pmatrix} -\left(1 - \frac{2GM_{Dark}^B}{c^2 r_B}\right) & 0 & 0 & 0 \\ 0 & \left(1 - \frac{2GM_{Dark}^B}{c^2 r_B}\right)^{-1} & 0 & 0 \\ 0 & 0 & r_B^2 & 0 \\ 0 & 0 & 0 & r_B^2 \sin^2\theta \end{pmatrix} \quad (6.7.30)$$

In the equation, M_{Dark} is the mass of dark matter sphere; G is gravitational constant; C is the speed of light. Take $x_0 = ct$, $x_1 = r$, $x_2 = \theta$, $x_3 = \varphi$.

Dark matter is a vacuum crack. Due to the existence of spin waves in space and time around a cosmic black hole, spin waves propagate in a vacuum crack, this will lead to the very slow rotation of dark matter in the universe. Therefore, the Schwarzschild solution can well describe the space-time curvature of dark matter.

The distribution of the crack is the distribution of the mass of the dark matter. The number of cracks is proportional to the strain caused by the vacuum bending. The total energy of the crack generated by the Big Bang has been discussed earlier.

$$W_2 = E_{Dark}^B(r) = \frac{\pi \lambda^2 (1 - \nu^2) P_m^2 r_0^2}{E_m (1 - \alpha)} \left[\left(\frac{r}{r_0}\right)^{2(1-\alpha)} - 1 \right] c^2 \quad (6.7.31)$$

$$E_{Dark}^B(r) = M_{Dark}^B(r) c^2$$

Through the above discussion, we can directly obtain the influence of cosmic dark matter on the space-time of the cosmic background.

It is worth noting that the earth we depend on is in the solar system. The solar system is in

Chapter 6   Dark Matter and Dark Energy in the Universe

the Milky way galaxy, and the center of the Milky Way Galaxy exists a huge black hole which mass is 4 million times more than the sun. Dark matter also exists around a black hole. It's also a background space. But as the whole galaxy and other galaxies are in the expansion of the universe, the Milky Way itself does not expand. As far as the observer is concerned, there is only one movement that is getting farther away from the center of the universe, so, when calculating the background space, only to think about existence the relatively separated cosmic background space.

Similarly, the dark matter in the center of the galaxy causes the space-time curvature of the Milky Way background, and its metric expression.

$$(g_{\mu\nu}^{M2}) = \begin{pmatrix} -\left(1 - \dfrac{2GM_{Dark}^M}{c^2 r_B}\right) & 0 & 0 & 0 \\ 0 & \left(1 - \dfrac{2GM_{Dark}^M}{c^2 r_B}\right)^{-1} & 0 & 0 \\ 0 & 0 & r_M^2 & 0 \\ 0 & 0 & 0 & r_M^2 \sin^2\theta \end{pmatrix} \quad (6.7.32)$$

For the dark matter in the Milky Way, dark matter originates from the Big Bang. Assuming that cracks have self-similarity, the distribution of dark matter is similar to that of the Big Bang. On the other hand, the number of cracks is proportional to the mass of the central region of the galaxy.

$$M_{Dark}^M(r) = \rho(r) M_M(r) \left(\dfrac{r}{r_0}\right)^{2(\alpha-1)} \quad (6.7.33)$$

M_M is the mass of the galaxy's central region; $\rho(r)$ is distribution coefficient of mass density in central region.

3. The calculation of the "cosmological constant Λ"

"Cosmological constant Λ" is actually a function: $\Lambda_{\mu\nu} = -\rho_\Lambda g_{\mu\nu}$. If you want to determine $\Lambda_{\mu\nu}$, the key is to solve the metric $g_{\mu\nu}^B$ and $g_{\mu\nu}^M$. The subscript B represents the residual cosmic black hole after the big bang; the subscript M represents the Milky way.

$$\Lambda_{\mu\nu} = g_{\mu\nu}^B - g_{\mu\nu}^M \quad (6.7.34)$$

Here, Einstein's cosmic constant is no longer a constant, but a slowly decreasing quantity. As the universe expands, $g_{\mu\nu}^B$ becomes smaller and smaller, eventually leading to accelerated expansion of the universe.

If there are black holes in both the center of the universe and the center of the galaxy, according to the breakdown of the vacuum, we can still get the conclusion that the universe accelerates expansion.

4. The large range of vacuum stress waves caused by big bang in cosmic background space-time

In the previous discussion, we know that the elastic stress waves produced by a vacuum explosion are propagated from the inside to the outside. The specific expression of the total energy of the elastic wave is as follows.

$$E_3 = \frac{\pi(1+\lambda^2)P_m^2 r_0^2}{2E_m(\alpha-1)}\left[(1-\frac{R_t}{r_0})^{2(\alpha-1)}\right] \qquad (6.7.35)$$

It can be roughly thought that this is an outward spherical wave. All strain wave energies are absorbed by cosmic matter, forming the kinetic energy of outward expansion.

5. The effect of the visible matter and gravitational wave on the background space

With the expansion of the universe, it can be seen that the distance between the materials is getting farther and farther. It further led to the decline of the gravity of the background space. But this effect is mainly due to the influence of adjacent galaxies. But because the visible matter is only less than 5% of the total universe, the effect is very small. We do not make any further discussion in this article. For gravitational waves, it can be seen that the gravitational wave energy transmitted between materials is even smaller. The influence on the space of the cosmic background can be ignored.

Chapter 7 Interaction and Super-unified Equation

The degrees of freedom of particles consist of different types of strains. These different types of degrees of freedom form a generalized multidimensional space. The generalized quantum field function is

$$\varphi(X) = \exp[-i([\varepsilon]_m g_{mn} X^n)] \qquad (7.0.1)$$

The intrinsic strain information of item i in quantum field $\varphi(X) = \exp[-i([\varepsilon]_n X_n)]$ is extracted by operator $\hat{\partial}_i$. If the quantum field itself is affected by the external field in space X_i, then the intrinsic strain $[\varepsilon_{\mu\nu}]_i$ in space X_i is changes and becomes $[\varepsilon_{\mu\nu}]'_i$, it is called the class i interaction between $\varphi(X_i)$ and $\varphi(X_i)$.

Superposition of multiple degrees of freedom in quantum space: a quantum field can be regard as a point with neighborhood, and the point with multiple strains. Each strain is a field. Each field is a degree of freedom which independent each other $X(X_1 \cdots X_n)$, a single field $[\varepsilon_{\mu\nu}]_i X_i$ (such as a weak field) constitute their own space it can be "bend" and to satisfy the gauge invariance.

An independent field strain matrix is $\varepsilon_{ji} = \partial_j u(\xi^i)$ in a single quantum field $\varphi(X)$. Different combinations of matrix $[\varepsilon_{\mu\nu}]$ constitute an independent degree of freedom in quantum field. As shown in Fig. 7.0.1, in one quantum field $\varphi(X)$ can overlap N kinds of field strain and each strain $[\varepsilon_{\mu\nu}]_i$ constitutes a space X^i, each space X^i can bend into X'^i. The N-dimensional $(X^1, X^2, \cdots, X^N)$ space of quantum field that is the superposition of variety field to produces the final result they are still in 4-dimensional space-time (the superposition of multiple fields in the same area allows the existence of parallel spaces).

$$\varphi(X) = \exp[-i([\varepsilon]_1 X^1 + \cdots + [\varepsilon]'_\alpha X'^\alpha + \cdots + [\varepsilon]_N X^N)] \qquad (7.0.2)$$

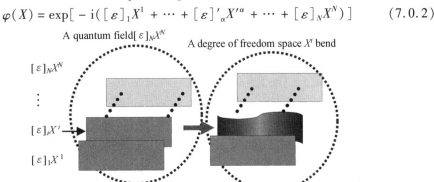

Fig. 7.0.1 Quantum intrinsic space

Considering the existence of an external strain field $\varphi(X^\alpha)$ in space X^α, there is an interaction in space X^α, then there is a coupling constant g_α between $\varphi(X^\alpha)$ and $\varphi(X^\alpha)$. Quantum field interaction will form a new quantum field. In order to realize this process, there must be two fields, one is Fermion field $\varphi(X)$ and the other is Boson field $\varphi(X^\alpha) = \exp(-i[\varepsilon^\phi]_\alpha X^\alpha)$ used to transfer interaction. The probability of interaction between these two fields is g_α. In this way, we obtain the interacting quantum field system. The quantum field system can be regard as a new quantum field $\varphi(X')$.

$$\varphi'(X) = \varphi(X)\exp\{-ig_\alpha[\varepsilon^\phi]_\alpha\}X^\alpha \tag{7.0.3}$$

It can be further expressed as

$$\varphi'(X) = \exp\{-i([\varepsilon]_1 X^1 + \cdots + ([\varepsilon]_\alpha + g_\alpha[\varepsilon^\phi]_\alpha)X^\alpha + \cdots + [\varepsilon]_N X^N)\} \tag{7.0.4}$$

$$D_\alpha \varphi'(X) = \gamma_\alpha([\varepsilon]_\alpha + g_\alpha[\varepsilon^\phi]_\alpha)\varphi'(X) \tag{7.0.5}$$

Interaction only exists in the same space. The above formula describes the interaction of a quantum field in a certain space X^α, most typically the electromagnetic interaction between photons and electrons.

If there are multiple interactions at the same time in space X^α, $\varphi(X)$ interacts with multiple Boson fields $\varphi(X^\alpha)$, $\psi(X^\alpha)$, $\cdots$, $\Phi(X^\alpha)$ at the same time, in space X^α. The coupling strength is g_α^ϕ, g_α^ψ, $\cdots$, g_α^Φ respectively in space X^α. Then

$$\gamma_\alpha D_\alpha \varphi'(X) = \gamma_\alpha([\varepsilon]_\alpha + g_\alpha^\phi[\varepsilon^\phi]_\alpha + g_\alpha^\psi[\varepsilon^\psi]_\alpha + \cdots + g_\alpha^\Phi[\varepsilon^\Phi]_\alpha)\varphi'(X) \tag{7.0.6}$$

Weak-electric interaction. The internal strain space is a weak-action space. The internal strain of Fermion field consists of intrinsic strain and electric flux line strain. The intrinsic strain constitutes the bare mass charge of Fermion and the electric flux line strain constitutes the charge of Fermion. They are two sides of a coin in the same space. Set the internal strain is ε, the intrinsic strain is ε_θ, and the electric flux line strain is ε_r. At one point in the intrinsic space p, the strain forms the spatial dimension to satisfying

$$\varepsilon^2 = \varepsilon_\theta^2 + \varepsilon_r^2 = \varepsilon^2 \sin^2\vartheta + \varepsilon^2 \cos^2\vartheta \tag{7.0.7}$$

ϑ is named Weinberg angle.

The space X_θ formed by the intrinsic strain to the strain ε_θ. In this space, the interacting boson field is $\varphi_\theta = \exp(-i\Delta\varepsilon_\theta X_\theta)$, and the coupling constant is g_t; the space X_r formed by the electric flux line strain ε_r. The interacting boson field in this space is $\varphi_r = \exp(-i\Delta\varepsilon_\theta X_\theta)$, and the coupling constant is g_y.

$$\Delta\varepsilon^2 = (\Delta\varepsilon_\theta)^2 + (\Delta\varepsilon_r)^2 = (g_\theta \Delta\varepsilon)^2 + (g_r \Delta\varepsilon)^2 = \Delta\varepsilon^2(g_\theta^2 + g_r^2) \tag{7.0.8}$$

$\Delta\varepsilon^2 = \Delta\varepsilon^2 \sin^2\vartheta + \Delta\varepsilon^2 \cos^2\vartheta$. Set: $G^2 = g_\theta^2 + g_r^2 = G^2 \sin^2\vartheta + G^2 \cos^2\vartheta$, then

$$g_\theta = G\sin\vartheta, \quad g_y = G\cos\vartheta \tag{7.0.9}$$

There are simultaneous interactions between space X_θ and space X_r, then

$$(\gamma_\theta D_\theta + \gamma_r D_r)\varphi'(X) = \gamma_\theta([\varepsilon_\theta] + g_\theta[\Delta\varepsilon_\theta])\varphi'(X) + \gamma_r([\varepsilon_r] + g_r[\Delta\varepsilon_r])\varphi'(X) \tag{7.0.10}$$

Since X_θ and X_r belong to weak action spaces, $X_\mu = X_\mu(X_\theta, X_r)$, they can be further

Chapter 7 Interaction and Super-unified Equation

written as
$$D_\mu = \partial_\mu + \Gamma_\mu(A) + \Gamma_\mu(B) \qquad (7.0.11)$$

We can get all the results of the gauge field theory directly by vacuum strain. In the theory of weak effects, the bare mass charge has 3-dimensional structure, the photon has only 1-dimensional structure. The covariant operator is expressed as

$$D_\mu = \partial_\mu + ig_t \frac{\sigma}{2} A_\mu + ig_y \frac{1}{2} B_\mu \qquad (7.0.12)$$

7.1 Electromagnetic interaction

7.1.1 The electromagnetic space of photon and electron in observation space

$\varphi_{\text{lepton}}(X) = \exp\{-i([\varepsilon]_\beta g_{\beta_\alpha} X^\alpha)\}$. Consider only lepton and photon in the observation space.

$$\varphi_e(X) = \exp\left\{-i\left(\begin{bmatrix} \varepsilon^e_{0,0} & 0 & 0 & 0 \\ 0 & \varepsilon^e_{1,1} & 0 & 0 \\ 0 & 0 & \varepsilon^e_{2,2} & 0 \\ 0 & 0 & 0 & \varepsilon^e_{3,3} \end{bmatrix} \widetilde{\gamma}_\mu \begin{pmatrix} x_0 & 0 & 0 & 0 \\ 0 & \xi_1 & 0 & 0 \\ 0 & 0 & \xi_2 & 0 \\ 0 & 0 & 0 & \xi_3 \end{pmatrix} + \right.\right.$$
Bare 4-momentum: $M_{\text{Bare}}c^2$, $M_{\text{Bare}}c$

$$\left.\left.\begin{bmatrix} A^e_{0,0} & 0 & 0 & 0 \\ 0 & A^e_{1,1} & 0 & 0 \\ 0 & 0 & A^e_{2,2} & 0 \\ 0 & 0 & 0 & A^e_{3,3} \end{bmatrix} \widetilde{\gamma}_\mu \begin{pmatrix} x_0 & 0 & 0 & 0 \\ 0 & x_1 & 0 & 0 \\ 0 & 0 & x_2 & 0 \\ 0 & 0 & 0 & x_3 \end{pmatrix}\right)\right\} \qquad (7.1.1)$$
Electromagnetic 4-momentum: $M_B c^2$, $M_E c$

$$\phi_\gamma(X) = \exp\left(-i\begin{bmatrix} A^\gamma_{0,0} & 0 & 0 & 0 \\ 0 & 0 & 0 & 0 \\ 0 & 0 & 0 & 0 \\ 0 & 0 & 0 & \varepsilon^\gamma_{3,3} \end{bmatrix} \widetilde{\gamma}_\mu \begin{pmatrix} x_0 & 0 & 0 & 0 \\ 0 & 0 & 0 & 0 \\ 0 & 0 & 0 & 0 \\ 0 & 0 & 0 & \xi_3 \end{pmatrix} + \right.$$
virtual photon   intrinsic space

$$\left.\begin{bmatrix} 0 & 0 & 0 & 0 \\ 0 & A^\gamma_{1,1} & 0 & 0 \\ 0 & 0 & A^\gamma_{2,2} & 0 \\ 0 & 0 & 0 & 0 \end{bmatrix} \widetilde{\gamma}_\mu \begin{pmatrix} x_0 & 0 & 0 & 0 \\ 0 & x_1 & 0 & 0 \\ 0 & 0 & x_2 & 0 \\ 0 & 0 & 0 & 0 \end{pmatrix}\right) \qquad (7.1.2)$$
electromagnetic wave   electromagnetic field space

7.1.2 The internal dimension of interaction

We know that the spaces of all the central regions of the quantum field are broken. In the process of interaction, the space of the region of interaction will also break up. Therefore, the space in the interaction region is no longer continuous and smooth. The space of rupture is no longer the original integer dimension, smaller than the original dimension of time and space. The amount of reduction in this dimension is called the anomalous dimension γ.

The stronger the interaction, the more serious the spatiotemporal rupture, so the anomalous dimension γ is proportional to the interaction strength.

Gravitational interaction does not destroy the space-time structure, so the anomalous dimension only exists in the quantum field interaction. According to the type of quantum field interaction, the spatial anomaly dimension of the interaction region can be divided into three categories: anomaly dimension of strong interaction anomaly γ_S, anomaly dimension of weak interaction γ_W and electromagnetic interaction anomaly dimension γ_E.

7.1.3 Virtual gauge particles in the inner space of quantum field interaction

The gauge field particles are the transfer interaction. In fact, only the virtual gauge field particles really transfer the interaction.

$$\phi_{\text{Virtual}}(X) = \exp\left\{-i \underbrace{g_i[A]}_{\text{virtual gauge field particles}} \breve{\gamma} \underbrace{(\xi^{1-\gamma_i})}_{\text{intrinsic space}}\right\}$$

Interacting virtual photons

$$\phi_{\text{Virtual}\gamma}(X) = \exp\left(-i \underbrace{\begin{bmatrix} A^{\gamma}_{0,0} & 0 & 0 & 0 \\ 0 & 0 & 0 & 0 \\ 0 & 0 & 0 & 0 \\ 0 & 0 & 0 & eA^{\gamma}_{3,3} \end{bmatrix}}_{\text{virtual photon}} \breve{\gamma}_\mu \underbrace{\begin{pmatrix} x_0 & 0 & 0 & 0 \\ 0 & 0 & 0 & 0 \\ 0 & 0 & 0 & 0 \\ 0 & 0 & 0 & \xi_3^{1-\gamma_E} \end{pmatrix}}_{\text{intrinsic space}}\right) +$$

$$\left(-i \underbrace{\begin{bmatrix} A^{\gamma}_{0,0} & 0 & 0 & 0 \\ 0 & A^{\gamma}_{1,1} & 0 & 0 \\ 0 & 0 & A^{\gamma}_{2,2} & 0 \\ 0 & 0 & 0 & 0 \end{bmatrix}}_{\text{electromagnetic wave}} \breve{\gamma}_\mu \underbrace{\begin{pmatrix} x_0 & 0 & 0 & 0 \\ 0 & x_1 & 0 & 0 \\ 0 & 0 & x_2 & 0 \\ 0 & 0 & 0 & 0 \end{pmatrix}}_{\text{electromagnetic field space}}\right) \quad (7.1.3)$$

The $\breve{\gamma}$ matrix here cannot be used for computation, but is used to annotate the characteristics of the coordinate space. Interacting virtual W^-:

$$\Phi_{\text{Virtual }W^-}(X) = \exp(-i)\left\{g_W \begin{bmatrix} W_{0,0} & W_{1,0} & W_{2,0} & W_{3,0} \\ W_{0,1} & W_{1,1} & (\Delta\gamma^W_{21} - \Delta\gamma'^W_{21}) + (\omega^W_{12} - \omega'^W_{12}) & 0 \\ W_{0,2} & (\Delta\gamma^W_{21} - \Delta\gamma'^W_{21}) + (\omega^W_{21} - \omega'^W_{21}) & W_{2,2} & 0 \\ W_{0,3} & 0 & 0 & W_{3,3} \end{bmatrix}\right.$$

$$\begin{pmatrix} \xi_0 & \xi_0 & \xi_0 & \xi_0 \\ \xi_1^{1-\gamma_W} & \xi_1^{1-\gamma_W} & \xi_1^{1-\gamma_W} & \xi_1^{1-\gamma_W} \\ \xi_2^{1-\gamma_W} & \xi_2^{1-\gamma_W} & \xi_2^{1-\gamma_W} & \xi_2^{1-\gamma_W} \\ \xi_3 & \xi_3 & \xi_3 & \xi_3 \end{pmatrix} +$$

Chapter 7 Interaction and Super-unified Equation

$$g_\gamma \begin{bmatrix} B_{0,0} & 0 & 0 & 0 \\ 0 & 0 & 0 & 0 \\ 0 & 0 & B_{2,2} & 0 \\ 0 & 0 & 0 & 0 \end{bmatrix} \begin{pmatrix} \xi_0 & 0 & 0 & 0 \\ 0 & 0 & 0 & 0 \\ 0 & 0 & \xi_2^{1-\gamma_E} & 0 \\ 0 & 0 & 0 & 0 \end{pmatrix}_\gamma + [A_{\mu,\mu}^{e^-}](X) \Bigg\} \quad (7.1.4)$$

The coupling constant of the expression indicates the action intensity of the virtual particle, and the spatial anomaly dimension represents the damage degree of the spatial dimension of the interaction region.

7.1.4 Electromagnetic interaction between photon and electron

When interacting, there will be a photon $\varphi_\gamma(X)$ interacting with each other to form Field functions after coupling success $\varphi'_e(X) = \varphi_e(X)\varphi_\gamma(X)$. Transfer momentum by virtual photon. Consider the coupling probability e, then

$$\varphi'_e(X) = \exp\Bigg\{-i\Bigg(\underbrace{\begin{bmatrix} \varepsilon_{00}^e & 0 & 0 & 0 \\ 0 & \varepsilon_{11}^e & 0 & 0 \\ 0 & 0 & \varepsilon_{22}^e & 0 \\ 0 & 0 & 0 & \varepsilon_{33}^e \end{bmatrix} \gamma_\mu \begin{pmatrix} x_0 & 0 & 0 & 0 \\ 0 & \xi_1 & 0 & 0 \\ 0 & 0 & \xi_2 & 0 \\ 0 & 0 & 0 & \xi_3 \end{pmatrix} + \begin{bmatrix} \varepsilon_{0,0}^\gamma & 0 & 0 & 0 \\ 0 & 0 & 0 & 0 \\ 0 & 0 & 0 & 0 \\ 0 & 0 & 0 & eA_{33}^\gamma \end{bmatrix} \breve{\gamma}_\mu \begin{pmatrix} x_0 & 0 & 0 & 0 \\ 0 & 0 & 0 & 0 \\ 0 & 0 & 0 & 0 \\ 0 & 0 & 0 & \xi_3^{1-\gamma_E} \end{pmatrix}}_{\text{The electrons interact with the virtual photon in the intrinsic space, and the electron obtains momentum } \varepsilon_{33}^\gamma. \text{ The probability is } e. \text{ The momentum absorbed by electrons does not destroy the intrinsic structure of electrons. This momentum is represented by photon jumping from velocity } v \text{ to } V \text{ along } x_3 \text{ axis. Set } e\varepsilon_{33}^\gamma = eA_{33}^\gamma} +$$

$$\underbrace{\begin{bmatrix} A_{0,0}^e + eA_{0,0}^\gamma & 0 & 0 & 0 \\ 0 & A_{1,1}^e + eA_{1,1}^\gamma & 0 & 0 \\ 0 & 0 & A_{2,2}^e + eA_{2,2}^\gamma & 0 \\ 0 & 0 & 0 & A_{3,3}^e \end{bmatrix} \breve{\gamma}_\mu \begin{pmatrix} x_0 & 0 & 0 & 0 \\ 0 & x_1 & 0 & 0 \\ 0 & 0 & x_2 & 0 \\ 0 & 0 & 0 & x_3 \end{pmatrix}}_{\text{Electron absorption photon fluctuation energy. The probability is } e}\Bigg)\Bigg\}$$

(7.1.5)

After experiencing the above process, the field function can be written directly as follows.

$$\varphi'_e(X) = \exp\Bigg(-i\begin{bmatrix} A_{0,0}^e + eA_{0,0}^\gamma & 0 & 0 & 0 \\ 0 & A_{1,1}^e + eA_{1,1}^\gamma & 0 & 0 \\ 0 & 0 & A_{2,2}^e(x) + eA_{2,2}^\gamma & 0 \\ 0 & 0 & 0 & A_{3,3}^e(x) + eA_{3,3}^\gamma \end{bmatrix} \breve{\gamma}_\mu \cdot \underbrace{\begin{pmatrix} x_0 & 0 & 0 & 0 \\ 0 & x_1 & 0 & 0 \\ 0 & 0 & x_2 & 0 \\ 0 & 0 & 0 & x_3 \end{pmatrix}}_{\text{Electromagnetic field space}}\Bigg)$$

(7.1.6)

That is

$$\varphi_e(X) \to \varphi'_e(X) = \exp\{-ie[A^\gamma_{\mu,\mu}](X_\mu)\}\varphi_e(X_\mu) \qquad (7.1.7)$$

It is known from the above formula.

$$X_\mu = \breve{\gamma}_\mu \begin{pmatrix} x_0 & 0 & 0 & 0 \\ 0 & x_1 & 0 & 0 \\ 0 & 0 & x_2 & 0 \\ 0 & 0 & 0 & x_3 \end{pmatrix} \qquad (7.1.8)$$

The change is $eA^\gamma_{\mu,\mu}$ in X_μ space. ϕ_γ is the gauge field for transmitting interactions. This is an operator specially used to extract information about the amount of change caused by the coupling of quantum fields.

$$\hat{\Gamma}_\mu \varphi'(X) = eA^\gamma_{\mu,\mu}\varphi'(X) \qquad (7.1.9)$$

For experimental observations, the momentum of electrons is obtained along the internal space of electrons ξ_3, and the momentum space of electrons changes. In the interior space, momentum space strain $[\varepsilon^e_\mu]$ and electromagnetic field spatial strain $[A^e_\mu]$ are integral and nondistinguishable. Therefore, it is easy to think that the momentum space will change after interacting with the gauge field 4-vector $[A^\gamma_{\mu,\mu}]$. Expressed as $[\varepsilon^e_\mu] \to [\varepsilon^e_\mu] + e[A^\gamma_{\mu,\mu}]$, It is further written as a form we know best.

$$p_\mu \to p_\mu + eA_\mu \qquad (7.1.10)$$

Using covariant derivative representation, it can be written as

$$D_\mu = \partial_\mu - \hat{\Gamma}_\mu$$

$$D_\mu \varphi'_e(x) = D_\mu \{\exp[-ieA^\gamma_{\mu,\mu}X_\mu]\varphi_e(x)\} = (\partial_\mu - \hat{\Gamma}_\mu)\{\exp[-ieA^\gamma_{\mu,\mu}X_\mu]\varphi_e(x)\}$$
$$= \exp[-ieA^\gamma_{\mu,\mu}X_\mu]\varphi_e(x)\partial_\mu\varphi_e(x) -$$
$$ieA^\gamma_{\mu,\mu}\exp[-ieA^\gamma_{\mu,\mu}X_\mu]\varphi_e(x) - \hat{\Gamma}_\mu\varphi'_e(x)$$

In order to keep the gauge invariance, requirements

$$-ieA^\gamma_{\mu,\mu}X_\mu \exp[-ieA^\gamma_{\mu,\mu}X_\mu]\varphi_e(X) - \hat{\Gamma}_\mu \varphi'_e(X) = 0 \qquad (7.1.11)$$

$$ieA^\gamma_{\mu,\mu}\varphi'_e(X) = \hat{\Gamma}_\mu \varphi'_e(X) \qquad (7.1.12)$$

Operator $D_\mu = \partial_\mu - ieA^\gamma_{\mu,\mu}$, then

$$\hat{\Gamma}_\mu \varphi'_e = ieA^\gamma_{\mu,\mu}\varphi'_e \qquad (7.1.13)$$

They are within the same space X_μ. We can understand that the covariant differential describes the lepton translation to produce a gauge field in the 4-dimensional space-time. The probability of translation to generate the gauge field is e (also known as the intensity of interaction). This gauge field is photons.

For the electromagnetic 4 vectors in the existing textbooks, we do the following regulations.

$$A_\mu \equiv A^\gamma_{\mu,\mu} \qquad (7.1.14)$$

The principal strain $A_{0,0}$ of electromagnetic field is understood as scalar potential with

Chapter 7 Interaction and Super-unified Equation

energy dimension and $A_{i,i}$ as vector potential with momentum dimension.

$U(1)$ local gauge symmetry in quantum space is as follows.

The $U(1)$ local gauge transformation:

$$\begin{cases} \psi(x) \to \psi'(x) = e^{-ieA(x)} \psi(x) \\ \bar{\psi}(x) \to \bar{\psi}'(x) = \bar{\psi}(x) e^{i\theta(x)} \end{cases} \quad (7.1.15)$$

Infinitesimal transformation $\theta(x) \ll 1$, get

$$\delta\psi(x) = -i\theta(x)\psi(x) \quad (7.1.16)$$

$$\delta(\partial_\mu \psi(x)) = \partial_\mu \delta\psi(x) = -i\theta(x)\partial_\mu \psi(x) - i(\partial_\mu \theta(x))\psi(x) \quad (7.1.17)$$

$$\delta\bar{\psi}(x) = \bar{\psi}(x) i\theta(x) \quad (7.1.18)$$

The resulting change of $\mathscr{L}_0$ is

$$\delta\mathscr{L}_0 = \frac{\partial_0 \mathscr{L}}{\partial \psi}\delta\psi + \frac{\partial_0 \mathscr{L}}{\partial \partial_\mu \psi}\delta\partial_\mu \psi + \delta\bar{\psi}\frac{\partial_0 \mathscr{L}}{\partial \bar{\psi}}$$

$$= 0 + \bar{\psi}i\gamma^\mu[-i\theta(x)\partial_\mu \psi(x) - i(\partial_\mu \theta(x))\psi(x) + i\theta(x)\psi(x)i\gamma^\mu \partial_\mu \psi(x)]$$

$$= -\bar{\psi}\gamma^\mu \psi \partial_\mu \theta(x) = -J^\mu \partial_\mu \theta(x) \quad (7.1.19)$$

Among

$$J^\mu = \bar{\psi}\gamma^\mu \psi \quad (7.1.20)$$

It is the electromagnetic flow.

$$\delta J^\mu = (\delta\bar{\psi})\gamma^\mu \psi + \bar{\psi}\gamma^\mu \delta\psi = 0 \quad (7.1.21)$$

(7.1.19) can be seen in the local gauge transformation (7.1.15), the Lagrangian density $\mathscr{L}_0$ is not invariable. In order to remain $\mathscr{L}_0$ invariant, the need to introduce the 4-vector gauge field $A_{\mu,\mu}(x)$, it directly coupling with the electromagnetic flow J^μ gives the interaction Lagrangian density.

$$\mathscr{L}_I = eJ^\mu A_{\mu,\mu} \quad (7.1.22)$$

$J^\mu A_{\mu,\mu}$ involved in the electromagnetic interaction, only $eJ^\mu A_{\mu,\mu}$ can change the electron momentum space, e is the optical coupling constant. For electronics, the movement of the center (momentum change) corresponds to the physical facts is electrons and photons interact, the absorption of a photon or emits a photon. And required to transform in (7.1.15), $A_{\mu,\mu}(x)$ is transformed into

$$\begin{cases} A_{\mu,\mu}(x) \to A'_{\mu,\mu}(x) = A_{\mu,\mu}(x) + \frac{1}{e}\partial_\mu \theta(x) \\ \delta A_{\mu,\mu}(x) = \frac{1}{e}\partial_\mu \theta(x) \\ \psi(x) \to \psi'(x) = e^{-i\theta(x)}\psi(x) \end{cases} \quad (7.1.23)$$

Then

$$\delta\mathscr{L}_I = e(\delta J^\mu)A_{\mu,\mu} + eJ^\mu \delta A_{\mu,\mu} = J^\mu \partial_\mu \theta(x) \quad (7.1.24)$$

which use the formula (7.1.21), combined (7.1.19) and (7.1.24) immediately get

$$\delta\mathscr{L}_0 + \delta\mathscr{L}_I = 0$$

$\mathscr{L}_0 + \mathscr{L}_I$ is invariant in the transform (7.1.18) and (7.1.23). As the new introduction of a gauge field A^μ, we need to construct its own free Lagrangian density $\mathscr{L}_g$; of course, we also require it is invariant in (7.1.23). Rotational strain of electromagnetic field in observation space

$$\Omega_{\mu\nu} = \frac{\partial A^\mu}{\partial x^\nu} - \frac{\partial A^\nu}{\partial x^\mu} = F_{\mu\nu} = \partial_\mu A_\nu - \partial_\nu A_\mu \tag{7.1.25}$$

That is the gauge field tensor. Quantum field in the electromagnetic interaction, transfer interaction is a gauge field particle, rather than the gauge field, gauge field Lagrangian density is $\mathscr{L}_g = -\frac{1}{4} F^{\mu\nu} F_{\mu\nu}$. Here the gauge field particles are photons.

$$\mathscr{L}_g = -\frac{1}{2}(\partial_\nu A_\mu)(\partial^\nu A^\mu) \tag{7.1.26}$$

The (7.1.15) and (7.1.23) in a local gauge transformation, the constant total Lagrangian density $\mathscr{L}$ can be written as

$$\begin{aligned}\mathscr{L} &= \mathscr{L}_0 + \mathscr{L}_g + \mathscr{L}_I \\ &= \overline{\psi}(i\gamma^\mu D_\mu - m)\psi - \frac{1}{2}(\partial_\nu A_\mu)(\partial^\nu A^\mu) \\ &= \overline{\psi}(i\gamma^n D_n)\psi - \frac{1}{2}(\partial_\nu A_\mu)(\partial^\nu A^\mu)\end{aligned} \tag{7.1.27}$$

In the existing theory $D_\mu = \partial_\mu + ieA_{\mu,\mu}$, it is called the covariant derivative. According to the previous strain theory analysis, we know that A_μ is the deformation quantity of the electromagnetic field along the x_μ in the observation space. Only strain $\partial_\mu A_\mu$ is the 4-momentum of the electromagnetic field has the dimension of energy and momentum. The $\partial_\mu\varphi = T_{\mu\mu}\varphi$ eigenvalue is the 4-momentum of the quantum field, and $\hat{\Gamma}_\mu\varphi = eA_\mu\varphi$ cannot get 4-momentum dimension. It must be $\hat{\Gamma}_\mu\varphi = eA_{\mu,\mu}\varphi$. $A_{\mu,\mu}$ is a strain, so that it matches the 4-dimensional momentum of interaction. Covariant derivatives are rewritten as follows:

$$D_\mu = \partial_\mu + ieA_{\mu,\mu} \tag{7.1.28}$$

$A_{\mu,\mu}$ is the principal strain of the electromagnetic field, A_μ in $(\partial_\nu A_\mu)(\partial^\nu A^\mu)$ is the deformation displacement of the electromagnetic field. In the original theory, it is wrong, it is necessary to pay special attention to it, and $A_{\mu,\mu} = \partial_\mu A_\mu$.

$$\mathscr{L}_0 + \mathscr{L}_g = \overline{\psi}[i\gamma^\mu(\partial_\mu + ieA_{\mu,\mu}) - m]\psi \tag{7.1.29}$$

The corresponding physical fact is that the translation of the electrostatic field of the electron ψ produces the gauge field A. The interaction between gauge field A and electron ψ leads to the translation of electrons. Both of them are opposite processes, which will cause the momentum k of electrons to change.

$$\psi(x) \to \psi'(x) = e^{-i[k - \Delta k(x)]x} \tag{7.1.30}$$

The amount of change is $\Delta k = eA_{\mu,\mu}(x)$. Conversion is not an easy thing, after effect by

Chapter 7 Interaction and Super-unified Equation

the gauge field $A_{\mu,\mu}$ (virtual photon); only 1/137 virtual photons are absorbed into momentum.

Direct calculation the covariant derivative D_μ and gauge field tensor $F_{\mu\nu}$ has the following relationship:

$$[D_\mu, D_\nu]\psi = ieF_{\mu\nu}\psi \qquad (7.1.31)$$

Using $U(1)$ localized gauge invariance, the above requirements to determine the interaction of system, (7.1.27) provisions of the electromagnetic interaction of spin particles and photons.

Here, we summarize the physical meaning of the operator.

$\partial_\mu \psi$: Eigenvalue equation, the eigenvalue $[\varepsilon]_\mu$ is obtained by the operator ∂_μ.

$D_\mu \psi$: Covariant eigenvalue operator. The eigenvalue $[\varepsilon]_\mu$ is obtained by the operator D_μ, and further available the eigenvalue $[\varepsilon]_\mu + eA_{\mu,\mu}$ after translation in space X_μ will result in the generation of gauge field $A_{\mu,\mu}$ in space X_μ.

$[D_\mu, D_\nu]\psi$: Covariant rotational eigenvalue operator, that is, following the translation of space X_μ and space X_ν, the rotation from space X_μ to space X_ν will result in ψ producing gauge field tensor $F_{\mu\nu}$.

7.2 Weak interaction field

The physical essence of Weak and electromagnetic has been considered to be different, because there are the following differences:

(1) Interaction intersity. The electromagnetic action coupling constant $\alpha = 1/137$, and the coupling constant of the weak interaction is $10^{-5}/M_p^2$.

(2) Force range is different. The electromagnetic effect is long-range force, and the weak force range less than 10^{-16} cm (Fig. 7.2.1).

Let us first qualitatively explain weak interaction. For intrinsic 4-momentum field, the displacement function u_r, u_θ, u_ϕ constitutes a weak field. The essence of weak interaction is unstable high-energy state fermions are decay separation (Fig. 7.2.2), back to the lower energy states, forming plurality of relatively stable low energy state fermions. That is like cell division, when the center into two, the weak interaction will be over, so the range force of weak interaction is very short. Internal mass field of fermions is static weak field.

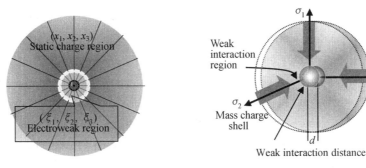

Fig. 7.2.1 Static charge tegion and electroweak region   Fig. 7.2.2 Weak interaction field

Consider μ-lepton decay. The weak interaction is restricted in the intrinsic space. The lepton intrinsic space radius is R, the lepton weak interaction split beginning from the central, when lepton in a high-mass state the intrinsic mass field stretching to the center (or compressed, for anti-particle), lepton field symmetry is not completely ball symmetry state and the neighborhood of the center had a very small deformation then the central region is in unstable state. The center in deformation limit will split that resulting in lepton decay. The splitting occurs when the center of two overlapping mass-charged spheres exchanges weak interacting bosons $W^{\pm}$, Z^0. The weak interaction coupling constant and electromagnetic coupling constant are consistent $g' \approx e$. The central area of the weak interaction less than process is less than 10^{-16} cm that is force rang of weak interaction. This central area can imagination as a droplet to split into two droplets, when the two droplets just tangent, the distance between the center of the sphere is the weak force range d. This can qualitatively explain why the distance between weak interaction is very short.

7.2.1 Interaction field function

1. Half-space of weak interaction

A single strain space region is defined as a half-space. Charged fermions intrinsic space is half-space, particles with fibrosis structure. The process of splitting the center of a particle into two can be understood as the transfer of W bosons through the fiber channel between the two centers. It splits into two particles (tensile state or compressed state), in other words, the weak decay process in the same space that is half-space, so weak interaction loss symmetry, parity non conservation.

$$\mu^- \to e^- \bar{\nu}_e \nu_\mu \tag{7.2.1}$$

By μ-lepton weak interactions to understand, the intrinsic space is how to split (Fig. 7.2.3). Figuratively speaking, the high-energy state lepton is like wearing a heavy coat of energy in an unstable state, only take off his coat can be return to a stable e^- electronic states. In order to maintain the conservation of angular momentum, the energy coat becomes two spin waves with opposite angular momentum. Lepton intrinsic mass field is static weak force field. The weak interaction is within intrinsic space, so the force range of weak interaction particularly short. The mass state of particles is the most stable in lowest energy level. The particle in a high-energy state that is corresponding lepton spin waves in high-energy state. This is an unstable state and ultimately back to the stability lower energy states. Fig. 7.2.3 shows this effect led to the μ decay into electronic, lepton released a spin wave ν_μ make polarization state energy to reduce, and taken away the energy of polarization state. Original lepton at low energy states without spin, apparently this state could not exist, so it released low-energy spin waves e^- to obtain an opposite spin to maintain the conservation of angular momentum. Eventually decay into stable low-energy state lepton, which is electronic.

Chapter 7   Interaction and Super-unified Equation

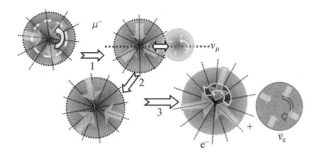

Fig. 7.2.3   The intrinsic space division by weak

2. The field functions of μ-lepton in space of weak interaction

$$\varphi_\mu(X) = \exp(-i)\left\{\left(\underbrace{\begin{bmatrix} \varepsilon_{\mu 0} & 0 & 0 & 0 \\ 0 & \varepsilon_{\mu 0} & 0 & 0 \\ 0 & 0 & \varepsilon_{\mu 0} & 0 \\ 0 & 0 & 0 & \varepsilon_{\mu 3} \end{bmatrix}}_{\text{Intrinsic radial strain of }\mu} + \underbrace{\frac{1}{2}\begin{bmatrix} 0 & \gamma_{01} & \gamma_{02} & \gamma_{03} \\ \gamma_{10} & 0 & \gamma_{12} & \gamma_{13} \\ \gamma_{20} & \gamma_{21} & 0 & \gamma_{23} \\ \gamma_{30} & \gamma_{31} & \gamma_{32} & 0 \end{bmatrix}}_{\text{Circumferential strain constitutes weak field}} + \underbrace{\frac{1}{2}\begin{bmatrix} 0 & \omega_{01} & \omega_{02} & \omega_{03} \\ \omega_{10} & 0 & \omega_{12} & \omega_{13} \\ \omega_{10} & \omega_{21} & 0 & \omega_{23} \\ \omega_{30} & \omega_{31} & \omega_{32} & 0 \end{bmatrix}}_{\text{Rotational strain, and spin}}\right)\underbrace{(1+\gamma_5)\begin{pmatrix} \xi_0 & \xi_0 & \xi_0 & \xi_0 \\ \xi_1 & \xi_1 & \xi_1 & \xi_1 \\ \xi_2 & \xi_2 & \xi_2 & \xi_2 \\ \xi_3 & \xi_3 & \xi_3 & \xi_3 \end{pmatrix}}_{\substack{\text{Weak interaction is self division}\\\text{and space is half space}}} + \right.$$

$$\underbrace{\text{Weak field of }\mu\text{, intrinsic four momentum, bare mass }M_Bc^2, M_Bc\text{ space}\Leftrightarrow(1+\gamma_5)}$$

$$\left.\underbrace{\begin{bmatrix} A^\mu_{0,0} & 0 & 0 & 0 \\ 0 & A^\mu_{1,1} & 0 & 0 \\ 0 & 0 & A^\mu_{2,2} & 0 \\ 0 & 0 & 0 & A^\mu_{3,3} \end{bmatrix}}_{\text{Electromagnetic 4-momentum: }M_Bc^2, M_Ec}\breve{\gamma}_\mu\begin{pmatrix} x_0 & 0 & 0 & 0 \\ 0 & x_1 & 0 & 0 \\ 0 & 0 & x_2 & 0 \\ 0 & 0 & 0 & x_3 \end{pmatrix}\right\} = \exp\{-i[\mu_{\alpha\beta}](\xi) + [A_\mu e^-](X)\} \quad (7.2.2)$$

$$\phi_{\text{Virtual }\gamma}(X) = \exp\left(-i\underbrace{\begin{bmatrix} B^\gamma_{0,0} & 0 & 0 & 0 \\ 0 & 0 & 0 & 0 \\ 0 & 0 & 0 & 0 \\ 0 & 0 & 0 & eB^\gamma_{3,3} \end{bmatrix}}_{\text{Virtual photon}}\underbrace{\begin{pmatrix} x_0 & 0 & 0 & 0 \\ 0 & 0 & 0 & 0 \\ 0 & 0 & 0 & 0 \\ 0 & 0 & 0 & \xi_3^{1-\gamma_E} \end{pmatrix}}_{\text{Intrinsic space}}\right) = \exp\{-i[B_{\mu\nu}](\xi_\gamma)$$

(7.2.3)

For the intrinsic space, the spin axis is ξ^3 and the rotation surface is $\xi^1-\xi^2$.

$$\varphi_{W^-}(X) = \exp(-i)\left\{\left(\underbrace{\begin{bmatrix} W_{0,0} & 0 & 0 & 0 \\ 0 & W_{1,1} & 0 & 0 \\ 0 & 0 & W_{2,2} & 0 \\ 0 & 0 & 0 & W_{3,3} \end{bmatrix}}_{\text{Radial strain }M_{0R}} + \underbrace{\frac{1}{2}\begin{bmatrix} 0 & \gamma^W_{01}+(\Delta\gamma^W_{01}-\Delta\gamma'^W_{01}) & \gamma^W_{02}+(\Delta\gamma^W_{02}-\Delta\gamma'^W_{02}) & \gamma^W_{03} \\ \gamma^W_{10}+(\Delta\gamma^W_{10}-\Delta\gamma'^W_{10}) & 0 & \gamma^W_{12}+(\Delta\gamma^W_{12}-\Delta\gamma'^W_{12}) & \gamma^W_{13} \\ \gamma^W_{20}+(\Delta\gamma^W_{20}-\Delta\gamma'^W_{20}) & \gamma^W_{21}+(\Delta\gamma^W_{21}-\Delta\gamma'^W_{21}) & 0 & \gamma^W_{23} \\ \gamma^W_{30} & \gamma^W_{31} & \gamma^W_{32} & 0 \end{bmatrix}}_{\text{Non spherically symmetric shear strain(Circumferential strain mass }M_{0C})} + \right.$$

$$\left.\underbrace{\frac{1}{2}\begin{bmatrix} 0 & \omega^W_{01} & \omega^W_{02} & 0 \\ \omega^W_{10} & 0 & \omega^W_{12}-\omega'^W_{12} & 0 \\ \omega^W_{20} & \omega^W_{21}-\omega'^W_{21} & 0 & 0 \\ 0 & 0 & 0 & 0 \end{bmatrix}}_{\substack{\text{Non spherically symmetric spin perturbation strain.}\\\text{Destroy heavy lepton stability}}}\right)\underbrace{(1+\breve{\gamma}_5)\begin{pmatrix} \xi_0 & \xi_0 & \xi_0 & \xi_0 \\ \xi_1 & \xi_1 & \xi_1 & \xi_1 \\ \xi_2 & \xi_2 & \xi_2 & \xi_2 \\ \xi_3 & \xi_3 & \xi_3 & \xi_3 \end{pmatrix}}_{\substack{\text{Weak interaction is self division}\\\text{and space is half space}}} + $$

· 323 ·

$$\left.\begin{array}{c}\begin{bmatrix} A^e_{0,0} & 0 & 0 & 0 \\ 0 & A^e_{1,1} & 0 & 0 \\ 0 & 0 & A^e_{2,2} & 0 \\ 0 & 0 & 0 & A^e_{3,3} \end{bmatrix} \tilde{\gamma}_\mu \begin{pmatrix} x_0 & 0 & 0 & 0 \\ 0 & x_1 & 0 & 0 \\ 0 & 0 & x_2 & 0 \\ 0 & 0 & 0 & x_3 \end{pmatrix}\end{array}\right\} \quad (7.2.4)$$

Spherical symmetric strain of electromagnetic field. The weak bosons are charged

$$= \exp(-i)([W_{\mu,\mu}] + \frac{1}{2}[\gamma^W + \Delta\gamma^W - \Delta\gamma'^W] + \frac{1}{2}[\omega^W - \omega'^W])(1+\gamma_5)(\xi^{1-\gamma_E}) + [A^e_{\mu,\mu}]\gamma_\mu(x)$$

$$= \exp(-i)\{[W_{\mu\nu}](\xi^{1-\gamma_E}) + [A^{e-}_\mu](X)\} \quad (7.2.5)$$

$\Delta\gamma^W_{ij} - \Delta\gamma'^W_{ij}$ superposition of small positive and negative shear strains constitutes strain fluctuation; $W_{i,i} \neq W_{j,j}(i \neq j)$, the superposition of positive and negative spin strains of $\frac{1}{2}[\omega^W - \omega'^W] = 0$ is shown to be zero, but it should be noted that some components are not necessarily zero.

The asymmetry of strain is the main factor that destroys the stability of high-energy leptons and causes the decay and separation of high-energy leptons. Therefore, the asymmetry of shear strain is the most important characteristic of weak-acting bosons. However, in order to obtain asymmetric strain, the effect of virtual photons will cause spherically symmetric strain to become aspherically symmetric strain. Transfer momentum by virtual photon. The virtual light will destroy the intrinsic structure of the high-energy μ-lepton and make the μ-lepton split and decay, so the weak effect is the result of the interaction of virtual photon and boson. Show a weak function

$$\begin{cases} \varphi'_\mu(X) = \varphi_\mu(X)\varphi_{\text{Virtual } W^-}(X)\varphi_{\text{Virtual }\gamma} \\ \varphi'_\mu(X) = \varphi_e(X)\varphi_\gamma(X)\varphi_{\bar{\nu}_e}(X)\phi_{\nu_\mu}(X) \end{cases} \quad (7.2.6)$$

$$\phi'_\mu(X) = \phi_\mu(X)\phi_{\text{Virtual } W^-}(X)\phi_{\text{Virtual }\gamma}$$

$$= \underbrace{\begin{bmatrix} \varepsilon_{W0} + g_\gamma\varepsilon_{\gamma 0} & 0 & 0 & 0 \\ 0 & \varepsilon_{\mu 1} + g_W W_{1,1} & 0 & 0 \\ 0 & 0 & \varepsilon_{\mu 2} + g_W W_{2,2} + g_\gamma\varepsilon_{\gamma 2} & 0 \\ 0 & 0 & 0 & \varepsilon_{\mu 3} + g_W W_{3,3} \end{bmatrix}}_{\text{Radial strain } M_{0R}} +$$

$$\underbrace{\frac{1}{2}\begin{bmatrix} 0 & \omega_{01} + g_W(\omega^W_{01} + \omega'^W_{01}) \times [\delta(i-2) + \delta(i-3)] & \omega_{02} + g_W(\omega^W_{20} + \omega'^W_{20}) \times [\delta(i-1) + \delta(i-3)] & \omega_{03} + g_W(\omega^W_{03} + \omega'^W_{03}) \times [\delta(i-1) + \delta(i-2)] \\ \omega_{10} + g_W(\omega^W_{10} + \omega'^W_{10}) \times [\delta(i-2) + \delta(i-3)] & 0 & \omega_{12} + g_W(\omega^W_{12} - \omega'^W_{12})\delta(i-3) & \omega_{13} + g_W(\omega^W_{13} - \omega'^W_{13})\delta(i-2) \\ \omega_{20} + g_W(\omega^W_{20} + \omega'^W_{20}) \times [\delta(i-3) + [\delta(i-1)] & \omega_{21} + g_W(\omega^W_{21} - \omega'^W_{21})\delta(i-3) & 0 & \omega_{23} + g_W(\omega^W_{23} - \omega'^W_{23})\delta(i-1) \\ \omega_{30} + g_W(\omega^W_{30} + \omega'^W_{30}) \times [\delta(i-2) + \delta(i-1)] & \omega_{31} + g_W(\omega^W_{31} - \omega'^W_{31})\delta(i-2) & \omega_{32} + g_W(\omega^W_{32} - \omega'^W_{32})\delta(i-1) & 0 \end{bmatrix}}_{\text{Spin. The lepton } \mu \text{ rotates along the } \xi_1 \text{ axis or } \xi_2. \; W^- \text{ only consider rotational strain along } \xi_3 \text{ axis}} +$$

Chapter 7　Interaction and Super-unified Equation

$$\frac{1}{2}\underbrace{\begin{bmatrix} 0 & \gamma_{01}+g_W(\gamma^W_{01}+\gamma'^W_{01})\times[\delta(i-2)+\delta(i-3)] & \gamma_{02}+g_W(\gamma^W_{20}+\gamma'^W_{20})\times[\delta(i-1)+\delta(i-3)] & \omega_{03}+g_W(\gamma^W_{03}+\gamma'^W_{03})\times[\delta(i-1)+\delta(i-2)] \\ \gamma_{10}+g_W(\gamma^W_{10}+\gamma'^W_{10})\times[\delta(i-2)+\delta(i-3)] & 0 & \gamma_{12}+g_W(\gamma^W_{12}-\gamma'^W_{12})\delta(i-3) & \omega_{13}+g_W(\gamma^W_{13}-\gamma'^W_{13})\delta(i-2) \\ \gamma_{20}+g_W(\gamma^W_{20}+\gamma'^W_{20})\times[\delta(i-3)+[\delta(i-1)] & \gamma_{21}+g_W(\gamma^W_{21}-\gamma'^W_{21})\delta(i-3) & 0 & \omega_{23}+g_W(\gamma^W_{23}-\gamma'^W_{23})\delta(i-1) \\ \gamma_{30}+g_W(\gamma^W_{30}+\gamma'^W_{30})\times[\delta(i-2)+\delta(i-1)] & \gamma_{31}+g_W(\gamma^W_{31}-\gamma'^W_{31})\delta(i-2) & \gamma_{32}+g_W(\gamma^W_{32}-\gamma'^W_{32})\delta(i-1) & 0 \end{bmatrix}}_{\text{Non spherically symmetric shear stain, only consider shear strain on }\xi^1-\xi^2\text{ surface}} \times$$

$$\left\{ (1+\gamma_5)\underbrace{\begin{pmatrix} \xi_0 & \xi_0 & \xi_0 & \xi_0 \\ \xi_1^{1-\gamma_W} & \xi_1^{1-\gamma_W} & \xi_1^{1-\gamma_W} & \xi_1^{1-\gamma_W} \\ \xi_2^{1-\gamma_W} & \xi_2^{1-\gamma_W} & \xi_2^{1-\gamma_W} & \xi_2^{1-\gamma_W} \\ \xi_3 & \xi_3 & \xi_3 & \xi_3 \end{pmatrix}}_{\substack{\text{Weak interaction is self division and}\\\text{space is half space}}} + \underbrace{\begin{bmatrix} A^e_{0,0} & 0 & 0 & 0 \\ 0 & A^e_{1,1} & 0 & 0 \\ 0 & 0 & A^e_{2,2} & 0 \\ 0 & 0 & 0 & A^e_{3,3} \end{bmatrix}}_{\text{Electromagnetic 4-momentum: e}} \overset{\smile}{\gamma_\mu} \begin{pmatrix} x_0 & 0 & 0 & 0 \\ 0 & x_1 & 0 & 0 \\ 0 & 0 & x_2 & 0 \\ 0 & 0 & 0 & x_3 \end{pmatrix} \right\} \quad (7.2.7)$$

The function $\delta(i-n)$ is introduced to ensure that the weak interaction exists only on two dimensions. The 2-dimensional surface is perpendicular to the rotation axis of the n.

γ_W is an anomalous dimension of one dimension in $\xi_1^{1-\gamma_W}$. The dimension of the interaction surface is equal to $\xi_2^{1-\gamma_W}\xi_1^{1-\gamma_W}$. The whole interaction dimension is equal to $\xi_1^{1-\gamma_w}\xi_2^{1-\gamma_w}\xi_3$.

The electromagnetic field only provides the power lines coupling channel for the interaction and determines the virtual photon interaction strength is g_γ. Weak interection is confined in the intrinsic space. The charge of the virtual $W^\pm$ gauge field is the lepton charge.

$\phi_{\text{Virtual }W^-}$ can be understood as the embryo state of W bosons, parasitic on μ-lepton, or it can be imagined that there is turbulent fluctuation outside the intrinsic space of μ-lepton (Fig. 7.2.4), which is the superposition of forward and backward fluctuations. This positive and negative strain superposition destroys the structure of μ-lepton.

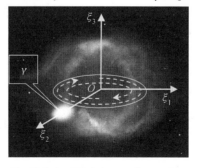

Fig. 7.2.4　Weak interaction surface

For W-boson, the superposition states of forward and reverse spin waves are coaxial, so the two superimposed spin waves can only exist on the 2-dimensional plane ξ^1-ξ^2 perpendicular to the rotational axis ξ^3. The shear strain on this plane also fluctuates due to the existence of spin waves, i.e. the shear strain and the reverse shear strain fluctuate each other. Superposition, therefore, makes the surface extremely unstable. If the surface interacts with the virtual photon, the structure of the μ-lepton will be destroyed and μ-lepton will decay from the high-energy state into the low-energy state. The 2-dimensional surface ξ^1-ξ^2 is called the weak interaction surface. Virtual photons can only exist on the surface above $\xi^1 - \xi^2$. $[\mu_{\alpha\beta}](\xi)$ is intrinsic strain of μ-lepton.

$$\varphi'_\mu(X) = \exp(-i)\left\{[\mu_{\alpha\beta}](\xi) + g_W \begin{bmatrix} W_{0,0} & W_{1,0} & W_{2,0} & W_{3,0} \\ W_{0,1} & W_{1,1} & (\Delta\gamma^W_{21} - \Delta\gamma^W_{21}) + (\omega^W_{12} - \omega'^W_{12}) & 0 \\ W_{0,2} & (\Delta\gamma^W_{21} - \Delta\gamma^W_{21}) + (\omega^W_{21} - \omega'^W_{21}) & W_{2,2} & 0 \\ W_{0,3} & 0 & 0 & W_{3,3} \end{bmatrix}(\xi) + \right.$$

$$\left. g_\gamma \begin{bmatrix} B_{0,0} & 0 & 0 & 0 \\ 0 & 0 & 0 & 0 \\ 0 & 0 & B_{2,2} & 0 \\ 0 & 0 & 0 & 0 \end{bmatrix}(\xi_\gamma) + [A^{e-}_{\mu,\mu}](X)\right\} \quad (7.2.8)$$

Symmetry is satisfied for virtual photon in 1-dimensional space structure.

$$B_{\mu\nu} = \frac{\partial B_\mu}{\partial \xi_\nu} - \frac{\partial B_\nu}{\partial \xi_\mu} = 0 \quad (7.2.9)$$

The weak action gauge field has four dimensional space-time structure. When there is no interaction, the strain is $W_{\mu\nu} = \frac{\partial W^\mu}{\partial x^\nu} - \frac{\partial W^\nu}{\partial x^\mu}$. When there is interaction, the interaction is caused by asymmetry. The degree of asymmetry is $[W_\mu, W_\nu]$. According to the conservation of strain, the asymmetric strain of the weak interaction gauge field eventually transfers to the intrinsic strain field of the particle after the end of the weak interaction. This is the interaction strain term $g_W[W_\mu, W_\nu]$. The gauge field can be written as

$$W_{\mu\nu} = \frac{\partial W^\mu}{\partial x^\nu} - \frac{\partial W^\nu}{\partial x^\mu} + g_W[W_\mu, W_\nu] \quad (7.2.10)$$

The interaction ends, field functions can be expressed as

$$\varphi'_\mu(X) = \varphi_e(X)\varphi_{\bar{\nu}_e}(X)\varphi_{\nu_\mu}(X) \quad (7.2.11)$$

Chapter 7 Interaction and Super-unified Equation

$$\varphi_e(X) = \exp(-i)\left\{\left(\underbrace{\begin{bmatrix} \varepsilon_{e0} & 0 & 0 & 0 \\ 0 & \varepsilon_{e1} & 0 & 0 \\ 0 & 0 & \varepsilon_{e2} & 0 \\ 0 & 0 & 0 & \varepsilon_{e3} \end{bmatrix}}_{\text{Intrinsic radial strain of }\mu.\ M_{0R}} + \frac{1}{2}\underbrace{\begin{bmatrix} 0 & \gamma_{01} & \gamma_{02} & \gamma_{03} \\ \gamma_{10} & 0 & \gamma_{12} & \gamma_{13} \\ \gamma_{20} & \gamma_{21} & 0 & \gamma_{23} \\ \gamma_{30} & \gamma_{31} & \gamma_{32} & 0 \end{bmatrix}}_{\text{Shear strain }M_{0C}} + \frac{1}{2}\underbrace{\begin{bmatrix} 0 & \omega_{01} & \omega_{02} & 0 \\ \omega_{10} & 0 & \omega_{12} & 0 \\ \omega_{10} & \omega_{21} & 0 & 0 \\ 0 & 0 & 0 & 0 \end{bmatrix}}_{\substack{\text{Rotational strain. Spin, only consider}\\ \text{rotation along }\xi_3\text{ axis }M_{0S}}}\right)\underbrace{(1+\breve{\gamma}_5)\begin{pmatrix} \xi_0 & \xi_0 & \xi_0 & \xi_0 \\ \xi_1 & \xi_1 & \xi_1 & \xi_1 \\ \xi_2 & \xi_2 & \xi_2 & \xi_2 \\ \xi_3 & \xi_3 & \xi_3 & \xi_3 \end{pmatrix}_e}_{\substack{\text{Weak interaction is self division}\\ \text{and space is half space}=X_\mu}}\right.$$

$$\underbrace{}_{\text{Weak field of }\mu,\text{Intrinsic four momentum, bare mass }M_{0B}c^2 = (M_{0R}+M_{0C}+M_{0S}),M_Bc\text{ space}\Leftrightarrow(1+\gamma_5)}$$

$$\left.+\underbrace{\begin{bmatrix} A^e_{0,0} & 0 & 0 & 0 \\ 0 & A^e_{1,1} & 0 & 0 \\ 0 & 0 & A^e_{2,2} & 0 \\ 0 & 0 & 0 & A^e_{3,3} \end{bmatrix}\breve{\gamma}_\mu\begin{pmatrix} x_0 & 0 & 0 & 0 \\ 0 & x_1 & 0 & 0 \\ 0 & 0 & x_2 & 0 \\ 0 & 0 & 0 & x_3 \end{pmatrix}}_{\text{Electromagnetic 4-momentum: }M_Ec^2,\ M_Ec}\right\} \quad (7.2.12)$$

$$\varphi_{\bar{\nu}_e} = \exp(-i)\left\{\underbrace{\begin{bmatrix} \varepsilon_{0/\!/} & 0 & 0 & 0 \\ 0 & 0 & 0 & 0 \\ 0 & 0 & 0 & 0 \\ 0 & 0 & 0 & -\varepsilon_3 \end{bmatrix}\begin{pmatrix} \xi_0 & 0 & 0 & 0 \\ 0 & 0 & 0 & 0 \\ 0 & 0 & 0 & 0 \\ 0 & 0 & 0 & x_3 \end{pmatrix}}_{\substack{\text{The strain structure similar virtual photon. Neutrinos}\\ \text{propagate at the speed of light and transmit momentum}}} + \underbrace{\begin{bmatrix} \varepsilon_{0\perp} & 0 & 0 & 0 \\ 0 & -\varepsilon_1 & 0 & 0 \\ 0 & 0 & -\varepsilon_2 & 0 \\ 0 & 0 & 0 & 0 \end{bmatrix}\begin{pmatrix} \xi_0 & 0 & 0 & 0 \\ 0 & \xi_1 & 0 & 0 \\ 0 & 0 & \xi_2 & 0 \\ 0 & 0 & 0 & 0 \end{pmatrix}}_{\text{Constitute 2-dimensional weak field radial strain part}} +$$

$$\frac{1}{2}\left(\underbrace{\begin{bmatrix} 0 & \Delta\gamma_{01} & \Delta\gamma_{02} & 0 \\ \Delta\gamma_{10} & 0 & -\Delta\gamma_{12} & 0 \\ \Delta\gamma_{20} & -\Delta\gamma_{21} & 0 & 0 \\ 0 & 0 & 0 & 0 \end{bmatrix}}_{\substack{\text{Constitute 2-dimensional}\\ \text{weak field shear strain part}}} + \underbrace{\begin{bmatrix} 0 & \omega'_{01} & \omega'_{02} & 0 \\ \omega'_{10} & 0 & -\omega'_{12} & 0 \\ \omega'_{20} & -\omega'^W_{21} & 0 & 0 \\ 0 & 0 & 0 & 0 \end{bmatrix}}_{\substack{\text{Two dimensional weak field}\\ \text{rotational strain part}}}\right)\cdot$$

$$\left.(1+\breve{\gamma}_5)\underbrace{\begin{pmatrix} 0 & \xi_0 & \xi_0 & 0 \\ \xi_1 & 0 & \xi_1 & 0 \\ \xi_2 & \xi_2 & 0 & 0 \\ 0 & 0 & 0 & 0 \end{pmatrix}_{\bar{\nu}_e}}_{\substack{\text{Weak interaction is self division}\\ \text{and space is half space}}}\right\} \quad (7.2.13)$$

$$\varphi_{\nu_\mu} = \exp\left(-i\left\{\underbrace{\begin{bmatrix} \varepsilon_{0/\!/} & 0 & 0 & 0 \\ 0 & 0 & 0 & 0 \\ 0 & 0 & 0 & 0 \\ 0 & 0 & 0 & \varepsilon_3 \end{bmatrix}\begin{pmatrix} \xi_0 & 0 & 0 & 0 \\ 0 & 0 & 0 & 0 \\ 0 & 0 & 0 & 0 \\ 0 & 0 & 0 & x_3 \end{pmatrix}}_{\substack{\text{The strain structure similar virtual photon. Neutrinos}\\ \text{propagate at the speed of light and transmit momentum}}} + \underbrace{\begin{bmatrix} \varepsilon_{0\perp} & 0 & 0 & 0 \\ 0 & \varepsilon_1 & 0 & 0 \\ 0 & 0 & \varepsilon_2 & 0 \\ 0 & 0 & 0 & 0 \end{bmatrix}\begin{pmatrix} \xi_0 & 0 & 0 & 0 \\ 0 & \xi_1 & 0 & 0 \\ 0 & 0 & \xi_2 & 0 \\ 0 & 0 & 0 & 0 \end{pmatrix}_{\nu_\mu}}_{\text{Constitute 2-dimensional weak field radial strain part}} +\right.\right.$$

$$\frac{1}{2}\left(\underbrace{\begin{bmatrix} 0 & \Delta\gamma_{01} & \Delta\gamma_{02} & 0 \\ \Delta\gamma_{10} & 0 & \Delta\gamma_{12} & 0 \\ \Delta\gamma_{20} & \Delta\gamma_{21} & 0 & 0 \\ 0 & 0 & 0 & 0 \end{bmatrix}}_{\text{Constitute 2-dimensional weak field shear strain part}} + \underbrace{\begin{bmatrix} 0 & \omega_{01} & \omega_{02} & 0 \\ \omega_{10} & 0 & \omega_{12} & 0 \\ \omega_{20} & \omega_{21} & 0 & 0 \\ 0 & 0 & 0 & 0 \end{bmatrix}}_{\text{Two dimensional weak field rotational strain part}} \right)\cdot$$

$$\left. (1+\breve{\gamma}_5)\underbrace{\begin{pmatrix} 0 & \xi_0 & \xi_0 & 0 \\ \xi_1 & 0 & \xi_1 & 0 \\ \xi_2 & \xi_2 & 0 & 0 \\ 0 & 0 & 0 & 0 \end{pmatrix}_{\nu\mu}}_{\text{Weak interaction is self division and space is half space}} \right\} \quad (7.2.14)$$

It can be abbreviated as

$$\varphi'_\mu(X) = \exp(-i)\{([\mu_{\alpha\beta}] + g_W[W_{\mu\nu}] + g_\gamma[B_{\mu\nu}])X_\mu + [A^{e^-}_\mu]X_\alpha\}$$
$$\varphi'_\mu(X) = \exp\{-i(g_W[W_{\mu\nu}] + g_\gamma[B_{\mu\nu}])X_\mu\}\varphi_\mu(X) \quad (7.2.15)$$

$$D_\mu = \partial_\mu - \hat{\Gamma}_\mu, \quad \partial_\mu = i\partial/\partial X_\mu$$

It is known from the above formula.

$$X_\mu = (1+\gamma_5)\begin{pmatrix} \xi_0 & \xi_0 & \xi_0 & \xi_0 \\ \xi_1 & \xi_1 & \xi_1 & \xi_1 \\ \xi_2 & \xi_2 & \xi_2 & \xi_2 \\ \xi_3 & \xi_3 & \xi_3 & \xi_3 \end{pmatrix} \quad (7.2.16)$$

$$D_\mu\varphi'_e(x) = (\partial_\mu - \hat{\Gamma}_\mu)\exp\{-i(g_W[W_{\mu\nu}] + g_\gamma[B_{\mu\nu}])X_\mu\}\varphi'_\mu(X) \quad (7.2.17)$$

$$\hat{\Gamma}_\mu\varphi'(X) = (g_W[W_{\mu\nu}] + g_\gamma[B_{\mu\nu}])\varphi'(X) \quad (7.2.18)$$

Weak action gauge fields also have W^+ and Z^0(Fig. 7.2.5).

Fig. 7.2.5 The structure of Z^0 bosons is a pair of positive and negative lepton coupling

Chapter 7 Interaction and Super-unified Equation

$$\phi_{W+}(X) = \exp\left(i\left\{\left(\underbrace{\begin{bmatrix} W_{0,0} & 0 & 0 & 0 \\ 0 & W_{1,1} & 0 & 0 \\ 0 & 0 & W_{2,2} & 0 \\ 0 & 0 & 0 & W_{3,3} \end{bmatrix}}_{\text{Radial strain } M_{0R}} + \right.\right.\right.$$

$$\frac{1}{2}\underbrace{\begin{bmatrix} 0 & \gamma^W_{01}+2\Delta\gamma^W_{01} & \gamma^W_{02}+2\Delta\gamma^W_{02} & \gamma^W_{03} \\ \gamma^W_{10}+2\Delta\gamma^W_{10} & 0 & \gamma^W_{12}+(\Delta\gamma^W_{12}-\Delta\gamma'^W_{12}) & \gamma^W_{13} \\ \gamma^W_{20}+2\Delta\gamma^W_{20} & \gamma^W_{21}+(\Delta\gamma^W_{21}-\Delta\gamma'^W_{21}) & 0 & \gamma^W_{23} \\ \gamma^W_{30} & \gamma^W_{31} & \gamma^W_{32} & 0 \end{bmatrix}}_{\text{Non spherically symmetric shear strain (Circumferential strain mass } M_{0C})} +$$

$$\frac{1}{2}\underbrace{\begin{bmatrix} 0 & \omega^W_{01} & \omega^W_{02} & 0 \\ \omega^W_{10} & 0 & \omega^W_{12}-\omega'^W_{12} & 0 \\ \omega^W_{20} & \omega^W_{21}-\omega'^W_{21} & 0 & 0 \\ 0 & 0 & 0 & 0 \end{bmatrix}}_{\substack{\text{Non spherically symmetric spin perturbation strain.} \\ \text{Destroy heavy lepton stability}}} (1-\breve{\gamma}_5)\underbrace{\begin{pmatrix} \xi_0 & \xi_0 & \xi_0 & \xi_0 \\ \xi_1 & \xi_1 & \xi_1 & \xi_1 \\ \xi_2 & \xi_2 & \xi_2 & \xi_2 \\ \xi_3 & \xi_3 & \xi_3 & \xi_3 \end{pmatrix}}_{\substack{\text{Weak interaction is self division} \\ \text{and space is half space}}} +$$

$$\left.\left.\underbrace{\begin{bmatrix} A^e_{0,0} & 0 & 0 & 0 \\ 0 & A^e_{1,1} & 0 & 0 \\ 0 & 0 & A^e_{2,2} & 0 \\ 0 & 0 & 0 & A^e_{3,3} \end{bmatrix}}_{\substack{\text{Spherical symmetric strain of electromagnetic} \\ \text{field. The weak bosons are charged}}} \breve{\gamma}_\mu \begin{pmatrix} x_0 & 0 & 0 & 0 \\ 0 & x_1 & 0 & 0 \\ 0 & 0 & x_2 & 0 \\ 0 & 0 & 0 & x_3 \end{pmatrix}\right\}\right) \quad (7.2.19)$$

$$\varphi_{Z^0}(X) = \exp(-i)\left\{\left(\underbrace{\begin{bmatrix} W_{0,0} & 0 & 0 & 0 \\ 0 & W_{1,1} & 0 & 0 \\ 0 & 0 & W_{2,2} & 0 \\ 0 & 0 & 0 & W_{3,3} \end{bmatrix}}_{\text{Radial strain } M_{0R}} + \frac{1}{2}\underbrace{\begin{bmatrix} 0 & \gamma^W_{01} & \gamma^W_{02} & \gamma^W_{03} \\ \gamma^W_{10} & 0 & \gamma^W_{12} & \gamma^W_{13} \\ \gamma^W_{20} & \gamma^W_{21} & 0 & \gamma^W_{23} \\ \gamma^W_{30} & \gamma^W_{31} & \gamma^W_{32} & 0 \end{bmatrix}}_{\substack{\text{Non spherically symmetric shear strain} \\ \text{(Circumferential strain mass } M_{0C})}} +$$

$$\frac{1}{2}\underbrace{\begin{bmatrix} 0 & \omega^W_{01} & \omega^W_{02} & 0 \\ \omega^W_{10} & 0 & \omega^W_{12} & 0 \\ \omega^W_{20} & \omega^W_{21} & 0 & 0 \\ 0 & 0 & 0 & 0 \end{bmatrix}}_{\substack{\text{Non spherically symmetric spin perturbation} \\ \text{strain. Destroy heavy lepton stability}}} (1+\breve{\gamma}_5)\underbrace{\begin{pmatrix} \xi_0 & \xi_0 & \xi_0 & \xi_0 \\ \xi_1 & \xi_1 & \xi_1 & \xi_1 \\ \xi_2 & \xi_2 & \xi_2 & \xi_2 \\ \xi_3 & \xi_3 & \xi_3 & \xi_3 \end{pmatrix}}_{\substack{\text{Weak interaction is self division} \\ \text{and space is half space. Forward} \\ \text{and backward coaxial rotation}}} +$$

$$\left.\begin{array}{l}
\underbrace{\begin{bmatrix} A^e_{0,0} & 0 & 0 & 0 \\ 0 & A^e_{1,1} & 0 & 0 \\ 0 & 0 & A^e_{2,2} & 0 \\ 0 & 0 & 0 & A^e_{3,3} \end{bmatrix} \breve{\gamma}_\mu \begin{pmatrix} x_0 & 0 & 0 & 0 \\ 0 & x_1 & 0 & 0 \\ 0 & 0 & x_2 & 0 \\ 0 & 0 & 0 & x_3 \end{pmatrix}}_{\text{Electromagnetic 4-momentum: } M_{EC}{}^2,\ M_{EC}} - \\[2em]
\underbrace{\begin{bmatrix} W'_{0,0} & 0 & 0 & 0 \\ 0 & W'_{1,1} & 0 & 0 \\ 0 & 0 & W'_{2,2} & 0 \\ 0 & 0 & 0 & W'_{3,3} \end{bmatrix}}_{\text{Radial strain } M_{OR}} - \frac{1}{2} \underbrace{\begin{bmatrix} 0 & \gamma'^W_{01} & \gamma'^W_{02} & \gamma'^W_{03} \\ \gamma'^W_{10} & 0 & \gamma'^W_{12} & \gamma'^W_{13} \\ \gamma'^W_{20} & \gamma'^W_{21} & 0 & \gamma'^W_{23} \\ \gamma'^W_{30} & \gamma'^W_{31} & \gamma'^W_{32} & 0 \end{bmatrix}}_{\substack{\text{Non spherically symmetric shear strain} \\ (\text{Circumferential strain mass } M_{OC})}} - \\[2em]
\frac{1}{2} \underbrace{\begin{bmatrix} 0 & \omega^W_{01} & \omega^W_{02} & 0 \\ \omega^W_{10} & 0 & \omega'^W_{12} & 0 \\ \omega^W_{20} & \omega'^W_{21} & 0 & 0 \\ 0 & 0 & 0 & 0 \end{bmatrix}}_{\substack{\text{Non spherically symmetric spin perturbation} \\ \text{strain. Destroy heavy lepton stability}}} (1 - \breve{\gamma}_5) \underbrace{\begin{pmatrix} \xi_0 & \xi_0 & \xi_0 & \xi_0 \\ \xi'_1 & \xi'_1 & \xi'_1 & \xi'_1 \\ \xi'_2 & \xi'_2 & \xi'_2 & \xi'_2 \\ \xi_3 & \xi_3 & \xi_3 & \xi_3 \end{pmatrix}}_{\substack{\text{Weak interaction is self division} \\ \text{and space is half space. Forward} \\ \text{and backward rotation coaxial}}} - \\[2em]
\underbrace{\begin{bmatrix} A^e_{0,0} & 0 & 0 & 0 \\ 0 & A^e_{1,1} & 0 & 0 \\ 0 & 0 & A^e_{2,2} & 0 \\ 0 & 0 & 0 & A^e_{3,3} \end{bmatrix} \breve{\gamma}_\mu \begin{pmatrix} x_0 & 0 & 0 & 0 \\ 0 & x_1 & 0 & 0 \\ 0 & 0 & x_2 & 0 \\ 0 & 0 & 0 & x_3 \end{pmatrix}}_{\text{Electromagnetic 4-momentum: } M_{EC}{}^2,\ M_{EC}}
\end{array}\right\} \quad (7.2.20)$$

There are three bosons that transmit weak interactions: W^+, W^-, Z^0. From the above discussion, we know that the unified theory of weak electricity can be obtained directly from the conservation of strain in the interaction field. The mass has been included in $[W_{\mu,\mu}]$ and $[\gamma_{\mu\nu}]$, and the Higgs mechanism is not involved in this process.

W^+, W^-, Z^0 whose internal structure is similar to heavy lepton, has turbulent fluctuation in the internal space, which is the superposition state of positive shear strain fluctuation and reverse shear strain fluctuation (Fig. 7.2.6). After the superposition of positive spin wave and reverse spin wave, the spin is zero, and the radial strain and shear strain still exist, which makes the static mass and radial strain stretch inward. There is charge e^-, the radial strain extrudes outward, showing charge e^+. The superposition of the positive reaction changes destroys its stability. After decay, turbulent fluctuations in the internal space change into spin wave ν and plane wave γ, leaving a strain state of relatively stable structure, i.e. e, μ, τ-lepton. For example: $W^+ \to e^+ + \nu + \gamma$; $W^+ \to e^+ + \nu$; $W^+ \to \mu^+ + \nu$; $W^+ \to \tau^+ + \nu$.

Chapter 7 Interaction and Super-unified Equation

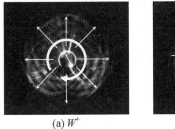
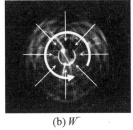
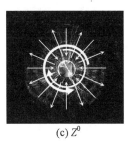

(a) W^+ (b) W^- (c) Z^0

Fig. 7.2.6 Schematic diagram of boson structure

7.2.2 Weak interaction $SU(2) \times U(1)$ localized gauge transformation

1. Covariant derivative in multiple degrees of freedom space

Considering a quantum field

$$\psi(X) = \exp\{-i([\varepsilon]_1 T^1 X^1 + \cdots + ([\varepsilon]_1^\alpha + \cdots + [\varepsilon]_n^\alpha) T^\alpha X^\alpha + \cdots + [\varepsilon]_{N+n} T^N X^N\}$$

There are $N+n$ strains and N spaces. T^α is the spatial structure matrix of space X^α. In space X^α, the coupling constant is $g_{1\alpha}, \cdots, g_{n\alpha}$, The covariant derivative of X^α is

$$D_\mu \psi = [\partial_\mu - i(g_1^\alpha [\varepsilon]_1^\alpha T_1^\alpha + g_2^\alpha [\varepsilon]_2^\alpha T_2^\alpha + \cdots + g_n^\alpha [\varepsilon]_n^\alpha T_n^\alpha)]\psi$$

$$D_\alpha = \partial_\alpha - \sum_{i=1}^{n} ig_i^\alpha [\varepsilon]_i^\alpha T_i^\alpha \tag{7.2.21}$$

$$D_\mu \varphi'_e(x) = (\partial_\mu - \hat{\Gamma}_\mu) \exp\left\{-i\left(g_W \begin{bmatrix} W_{0,0} & 0 & 0 & 0 \\ 0 & W_{1,1} & 0 & 0 \\ 0 & 0 & W_{2,2} & 0 \\ 0 & 0 & 0 & W_{3,3} \end{bmatrix} + g_\gamma \begin{bmatrix} W_{0,0} & 0 & 0 & 0 \\ 0 & 0 & 0 & 0 \\ 0 & 0 & 0 & 0 \\ 0 & 0 & 0 & B_{3,3} \end{bmatrix}\right) X_\mu\right\} \varphi'_\mu(X)$$

2. $SU(2) \times U(1)$ localized gauge transformation

The intrinsic strain region of charged lepton is the weak electric mixing region. Weak interacting has a weak interaction gauge field (A_μ, B_μ). The elements of group $U(1)$ are $\exp[-i\theta(Y/2)]$, corresponding gauge fields $B_\mu(x)$. The elements of group $SU(2)$ are $\exp[-i\theta T]$. The gauge group generators $T(T^i(x), i=1,2,3)$ constitute 3D space. The corresponding gauge field is A_μ and components are $A_\mu^i(x)$, $i=1,2,3$. The group element in $SU(2) \times U(1)$ is

$$U(\theta) = \exp\left[-i\theta \cdot T - i\vartheta \frac{Y}{2}\right] = \exp[-i\theta \cdot T] \exp\left[-i\vartheta \frac{Y}{2}\right] \tag{7.2.22}$$

$\psi(x) \to \psi'(x) = U\psi(x)$. Covariant derivative

$$D_\mu \psi = \left(\partial_\mu + ig_t \frac{\sigma}{2} A_\mu + ig_y \frac{1}{2} B_\mu\right) \psi \tag{7.2.23}$$

Gaug field tensor is
$$F^k_{\mu\nu} = \partial_\mu A^k_\nu - \partial_\nu A^k_\mu - g\varepsilon^{ijk}A^i_\mu A^j_\nu \; ; \; B_{\mu\nu}(x) = \partial_\mu B_\nu - \partial_\nu B_\mu \qquad (7.2.24)$$
Gauge field $A^k_\mu(x)$ with self-interaction, $B_\mu(x)$ is no interaction.

7.3 Higgs mechanism

Higgs-field strain function

$$\Phi_{\text{Higgs}} = \exp\left\{ -ig_W \begin{bmatrix} W_{0,0} & 0 & 0 & 0 \\ 0 & W_{1,1} & 0 & 0 \\ 0 & 0 & W_{2,2} & 0 \\ 0 & 0 & 0 & W_{3,3} \end{bmatrix} \begin{bmatrix} \xi^0 & 0 & 0 & 0 \\ 0 & \xi_1 & 0 & 0 \\ 0 & 0 & \xi_2 & 0 \\ 0 & 0 & 0 & \xi_3 \end{bmatrix} - ig_\gamma \begin{bmatrix} B_{0,0} & 0 & 0 & 0 \\ 0 & 0 & 0 & 0 \\ 0 & 0 & 0 & 0 \\ 0 & 0 & 0 & B_{3,3} \end{bmatrix} \begin{bmatrix} \xi^0 & 0 & 0 & 0 \\ 0 & 0 & 0 & 0 \\ 0 & 0 & 0 & 0 \\ 0 & 0 & 0 & 0 \end{bmatrix} \right\}$$

Goldstone particles move along T^α to produce Higgs particles, along Y move to make Higgs particles to broken along Y, two combined actions make Higgs particle, whose breaking mechanism is called the Higgs mechanism. The Higgs particle is a combination of virtual W bosons and virtual photons.

7.3.1 Higgs field

Consider the existence a ground state of quantum field Φ_0, it interact with gauge field θ, Made the gauge transformation, we get

$$\Phi_0 \to \Phi'_0 = \Phi_0 \exp[-i\theta \cdot T] \qquad (7.3.1)$$

$$\mathscr{L}_{\phi g} = \frac{1}{2}D_\mu \Phi_0^+ D^\mu \Phi_0 = \frac{1}{2}(\partial_\mu + i\theta \cdot T)\Phi_0^+(\partial_\mu - i\theta \cdot T)\Phi_0$$

$$= \frac{1}{2}\partial_\mu \Phi_0^+ \partial_\mu \Phi_0 - \frac{1}{2}(\theta \cdot T)^2 \Phi_0^+ \Phi_0 \qquad (7.3.2)$$

Obtain mass $(\theta \cdot T)^2$, set $\Phi'_0 = \phi_0$. The form of field Φ_0

$$\phi = \Phi \exp\{-i\theta \cdot T\} \qquad (7.3.3)$$

This can be understood as there is no breaking of axisymmetric convergence wave on the hypersurface T(Fig. 7.3.1). θ moving along T^k to constitute a convergence wave makes the vacuum cylinder to produce a convergence wave, makes the vacuum to generate a cylinder of convergence wave. The cylindrical wave perpendicular to the fluctuations hyper-plane, at which time, the vibration of the cylindrical wave constitute a no breaking of the field ϕ_0, with rotation invariance. There is another situation for hypersurface wave, and this is spread out by the vibration of the center of spherical wave, that is T space diffusion wave of hypersurface, which spread the vibration source is a Higgs field ϕ_0^+, ϕ_0 and ϕ_0^+ have the structure of the coupling, both a complete ϕ field, spin is 0.

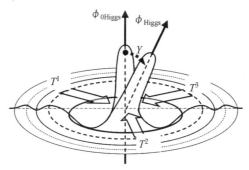

Fig. 7.3.1 Move along T^k, form the convergence wave of hypersurface, constitute a source of vibration in a vacuum, is a virtual mass field

For the description of Higgs scalar field ϕ, the Lagrange density

$$L_\phi = \frac{1}{2}\partial_\mu \phi^+ \partial \phi - \frac{\mu^2}{2}\phi^+ \phi - \frac{\lambda}{4}(\phi^+ \phi)^2 \qquad (7.3.4)$$

The same as the usual scalar field, and $\lambda > 0$, with the self-interaction. The (7.3.4) transformation of reflection $\phi(x) \to -\phi(x)$ in the internal space is invariant. Its interior has reflection symmetry. Higgs field there is a minimum, there are

$$\frac{dV}{d\phi^+} = \mu^2 \phi + \lambda \phi(\phi^+ \phi) = 0 \qquad (7.3.5)$$

Available, the $V(\phi)$ minimum in

$$|\phi_0|^2 = -\mu^2/\lambda = v^2 \qquad (7.3.6)$$

There are infinite numbers of solutions. Contrast (7.3.2) and (7.3.4):

$$\mu^2 = (\theta \cdot T)^2 < 0 \qquad (7.3.7)$$

That is the virtual mass, virtual mass particles may be superluminal motion, which is the features of source of the disturbance.

By $V(\phi) = \frac{\mu^2}{2}\phi^2 + \frac{\lambda}{4}\phi^4$ and $V_{\min}(\phi) = -\frac{\mu^4}{4}$, these conditions mean field vacuum expectation value is $|\langle 0 | \phi | 0 \rangle|^2 = v^2$, Satisfy this condition of vacuum $|0\rangle$ have an unlimited number, corresponding to $\langle 0 | \phi | 0 \rangle$, the circle radius is v on the complex plane. Now the vacuum is infinitely degenerate, only one is the physical vacuum state, this field of vacuum expectation value non-zero is the Higgs field. Higgs field in the complex plane:

$$i\phi_{01} = iv\sin\alpha, \quad \phi_{02} = v\cos\alpha \qquad (7.3.8)$$

$$\phi_0 = i\phi_{01} + \phi_{02} = v\exp i\alpha \qquad (7.3.9)$$

α can be any value with rotation invariance. This is extremely important for the Higgs field (Fig. 7.3.2), where the dashed circle (the bowl depression) exists $\phi_{0\text{Higgs}}$ field, the entire ring are likely to exist, no observable effects.

Contrast (7.3.6), (7.3.3) and (7.3.7), we can know

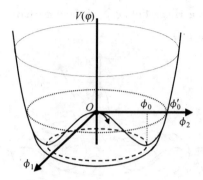

Fig. 7.3.2 Higgs field vacuum state structure

$$\Phi_0 = v = \sqrt{-\theta^2/\lambda} \ ; \ \alpha = \theta \cdot T \qquad (7.3.10)$$

Get

$$\phi_{0\text{Higgs}} = \sqrt{-\frac{\theta^2}{\lambda}} \exp[-i\theta \cdot T] \qquad (7.3.11)$$

If broken it will be eaten Goldstone particles, into a real mass field ϕ_{Higgs}.

$\phi = \pm\sqrt{-2\mu^2/\lambda}$. When $\phi_0 = 1$, $V(\phi) = 0$. Especially $\phi_0 = 1, \mu^2 = \frac{\lambda}{2} = 0$, the gauge field A_μ without interaction. $E = V = 0$, on $\phi_0 = 1$ ring, Higgs particles degenerate into Stone particles. Goldstone particle are massless scalar field, remember $\phi_0 = \exp(-i_0)$, it is the scalar particles which mass and momentum are zero. The experiments can only be carried out on the basis of the average exctited starte. The various particles observed in the experiment are fields produced by vacuum excitation.

7.3.2 $SU(2) \times U(1)$ localized gauge transformation

Unified gauge field theory of weak used in the $SU(2) \times U(1)$ group, in which the weak isospin group is $SU(2)$, the weak hypercharge group is $U(1)$. The elements of group $U(1)$ are $\exp[-i\vartheta \frac{Y}{2}]$, the super-charged Y is the generator, there is a corresponding gauge fields $B_\mu(x)$. The elements of group $SU(2)$ are $\exp[-i\theta \cdot T]$, the gauge group generators $T(T^i, i = 1,2,3)$ constitute isospin space. The corresponding gauge field is $A_\mu(x)$, the components are $A_\mu^i(x), i = 1,2,3$. $SU(2) \times U(1)$ The group element is

$$U(\theta) = \exp[-i\theta \cdot T - i\vartheta \frac{Y}{2}] = \exp[-i\theta \cdot T] \exp[-i\vartheta \frac{Y}{2}] \qquad (7.3.12)$$

$\psi(x) \to \psi'(x) = U\psi(x)$, Covariant derivative

$$D_\mu \psi = \left(\partial_\mu + ig_t \frac{\sigma}{2} A_\mu + ig_y \frac{1}{2} B_\mu\right)\psi \qquad (7.3.13)$$

Gauge field tensor $F_{\mu\nu}^k = \partial_\mu A_\nu^k - \partial_\nu A_\mu^k - g\varepsilon^{ijk} A_\mu^i A_\nu^j$, and $B_{\mu\nu}(x) = \partial_\mu B_\nu - \partial_\nu B_\mu$. Gauge field A_μ^k with self-interaction, B_μ is no interaction.

Chapter 7 Interaction and Super-unified Equation

$$A_\mu^a \to A'^a_\mu = A_\mu^a + f^{abc}\theta^b(x)A_\mu^c + \frac{1}{g_t}\partial_\mu\theta^a(x), g_t\delta A_\mu(x) = \partial_\mu\theta^\alpha(x) \quad (7.3.14)$$

$$B_\mu \to B'_\mu = B_\mu + \frac{1}{g_y}\partial_\mu\theta(x), \; g_y\delta B_\mu(x) = \partial_\mu\theta(x) \quad (7.3.15)$$

(7.3.13) the physical picture is: gauge field A_μ acting on ψ and A_μ moving along $\frac{\sigma}{2}$ by the scale factor g_t, then the amount of movement $\theta = g_t\frac{\sigma}{2}A_\mu$, ψ change makes isospin space ψ change:

$$\begin{aligned}\psi = \exp[-ikx] \to \psi' &= U(\theta)\exp[-ikx] \\ &= \exp[-ikx + i\theta \cdot T] \\ &= \exp[-ikx + ig_t A_\mu \frac{\sigma}{2}]\end{aligned} \quad (7.3.16)$$

T^k is isospin space frame of particle space representation: $T^k = \frac{1}{2}\sigma^k$. σ^k is pauli matrices, contrast (7.3.16) and (7.3.13), define $\theta = g_t A_\mu$, the effect of making ψ' to obtain virtual rest mass, form is not breaking Higgs field $\phi_{0Higgs} = \sqrt{-\frac{\mu^2}{\lambda}}\exp\left[ig_t A_\mu\frac{\sigma}{2}\right]$. On the other hand, B_μ acting on ψ, resulting in the hypercharge space of ψ to change, B_μ according the scale factor g_y moves along the frame $Y/2$, $\theta(x) = g_y B_\mu$, i.e.

$$\psi = \exp[-ikx] \to \psi' = U(\theta)\exp[-ikx] = \exp[-ikx + ig_y\frac{Y}{2}B_\mu] \quad (7.3.17)$$

The generators determine the direction of breaking, making ψ' with virtual photon coupling. It is worth mentioning that only A_μ, B_μ combined effects, weak action of the ψ can be achieved, to obtain real mass and real photons. This is because only when exist group element of $SU(2)$, the vacuum is not broken, the virtual mass have no observability. To join the group $U(1)$ element, breaking to obtain a specific direction, making the vacuum occurrence breaking. The $SU(2) \times U(1)$ local gauge transformation make the ψ obtain mass.

From the above analysis, we get to this conclusion: the weak gauge groups elements $U(\theta)$ interact with ψ resulting in ψ obtain mass. In other words, $U(\theta)$ constitutes a special vacuum disturbance source, making changes in the vacuum structure of ψ obtain mass, this is the Higgs particle characteristic, so we think that $U(\theta)$ is the Higgs field directly.

7.3.3 Obtain mass by Higgs of breaking

Consider the Higgs field in the internal space have n real component (a complex component contains two real components). In order, $\phi(x) = \begin{vmatrix}\phi_1(x)\\ \phi_2(x)\\ \vdots\\ \phi_n(x)\end{vmatrix}$, expansion into a particle of space, $\phi_n(x)$ to express the n kinds of particles, then

$$\phi^+(x)\phi(x) = \phi(x)\phi(x) = \phi_1^2(x) + \phi_2^2(x) + \cdots + \phi_n^2(x)$$

Obviously, Higgs field rotation is constant in n dimension interior space. Its internal space has rotational symmetry, that is spherical symmetry. Higgs field on

$$\phi_0^+ \phi_0 = \phi_{01}^2 + \phi_{02}^2 + \cdots + \phi_{0n}^2 = \phi_0^2 = v \qquad (7.3.18)$$

Take the minimum energy state. In the internal space of n dimensional particles

$$\phi^+ \phi = \phi_1^2 + \phi_2^2 + \cdots + \phi_n^2 = \phi^2 = 1 \qquad (7.3.19)$$

$\phi_0 = 1$ is the radius of a sphere, there are mass of generators $\phi_{01}, \phi_{02}, \cdots, \phi_{0n}$, which take the vacuum value. If we take $\phi_1(x), \phi_2(x), \cdots, \phi_n(x)$ to become n dimensional space axes (as pictures 1), to take a vacuum in the ϕ_n an axis: $\phi_0 = \langle |\phi(x)|\rangle_0 = \begin{vmatrix} 0 \\ \vdots \\ 1 \end{vmatrix}$, obviously, such a vacuum will change under $SO(n)$ group rotation. Because only one direction, the breaking generator K_i can be projection onto the mass of the generator v. Radial quantum field excitation corresponds to a mass of particles, and the circle corresponding to the quantum excitation of zero mass particles (because the mass equal $V(\phi)$) in the direction of the second derivative. Consider the radial excitation corresponds to the amplitude can be written as $v + \rho(x)$, the circle quantum excitation corresponds to the rotation, to determine the breaking direction, expressed as, $\exp[i\theta T + i\vartheta Y/2]$ and thus the full expression for the Higgs field

$$\phi_{\text{Higgs}} = \left(\sqrt{-\frac{\mu^2}{\lambda}} + \rho(x)\right) \exp\left[ig_t A_\mu \frac{\sigma}{2} + i\vartheta Y/2\right] \qquad (7.3.20)$$

With the virtual mass, energy state is uncertain. $\phi_{\text{Higgs}} = \phi_{\text{0Higgs}} \exp[-i\vartheta Y/2]$ which along $Y/2$ is the direction of the breaking Higgs field, the convergent cylindrical wave tilt along the $Y/2$ direction will be down, when the convergent cylindrical wave falls down, the mass of ϕ_n is obtained which brings together the state of motion in the tilt cylinder wave form the breaking properties of Higgs particles. Thus the state of motion in the tilt of convergence wave to constitute the Higgs particles, which have the breaking characteristics. From the vertical to the fallen state of the Higgs particle is the lifetime. The cylindrical height (amplitude) of convergent wave determines the strength of the vibration point, which is a scalar field with energy and momentum, and did not form a physical particle, its features like virtual pair of electrons, hence have virtual mass. Convergence wave of Higgs field is form a vacuum disturbance source. Consider the symmetry breaking Higgs field, and Φ is zero mass scalar field, at $V(\phi) = 0$ another point, when $|\phi_0'| = 1, \mu^2 = \frac{\lambda}{2} = 0$, to take $|\Phi| = 1$, then the gauge field $U(1)$ coupled with Φ:

$$\mathcal{L}_{\phi g} = \frac{1}{2} D_\mu \Phi^+ D^\mu \Phi = \frac{1}{2} \left| 0 \quad \phi_{\text{0Higgs}}(x) \right| \left(\overleftarrow{\partial}_\mu + \frac{i}{2} g_y B_\mu\right) \cdot \left(\overrightarrow{\partial}_\mu - \frac{i}{2} g_y B_\mu\right) \left| \begin{array}{c} 0 \\ \phi_{\text{0Higgs}}(x) \end{array} \right|$$

$$= \frac{1}{2} \left| 0 \quad \sqrt{-\frac{\mu^2}{2\lambda}} + v \right| \left(\overleftarrow{\partial}_\mu + \frac{i}{2} g_t \sigma \cdot A_\mu + \frac{i}{2} g_y B_\mu\right) \cdot$$

Chapter 7   Interaction and Super-unified Equation

$$\left(\overleftarrow{\partial}_\mu - \frac{i}{2}g_t \sigma \cdot A_\mu - \frac{i}{2}g_y B_\mu\right) \begin{vmatrix} 0 \\ \sqrt{-\frac{\mu^2}{2\lambda}} + v \end{vmatrix}$$

$$= \frac{1}{2}\partial_\mu \phi \cdot \partial^\mu \phi - \frac{g_t^2}{8}\phi^2(A_\mu^1 - iA_\mu^2)(A_\mu^1 + iA_\mu^2) - \frac{1}{8}\phi^2(g_y B_\mu - g_t A_\mu^3)^2 \quad (7.3.21)$$

Here $\vartheta Y/2 = \frac{1}{2}g_y B_\mu$ to determine the breaking direction; $\phi = \begin{vmatrix} 0 \\ v + \rho(x) \end{vmatrix}$, as a $U(1)$ gauge transformation Higgs particle disappear after breaking, ϕ obtain mass, the mass is

$$M_{\text{Goldstone}} = \frac{g_t^2}{8}(A_\mu^1 - iA_\mu^2)(A_\mu^1 + iA_\mu^2) - \frac{1}{8}(g_y B_\mu - g_t A_\mu^3)^2 \quad (7.3.22)$$

From the experimental observation of the point of view, mass is relative to the vacuum state, at $V(\phi) = 0$ (Fig. 7.3.2).

$$\phi' = v + \rho(x) = \sqrt{-2\mu^2/\lambda}$$

This field in the vacuum ground state, we can understand the negative energy state ϕ_0 radial excitation into 0-energy particle, which in vacuum ground state. The particles in the vacuum ground state are Goldstone particles. The particle is very unstable, the gauge field will be "eaten", with the gauge field to obtain the mass of observable effects, A_μ and B_μ recombined to form real particles. To introduce the complex field

$$W_\mu^- = \frac{1}{\sqrt{2}}(A_\mu^1 + iA_\mu^2), W_\mu^+ = \frac{1}{\sqrt{2}}(A_\mu^1 - iA_\mu^2), Z_\mu = \frac{1}{\sqrt{g_t^2 + g_y^2}}(g_t A_\mu^3 - g_y B_\mu)$$

$$(7.3.23)$$

Each generator corresponds to a standard particle, which has three standard particles $W^\pm$, Z^0. (7.3.20) expression gives the gauge field and the composition of the relationship between observed particles $W_\mu^\pm$ and Z_μ. Take them into (7.3.19), we get a very simple expression:

$$\mathscr{L}_{\phi g} = -\frac{g_t^2}{4}v^2 W_\mu^+ W^{-\mu} - \frac{g_t^2 + g_y^2}{8}v^2 Z_\mu Z^\mu \quad (7.3.24)$$

It can be seen: W_μ^+, W_μ^-, Z_μ are the eigenstates of mass field, the corresponding mass of the particle is

$$m_W = \frac{1}{2}g_t v, \quad m_Z = \frac{1}{2}\sqrt{g_t^2 + g_y^2}\, v \quad (7.3.25)$$

Three of the four gauge particles have mass. W_μ^+, W_μ^-, Z_μ is a linear combination by A_μ, B_μ. There is a zero mass particle is the photon, A_μ composition by A_μ, B_μ, its relationship is

$$A_\mu = A_\mu^3 \sin\theta_W + B_\mu \cos\theta_W \quad (7.3.26)$$

This is the photon field. By (7.3.22) we can see, Higgs particles combination by the gauge field (A_μ, B_μ), which means that Higgs particles resulting in the mass of other particles, the gauge field must exist (A_μ, B_μ) background.

Here Goldstone particle has been "eaten" by Higgs, to obtain virtual mass, Goldstone particles disappear. Virtual mass Higgs has been "eaten" by recombination of the gauge field, by recombination of mass $W^{\pm}, Z^0$ and massless gauge fields A_μ. Gauge field component A^i_μ, B_μ can't appear alone. Goldstone particles and the gauge field A^i_μ, B_μ does not appear, which is what we want to get results.

7.3.4 Lepton field of $SU(2) \times U(1)$ localized gauge transformation

Lepton fields are $L = \begin{vmatrix} \nu_L \\ l_L \end{vmatrix}, l_R$, are the $SU(2)$ doublet and singlet state. Generators $T = \dfrac{\sigma}{2}$ and $T = 0, \dfrac{Y}{2} = -\dfrac{1}{2}$ and $\dfrac{Y}{2} = -1$. Lepton fields under $SU(2) \times U(1)$ localized gauge transformation

$$\mathcal{L}_{lg} = \bar{L}(i\gamma^\mu D_\mu)L + \bar{l}_R(i\gamma^\mu D'_\mu)l_R = \bar{L}\left[i\gamma^\mu(\partial_\mu - ig_t\dfrac{\sigma}{2}A + g_y\dfrac{1}{2}B)\right]L - \bar{L}[i\gamma^\mu(\partial_\mu + g_y B)]L \tag{7.3.27}$$

(7.3.27) instructions the lepton interaction with gauge field, then provides a gauge field (A_μ, B_μ) background, only in this background space, Higgs field is possible. the first of (7.3.27) give the conditions of Higgs particle generation, and thus the first of the lepton mass can be obtained; the second Higgs particles have not created the conditions, the Higgs particles with lepton does not exist interaction, naturally have not rest mass.

$$L(x) \to L'(x) = e^{-i\theta \cdot M} e^{-i\theta \cdot x} L(x) \ ; \ l_R(x) \to l'_R(x) = e^{-i\theta \cdot x} l_R(x) \tag{7.3.28}$$

Lepton field rest mass is zero, Higgs field is written as $\Phi_H = \begin{vmatrix} 0 \\ U(\theta(x)) \end{vmatrix}$, where the Higgs field has advanced in the gauge field (A_μ, B_μ) background space to obtain a virtual mass, and lepton interaction with Higgs field:

$$\mathcal{L}_{l\phi} = -G_l(\bar{L}\phi l_R + \bar{l}_R \phi \bar{L}) = -G_l\left(\begin{vmatrix} \bar{\nu}_L & \bar{l}_L \end{vmatrix} \begin{vmatrix} \phi^+ \\ \phi^0 \end{vmatrix} l_R + \bar{l}_R \begin{vmatrix} \phi^+ & \phi^0 \end{vmatrix} \begin{vmatrix} \nu_L \\ l_L \end{vmatrix}\right)$$

$$\mathcal{L}_{l\phi} = -G_l\left(\begin{vmatrix} \bar{\nu}_L & \bar{l}_L \end{vmatrix} \begin{vmatrix} 0 \\ U(\theta(x)) \end{vmatrix} l_R + \bar{l}_R \begin{vmatrix} 0 & U(\theta(x)) \end{vmatrix} \begin{vmatrix} \nu_L \\ l_L \end{vmatrix}\right) = -G_l U(\theta(x)) \bar{l}l \tag{7.3.29}$$

Get the mass of lepton

$$m_l = G_l U(\theta(x)) \tag{7.3.30}$$

$U(\theta(x))$ is not a constant, determined by the gauge field strength, which makes the existence of different generation of lepton. Mass of change making 4-momentum space changes, L lepton field to obtain mass and momentum, right-handed lepton fields l_R obtain momentum. Higgs particle is convergence wave in the breaking, constitute the cylinder of convergence wave, cylinder diameter to meet

$$m_e c \cdot \phi_e = m_\mu c \cdot \phi_\mu = m_\tau c \cdot \phi_\tau = 2\hbar \tag{7.3.31}$$

Lepton mass derived from 3-D spherically symmetric strain of field, deformation of field is

Chapter 7 Interaction and Super-unified Equation

$e\mathcal{H}/\Phi$. Neutrino 2-dimensional structure makes the neutrino mass is 0, so the neutrino and leptons in mass space projection is

$$\begin{pmatrix} \text{Neutrino} \\ \text{Lepton} \end{pmatrix} \xrightarrow{\text{projection in rest mass space}} \begin{pmatrix} 0 \\ v \end{pmatrix} = \begin{pmatrix} 0 \\ e\mathcal{H}/\Phi \end{pmatrix} \quad (7.3.32)$$

Here Φ is the lepton intrinsic space diameter.

7.3.5 Higgs particle mass

The Higgs particle is a source of disturbance, the convergent wave in a vacuum is no charge before rupture (Fig. 7.3.1). The convergent wave is Z^0 without rupture, the disturbance source is pushed down along the $U(1)$ and the remain is Z^0 particles, therefore, the simplest way to infer the mass of Higgs particles is Z^0 particles plus energy along the $U(1)$.

$M_{\text{Goldstone}} = \frac{g_t^2}{8}(A_\mu^1 - iA_\mu^2)(A_\mu^1 + iA_\mu^2) - \frac{1}{8}(g_y B_\mu - g_t A_\mu^3)^2$ can be restated as $M_{\text{Goldstone}} = g^2 \Phi(A_\mu^1, A_\mu^2, A_\mu^3, B_\mu)$ in space of $SU(2) \times U(1)$, where g is the coupling strength in the $SU(2) \times U(1)$ weak interaction space, $\Phi(A_\mu^1, A_\mu^2, A_\mu^3, B_\mu)$ is the source of disturbance in the space before the rupture, its projection in $SU(2)$ space is called Z^0 particle, the coupling strength is g. The projection in $U(1)$ space is the γ particle, the coupling strength is g_y, and the following expression is thus obtained

$$M_{Z^0} = (g\cos\theta_w)^2 \Phi(A_\mu^1, A_\mu^2, A_\mu^3, B_\mu) \quad \text{in } SU(2) \text{ space} \quad (7.3.33)$$

$$M_\gamma = (g\sin\theta_w)^2 \Phi(A_\mu^1, A_\mu^2, A_\mu^3, B_\mu) \quad \text{in } U(1) \text{ space} \quad (7.3.34)$$

(7.3.33) can be obtained

$$M_{\text{Goldstone}} = \frac{Z^0}{\cos^2\theta_w} \quad (7.3.35)$$

According to $M_{Z^0} = (91.187 \pm 0.007)$ GeV, $\theta_W = 28.47° - 28.16°$, we can calculate $M_{\text{Goldstone}} \approx 118.0$ GeV, the actual measurement value is $M_{\text{Goldstone(Experimental value)}} = 125$ GeV, there is a 5.6% error.

As the rest mass of charged particles cannot be zero. Higgs particle is the disturbance source, which has no rest mass. The disturbance source $U(x) \leq \phi_{0\max}$, that is the quantum fluctuations in part of the bowl produce virtual mass, at $U(x) = \phi_{0\max}$, degenerate into Goldstone particles, $U(x) > \phi_{0\max}$. Goldstone particles are excited into Higgs particles, at $\phi_{0\max} > 1$, vacuum breaking, Higgs particles obtained the mass in the gauge field background, the mass will be turned into mass bosons $W^\pm, Z^0$ particles.

Before breaking, Higgs particles have the most typical features: ①Without charge; ②spin 0; ③The rest mass >118 GeV. In short, the Goldstone particle is a basic vacuum unit h_f. The maximum spherically symmetric strain that it can bear is $M_{\text{Higgs}} C$, beyond which the vacuum breaks and produces fermions with rest mass.

7.4 Calculation of Weinberg angle θ_w

7.4.1 The physical image of the Weinberg angle θ_w

In $SU(2) \times U(1)$ electro weak unified theory $W^\pm$, Z^0 and γ in the same position they are quantum of gauge field, see in Table 7.4.1. A_μ^1, A_μ^2, A_μ^3 and B_μ (can be recombined into $W^\pm$, Z^0 and γ), and electro weak unified coupling constant are g_y and g_t. The weak interation process is the separation of spherical shear strain $[\gamma_{\mu\nu}]$ region.

Coupling constants g_y and g_t in Table 7.4.1 (cannot be measured directly) their relationship is expressed as following Fig. 7.4.1. g project to two-axis y and t_3 (that is σ_3) respectively are g_y and g_t.

Table 7.4.1 Electro weak unified theory's relevant parameters and description

Electro weak unified $SU(2) \times U(1)$	Coupling constants	Field's current	transfer quantum	
	g_y	B_μ	B_μ	γ
	g_t	$A_\mu^i(x)$	$A_\mu^1, A_\mu^2, A_\mu^3$	$W^\pm, Z^0$

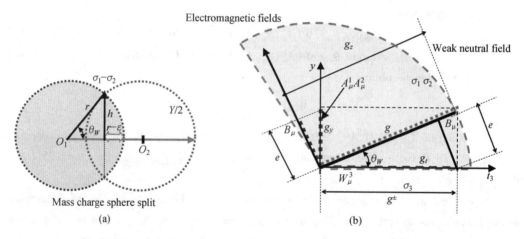

Fig. 7.4.1 The relationship between weak coupling constant e, g and g_y, g_t

$$g_t = g\cos\theta_W \quad (7.4.1)$$
$$g_y = g\sin\theta_W \quad (7.4.2)$$
$$g_y/g_t = \tan\theta_W \quad (7.4.3)$$

Take the weak neutral coupling constant $g_z = g$, the coupling constant of the weak charged $g^\pm = g_t$, a group axis of electromagnetic field and the weak neutral field and y-axis, σ_3-axis their relative rotation angle is θ_W (Fig. 7.4.1). The secondary projection of g can be written as

$$g\sin\theta_W\cos\theta_W = e \quad (7.4.4)$$
$$g_t = e/\sin\theta_W = g^\pm \quad (7.4.5)$$

Chapter 7 Interaction and Super-unified Equation

$$g_y = e/\cos\theta_W = g' \tag{7.4.6}$$

When $\vartheta_W = 90°$ (ϑ_W denotes the variable Weinberg angle), the particle is not split (Fig. 7.4.2 (a)), and this is the starting point. Consider the particles split along the direction g_t, $g_t = e/\sin\vartheta_W$, $g_t = e$. The coupling constant of single lepton along the split direction is e reduced to pure electromagnetic interaction. That is showed that the coupling strength of single lepton split along the vertical direction is $g_y = e/\cos 90°$, namely $g_y = \infty$, particles along the vertical direction of the split (regarded as the center of the two-particle coincidence, the coupling strength is ∞). $\vartheta_W = 0°$ shows that the particle has completed the split. Weak interaction is end. $g_t = e/\sin 0°$, $g_t = \infty$ shows that particles coupling strength in the split direction is ∞, no split. For free particles $g_y = e/\cos 0°$, the weak interaction coupling constant degradation pure electromagnetic coupling constant e. $\vartheta_W = 90°$ and $\vartheta_W = 0°$ are both in a state of free particles.

θ_W is a parameters which contact weak and electromagnetic interaction coupling constant (e, g). In addition, the relationship between the weak unified field W_μ^3, B_μ with weak neutral field Z_μ^0 and electromagnetic field A_μ indicates by Fig. 7.4.2.

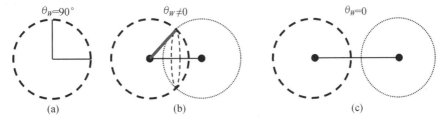

Fig. 7.4.2 Weak unified action, the coupling constant relationship between e and g_t

Still taken y, σ_3 as the two axises of the weak unified field (y corresponds to the projection of $\sigma_1\sigma_2$ 2-dimensional surface on the y shaft, and the σ_3 corresponding shaft x, and weak interaction field having a 3-dimensional structure), while the electromagnetic field and weak neutral field taken two mutually perpendicular axis. There is a rotation angle θ_W between two groups of axis, the projection of A_μ^3 and B_μ on the axis of electromagnetic field and weak neutral field are

$$A_\mu = B_\mu\cos\theta_W + A_\mu^3\sin\theta_W \tag{7.4.7}$$

$$Z_\mu^0 = -B_\mu\sin\theta_W + A_\mu^3\cos\theta_W \tag{7.4.8}$$

In weak unified theory $SU(2)\times U(1)$, electromagnetic A_μ and weak neutral field Z_μ^0 is not the basic field that is mixture field by the basic field W_μ^3 and B_μ. Mixed mode and A_μ, Z_μ^0 are decided by θ_W.

7.4.2 Correspondence between particles space the weak interaction intrinsic space

In Table 7.4.2, the σ form a 3-dimensional particle space, and the base of coordinates e_i form lepton intrinsic space. Field strain $\varepsilon_{xx}, \varepsilon_{yy}, \varepsilon_{zz}$ along e_x, e_y, e_z constituting the intrinsic leptons space of weak force field.

The basic conditions of mass generation is exists spherical strain $\varepsilon_{xx} = \varepsilon_{yy} = \varepsilon_{zz}$. There are only three dimensions inward (or outward) to form spherical strain can be production the mass of leptons. If only 2-dimensional strain $\varepsilon_{xx} = \varepsilon_{yy}$ then formation of the neutrino lepton. If only exists 1-dimensional strain, the quantum wave is formed in which photons.

In the interaction (field splitting), A_μ^i is the bosons which transmission interacting, because A_μ^i only exists in the internal the particle space, we can understand the particles A_μ^i as a v virtual particle. We can use of the corresponding between the intrinsic space and particle space, the intrinsic space into particles space.

Table 7.4.2 Particles space and lepton intrinsic space corresponding

		SU(2)	U(1)
Particles space	Generators	$\sigma_1, \sigma_2, \sigma_3$	$Y/2$
	Gauge field	$A_\mu^1(x), A_\mu^2(x), A_\mu^3(x)$	$B_\mu(x)$
Lepton intrinsic space		3-dimensional lepton coordinate space	1-dimensional coordinate space
	Coordinate base	e_x, e_y, e_z	x
	Field strain	$\varepsilon_{xx}, \varepsilon_{yy}, \varepsilon_{zz}$	ε_x

Particle space $B_\mu(x)$ corresponding ε_x, we should know that particle $B_\mu(x)$ field deformation in only one direction which is photon of no rest mass. ε_{ii} corresponds to A_μ^i a single A_μ^i in intrinsic field can only constitute a deformation in one direction, such particles cannot generate a rest mass, only when the three directions $A_\mu^1, A_\mu^2, A_\mu^3$ at the same time spread to the center of the sphere in intrinsic space can be formation of a rest mass of the lepton.

1. The nature of coupling constants g and θ_W

With the above image, we can calculate θ_W. Weak force field is the front end of the electric flux line, and thus the coupling constant of electromagnetic g_e and coupling constant of weak interaction g have same essence. Their performance are different that is due to: first, the space of effect are different, coupling constant of weak interaction g is the coupling constant of electromagnetic g_e which within the division region of particle center (average); and second, electromagnetic interaction in normal space, and the weak interaction exists only in half-space.

The fiber field is divided into two parts: part of the spherically symmetric of tension (or compression) g_t to the center and part of the deformation of non-spherically symmetric g_y. From a point of view of weakly interacting boson W transfer, there are two types of field fiber channel, spherically symmetric and non-spherical symmetry. After decomposition, we can be regard g as similar to a vector, composition by g_y and g_t (Fig. 7.4.1). First, let's explain why g_y perpendicular to g_t. g_t is the part of spherically symmetric stretched (or compressed) $g_t = g_{tx} + g_{ty} + g_{tz}$. g_y is a portion, which along the weak interaction bosons transmission direction, g_y is given y-axial direction, g_y and g_t are independent of each other, namely $g_y \perp g_t$.

2. To work out Weinberg angle

Consider the Fig. 7.4.2 center area of the weak interaction. Imagine a very small sphere

Chapter 7 Interaction and Super-unified Equation

split into two spheres. The distance d between the centers of the sphere is maximum range of the weak interaction when two spherical shells tangent. Sphere surface area is $4\pi r^2$. g_y is the intersection area of two spheres. In other words, two spheres beginning division, g_y is the maximum, and after the end of the split, which in a low-energy polarized state, g_y is a minimum. The interaction surface of two spheres intersect perpendicular to the interaction direction σ_3, the greater the intersecting interaction surface, the greater the probability along g_y coupled, i.e. $g_y \propto$ intersect interaction surface.

$$g_y \propto 2\int_{-\vartheta}^{\vartheta} r\cos\theta d\theta \int_{-r}^{r} dr \tag{7.4.9}$$

Multiplied by 2 is because there are two effects field overlap.

g_t is corresponding to the center of the sphere surface. The larger the surface area is spherically symmetric field fiber quantity is more. W boson along the ball the symmetric field fiber transmission has higher the probability, i.e. $g_t \propto$ the surface area of the center ball.

$$g_t \propto \int_{-\pi}^{\pi} r d\theta \int_{-r}^{r} dr \tag{7.4.10}$$

May be learned by Formula (7.4.9) and (7.4.10)

$$\frac{g_y}{g_t} = \frac{2\int_{-\theta}^{\theta} r\cos\vartheta d\vartheta \int_{-r}^{r} dr}{\int_{-\pi}^{\pi} r d\theta \int_{-r}^{r} dr} = \frac{2\int_{0}^{\theta} \cos\theta d\theta}{\pi} \tag{7.4.11}$$

Before splitting (Fig. 7.4.2), at the moment $\varepsilon = r$, O_1 and O_2 completely overlap, when completely split at $\varepsilon = 0$ the end of the weak interaction. The distance between O_1 and O_2 is $2r$ that is weak interaction process. The process average value g_y/g_t of completely overlapping to the complete separation it is corresponding ϑ_W, then can be obtained value ϑ_W in this process.

$$h = \sqrt{r^2 - (r-\varepsilon)^2} = \sqrt{2r\varepsilon - \varepsilon^2}$$

The two sphere intersection area:

$$S = \pi h^2 = \pi(2r\varepsilon - \varepsilon^2) \tag{7.4.12}$$

Here, ε is the distance which from the intersect planes to sphere center O_1, ε is the distance between the intersecting planes along the $Y/2$ axis to the spherical O_1 surface. When O_1 and O_2 completely overlaps ($\varepsilon = r$) becomes completely split ($\varepsilon = 0$), the average value of the intersecting surface of two mass-charged spere is

$$S = \frac{\int_0^r \pi(2r\varepsilon - \varepsilon^2)d r}{\int_0^r dr} = 0.667\pi r^2 \tag{7.4.13}$$

By $S = \pi(2r\bar{\varepsilon} - \bar{\varepsilon}^2)$:

$$\bar{\varepsilon}^2 - 2\bar{\varepsilon}r + 0.667r^2 = 0$$

The solution was $\bar{\varepsilon}_1 = 0.423r, \bar{\varepsilon}_2 = 1.155r$ (because of greater than r, unreasonable rounding) $\theta = \arccos\left(\frac{r-\bar{\varepsilon}}{r}\right) = \arccos(1 - 0.423) = 54.76°$, by $g_y/g_t = \tan\theta_W$ and (7.4.11):

$$\theta_W = \arctan\left[\frac{2\int_0^\theta \cos\theta d\theta}{\pi}\right] = \arctan\left(\frac{2\sin 54.76°}{\pi}\right) = 27.47° \qquad (7.4.14)$$

This value has nothing to do with r. r is a variable in a weak interaction in the process. Here θ_W is Weinberg angle. Theoretical calculations: $\sin^2\theta_W = 0.213$ which is close to the experimental value $\sin^2\theta_W = 0.2292 \pm 0.0013$.

7.4.3 The vertex of fermions interaction

1. Fermi V-A theory

Fermi proposed constant amplitude $\mu = G(\bar{u}_n\gamma^\mu u_p)(u_{\nu_e}\gamma_\mu u_e)$ of β decay which inspired by the current structure of the electromagnetic interaction. Particle interaction vertex γ^μ is symmetric space vertex and Parity conservation. That the $V-A$ theory the four-fermions $\psi_a, \psi_b, \psi_c, \psi_d$ interaction Lagrangian.

$$\mathscr{L}_w \sim \bar{\psi}_a(x)\Gamma\psi_b(x) \cdot \bar{\psi}_c(x)\Gamma'\psi_d(x)$$

That is the form of direct coupling, known as the four fermions direct coupling theory. Here matrix Γ and matrix Γ' respectively constitutes the vertex space of the interaction fermions: ψ_a, ψ_b and ψ_c, ψ_d.

In 1956, Yang Zhenning and Li Zhendao proposed the weak interactions in parity is not conservation. Shortly, Wu Jianxiong use the ingenious experiments confirm their theory. Series of experiments found that β decay destruction parity showed that interaction vertex γ_μ necessary to make such a replacement $\gamma_\mu \to \gamma_\mu(1+\gamma_\mu)$. Now, we know that this is caused by the properties of half-space. γ_μ corresponding to representations of the symmetric space, i.e. parity conservation. Particle interaction vertex $\gamma^\mu(1+\gamma^5)$ is the particle space representations of the half-space, non-conservation of parity. After substitution, proton p and lepton μ weak decay constant amplitude.

$$\mu(p \to ne^+\nu_e) = \frac{G}{\sqrt{2}}[\bar{u}_n\underbrace{\gamma_\mu(1+\gamma_5)u_p}_{\text{Half-space vertex}}][\bar{u}_{\nu_e}\underbrace{\gamma_\mu(1+\gamma_5)u_e}_{\text{Half-space vertex}}] \qquad (7.4.15)$$

$$\mu(\mu^- \to e^-\bar{\nu}_e\nu_\mu) = \frac{G}{\sqrt{2}}[\bar{\psi}_e\underbrace{\gamma_\mu(1+\gamma_5)\psi_{\nu_e}}_{\text{Half-space vertex}}][\bar{\psi}_{\bar{\nu}_\mu}\underbrace{\gamma_\mu(1+\gamma_5)\psi_\mu}_{\text{Half-space vertex}}] \qquad (7.4.16)$$

The factor $1/\sqrt{2}$ in formula is the convention. After comprehensive experimental results obtained

$$-\mathscr{L}_w = \frac{G}{\sqrt{2}}J_\mu^+ J^\mu$$

$$J_\mu = \bar{\nu}\gamma_\mu(1+\gamma_5)l + g_0\bar{u}\gamma_\mu(1+\gamma_5)d + g_1\bar{u}\gamma_\mu(1+\gamma_5)s \qquad (7.4.17)$$

Here, using particle symbol such as leptons l, ν and quarks u, d, s to express corresponding the particles field, such as $\psi_l(x), \psi_\nu(x), \psi_u(x), \psi_d(x), \psi_s(x)$. The above formula called Fermi suitable V-A theory. It is calculated the process $\mu^- \to e^- + \bar{\nu}_\mu + \nu_\mu$ by $\frac{G}{\sqrt{2}}\bar{e}\gamma_\mu(1+\gamma_5)\nu_e$.

$\bar{\nu}_\mu \gamma_\mu (1+\gamma_5) \mu$ in formula (7.4.15). Comparison the calculation results and experimental, obtain the weak interaction coupling constant

$$G = 1.01 \times 10^{-5}/m_p^2 \approx 0.71 \times 10^{-5} \qquad (7.4.18)$$

Among them m_p is the proton mass.

Fermi theory successfully explained some phenomena of the low-energy weak interaction, but at high energies ($\geqslant 300$ GeV) would undermine the unitary limit theorem. Taking into account the similarity of the electromagnetic interactions to envisage weak interactions through the exchange spin 1 charged vector weak bosons $W^\pm$. Weak bosons are similar to the photon in electromagnetic action or gluon in color action. Decay of μ^- is achieved by exchanging the bosons W^- (Fig. 7.4.3), the form of the amplitude is

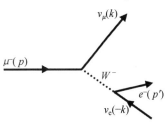

Fig. 7.4.3 μ^- Decay

$$\mu = \left[\frac{g}{\sqrt{2}} u_{\nu_\mu} \gamma^\sigma \frac{1}{2}(1-\gamma^5) u_\mu\right] \frac{1}{M_M^2 - q^2} \left[\frac{g}{\sqrt{2}} \bar{u}_e \gamma_\sigma \frac{1}{2}(1-\gamma^5) u_{\nu e}\right] \qquad (7.4.19)$$

The above equation has been summation for spin of weak Bose propagator. Here $g/\sqrt{2}$ is a dimensionless coupling constant, Weak bosons carry the momentum q, factor $1/\sqrt{2}$ and $1/2$ to comply with the contract definition. Contrast with the photon, weak boson must have mass, experimentally confirmed = (80.33 ± 0.15) GeV. As shown in Fig. 7.4.4, four fermions point interaction coupling constant G and weak interactions coupling constants g, which between the intermediate boson ($W^\pm$) and lepton that is the description of the two constants in the same weak process. Consider (7.4.15) or (7.4.17) type, in the low momentum exchange ($q^2 \to 0$), should be

$$\lim_{q^2 \to 0} \frac{g'^2}{q^2 + M_W^2} \equiv \frac{G}{\sqrt{2}} \qquad (7.4.20)$$

$$M_W^2 = \frac{\sqrt{2} g'^2}{G} \qquad (7.4.21)$$

Weak field and electromagnetic interaction field with the fiber structure. Weak and electromagnetic effects are the fiber field in different regions, and thus the size of the coupling constant g' of the weak interaction and electromagnetic coupling constants e should be equal, by the (7.4.20) can estimate the mass of the weak interaction intermediate boson.

$$M_W^2 = \frac{\sqrt{2} e^2}{G} \qquad (7.4.22)$$

Here, $e^2 = 4\pi/137$, $G = 10^{-5}/M_{p2}$ (natural units $= c = 1$) and substituting into the above equation, we can estimate $M_W \approx 90$ GeV this value is consistent with the experimental value. If $q^2 \ll M_{W2}$ (such as μ-decay), formula (7.4.19) return the formula (7.4.17) and have

$$\frac{G}{\sqrt{2}} = \frac{g^2}{8M_W^2} \qquad (7.4.23)$$

In this limit, the propagator disappears between the weak current, weak current interact approximation at one point (note $g^2 = 8g'^2$). Why the weak interaction is weak $g \ll e$, because of M_{W2} is great. If g' with e are approximation and the energy reaches the M_W or higher, the strength of the weak interaction will become quite same as the strength of electromagnetic interaction. $g' \approx e$ can be considered the unified of weak and electromagnetic interaction.

Weak-interaction carrier particles (W and Z bosons) have large masses, so their lifetime is very short: the average lifetime is about 3×10^{-25} s. The coupling constants of the weak interaction are between 10^{-7} and 10^{-6}, whereas the coupling constants of the strong interaction are about 1, and the weak interaction is weak. The action distance of weak phase is very short (about 10^{-17}—10^{-16} m). At a distance of about 10^{-18} m, the strength of weak interaction is approximately the same as that of electromagnetic field, but at a distance of about 3×10^{-17} m, the strength of weak interaction is 10,000 times weaker than that of electromagnetic field.

2. The vertex of incomplete half-space

Protons p composed by uud quarks and neutrons n composed by udd quarks, in (7.4.19):

$$\bar{u}\gamma_\mu(1 + \gamma_5)d \sim \bar{p}\gamma_\mu(1 + a\gamma_5)n \qquad (7.4.24)$$

Due to quark compose hadrons, the vector shape factor and axial vector shape factor is not the same, therefore $\alpha \neq 1$, and deviate from the V–A form, which indicates that the vertex is not completely half space. People use

$$\frac{G}{\sqrt{2}}\bar{e}\gamma_\mu(1 + \gamma_5)\nu_e \cdot g_0\bar{p}\gamma^\mu\gamma_\mu(1 + \alpha\gamma_5)n \qquad (7.4.25)$$

To calculate the β decay process: $n \to p + e^- + \bar{\nu}_e$, the calculated results and experimental comparison identified

$$g_0 \leqslant 1, \alpha \approx 1.2 \qquad (7.4.26)$$

3. Coupling constant of Quark weak interaction

Hyperons Λ are composed by u, d and s quarks, the proton are composed by uud quarks in the formula (7.4.24).

$$\bar{u}\gamma_\mu(1 + \gamma_5)s \sim \bar{p}\gamma_\mu(1 + \alpha\gamma_5)\Lambda$$

People use

$$\frac{G}{\sqrt{2}}\bar{e}\gamma_\mu(1 + \gamma_5)\nu_e \cdot g_1\bar{p}\gamma^\mu(1 + \alpha\gamma_5)\Lambda$$

To calculate the process: $\Lambda \to p + e^- + \bar{\nu}_e$, the calculation results and experimental comparison can be obtained

$$g_1 \sim 10^{-2} \qquad (7.4.27)$$

4. Cablbbo angle

Considering the hadrons quark structure, people can use universal Fermi theory of V–A (7.4.24) to explain the weak interaction experience rule. However, represented the g_0, g_1 by (7.4.26), (7.4.27) is not very clear. In 1963, after to analysis and comprehensive experimental results, Cablbbo point out

Chapter 7 Interaction and Super-unified Equation

$$g_0^2 + g_1^2 \approx 1 \qquad (7.4.28)$$

In addition, the introduction of Cablbbo angle θ_c, make

$$g_0 = \cos\theta_c, \quad g_1 = \sin\theta_c \qquad (7.4.29)$$

So, make two undetermined constants g_0, g_1 reduced to a certain constant θ_c.

Use (7.4.29), (7.4.24) can be written as

$$-\mathscr{L}_w = \frac{G}{\sqrt{2}} J_\mu^+ J^\mu$$

$$J_\mu = \bar{\nu}\gamma_\mu(1+\gamma_5)l + \bar{u}\gamma_\mu(1+\gamma_5)d_\theta \qquad (7.4.30)$$

u, d, s are the quark mass Eigen states; d_θ is quark weak Eigen states.

5. Left hand current

Weak interaction flow represented by the formula (7.4.30) is V–A form that very plainly expressed vertex $\gamma_\mu(1+\gamma_5)$ in half-space, γ_μ is the vertex in ordinary space. In order to make the theory of representations having the gauge symmetry, here to rewrite it into the left hand vector current form. According to the single power of chiral operator and the anticommutation of γ_5 with γ_μ, there is

$$\gamma_0\gamma_\mu \frac{1+\gamma_5}{2} = \gamma_0\gamma_\mu \frac{1+\gamma_5}{2} \cdot \frac{1+\gamma_5}{2} = \frac{1+\gamma_5}{2}\gamma_0\gamma_\mu \frac{1+\gamma_5}{2}$$

And because

$$\psi_L = \frac{1+\gamma_5}{2}\psi, \quad \overline{\psi}_L = \psi_L^+ \gamma_0 = \psi^+ \frac{1+\gamma_5}{2}\gamma_0$$

Therefore, the formula (7.4.30) can be rewritten as

$$J_\mu = 2J_{l\mu}$$
$$J_{l\mu} = \bar{\nu}_L\gamma_\mu l_L + \bar{u}_L\gamma_\mu d_{L\theta} \qquad (7.4.31)$$
$$d_{L\theta} = d_L\cos\theta_c + s_L\sin\theta_c$$

7.4.4 Weak action

1. Self-acting

Gauge field exist the self-acting, they are

$$\mathscr{L}_{si} = -g\partial_\mu A_\nu \cdot A^\mu \times A^\nu - \frac{g^2}{4}(A_\mu \times A_\nu)\cdot(A^\mu \times A^\nu) \qquad (7.4.32)$$

2. The interaction

Because the Higgs field does not exist, therefore only need to consider the interaction of leptons and gauge field

$$\mathscr{L}_{li} = \bar{L}\left(\frac{g_t}{2}\sigma\cdot\gamma_\mu A - \frac{g_y}{2}B\right)L - g_y\bar{l}_R\gamma_\mu B l_R$$

$$= \begin{vmatrix}\bar{\nu}_L & \bar{l}_L\end{vmatrix}\gamma_\mu\left(\frac{g_t}{2}\begin{vmatrix} A_\mu^3 & A_\mu^1 - iA_\mu^2 \\ A_\mu^1 + iA_\mu^2 & -A_\mu^3 \end{vmatrix} - \frac{g_y}{2}\gamma_\mu B\right)\begin{vmatrix}\nu_L \\ l_L\end{vmatrix} - g_y\bar{l}_R\gamma_\mu l_R B_\mu$$

$$= |\bar{\nu}_L \quad l_L| \frac{\gamma_\mu}{2} \begin{vmatrix} g_t A_\mu^3 - g_y B_\mu & g_t(A_\mu^1 - iA_\mu^2) \\ g_t(A_\mu^1 + iA_\mu^2) & -g_t A_\mu^3 - g_y B_\mu \end{vmatrix} \begin{vmatrix} \nu_L \\ l_L \end{vmatrix} - g_y l_R \gamma_\mu l_R B_\mu$$

$$= |\bar{\nu}_L \quad l_L| \frac{\gamma_\mu}{2} \begin{vmatrix} \sqrt{g_t^2 - g_y^2} Z_\mu & \sqrt{2} g_t W_\mu^+ \\ \sqrt{2} g W_\mu^- & -\frac{2g_t g_y}{\sqrt{g_t^2 + g_y^2}} A_\mu + \frac{g_y^2 - g_t^2}{\sqrt{g_t^2 + g_y^2}} Z_\mu \end{vmatrix} \begin{vmatrix} \nu_L \\ l_L \end{vmatrix} -$$

$$l_R \gamma_\mu l_R \left(\frac{g_t g_y}{\sqrt{g_t^2 + g_y^2}} A_\mu - \frac{g_y^2}{\sqrt{g_t^2 + g_y^2}} Z_\mu \right) \tag{7.4.33}$$

Use of the relationship between A_μ, B_μ and $W_\mu^\pm$, Z_μ, A_μ. Multiply the above matrix to get

$$\mathscr{L}_{li} = \frac{g_t}{\sqrt{2}} (W_\mu^+ \bar{\nu}_L \gamma^\mu l_L - W_\mu^- \bar{l}_L \gamma^\mu \nu_L) +$$

$$\sqrt{g_t^2 + g_y^2} Z_\mu \left(\frac{\bar{\nu}_L \gamma^\mu \nu_L - \bar{l}_L \gamma^\mu l_L}{2} + \sin^2 \theta_W \bar{l} \gamma^\mu l \right) -$$

$$\frac{g_t g_y}{\sqrt{g_t^2 + g_y^2}} A_\mu \bar{l} \gamma^\mu l \tag{7.4.34}$$

Back to the eigenstates of lepton mass, the above equation becomes

$$\mathscr{L}_{li} = \frac{g_t}{2\sqrt{2}} [W_\mu^+ \bar{\nu} \gamma^\mu (1 + \gamma_5) l + W_\mu^- \bar{l} \gamma^\mu (1 + \gamma_5) \nu] +$$

$$\frac{\sqrt{g_t^2 + g_y^2}}{4} Z_\mu [\bar{\nu} \gamma^\mu (1 + \gamma_5) \nu + \bar{l} \gamma^\mu (4 \sin^2 \theta_W - 1 - \gamma_5) l] - \frac{g_t g_y}{\sqrt{g_t^2 + g_y^2}} A_\mu \bar{l} \gamma^\mu l$$

The apex angle factor

(a) $\dfrac{ig_t}{2\sqrt{2}} \gamma_\mu (1 + \gamma_5)$; (b) $\dfrac{i\sqrt{g_t + g_y'}}{4} \gamma_\mu (1 + \gamma_5)$; \hfill (7.4.35)

(c) $\dfrac{i\sqrt{g_t + g_y'}}{4} \gamma_\mu (4 \sin^2 \theta_W - 1 - \gamma_5)$; (d) $-\dfrac{ig_t g_y}{\sqrt{g_t^2 + g_y^2}} \gamma_\mu = -ie\gamma_\mu$.

3. Characteristic parameters

Original parameters v, g_t, g_y are introduced in the theory. They determine the mass of intermediate bosons and leptons.

$$m_W = \frac{1}{2} g_t v, \quad m_Z = \frac{1}{2} \sqrt{g_t^2 + g_y^2} \, v, \quad m_l = g_t v \tag{7.4.36}$$

Also various interaction vertex intensity determined by g_t, g_y. 3 parameters are v, g_t, g_y, because

$$e = \frac{g_t g_y}{\sqrt{g_t^2 + g_y^2}}, \tan\theta_W = \frac{g_y}{g_t} \qquad (7.4.37)$$

are replaced by v, e, θ_W. e in addition, θ_W can be counted out in the vacuum field theory, thus g_t, g_y is no longer a parameter. v is projection value in leptons mass space and thus nor is the parameter. So there are no parameters exist in the weak interaction.

7.5 Weak interaction of quark

We look at the characteristics of the weak interactions. For example, the neutron decay, $n \to p + e^- + \bar{\nu}$, is allowed by the conservation of energy, and baryon number conservation. Neutron lifetime of approximately 15min, it is far greater than the life of the particles of decay by the electromagnetic action. For example, $\pi^0 \to \gamma\gamma$ this decay is the electromagnetic effect, life is about 10^{-16} s.

$\tau_e/\tau_s \approx (10^4 - 10^6)$ time. Because $\alpha_s^2/\alpha^2 = (10^4 - 10^6)$, the life of particles decay through the electromagnetic interaction than the particles through strong interaction decay characteristic time, $\tau_e/\tau_s \approx (10^4 - 10^6)$. Only α, α_s two scaling cannot explain the ratio 10^{-16} s much longer life. The charged mesons $\pi^- \to e^- \bar{\nu}$ life is 10^{-12} s. The series of measurements to the life of the strange particles are 10^{-10} s. From the

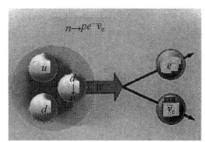

Fig. 7.5.1 Particle decay

point of view of energy to see the weak decay $\Sigma \to n + \pi$ and strong decay $\Delta \to n + \pi$ is almost the same, their life was a difference of 13 orders of magnitude. Therefore, there is a new scale α_w used to describe such a weak interaction.

$$\tau(\Delta \to \pi + n)/\tau(\Sigma \to \pi + n) = 10^{-23} \text{s}/10^{-10} \text{s} = (\alpha_w/\alpha_s)^2, \alpha_w \approx 10^{-6}$$

It is the quantum probability of emission or absorption of weakly interacting ($\alpha_s \approx 1$ compared with $\alpha \approx 10^{-2}$).

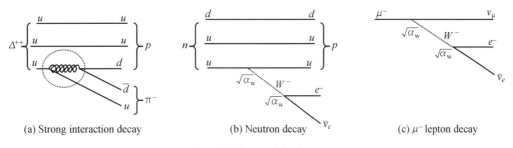

(a) Strong interaction decay (b) Neutron decay (c) μ^- lepton decay

Fig. 7.5.2 Particle decay

A novel characteristic of the weak interaction, it can be turned u quarks into d quarks, and make μ into e, that is, the weak interaction can change the "flavor" of quarks and leptons. Now the experiment has been found to transfer of weakly interacting has positively, negatively charged and neutral.

Weak field fiber is the starting part of electric flux line (Fig. 7.5.2(b) and Fig. 7.5.2(c)), the coupling strength of W is the same as that of electromagnetic interaction, which means that the probability of emitted W is substantially the same as the probability of emitted photons, the slow decay rate of the weak interaction is caused by great W mass. Therefore, the possibility of W transfer is less than that of photon exchange. It is difficult to transmit because of its high quality. The large mass of M_W decreases the coupling strength of the weak interacting intermediate bosons. The effective coupling strength $\alpha_w = \alpha/(M_w/M_p)^2 \approx 10^{-6}$.

7.5.1 Cabbibo angle CKM matrix

To understand the source of Cabbibo angle, it is necessary to look at the structure of quarks. Like leptons, quark instability in the background field will form the center of weak interaction. The center of weak interaction is a basic unit field, weak interaction center and quark lepton center coupling is forming weak interaction. The weak interactions of quarks and leptons are essentially the same, but there are also differences. The biggest difference is that the quark with strings, lepton with no strings. Chord existence is impact the weak interaction coupling constant. Now let's look at how quark strings affect the coupling constants of weak interactions.

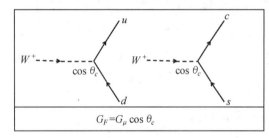 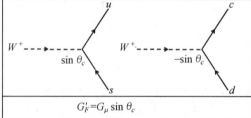

When the string is exists, it's connected to the center and the center bare, then the center decay impossible along the chord direction, only along the other direction to passed the bosons of weak interaction. The cross section S string of chord would block very small part of the area S_{sphere} of the ball, the cover part cannot induce weak interaction thus reducing the coupling strength $G_F < G_\mu$. Weak interaction transfer $W^\pm$ boson. Weak interaction coupling strength proportional to the interaction surface: $G \propto s_1$. After Consider shielding effect, the coupling constant G_F is

$$G_F = \frac{S - s_1}{S} G_\mu \tag{7.5.1}$$

By $G_\mu = (1.458 \pm 0.003) \times 10^{-49}$ erg·cm³, $G_F = (1.415 \pm 0.003) \times 10^{-49}$ erg·cm³. Chord is very fine, with minimal impact on the center of the weak interaction, more careful consideration is as follows.

Chapter 7 Interaction and Super-unified Equation

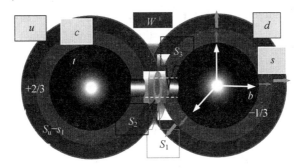

Fig. 7.5.3 The field essence of Cablbbo angle

The first generation quarks have one layer to reach the strain limit, the second-generation quarks have two layers to reach the strain limit, and the third generation quarks have three layers to reach the strain limit. When the second and third generation quarks interact with $W^\pm$ bosons, the string of quarks constitutes the only channel through which the interaction with $W^\pm$ bosons can take place. For this channel, the cross-sectional area of quark string directly determines the probability of interaction. The stronger the quark contraction is, the smaller the cross-sectional area of the string is, and the lower the probability of interaction with $W^\pm$ bosons is. The effect of bosons transfor varies with the cross-sectronal orea of the string.

$u \to d$, $c \to s$, $t \to b$: In the same generation of quarks, the fibers overall outer surface, excluding the shielding part of the chord section, can form channels for transferring weak-acting bosons.

$$\begin{cases} U_{ud} = [(S_u - s_1)/S_u]^{1/2}[(S_d - s_1)/S_d]^{1/2} = 0.9745 - 0.9757 \\ U_{cs} = [(S_c - s_1)/S_c]^{1/2}[(S_s - s_1)/S_s]^{1/2} = 0.9736 - 0.9750 \\ U_{tb} = [(S_t - s_1)/S_t]^{1/2}[(S_b - s_1)/S_b]^{1/2} = 0.9989 - 0.9993 \end{cases} \quad (7.5.2a)$$

The weak interaction betweenu quarks and different generations of s and b quarks

$$U_{us} = [(s_1 - s_2)/S_u]^{1/2}[(s_2 - s_3)/S_s]^{1/2} = 0.219 - 0.224 \quad (7.5.2b)$$

$$U_{ub} = [(s_1 - s_2)/S_u]^{1/3}[(s_2 - s_3)/S_c]^{1/3}[s_3/S_b]^{1/3} = 0.002 - 0.005 \quad (7.5.2c)$$

$u \to b$: To cross the three level channels, the three probabilitys multiplies. The $s_1 - s_2$ constitutes the exit channel of u quarks; $s_2 - s_3$ constitutes a channel to enter s quarks; s_3 enters the b quark channel.

The weak interaction betweenc quarks and different generations of d and b quarks

$$U_{cd} = [(s_2 - s_3)/S_c]^{1/2}[(s_1 - s_2)/S_d]^{1/2} = 0.218 - 0.224 \quad (7.5.2d)$$

$$U_{cb} = [(s_2 - s_3)/S_c]^{1/2}[s_3/S_b]^{1/2} = 0.036 - 0.046 \quad (7.5.2e)$$

The $s_2 - s_3$ constitutes the exit channel of c quarks; $s_1 - s_2$ constitutes a channel to enter d quarks; s_3 enters the b quark channel.

The weak interaction betweent quarks and different generations of d and s quarks

$$U_{td} = [s_3/S_t]^{1/3}[(s_2 - s_3)/S_c]^{1/3}[(s_1 - s_2)/S_d]^{1/3} = 0.004 - 0.014 \quad (7.5.2f)$$

$$U_{ts} = [s_3/S_t]^{1/2}[(s_2 - s_3)/S_s]^{1/2} = 0.034 - 0.046 \quad (7.5.2g)$$

$t \to d$: To cross the three level channels, then the three probabilitys multiplies. The s_3 constitutes the exit channel of t quarks; s_1-s_2 constitutes a channel to enter d quarks; s_2-s_3 enters the s quark channel.

We can write the CKM matrix

$$U = \begin{bmatrix} U_{ud} & U_{us} & U_{ub} \\ U_{cd} & U_{cs} & U_{cb} \\ U_{td} & U_{ts} & U_{tb} \end{bmatrix} = \begin{bmatrix} 0.9745-0.9757 & 0.219-0.224 & 0.002-0.005 \\ 0.218-0.224 & 0.9736-0.9750 & 0.036-0.046 \\ 0.004-0.014 & 0.034-0.046 & 0.9989-0.9993 \end{bmatrix}$$

(7.5.3)

Considering all surface weak interactions including chord cross sections, the total probability is 1, then the following relations exist.

$$U_{ud}^2 + U_{us}^3 + U_{ub}^3 = 1 \ ; \ U_{cd}^2 + U_{cs}^2 + U_{cb}^2 = 1 \ ; \ U_{td}^3 + U_{ts}^2 + U_{tb}^2 = 1 \quad (7.5.4)$$

Because it is the multiplication of three pobabilities, the value is very small, U_{ub}, U_{td} will be approximated to zero.

$$U_{ud}^2 + U_{us}^2 = 1 \ ; \ U_{ts}^2 + U_{tb}^2 = 1 \quad (7.5.5)$$

The theory of quark chord channel includes Cabbibo angle and KMC matrix, and Cabbibo angle is an approximate expression of the theory.

Cabbibo angle: The above and (7.5.1) can be represented by a trigonometric function, written as

$$G_\mu = \sqrt{G_F^2 + G_F'^2} \quad (7.5.6)$$

$$G_F = G_\mu \cos\theta_c^v \ ; \ G'_F = G_\mu \sin\theta_c^v \quad (7.5.7)$$

This is the expression of well-known the Cabbibo's angle $G'_F/G_F = \tan\theta_c^v$, $\theta_c^v = 0.247 \pm 0.008$. Here by the way, the coupling constant is $G_{G-T} = (1.26 \pm 0.02)G_F$ in the allowed transition of Gamow-Taylor-type (Actually, that is pseudo-vector flow A_μ plays a role). Experimentally by comparing two pure Gamow-Taylor transition probabilities can be obtained the ratio of the two coupling constants G'_{G-T} and G_{G-T}.

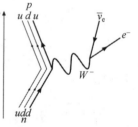

Fig. 7.5.4 Weak interaction space

$$\theta_c^A = \arctan\left(\frac{G'_{G-T}}{G_{G-T}}\right) = 0.226 \pm 0.005 \quad (7.5.8)$$

Knowable $\dfrac{G'_{G-T}}{G_{G-T}} \approx \dfrac{G'_F}{G_F}$. Show that the string of quark to weak interaction is consistent. There are 26% differences between G_{G-T} and G_F, it has should be caused by the different of background field of quarks.

7.5.2 Intrinsic strain of quark with hadronic weak action

$$n \to p + e^- + \bar{\nu}_e$$

Chapter 7   Interaction and Super-unified Equation

$$\varphi_d{}'(X) = \exp(-\mathrm{i})\left\{\frac{1}{3}\underbrace{\begin{bmatrix} \varepsilon_{00} & \varepsilon_{01} & \varepsilon_{02} & \varepsilon_{03} \\ \varepsilon_{10} & \varepsilon_{11} & \varepsilon_{12} & \varepsilon_{13} \\ \varepsilon_{20} & \varepsilon_{21} & \varepsilon_{22} & \varepsilon_{23} \\ \varepsilon_{30} & \varepsilon_{31} & \varepsilon_{32} & \varepsilon_{33} \end{bmatrix}^d}_{\text{Intrinsic strain of } d} \underbrace{(1+\breve{\gamma}_5)\begin{pmatrix} \xi_0 & \zeta_0 & \zeta_0 & \zeta_0 \\ \zeta_1 & \zeta_1 & \zeta_1 & \zeta_1 \\ \zeta_2 & \zeta_2 & \zeta_2 & \zeta_2 \\ \zeta_3 & \zeta_3 & \zeta_3 & \zeta_3 \end{pmatrix}^d}_{\text{Weak action space is half space}(\gamma_\mu+1)} +\right.$$

$$\frac{1}{3}\underbrace{\begin{bmatrix} A_{0,0} & A_{0,1} & A_{0,2} & A_{0,3} \\ A_{1,0} & A_{1,1} & A_{1,2} & A_{1,3} \\ A_{2,0} & A_{2,1} & A_{2,2} & A_{3,2} \\ A_{3,0} & A_{3,1} & A_{3,2} & A_{3,3} \end{bmatrix}^d}_{\text{Electromagnetic 4-momentum}} \breve{\gamma}_\mu \begin{pmatrix} \xi_0 & \zeta_0 & \zeta_0 & \zeta_0 \\ \zeta_1 & \zeta_1 & \zeta_1 & \zeta_1 \\ \zeta_2 & \zeta_2 & \zeta_2 & \zeta_2 \\ \zeta_3 & \zeta_3 & \zeta_3 & \zeta_3 \end{pmatrix}^d +$$

$$\left(\begin{bmatrix} \varepsilon_{00} & \varepsilon_{01} & \varepsilon_{02} & \varepsilon_{03} \\ \varepsilon_{10} & \varepsilon_{11} & \varepsilon_{12} & \varepsilon_{13} \\ \varepsilon_{20} & \varepsilon_{21} & \varepsilon_{22} & \varepsilon_{23} \\ \varepsilon_{30} & \varepsilon_{31} & \varepsilon_{32} & \varepsilon_{33} \end{bmatrix}^{\text{Background}} + \right.$$

$$\left.\begin{bmatrix} \varepsilon_{00} & 0 & 0 & 0 \\ 0 & \delta(i-1)\varepsilon_{11} & 0 & 0 \\ 0 & 0 & \delta(i-2)\varepsilon_{22} & 0 \\ 0 & 0 & 0 & \delta(i-3)\varepsilon_{33} \end{bmatrix}^L \begin{pmatrix} \xi_0 & 0 & 0 & 0 \\ 0 & \xi_1 & 0 & 0 \\ 0 & 0 & \xi_2 & 0 \\ 0 & 0 & 0 & \xi_3 \end{pmatrix}\right) +$$

$$g_W \begin{bmatrix} W_{0,0} & W_{0,1} & W_{0,2} & W_{0,3} \\ W_{1,0} & W_{1,1} & (\Delta\gamma_{21}^W - \Delta\gamma_{21}^W) + (\omega_{12}^W - \omega{'}_{12}^W)W_{1,3} \\ W_{2,0}(\Delta\gamma_{21}^W - \Delta\gamma_{21}^W) + (\omega_{21}^W - \omega{'}_{21}^W) & W_{2,2} & W_{2,3} \\ W_{3,0} & W_{3,1} & W_{3,2} & W_{3,3} \end{bmatrix}^{W^-}(\xi) +$$

$$g_\gamma \begin{bmatrix} B_{0,0} & 0 & 0 & 0 \\ 0 & 0 & 0 & 0 \\ 0 & 0 & B_{2,2} & 0 \\ 0 & 0 & 0 & 0 \end{bmatrix}^{W^-}\begin{pmatrix} \xi_0 & \zeta_0 & \zeta_0 & \zeta_0 \\ \zeta_1 & \zeta_1 & \zeta_1 & \zeta_1 \\ \zeta_2 & \zeta_2 & \zeta_2 & \zeta_2 \\ \zeta_3 & \zeta_3 & \zeta_3 & \zeta_3 \end{pmatrix}^d (\xi) +$$

$$\left.g_W \begin{bmatrix} A_{0,0} & A_{0,1} & A_{0,2} & A_{0,3} \\ A_{1,0} & A_{1,1} & A_{1,2} & A_{1,3} \\ A_{2,0} & A_{2,1} & A_{2,2} & A_{3,2} \\ A_{3,0} & A_{3,1} & A_{3,2} & A_{3,3} \end{bmatrix}^{W^-}\begin{pmatrix} \xi_0 & \zeta_0 & \zeta_0 & \zeta_0 \\ \zeta_1 & \zeta_1 & \zeta_1 & \zeta_1 \\ \zeta_2 & \zeta_2 & \zeta_2 & \zeta_2 \\ \zeta_3 & \zeta_3 & \zeta_3 & \zeta_3 \end{pmatrix}^d \right\} \quad (7.5.9)$$

At the end of the interation, the field function can be expressed as

$$\varphi_u(X) = \exp\left\{(-\mathrm{i})\left[-\frac{2}{3}\underbrace{\begin{bmatrix} \varepsilon_{e0} & 0 & 0 & 0 \\ 0 & \varepsilon_{e1} & 0 & 0 \\ 0 & 0 & \varepsilon_{e2} & 0 \\ 0 & 0 & 0 & \varepsilon_{e3} \end{bmatrix}}_{\text{Intrinsic radial strain of } u} + \frac{1}{2}\underbrace{\begin{bmatrix} 0 & \gamma_{01} & \gamma_{02} & \gamma_{03} \\ \gamma_{10} & 0 & \gamma_{12} & \gamma_{13} \\ \gamma_{20} & \gamma_{21} & 0 & \gamma_{23} \\ \gamma_{30} & \gamma_{31} & \gamma_{32} & 0 \end{bmatrix}}_{\text{Shear strain}} + \right.\right.$$

· 353 ·

$$\left.\frac{1}{2}\begin{bmatrix} 0 & \omega_{01} & \omega_{02} & 0 \\ \omega_{10} & 0 & \omega_{12} & 0 \\ \omega_{10} & \omega_{21} & 0 & 0 \\ 0 & 0 & 0 & 0 \end{bmatrix}\right)(\breve{\gamma}_\mu - 1)\begin{pmatrix} \xi_0 & \zeta_0 & \zeta_0 & \zeta_0 \\ \zeta_1 & \zeta_1 & \zeta_1 & \zeta_1 \\ \zeta_2 & \zeta_2 & \zeta_2 & \zeta_2 \\ \zeta_3 & \zeta_3 & \zeta_3 & \zeta_3 \end{pmatrix}_u -$$

<u>Rotational strain. Spin, only consider rotation along ξ_3 axis</u>   <u>Strain changes from negative strain to normal strain; space from $(1-\gamma_5)$ to $(1+\gamma_5)$</u>

$$\frac{2}{3}\begin{bmatrix} A^u_{0,0} & A^u_{0,1} & A^u_{0,2} & A^u_{0,3} \\ A^u_{1,0} & A^u_{1,1} & A^u_{1,2} & A^u_{1,3} \\ A^u_{2,0} & A^u_{2,1} & A^u_{2,2} & A^u_{3,2} \\ A^u_{3,0} & A^u_{3,1} & A^u_{3,2} & A^u_{3,3} \end{bmatrix}\begin{pmatrix} \xi_0 & \xi_0 & \xi_0 & \xi_0 \\ \xi_1 & \xi_1 & \xi_1 & \xi_1 \\ \xi_2 & \xi_2 & \xi_2 & \xi_2 \\ \xi_3 & \xi_3 & \xi_3 & \xi_3 \end{pmatrix}_u +$$

<u>Electromagnetic 4-momentum</u>

$$\begin{bmatrix} \varepsilon_{00} & \varepsilon_{01} & \varepsilon_{02} & \varepsilon_{03} \\ \varepsilon_{10} & \varepsilon_{11} & \varepsilon_{12} & \varepsilon_{13} \\ \varepsilon_{20} & \varepsilon_{21} & \varepsilon_{22} & \varepsilon_{23} \\ \varepsilon_{30} & \varepsilon_{31} & \varepsilon_{32} & \varepsilon_{33} \end{bmatrix}^B \begin{pmatrix} \xi_0 & \xi_0 & \xi_0 & \xi_0 \\ \xi_1 & \xi_1 & \xi_1 & \xi_1 \\ \xi_2 & \xi_2 & \xi_2 & \xi_2 \\ \xi_3 & \xi_3 & \xi_3 & \xi_3 \end{pmatrix}_u +$$

$$\left.\begin{bmatrix} \varepsilon_{00} & 0 & 0 & 0 \\ 0 & \delta(i-1)\varepsilon_{11} & 0 & 0 \\ 0 & 0 & \delta(i-2)\varepsilon_{22} & 0 \\ 0 & 0 & 0 & \delta(i-3)\varepsilon_{33} \end{bmatrix}^L \begin{pmatrix} \xi_0 & 0 & 0 & 0 \\ 0 & \xi_1 & 0 & 0 \\ 0 & 0 & \xi_2 & 0 \\ 0 & 0 & 0 & \xi_3 \end{pmatrix}\right]\right\}$$

(7.5.10)

$$\varphi_e(X) = \exp\left\{-i\left(\begin{bmatrix} \varepsilon_{e0} & 0 & 0 & 0 \\ 0 & \varepsilon_{e1} & 0 & 0 \\ 0 & 0 & \varepsilon_{e2} & 0 \\ 0 & 0 & 0 & \varepsilon_{e3} \end{bmatrix} + \frac{1}{2}\begin{bmatrix} 0 & \gamma_{01} & \gamma_{02} & \gamma_{03} \\ \gamma_{10} & 0 & \gamma_{12} & \gamma_{13} \\ \gamma_{20} & \gamma_{21} & 0 & \gamma_{23} \\ \gamma_{30} & \gamma_{31} & \gamma_{32} & 0 \end{bmatrix}\right.\right. +$$

<u>Intrinsic radial strain of u</u>   <u>Shear strain</u>

$$\left.\frac{1}{2}\begin{bmatrix} 0 & \omega_{01} & \omega_{02} & 0 \\ \omega_{10} & 0 & \omega_{12} & 0 \\ \omega_{10} & \omega_{21} & 0 & 0 \\ 0 & 0 & 0 & 0 \end{bmatrix}\right)(\breve{\gamma}_\mu + 1)\begin{pmatrix} \xi_0 & \xi_0 & \xi_0 & \xi_0 \\ \xi_1 & \xi_1 & \xi_1 & \xi_1 \\ \xi_2 & \xi_2 & \xi_2 & \xi_2 \\ \xi_3 & \xi_3 & \xi_3 & \xi_3 \end{pmatrix}_e +$$

<u>Rotational strain. Spin, only consider rotation along ξ_3 axis</u>   <u>Strain changes from negative strain to normal strain; space from $(1-\gamma_5)$ to $(1+\gamma_5)$</u>

$$\left(\begin{bmatrix} A^u_{0,0} & A^u_{0,1} & A^u_{0,2} & A^u_{0,3} \\ A^u_{1,0} & A^u_{1,1} & A^u_{1,2} & A^u_{1,3} \\ A^u_{2,0} & A^u_{2,1} & A^u_{2,2} & A^u_{3,2} \\ A^u_{3,0} & A^u_{3,1} & A^u_{3,2} & A^u_{3,3} \end{bmatrix} + \begin{bmatrix} \varepsilon_{00} & \varepsilon_{01} & \varepsilon_{02} & \varepsilon_{03} \\ \varepsilon_{10} & \varepsilon_{11} & \varepsilon_{12} & \varepsilon_{13} \\ \varepsilon_{20} & \varepsilon_{21} & \varepsilon_{22} & \varepsilon_{23} \\ \varepsilon_{30} & \varepsilon_{31} & \varepsilon_{32} & \varepsilon_{33} \end{bmatrix}^B\right)\begin{pmatrix} x_0 & x_0 & x_0 & x_0 \\ x_1 & x_1 & x_1 & x_1 \\ x_2 & x_2 & x_2 & x_2 \\ x_3 & x_3 & x_3 & x_3 \end{pmatrix}_B$$

<u>Electromagnetic 4-momentum</u>

(7.5.11)

Chapter 7 Interaction and Super-unified Equation

$$\phi_{\bar{\nu}_e} = \exp(-i) \left\{ \underbrace{\begin{bmatrix} \varepsilon_{0/\!/} & 0 & 0 & 0 \\ 0 & 0 & 0 & 0 \\ 0 & 0 & 0 & 0 \\ 0 & 0 & 0 & -\varepsilon_3 \end{bmatrix} \begin{pmatrix} \xi_0 & 0 & 0 & 0 \\ 0 & 0 & 0 & 0 \\ 0 & 0 & 0 & 0 \\ 0 & 0 & 0 & x_3 \end{pmatrix}}_{\text{The strain structure similar virtual photon. Neutrinos propagate at the speed of light and transmit momentum}} + \right.$$

$$\underbrace{\begin{bmatrix} \varepsilon_{0\perp} & 0 & 0 & 0 \\ 0 & -\varepsilon_1 & 0 & 0 \\ 0 & 0 & -\varepsilon_2 & 0 \\ 0 & 0 & 0 & 0 \end{bmatrix} \begin{pmatrix} 0 & 0 & 0 & 0 \\ 0 & \xi_\mu & 0 & 0 \\ 0 & 0 & \xi_2 & 0 \\ 0 & 0 & 0 & 0 \end{pmatrix}_{\bar{\nu}_e}}_{\text{Constitute 2-dimensional weak field radial strain part}} +$$

$$\frac{1}{2} \left(\underbrace{\begin{bmatrix} 0 & \Delta\gamma_{01} & \Delta\gamma_{02} & 0 \\ \Delta\gamma_{10} & 0 & -\Delta\gamma_{12} & 0 \\ \Delta\gamma_{20} & -\Delta\gamma_{21} & 0 & 0 \\ 0 & 0 & 0 & 0 \end{bmatrix}}_{\text{Constitute 2-dimensional weak field shear strain part}} + \underbrace{\begin{bmatrix} 0 & \omega'_{01} & \omega'_{02} & 0 \\ \omega'_{10} & 0 & -\omega'_{12} & 0 \\ \omega'_{20} & -\omega'^{W}_{21} & 0 & 0 \\ 0 & 0 & 0 & 0 \end{bmatrix}}_{\text{Two dimensional weak field rotational strain part}} \right) \cdot$$

$$\left. (1+\breve{\gamma}_5) \underbrace{\begin{pmatrix} 0 & \xi_0 & \xi_0 & 0 \\ \xi_1 & 0 & \xi_1 & 0 \\ \xi_2 & \xi_2 & 0 & 0 \\ 0 & 0 & 0 & 0 \end{pmatrix}_{\bar{\nu}_e}}_{\text{Weak interaction is self division and space is half space}} \right\} \qquad (7.5.12)$$

It can be abbreviated as

$$\varphi'_d(X) = \exp\left\{(-i)\left(\frac{2}{3}[\alpha_{\alpha\beta}] + g_W[W^-_{\mu\nu}] + g_\gamma[B_{\mu\nu}]\right)X_\mu\right\} \qquad (7.5.13)$$

$$\varphi'_d(X) = \exp\{-i(g_W[W_{\mu\nu}] + g_\gamma[B_{\mu\nu}])X_u\}\phi_d(X) \qquad (7.5.14)$$

$$D_\mu = \partial_\mu - \hat{\Gamma}_\mu;\ \partial_\mu = i\partial/\partial X_\mu \qquad (7.5.15)$$

7.5.3 CP symmetry of lepton and hadron decay

The weak interaction of leptons violates CP symmetry. The weak interaction of hadrons does not violate CP symmetry.

This is because the whole weak interaction process of leptons is completed in the half-direction space, thus losing the CP symmetry (Fig. 7.5.5). For the weak interaction of hadrons, d quarks have a charge of $-e/3$, which is the tensile strain (positive matter), and d quarks are in the positive half-direction space. After the weak interaction, d quark becomes u quark with $+2/3$ charge. It is a compressive strain (antimatter). u quark is in the negative half space. The weak interaction process of quarks transforms from positive half space to negative half space. Considering that the internal space of nucleon is tensile strain (positive space) due to the stretching effect of strings, the quark background space migrates along the positive direction, and the symmetry axis after migration is O'. The spatial phase of d-quark becomes u-quark symmetrically relative to O'-axis, and changes from positive half-space $(1+\gamma_5)$ to negative half-space $(1-\gamma_5)$, so CP is conserved.

Fig. 7.5.5 weak interaction space
(Spin angular momentum and spatial properties)

7.5.4 The expression of Gauge theory of quark weak unified

Next, quarks introduce GWS gauge theory to describe the weak interaction process of hadron. The gauge group is still $SU(2) \times SU(1)$ group. Corresponding generators are still isospin T_1, T_2, T_3 and hypercharge $Y/2$. Corresponding gauge fields are still A_μ and B_μ. Similar to leptons, quarks projection value in vacuum mass space is $\phi_m(x) = (0, v)$. Generator $T_+ = \frac{1}{2}(T_1 + iT_2), T_- = \frac{1}{2}(T_1 - iT_2), T_3 - \frac{Y}{2}$ corresponds to the intermediate boson field W_μ^-, W_μ^+, Z_μ. Maintaining the generator of symmetry $Q = T_3 - Y/2$ corresponding with photon field A_μ.

1. Quark field

Consider the situation of the quark u, d, c, s. According to the results of phenomenological analysis that involved in weak interaction quark eigenstates are u_L, c_L, and $d_{L\theta} = d_L \cos\theta_c + s_L \sin\theta_c, s_{L\theta} = s\cos\theta_c - d\sin\theta_c$. θ_c is Cabbibo's angle. For the group $SU(2)$, $u_L, d_{L\theta}, c_L$ and $s_{L\theta}$ were composed of double state

$$q_L = \begin{vmatrix} u_L \\ d_{L\theta} \end{vmatrix} ; \quad q'_L = \begin{vmatrix} c_L \\ s_{L\theta} \end{vmatrix} \tag{7.5.21}$$

Decided the generator of transform of nature is $\sigma/2$. For $U(1)$ group, hypercharge

$$\frac{Y}{2} = Q - T_3 = \begin{vmatrix} \frac{2}{3} \\ -\frac{1}{3} \end{vmatrix} - \begin{vmatrix} \frac{1}{2} \\ -\frac{1}{2} \end{vmatrix} = \frac{1}{6} \tag{7.5.22}$$

Therefore, under the gauge transformation of $SU(2) \times U(1)$ group, its transformation law

Chapter 7 Interaction and Super-unified Equation

$$q_L(q'_L) \to \exp\left[-i\theta \cdot \frac{\tau}{2} - i\frac{\theta}{6}\right] q_L(q'_L) \tag{7.5.23}$$

Accordingly, their covariant derivative is

$$D_\mu = \partial_\mu - ig_t \frac{\sigma}{2} \cdot A_\mu - i\frac{g_y}{6} B_\mu \tag{7.5.24}$$

The quarks have mass. They don't only have left state $u_L, d_{L\theta}, c_L, s_{L\theta}$. There are right-handed state $u_R, d_{R\theta}, c_R, s_{R\theta}$. Under the $SU(2) \times U(1)$ group, the left-handed state has the nature which described by above, the right hand state? They do not appear in the phenomenological theory of weak interactions, thus the natural assumption is: they are $SU(2)$ a single state. In the group $U(1)$, their hypercharge is $Y/2 = Q$, i.e.

$$\text{for } u_R \text{ and } c_R \text{ has } \frac{Y}{2} = Q = \frac{2}{3}; \text{ to } d_{R\theta} \text{ and } s_{R\theta} \text{ has } \frac{Y}{2} = Q = -\frac{1}{3} \tag{7.5.25}$$

Therefore, in $SU(2) \times U(1)$ group, their conversion rule:

$$u_R(c_R) \to \exp\left[-i\frac{2}{6}\theta\right] u_R(c_R) \tag{7.5.26}$$

$$d_R(s_R) \to \exp\left[i\frac{1}{3}\theta\right] d_{R\theta}(s_{R\theta}) \tag{7.5.27}$$

Correspondingly, the covariant derivative is

$$D_\mu = \partial_\mu - ig_y \frac{2}{3} B_\mu \tag{7.5.28}$$

$$D_\mu = \partial_\mu + ig_y \frac{1}{3} B_\mu \tag{7.5.29}$$

Quark field is a spin 1/2 fermion field. Lagrange density has the form of the Dirac theory

$$\begin{aligned}\mathcal{L}_{qg} &= i\bar{q}_L \gamma_\mu D q_L + i\bar{q}'_L \gamma_\mu D q'_L + i\bar{u}_R \gamma_\mu D u_R + i\bar{d}_{R\theta} \gamma_\mu D d_{R\theta} + i\bar{c}_R \gamma_\mu D c_R + i\bar{s}_{R\theta} \gamma_\mu D s_{R\theta} \\ &= i\begin{vmatrix} \bar{u}_L & \bar{d}_{L\theta} \end{vmatrix} \left(\gamma_\mu \partial - i\frac{g_t}{2}\sigma \cdot \gamma_\mu A - i\frac{g_y}{6}\gamma_\mu B\right) \begin{vmatrix} u_L \\ d_{L\theta} \end{vmatrix} + i\bar{u}_R\left(\gamma_\mu \partial - i\frac{2}{3}g_y B\right) u_R + \\ &\quad i\bar{d}_{R\theta}\left(\gamma_\mu \partial + i\frac{1}{3}g_y B\right) d_{R\theta} + i\bar{c}_R\left(\gamma_\mu \partial - i\frac{2}{3}g_y B\right) c_R + i\bar{s}_{R\theta}\left(\gamma_\mu \partial + i\frac{1}{3}g_y B\right) s_{R\theta}\end{aligned} \tag{7.5.30}$$

2. Quark of the vertex angle and propagator

In the above existed quarks and gauge field interaction term:

$$\begin{aligned}\mathcal{L}_{qg} &= \begin{vmatrix}\bar{u}_L & \bar{d}_{L\theta}\end{vmatrix}\left(\frac{g_t}{2}\sigma \cdot \gamma_\mu A + \frac{g_y}{6}\gamma_\mu B\right)\begin{vmatrix} u_L \\ d_{L\theta}\end{vmatrix} + \bar{u}_R \frac{2}{3}g_y \gamma_\mu B u_R - \bar{d}_{R\theta}\frac{g_y}{3}\gamma_\mu B d_{R\theta} + (u \to c, d \to s) \\ &= \begin{vmatrix}\bar{u}_L & \bar{d}_{L\theta}\end{vmatrix}\frac{\gamma^\mu}{2}\begin{vmatrix} g_t A_\mu^3 + \frac{g_y}{3}B_\mu & g_t(A_\mu^1 - iA_\mu^2) \\ g_t(A_\mu^1 + iA_\mu^2) & -g_t A_\mu^3 + \frac{3}{}B_\mu \end{vmatrix}\begin{vmatrix} u_L \\ d_{L\theta}\end{vmatrix} + \\ &\quad \frac{2}{3}g_y \bar{u}_R \gamma^\mu u_R B_\mu - \frac{g_y}{3}\bar{d}_{R\theta}\gamma^\mu d_{R\theta} B_\mu + (u \to c, d \to s)\end{aligned}$$

$$= |\bar{u}_L \bar{d}_{L\theta}| \frac{\gamma^\mu}{2} \begin{vmatrix} \dfrac{g_t^2 - \dfrac{g_y^2}{3}}{\sqrt{g_t^2 + g_y^2}} Z_\mu + \dfrac{4}{3} \dfrac{g_y g_t}{\sqrt{g_t^2 + g_y^2}} A_\mu & \sqrt{2} g_t W_\mu^+ \\ \sqrt{2} g_t W_\mu^- & -\dfrac{g_t^2 - \dfrac{g_y^2}{3}}{\sqrt{g_t^2 + g_y^2}} Z_\mu - \dfrac{2}{3} \dfrac{g_y g_t}{\sqrt{g_t^2 + g_y^2}} A_\mu \end{vmatrix} \cdot$$

$$\begin{vmatrix} u_L \\ d_{L\theta} \end{vmatrix} + \frac{2}{3} \left(\frac{g_y g_t}{\sqrt{g_t^2 + g_y^2}} A_\mu - \frac{g_y^2}{\sqrt{g_t^2 + g_y^2}} Z_\mu \right) \bar{u}_R \gamma^\mu u_R -$$

$$\frac{1}{3} \left(\frac{g_y g_t}{\sqrt{g_t^2 + g_y^2}} A_\mu - \frac{g_y^2}{\sqrt{g_t^2 + g_y^2}} Z_\mu \right) \bar{d}_{R\theta} \gamma^\mu d_{R\theta} + (u \to c, d \to s)$$

$$= \frac{g_t}{\sqrt{2}} (W_\mu^+ \bar{u}_L \gamma^\mu d_{L\theta} + W_\mu^- \bar{d}_{L\theta} \gamma^\mu u_L) + \frac{2}{3} e A_\mu \bar{u} \gamma^\mu u - \frac{1}{3} e A_\mu \bar{d} \gamma^\mu d_3 +$$

$$\frac{1}{2} \sqrt{g_t^2 + g_y^2} Z_\mu \left(\bar{u}_L \gamma^\mu u_L - \frac{4}{3} \sin^2 \theta_W \bar{u}_L \gamma^\mu u_L \right) -$$

$$\frac{1}{2} \sqrt{g_t^2 + g_y^2} Z_\mu \left(\bar{d}_L \gamma^\mu d_L - \frac{2}{3} \sin^2 \theta_W \bar{d} \gamma^\mu d \right) + (u \to c, d \to s)$$

$$= \frac{g_t}{2\sqrt{2}} [W_\mu^+ \bar{u} \gamma^\mu (1 + \gamma_5) d_\theta + W_\mu^- \bar{d}_\theta \gamma^\mu (1 + \gamma_5) u] + \frac{2}{3} e A_\mu \bar{u} \gamma^\mu u - \frac{1}{3} e A_\mu \bar{d} \gamma^\mu d +$$

$$\frac{1}{4} \sqrt{g_t^2 + g_y^2} Z_\mu \bar{u} \gamma^\mu \left(1 + \gamma_5 - \frac{8}{3} \sin^2 \theta_W \right) u -$$

$$\frac{1}{4} \sqrt{g_t^2 + g_y^2} Z_\mu \bar{d} \gamma^\mu \left(1 + \gamma_5 - \frac{4}{3} \sin^2 \theta_W \right) d + (u \to c, d \to s) \qquad (7.5.36)$$

The corresponding vertex angles:

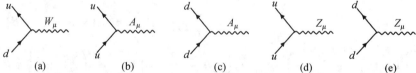

(a) (b) (c) (d) (e)

The corresponding vertex factor:

(a) $\dfrac{g_t}{2\sqrt{2}} \gamma^\mu (1 + \gamma_5)$;

(b) $i \dfrac{2}{3} e \gamma^\mu$;

(c) $-i \dfrac{1}{3} e \gamma^\mu$;

(d) $\dfrac{i}{4} \sqrt{g_t^2 + g_y^2} \gamma^\mu \left(1 + \gamma_5 - \dfrac{8}{3} \sin^2 \theta_W \right)$;

(e) $-\dfrac{i}{4} \sqrt{g_t^2 + g_y^2} \gamma^\mu \left(1 + \gamma_5 - \dfrac{4}{3} \sin^2 \theta_W \right)$.

$$L_q^0 = \bar{q}_i (i \partial - m_i) q_i \quad (i = u, d, c, s) \qquad (7.5.27)$$

Chapter 7 Interaction and Super-unified Equation

They determine the quark propagator:

$$\bullet\xrightarrow[q_i]{p}\bullet \qquad \frac{i}{p-m_i}$$

Collectively, the formula (7.5.9), (7.5.10) and (7.5.11) together and the corresponding graphics and factors constitute a GWS weak unified gauge theory of quarks and leptons. Use them can be calculated various weak process.

7.6 Super-unified field equation

According to the principle that all kinds of strains in vacuum satisfy the strain conservation $\sum_{\mu=0}^{4}\partial_\mu\varphi=0$, see (2.2.2), and consider the interaction. On this basis, if the light cone condition is satisfied in observation space or in high-speed motion state, this means that the strain limit is reached in the observation space. $\varepsilon_{12}+\varepsilon_{22}+\varepsilon_{32}=\varepsilon_{02}$. $\varepsilon_\mu=\partial u_\mu/\partial\xi_\mu$. That is $\Box\phi_\mu=0$, see (2.2.1b). The superunified field equation can be further written as

$$\sum_{n=i=0,l=0}^{N,1}\hat{\partial}_{n;i}^{1+\delta(i)\delta(l)}\Phi_a(X)=0 \tag{7.6.1}$$

Here, i is spatial dimension of interaction; n is generalized space dimension.

Provisions:

(1) $l=1$. Quantum field does not reach the strain limit along the propagation direction (in the case of low velocity).

$$\sum_{n=i=0}^{N}\hat{\partial}_{n;i}\Phi_a(X)=0 \tag{7.6.2}$$

(2) $l=0$. Quantum field reaches strain limit along propagation direction (high-speed relativistic case), and considering that the described free quantum field does not interact, take $i=0$.

$$\sum_{\mu=0}^{N}\partial_\mu^2\Phi_a(X)=0 \tag{7.6.3}$$

Here, $\Phi_a(X)=A_a\exp\{-i[\varepsilon_n]g_{nm}(X^m)\}$; when $i=0$, $\hat{\partial}_{n;0}=\hat{\partial}_n$; $g_{nm}=g_{\mu\nu}$, when $0\leq m,n\leq 4$.

Now let's look at a more specific formulation of the superunified field equation.

1. Schrodinger equation

Schrodinger equation is the simplest equation describing the low-speed motion of electrons in observation space. From the discussion in Chapter 3, we know that the field function of electrons is extremely complex, but the intrinsic strain in the observation space is not observable. Besides, without considering the electromagnetic effect, then there is only the observable space-time strain, and we can write the field function $\Phi_a(X)=A_a\exp\{-i(\varepsilon_0 X^0-\varepsilon_1 X^1-\varepsilon_2 X^2-\varepsilon_3 X^3)\}$. That is

$$\varphi(X) = \exp\left\{-i\left[\begin{bmatrix}E & 0 & 0 & 0\\0 & 0 & 0 & 0\\0 & 0 & 0 & 0\\0 & 0 & 0 & 0\end{bmatrix}\begin{pmatrix}x^0 & 0 & 0 & 0\\0 & 0 & 0 & 0\\0 & 0 & 0 & 0\\0 & 0 & 0 & 0\end{pmatrix} - \begin{bmatrix}0 & 0 & 0 & 0\\0 & \varepsilon_1 & 0 & 0\\0 & 0 & 0 & 0\\0 & 0 & 0 & 0\end{bmatrix}\begin{pmatrix}0 & 0 & 0 & 0\\0 & x^1 & 0 & 0\\0 & 0 & 0 & 0\\0 & 0 & 0 & 0\end{pmatrix} - \right.\right.$$

$$\left.\left.\begin{bmatrix}0 & 0 & 0 & 0\\0 & 0 & 0 & 0\\0 & 0 & \varepsilon_2 & 0\\0 & 0 & 0 & 0\end{bmatrix}\begin{pmatrix}0 & 0 & 0 & 0\\0 & 0 & 0 & 0\\0 & 0 & x^2 & 0\\0 & 0 & 0 & 0\end{pmatrix} - \begin{bmatrix}0 & 0 & 0 & 0\\0 & 0 & 0 & 0\\0 & 0 & 0 & 0\\0 & 0 & 0 & \varepsilon_3\end{bmatrix}\begin{pmatrix}0 & 0 & 0 & 0\\0 & 0 & 0 & 0\\0 & 0 & 0 & 0\\0 & 0 & 0 & x^3\end{pmatrix}\right]\right\}$$

Without considering the spatial structure, the field equation can be abbreviated as

$$\varphi_a(X) = \varphi_0 \exp[-i(E \cdot t - p_1 x^1 - p_2 x^2 - p_3 x^3)]$$

The space-time strain conversion coefficient isv at low speed, $x^0 = vt$, $pv = E$. From (7.6.1), there is no interaction, take item i does not exist, so the covariant derivative degenerates to the ordinary derivative. We can get

$$(\hat{\partial}_0 + \hat{\partial}_1 + \hat{\partial}_2 + \hat{\partial}_3)\varphi_0 \exp[-i(E \cdot t - p_1 x^1 - p_2 x^2 - p_3 x^3)] = 0$$

$$\hat{\partial}_n = -i\partial/\partial X_n; \hat{\partial}_0 = \frac{1}{v}\hat{\partial}_t; \hat{\partial}_t \varphi_a = E\varphi_a$$

then

$$\sum_{n=0}^{3} \partial_n \varphi_a(x) = 0 \Rightarrow [(1/v)E\varphi_a - p_1 - p_2 - p_3]\varphi_a = 0, \text{ i. e.}$$

$$E\varphi_a = v(p_1 + p_2 + p_3)\varphi_a \tag{7.6.4}$$

This is the Schrodinger equation in the form of superunified field equation. Next, let's write the equation as a familiar form. In the framework of Newtonian mechanics, $E = p^2/2m$, $E_i = p_i^2/2m$. Then it can be written in the following form:

$$E\varphi_a = (E_1 + E_2 + E_3)\varphi_a = \frac{1}{2m}(p_1^2 + p_2^2 + p_2^2)\varphi_a$$

According to

$$\hat{\partial}_i \varphi_a = p_i \varphi_a, \hat{\partial}_\mu = -i\frac{\partial}{\partial x^\mu}, \nabla^2 = \frac{\partial^2}{\partial(x^1)^2} + \frac{\partial^2}{\partial(x^2)^2} + \frac{\partial^2}{\partial(x^3)^2}$$

Eigenvalues are equivalent to operators, then we use operators to represent it.

$$\hat{\partial}_0 \varphi_a = \frac{1}{2m}(\hat{\partial}_1^2 + \hat{\partial}_2^2 + \hat{\partial}_2^2)]\varphi_a \Rightarrow i\partial_0 \varphi_a = -\frac{1}{2m}\nabla^2 \varphi_a$$

The quantum field functions of the new theory differ from those of the existing quantum theory as follows:

$$\Phi_a(X) = \exp\{-i[\varepsilon_n]g^{nm}X_m\} \Leftrightarrow \varphi_a(x) = A_a \exp\left\{-i\frac{i}{\hbar}[p_\mu]\eta^{\mu\nu}x_\nu\right\}$$

Finally, we get the familiar Schrodinger equation:

$$i\frac{\partial \varphi_a(r,t)}{\partial t} = -\frac{\hbar^2}{2m}\nabla^2 \varphi_a(r,t) \tag{7.6.5}$$

Chapter 7 Interaction and Super-unified Equation

2. Dirac equation

Dirac equation is used to describe charged fermions with spin 1/2. Multiple degrees of freedom appear in the observation space.

$$\psi(X) = u\exp\{i[P_0X^0 - p_1X^1 - p_2X^2 - p_3X^3 - m_0X^4]\} \quad (7.6.6)$$

$$X_0 = \breve{\gamma}_0\begin{pmatrix}x_0 & 0 & 0 & 0\\ 0 & 0 & 0 & 0\\ 0 & 0 & 0 & 0\\ 0 & 0 & 0 & 0\end{pmatrix}; X_1 = \breve{\gamma}_1\begin{pmatrix}0 & 0 & 0 & 0\\ 0 & x_1 & 0 & 0\\ 0 & 0 & 0 & 0\\ 0 & 0 & 0 & 0\end{pmatrix}; X_2 = \breve{\gamma}_2\begin{pmatrix}0 & 0 & 0 & 0\\ 0 & 0 & 0 & 0\\ 0 & 0 & x_2 & 0\\ 0 & 0 & 0 & 0\end{pmatrix};$$

$$X_3 = \breve{\gamma}_3\begin{pmatrix}0 & 0 & 0 & 0\\ 0 & 0 & 0 & 0\\ 0 & 0 & 0 & 0\\ 0 & 0 & 0 & x_3\end{pmatrix}; X_4 = \breve{\gamma}_4\begin{pmatrix}x_0 & 0 & 0 & 0\\ 0 & \xi_1 & 0 & 0\\ 0 & 0 & \xi_2 & 0\\ 0 & 0 & 0 & \xi_3\end{pmatrix}$$

$$\gamma = \gamma_1 + \gamma_2 + \gamma_3;\ \gamma_0 = I;\ \gamma_4 = \beta;\ \hat{E} = i\frac{\partial}{\partial X_0};\ \hat{p}_i = -i\frac{\partial}{\partial X_i};$$

$$\hat{m} = -i\frac{\partial}{\partial X_4};\ \psi(X) = \begin{pmatrix}\psi_1(x)\\ \vdots\\ \psi_4(x)\end{pmatrix}$$

By (7.6.1), Dirac equation can be obtained directly in the form of superunified field equation.

$$\sum_{n=0}^{4}\hat{\partial}_n\psi(X) = 0 \quad (7.6.7)$$

According to the above formula, we can see that

$$(\hat{\partial}_0 - \hat{\partial}_1 - \hat{\partial}_2 - \hat{\partial}_3 - \hat{\partial}_4)\psi(X) = (\gamma_0P_0 - \gamma_1p_1 - \gamma_2p_2 - \gamma_3p_3 - \gamma_4m)\psi(X) = 0 \quad (7.6.8)$$

We can get a more familiar formula: $(i\gamma^\mu\partial_\mu)\psi(x) = 0$.

3. Klein-Gordon equation

Particles have only mass and no charge and Spin, there are no-multiple degrees of freedom in the observation space. Then the spatial structure γ_iX_i with γ_i matrix degenerates to the ordinary coordinate space x_i.

$$\phi(x) = u\exp i[P_0t^0 - p_1x^1 - p_2x^2 - p_3x^3 - m_0x^4]$$

The field equation of particles is a real scalar field as

$$\sum_{n=0}^{4}\hat{\partial}_n\phi(x) = (P_0 - p_1 - p_2 - p_3 - m)\psi(x) = 0 \quad (7.6.9)$$

Under relativistic conditions: $E^2 = (pc)^2 + (mc^2)^2$. The space-time strain conversion coefficient is c, $x_0 = ct$. The strain limit is reached in observation space-time. The strain limit conditions of the superunified field equation are satisfied. The scalar field equation of particles moving at high speed can be written directly.

$$\sum_{\mu=0}^{4} \partial_\mu^2 \phi(x) = 0 \qquad (7.6.10)$$

Further written as the well-known K–G equation: $(\Box - m^2)\phi(x) = 0$.

4. Electromagnetic wave equation

The propagation velocity of electromagnetic wave is c, which satisfies the optical cone condition. Therefore, the electromagnetic wave satisfies two conditions of the superunified field. The electromagnetic wave field function is $A^t(x,t)$. The field equation of the electromagnetic wave can be written as follows.

$$\begin{cases} \partial_\mu A^t(x) = 0 \\ \Box A^t(x) = 0 \end{cases} \qquad (7.6.11)$$

$$A^t(x,t) = \sum_k \{ c_k(t) e^{ikx} + c_k^*(t) e^{-ikx} \} \qquad (7.6.12)$$

5. Gravitational wave equation

The propagation velocity of Gravitational wave is c, which satisfies the optical cone condition. According to the superunified field, the field equation of gravitational wave can be written directly.

$$\Box h_{\mu\nu} = 0 \qquad (7.6.13)$$

$$h_{\mu\nu}(x) = e_{\mu\nu} \exp(ik_\lambda x^\lambda) + e_{\mu\nu}^* \exp(-ik_\lambda x^\lambda)$$

6. The equation of gravitational field

$$\Phi_G(X) = \exp\{-i([T_{\mu\nu}](X_T) + \frac{1}{2}\kappa'R[g_{\mu\nu}(r,\theta,\varphi)](X_R))\} \qquad (7.6.14)$$

$$\frac{1}{2}R[g_{\mu\nu}] = G_{\mu\nu}, \kappa' = \frac{1}{\kappa}; \hat{\partial}_{G;i} = \hat{\partial}_G + \hat{\Gamma}_G = i\left(\frac{\partial}{\partial X_T} - g_i\frac{\partial}{\partial X_R}\right); \hat{\Gamma}_G = -i\frac{1}{\kappa'}\frac{\partial}{\partial X_R} \Rightarrow$$

$$(\hat{\partial}_G + \hat{\Gamma}_G)\Phi_G(X) = ([T_{\mu\nu}] - \frac{1}{2\kappa}[G_{\mu\nu}])\Phi_G(X) = 0$$

We can get the equation of gravitational field directly.

$$\kappa[T_{\mu\nu}]_p - [G_{\mu\nu}]_{p'} = 0$$

7. Electromagnetic field equation of electron-photon interaction

$\hat{D}_\mu = \hat{\partial}_\mu - \hat{\Gamma}_\mu = \partial_\mu - ieA^\gamma_{\mu,\mu}$. According to $\sum_{n=i=0}^{N,m} \hat{\partial}_{n;i}\Phi_a(X) = 0$, take $N=4$, $i=1$, get $\hat{\partial}_{n;i}\varphi(x) = 0$. We can get the equation of electromagnetic interaction field $(\gamma_n\partial_n - ieA^\gamma_{\mu,\mu})\varphi(x) = 0$, namely

$$(\gamma_\mu\partial_\mu + m - ieA^\gamma_{\mu,\mu})\phi(x) = 0 \qquad (7.6.13)$$

Chapter 7 Interaction and Super-unified Equation

8. Weak interaction field equation

$$\hat{D}_\mu \psi = (\hat{\partial}_\mu + \hat{\Gamma}_{\mu 1} + \hat{\Gamma}_{\mu 2}) = \left(\partial_\mu + ig_t \frac{\sigma}{2} A_\mu + ig_y \frac{1}{2} B_\mu \right) \psi = 0$$

9. Strong interaction field equation

See (4.4.16) and (4.4.17), available

$$D_{v,kl} F_l^{\mu\nu} = g j_k^\mu \text{ (Gluon wave equation)}$$

$$i\gamma^\alpha D_\alpha q - mq = 0 \text{ (Quark field equation)}$$

$$D_\mu = \partial_\mu + ig \Lambda_{\mu,k}(x) \frac{\lambda_k}{2}$$

Chapter 8 Supplementary Notes on Some Basic Issues

8.1 Basic idea of quantum field

According to the physical properties of besic particles, the classification table is as Table 8.1.1.

Table 8.1.1 Classification of fundamental particles

Hadron	Baryon (By the quarks)	Nuclear	Proton: p			Antiparticle
			Neutron: n			
		Hyperon	$\Lambda^0, \Omega, \Sigma^+, \Sigma^-, \Sigma^0$			
		Particle resonance	$\Delta^-\ \Delta^0\Delta^+\ \Delta^{++}, \Sigma^{*+}\Sigma^{*-}\Sigma^{0+}, \Xi^{*-}\Xi^{*0}\Omega$			
	Meson		$\pi^+\pi^-\pi^0, K^+K^-K^0, \eta^+\eta^-\eta^0$			
Lepton			e	μ	τ	
			ν_e	ν_μ	ν_τ	
Propagator			Strong: gluon and meson	Weak force: intermediate boson W^+, W^-, Z^0	Electromagnetic force: photon	Gravity: graviton

According to the statistical properties of particles, there are two types of particles: Fermion is obeyed by Fermi statistics; Boson is obeyed by Bose-Einsten statistics. Boson can transfer interaction, such as photons.

The most basic particle is Fermion of 1/2 spin.

(1) Quark $\begin{pmatrix} u \\ d \end{pmatrix} \begin{pmatrix} c \\ s \end{pmatrix} \begin{pmatrix} t \\ b \end{pmatrix}$;

(2) Lepton $\begin{pmatrix} \nu_e \\ e \end{pmatrix} \begin{pmatrix} \nu_\mu \\ \mu \end{pmatrix} \begin{pmatrix} \nu_\tau \\ \tau \end{pmatrix}$.

$m_e \approx 0.511$ MeV, $m_\mu \approx 105.66$ MeV, $m_\tau \approx 1,784.1$ MeV.

Gauge bosons of 1 spin.

The transfer electromagnetic interaction gauge bosons are photons, γ.

Transfer strong interaction gauge bosons are gluons g^α ($\alpha = 1, 2, \cdots, 8$), with color freedom, $SU(3)$ group.

Transfer weak interaction gauge bosons are W^+, W^-, Z^0.

Chapter 8 Supplementary Notes on Some Basic Issues

$m_W = (80.419 \pm 0.056)\,\text{GeV};\ m_Z = (91.1882 \pm 0.0022)\,\text{GeV}$

$W^+ \to e^+ v_e \mu^+ v_\mu \tau^+ v_\tau \overline{cs},\ Z^0 \to e^+ e^-, \mu^+ \mu^-, \tau^+ \tau^- \cdots$

The motion characteristics of elementary particles reveal the internal information of matter, so it is very important to study their regularity. At present, the most successful basic particle theory is the standard model theory, but unfortunately, the theory has no knowledge of the internal structure of particles. The basic particle information we can know is limited to the experimental observation space, so the whole quantum theory system is constructed in the observation space.

8.1.1 Feynman path integral

Theory of physics is an experimental science, how through the experiment method to shown the intrinsic structure of the particles, it is necessary briefly described here. The basic process is as follows.

Establish the basic properties of the vacuum→Establishment particle model→Obtains the nature of the particle's intrinsic→Intrinsic space 4-momentum and mass→Lagrangian quantum field→The nature of Feynman path integral→Experimental observation of the total probability amplitude.

Probability amplitude of quantum mechanics: Particles from a to b in the process, the probability that from (t_a, x_a) to (t_b, x_b) which is the absolute square of the probability amplitude

$$P(b,a) = |K(b,a)|^2 \tag{8.1.1}$$

The probability amplitude $\varphi[x(t)]$ is the sum of contribution of each path.

$$K(b,a) = \sum_{\text{All paths from }a\text{ to }b} \varphi[x(t)] \tag{8.1.2}$$

The contribution of a single path has an effect proportional to the amount of S phase.

$$\varphi[x(t)] = \text{const}\, e^{(i/h)S[x(t)]} \tag{8.1.3}$$

$$S[a,b] = \int_{t_a}^{t_b} L(\hat{x}, x, t)\,dt \tag{8.1.4}$$

Specifically, we can further look at the path integral form of quantum mechanics.

Dynamics of quantum mechanical systems described by the Hamiltonian $\mathcal{H}(p,q)$, the time evolution of the system depends on the Schrodinger equation.

$$i\frac{\partial \psi(t)}{\partial t} = \mathcal{H}(p,q)\psi(t) \tag{8.1.5}$$

Formal solution of this equation can be expressed as

$$\psi(t) = U(t'',t')\psi(t) \tag{8.1.6}$$

And

$$U(t'',t') = \exp[-i\mathcal{H} \cdot (t'' - t')] \tag{8.1.7}$$

It is the evolution operator. The probabilistic amplitude of the change of system state from q' to q'' is expressed as

$$\langle q''| U(t'',t')| q'\rangle = \langle q''| \exp[-i\mathcal{H}\cdot(t''-t')]| q'\rangle \equiv \langle q'',t''| q',t'\rangle \tag{8.1.8}$$

where $|q\rangle$ and $|q,t\rangle$ are the coordinates of the eigenvectors in Schrodinger and Heisenberg picture. Consider the Feynman path integral, the field of action can be roughly described as

$$\langle q'',t''| q',t'\rangle = N\int e^{i\int L[\psi]} D\psi \tag{8.1.9}$$

Then the action of the quantum field can be re-expressed as

$$\langle q'',t''| q',t'\rangle = N\int e^{i\int \frac{1}{2}[(\varepsilon^0 x_0)^2 - (\varepsilon^1 x_1)^2 - \cdots - (\varepsilon^N x_N)^2]\varphi(x)} D\psi$$

$$= N\int e^{i\int \frac{1}{2}\eta^{\mu\nu}\partial_\mu\varphi(x)\partial_\nu\varphi(x)} D\psi \tag{8.1.10}$$

For the anti-matter, the action of the quantum field can be expressed as

$$\langle q'',t''| q',t'\rangle = N\int e^{i\int \frac{1}{2}\eta^{\mu\nu}\partial_\mu\overline{\varphi}(x)\partial_\nu\overline{\varphi}(x)} D\psi \tag{8.1.11}$$

The basic idea of the vacuum to do a general introduction, thus solving the Lagrangian density L is the core to determine the quantum field equations of motion. For the type of $\exp(i\int L[\psi])$. Consider the existence of j-particles, which can be roughly written as

$$\exp(\sum_N L_j[\psi]) = \exp(iL_1[\psi])\cdot \exp(iL_2[\psi])\cdots\exp(iL_j[\psi])\cdots\exp(iL_N[\psi]) \tag{8.1.12}$$

For each single particle, $\exp(i\int L[\psi])$ is interpreted as the field function ψ that constitutes by deformation field.

8.1.2 Non-relativistic perturbation theory

At low energy, the particles moving in the interaction potential $V(x,t)$ obey the Schrodinger equation.

$$[\mathcal{H}_0 + V(x,t)]\psi = i\frac{\partial\psi}{\partial t}$$

Solving equations

$$\psi = \sum_n a_n(t)\varphi_n(x)e^{-iE_n t}$$

Because $\psi\psi^* = |a_n|^2$ is the probability amplitude of the interaction wave function, the expression of $a_n(t)$ is obtained.

$$i\left(\sum_n \frac{da_n}{dt}\varphi_n(x)e^{-iE_n t} = \sum_n V(x,t)a_n\varphi_n(xe^{-iE_n t}\times\varphi_f^*(x))\right)$$

$$\frac{da_f}{dt} = -i\left(\sum_n a_n(t)\int d^3x\varphi_f^*V(x,t)\varphi_n(x)e^{i(E_f-iE_n)t}\right)$$

$$a_n\left(-\frac{T}{2}\right) = \begin{cases} 1, n=i \\ 0, n\neq i \end{cases}$$

Chapter 8 Supplementary Notes on Some Basic Issues

$$\frac{\mathrm{d}a_f}{\mathrm{d}t} = -\mathrm{i}\int \mathrm{d}^3 x \varphi_f^* V(x,t)\varphi_n(x) \mathrm{e}^{\mathrm{i}(E_f - \mathrm{i}E_n)t}$$

Integral:

$$a_f(t) = -\mathrm{i}\int_{-T/2}^{t} \mathrm{d}t' \int \mathrm{d}^3 x \varphi_f^* V(x,t)\varphi_i(x) \mathrm{e}^{\mathrm{i}(E_f - \mathrm{i}E_i)t'}$$

$$S_{fi} \equiv a_f(T/2) = -\mathrm{i}\int_{-T/2}^{T/2} \mathrm{d}t \int \mathrm{d}^3 x [\varphi_f \mathrm{e}^{-\mathrm{i}E_f t}]^* V(x,t)[\varphi_i(x)\mathrm{e}^{-\mathrm{i}E_i t}]$$

$$= \int \mathrm{d}^3 x \varphi_f^*(x) V(x,t) \varphi_i(x)$$

The above formula has a better approximation only when $a_f(t) \ll 1$ is used. $|S_{fi}|^2$ is interpreted as the probability of scattering from the initial state i to the final state f. Consider the simplest case, where V is time-independent, that is $V(x,t) = V(x)$.

$$S_{fi} = -\mathrm{i}\underbrace{\int_{-\infty}^{+\infty} \mathrm{d}^3 x \varphi_f^*(x) V(x) \varphi_i(x)}_{=V_{fi}} \int_{-\infty}^{+\infty} \mathrm{d}t \mathrm{e}^{\mathrm{i}(E_f - E_i)t} = -\mathrm{i}V_{fi} \cdot 2\pi \underline{\delta(E_f - E_i)}$$

$$\uparrow$$
Conservation of energy

At low energies, particles do not produce or annihilate, which belongs to the category of quantum mechanics. Now consider the case of high energy, the existence of particle production and annihilation that is, the category of quantum electrodynamics. In quantum electrodynamics, the expression of matrix elements of S transformation can be written directly by using Feynman rule.

Since $|S_{fi}|^2$ is interpreted as the probability of particle scattering from initial state i to final state f, then all the calculations of quantum theory can finally be expressed by $|S_{fi}|^2$. Under strict expermental conditions, the particle accelerator verifies the distribution of the measured particles. The particle conforms to $|S_{fi}|^2$, it shows that the theoretical reasoning is consistent with the calculation. In this way, a huge bridge has been built so that the quantum theory of the standard model can be verified by experiments.

The disadvantage standard model theory is that there are too many parameters in point-like particle theory. The standard model contains 19 free parameters, which can only be determined by experiments.

8.2 Quantum fluctuations and space-time rupture

8.2.1 Running coupling constant

When the electronic interaction, electronic inevitably there is a disturbance, electronic sudden emission of a photon, the photon may be transformed into a pair of virtual electron pairs then annihilation into a photon absorbed by electrons. Range of interaction is closer, disturbance is stronger and this effect increases.

The presence of the charge lead to fluctuations of vacuum quantum field (Fig. 8.2.1), more closely, then greater the fluctuation. Vacuum is understood as electrically neutral boson condensed matter, the fluctuation of the quantum field lead to the vacuum polarization effects (provisions the hole with a negative charge, stimulate "sea level" with a positive charge), the effect constitutes charge screening. The electron is surrounded by opposite charges attract each other, position tend to original electronic. This is called the electron screening in QED. Therefore, the electrons are surrounds by polarized charge cloud and the positive charge near the electron. In this way, the negative charge of the original electron is shielded. If you want to test charge by the Coulomb force to determine the original charge of the electron, then it related to the location of the test charge (Fig. 8.2.2). If the test charge move toward the charge, it is necessary to pass through the shielding cloud of the electron charge. Test charge is closer to the electron, there are more electric flux lines intersect, the channel of interaction is increased, interaction becomes stronger, coupling constant, which is measured by the charge to the greater, the coupling constant becomes large, i.e. the charge which is measured to the greater. The measured charge value dependent on the probe when the electron is charged, which is calculated by considering all the possible the con figuration of charge cloud these surrounded the electronic. Consider the correction of charge vacuum polarization circle in the photon propagator. This cycle is repeated in the higher order diagrams. The running coupling constant $\alpha(Q^2) \equiv e^2(Q^2)/4\pi$. Running coupling constant describes the relationship of distance dependence between two effective charged particles.

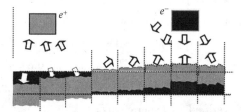

Fig.8.2.1 Charge resulting quantum field vacuum fluctuations that changes with distance schematic (Quantum field vacuum fluctuation caused by charge. The closer the greater the ups and downs, which constitutes a charge shielding effect)

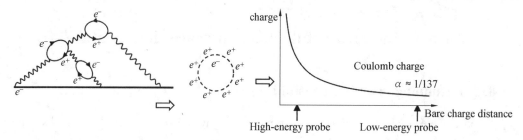

Fig. 8.2.2 Charge screening

Chapter 8 Supplementary Notes on Some Basic Issues

8.2.2 The strong interaction coupling constant in close range

Similar to QED, we can be calculated as (Q^2) of the quark color charge. If we do not consider the interaction between the gluon, the color screening and electrical shielding is very similar. However, due to the self-interaction between the gluon, coupling constant $\alpha_S(Q^2)$ in QCD dependence to Q^2 is completely different in QED.

String is channel of the strong interaction transfer gluon, thus gluon itself with color. When using the test color charge to measure the original color charge, the measurement itself will produce disturbance to the original color charge, and disturbance make the background vacuum of the measured quark polarization, inspired virtual quark pair, virtual quark emitting gluon. Quark vibration produce gluon, gluon vibration generated on gluon pair. This effect will continue to occur formation of quantum fluctuations and around the measured quark becomes very complex, virtual quark generation and disappear, gluons and string intertwined resulting in the original color charge becomes like a maze. Now consider a color charge, such as the red color charge. Disturbance of the color charge, so that the vacuum polarization, red color charge was priority surrounded by red color charge (Fig. 8.2.3). When the test color charge closer to the measured color charge the test color charge into the ball with mainly red charge, the quantity of the measured red charge reduction. This effect is red anti-shielding, it has the nature of "Asymptotic freedom".

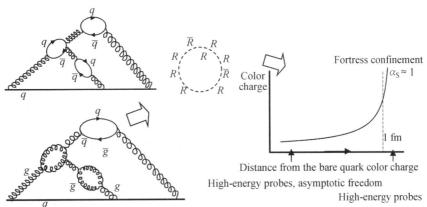

Fig. 8.2.3 Color charge shielding

"Progressive freedom" can be considered from two aspects.

(1) From the momentum point of view, we can make such an analogy, assuming that the faster the bullet passes through a glass plate, the smaller the hole, the smaller the vibration of the whole glass.

(2) On the contrary, the slower the speed the greater the vibration of the whole glass and the greater the impact force. From the point of view of the string, the strong interaction field is a space full of strings. The closer the two hadrons are, the faster the speed is, the less the pos-

sibility of the strings intertwining. On the contrary, the smaller the momentum of the detected particles, the farther the distance and the slower the speed are, the greater the possibility of the strings intertwining. We can also make this analogy: the slower a small insect passes through a spider web, the easier it is to be caught by the spider web. Corresponding physical image is the interaction distance closer the strong coupling constant is smaller. Whether the quantum fluctuations more complex on background vacuum cause by of quark polarization, its rules are simple, just repeat the simple effect, so we can determine the coupling constant $\alpha_S(Q^2)$ in QCD dependent on Q^2. This requires the calculation $I(q^2)$ in QCD. Gluon propagator diagram as follows.

Additional items from the gluon self-acting, C and T denote the Coulomb gluon and transverse gluons. On the diagram the first circle fluctuations into a pair of virtual $q\bar{q}$. Second is Gluon itself interact, making theory of the strong interaction is no-able. Gluon coupling make QCD with color anti-shielding. The running coupling constant of the QCD can be calculated by above formula (refer to any one the QCD theory books).

$$\alpha_S(Q^2) = \frac{\alpha_S(\mu^2)}{1 + \frac{\alpha_S(\mu^2)}{12\pi}(33 - 2n_f)\ln\frac{Q^2}{\mu^2}}$$

In QCD theory, $\alpha_S(Q^2)$ increases with the Q^2.

8.2.3 Hardening vacuum background field lead to a unified interaction coupling constant

For interacting quantum field, interaction is stronger, the disturbance of vacuum is stronger, the vacuum hardening is harder. The vacuum hardening strict proportional to the energy of interaction, with the increase of energy, the background field in the interaction region becomes harder and harder.

Background field is completely harden and loses elasticity, the a localized vacuum reaches the deformation limit, at this time all the deformation no longer is the elastic deformation, quantum field and gravitational field these based on the elastic deformation will be completely dissipated, interaction to achieve unification, in this case, physics also disappeared. This situation is similar to the center of the black hole area.

Physics is the product of strain in local vacuum, which is elastic based on background vacuum. With the increase of interaction energy, the hardening of background field reaches the deformation limit, the background vacuum loses elasticity, and the coupling constants of interaction operation must be classified as the same point (Fig. 8.2.4), at which the whole physics disappears.

Chapter 8　Supplementary Notes on Some Basic Issues

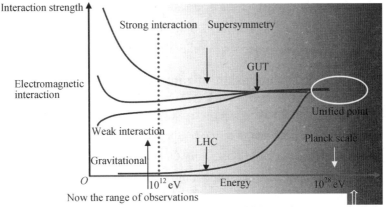

Fig. 8.2.4　Unified interaction

8.2.4　The concept of neighborhood of quantum field

Quantum field anomalous dimension as a mathematical tool exists in quantum field theory, there is no actual physical properties. However, in fact, the anomalous dimension of the quantum field does exist in the interacting quantum field, which has the actual physical significance and physical image. The content of this chapter is to study anomalous dimension of quantum field from the point of view of the vacuum.

Interacting quantum field:

$$\varphi_R(x) = \sum_k [a_R(k)f_k(x) + a_R^+(k)f_k^*(x)], f_k(x) = \frac{1}{(2\pi)^D}\exp(-ikx)$$

Here, $D = 4$. In the perturbation theory, the equation of nonlinear interaction of Bose quantum field is written as the form of renormalization.

$$\varphi_R(x) = Z_B^{-1/2}(e_R)\varphi_b(x) \quad (\text{QED}) \tag{8.2.1}$$

Bose in QCD has a similar expression.

$$\varphi_R(x) = Z_B^{-1/2}(g_R)\varphi_b(x) \quad (\text{QCD}) \tag{8.2.2}$$

For spin $J = 1/2$ Dirac field:

$$\varphi_N(x) = Z_F^{-1/2}(e_R)\varphi_b(x) \quad (\text{QED}) \tag{8.2.3}$$

$$\varphi_R(x) = Z_F^{-1/2}(g_R)\varphi_b(x) \quad (\text{QCD}) \tag{8.2.4}$$

Here, $\varphi_R(x)$ is the quantum field of renormalization constant; $\varphi_b(x)$ is the bare quantum field; $Z_B^{-1/2}$ and $Z_F^{-1/2}$ are the renormalization constant.

Quantum field $\varphi_R(x)$ and $\varphi_b(x)$ have neighborhood and measure, we can establish a coordinate frame in this geometry, the coordinate frame having fractal dimensional structure. Quantum field $\varphi_R(x)$ in the interaction with the fractal dimension bubble geometry, $\varphi_b(x)$ with 4-dimensional structure, $Z_{B,F}^{-1/2}\varphi_b(x)$ is understood as (Cantor segment $0 < \sqrt{Z} < 1$) $\varphi_b(x)$ projected on the fractal dimension coordinates based it with the bubble structure. The result makes the renormalization constant has the significance of the geometric segment, which is similar to the Cantor set, of course, the Cantor set is just a special case. The segment geometry of renormalization (constant) can be very complex, but since we cannot be observed intrinsic space of the quantum field, we are only interested in the dimensions of the segment.

8.2.5 Anomalous dimension of quantum field

By the definition of the fractal dimension: $N(r) = r^{-D}$ i.e. $D = \log N(r) / \log N(1/r)$. The basic unit of the size of the vacuum of space by the vacuum, divided into an infinite number of small size l^3 of the cube is the basic unit, l^3 is the volume of vacuum basic unit. To mark by $\alpha = 1, 2, \cdots$. The fractal dimension of distribution is formed.

$$D = \frac{\log\left(\frac{N_{m+1}}{N_m}\right)}{\log\left(\frac{N_m}{N_{m+1}}\right)} = \frac{\log(7P + 1)}{\log 2}$$

This piece of the vacuum there is a strict fractal structure, which always can be divided. After split infinite times vacuum material there is always an element of most basic. Vacuum with particle structure, probability that can be divided is 1, namely $p = 1$, then

$$D = \frac{\log 8}{\log 2} = 3$$

The manifold dimensions is $3D$.

If the vacuum is disturbed then structure will changes, such as field quality defect (local), then $0 < p < 1$, $D < 3$, in quantum field, namely the interaction region, the spatial dimension of vacuum will be smaller than the 3 dimension. Interacting quantum field exists resulting the vacuum basic unit is plastic deformation and occurrence defects, resulting in dimension from 3 dimension into the 1 or 0 dimensions, so the dimension of neighborhood of point less than 3 dimension, there anomalous dimension (Fig. 8.2.5). The anomalous dimension increases by increasing the disturbance.

8.2.6 Renormalization constant

Quantum electrodynamics, quantum field renormalization calculate the second-order perturbation series, the renormalization constant divergence (∞) expression is

$$Z_B^{-1/2}(e_R, \sigma) = 1 + \frac{\alpha_e}{6\pi} \ln\left(\frac{\sigma^2 + m^2}{m^2}\right)\bigg|_{\sigma \to \infty} \quad \text{(QED)} \quad (8.2.5)$$

$$Z_F^{-1/2}(e_R, \sigma) = 1 - \frac{\alpha_e}{6\pi} \ln\left(\frac{\sigma^2 + m^2}{m^2}\right)\bigg|_{\sigma \to \infty} \quad \text{(QED)} \quad (8.2.6)$$

Mass renormalization ($Z_m^{-1/2}(e_R, \sigma) m = m_R$) and charge renormalization ($Z_e^{-1/2}(e_R, \sigma) e = e_R$), m_R, e_R corresponds to the experimental values. Mass renormalization constant:

$$Z_m^{-1/2}(e_R, \sigma) = 1 + \frac{3}{4\pi} \alpha_e \ln\left(\frac{\sigma^2 + m^2}{m^2}\right)\bigg|_{\sigma \to \infty} \quad \text{(QED)} \quad (8.2.7)$$

Chapter 8 Supplementary Notes on Some Basic Issues

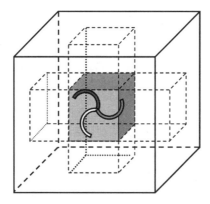

Fig. 8.2.5 Schematic diagram of the anomalous dimension
(Plastic deformationof a basic unit in vacuum resulting in 3-dimensional structure into a 1-dimensional structure, so that the focal areas of the vacuum dimension is less than 3-dimensional)

Here, σ is the 4-momentum of the virtual particles, not measurable physical quantities. By covering the definition of measure and Hausdorff measures-dimension definition (8.2.7), (8.2.8), we obtain

$$\sigma \frac{\partial}{\partial \sigma} Z_B^{-1/2}(e_R, \sigma) = \gamma_B(e_R) Z_B^{-1/2}(e_R, \sigma) \qquad (8.2.8)$$

By (8.2.5) and (8.2.8), obtain Bose quantum field anomalous dimension in QED is

$$\gamma_B(e_R) = \frac{1}{3\pi} \alpha_e \quad (\text{QED}) \qquad (8.2.9)$$

By (8.2.6), (8.2.7) and (8.2.8) obtain the anomalous dimensions of the corresponding quantum field.

$$\gamma_F(e_R) = -\frac{1}{2\pi} \alpha_e(Q^2) \quad (\text{QED}) \qquad (8.2.10)$$

$$\gamma_m(e_R) = \frac{3}{2\pi} \alpha_e(Q^2) \quad (\text{QED}) \qquad (8.2.11)$$

The spin top vertex renormalization $\gamma_\mu(e_R, \sigma) = Z_\nu^{-1} \gamma_\mu$, here, γ_μ is Dirac matrices. Defined spin anomalous dimension:

$$\gamma_\nu(e_R) = \frac{\alpha_e(Q^2)}{2\pi} \quad (\text{QED}) \qquad (8.2.12)$$

8.2.7 Running coupling constant

1. In QED

Above $\alpha(Q^2) \equiv e^2(Q^2)/4\pi$ is called the running coupling constant. Here, $e^2(Q^2) = e_{02}/[1+I(q^2)]$. In the limit of large $Q^2 \equiv -q^2$, in order to eliminate the $\alpha(Q^2)$ obvious dependence to truncated M, you can choose a renormalization or reference momentum values μ.

For large Q^2, there is $\alpha(Q^2) = \dfrac{\alpha(\mu^2)}{1 - \dfrac{\alpha(\mu^2)}{3\pi}\ln\left(\dfrac{Q^2}{\mu^2}\right)}$, running coupling constant $\alpha(Q^2)$ describes the distance dependence relationship between two charged particles of interaction. When Q^2 increase the photon, see more and more charge until the coupling constant becomes infinite. However all Q^2 can be achieved in experiments, $\alpha(Q^2)$ change is very small. $\alpha(Q^2)$ very slow increase from 1/137 with Q^2 increase. For the electromagnetic interaction, the interaction distance is short then interaction is strong, the more serious destruction of the dimension of the vacuum, corresponding to anomalous dimension bigger.

Physical reality of Anomalous dimension of quantum fields $\gamma_{BF}(e_R)$ and $\gamma_m(e_R)$ need the new physical hypothesis, this hypothesis must be a large number experimental support. The current experimental support the existence of anomalous dimensions of the quantum field, such as charge · charge correlation experiment.

2. In QCD

QCD running coupling constant $\alpha_S(Q^2) = \dfrac{\alpha_S(\mu^2)}{1 + \dfrac{\alpha_S}{12\pi}(33 - 2n_f)\ln\left(\dfrac{Q^2}{\mu^2}\right)}$ in QCD theory, $\alpha_S(Q^2)$ decreases with increasing Q^2. Interactions becomes small in the small distance. This theory is known as asymptotic freedom.

Strong interaction quantum field anomalous dimension $\gamma_{BF}(g_R)$ exists in the process of N–hadrons production. In CMS energy range $\sqrt{S} = (2\text{—}1\,800)\,\text{GeV}$, $-\gamma_B(g_R)$ gradually become smaller, because the few experimental data $-\gamma_F(g_R)$, and the error is large, the preliminary conclusion is

$$-\gamma_B(g_R) = 0.060 - 0.030 \quad \sqrt{S} = (4\text{—}1\,800)\,\text{GeV} \qquad (8.2.13)$$
$$-\gamma_F(g_R) = 0.45 \pm 0.05 \qquad (8.2.14)$$

By (8.2.16), $-\gamma_B(g_R)$ within the CMCS energy $\sqrt{S} = (2\text{—}1\,800)\,\text{GeV}$ becomes smaller gradually show that the strong interaction becoming weaker in small distance, the nuclear intrinsic space vacuum disturbance smaller, the corresponding intrinsic space the anomalous dimension smaller[①].

8.3 The difference between vacuum unified theory and string/M theory

The goal of superstring theory and vacuum super–unified theory is to establish a unified

---

① Note: My tutor Zhao Shusong and my classmate Zhao Xi took the lead in conducting long-term and in-depth research on anomalous dimensions, and obtained fruitful research results. See in detail the academic monograph "Principles of Strong Interaction Quantum Field" (Zhao Xi, Zhao Shusong, Science Press, May 2012).

Chapter 8 Supplementary Notes on Some Basic Issues

physical theory. Here, the author gives a very unprofessional introduction to superstring theory, and compares it with the basic characteristics of vacuum super-unified theory.

8.3.1 The basic idea of Vacuum unified theory and string theory

1. Space-time of string theory and the Vacuum unified theory

String theory and vacuum superunification theory have different understandings of space-time, which leads to a great distance between the two theories. But we live in 4-dimensional space-time, which is obviously inconsistent with the 11-dimensional space-time of string theory. In order to conform to the real world, 11-dimensional space is Calabi-Yau space, which hides independently at every point of 4-dimensional space-time (Fig. 8.3.1). We can't see it, but the string theory regard as it exists. The basic assumption in string theory concerns as it is that all fundamental particles are composed by constantly vibrating chords; time and space must allow some kind of supersymmetry.

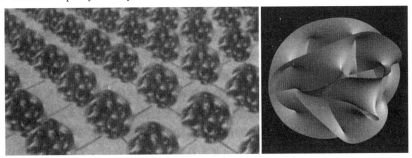

(Left: According to string theory, extra dimension curling into a Calabi-Yau space in universe;
Right: an example of Calabi-Yau space)

Fig. 8.3.1 Space in string theory

The added dimension space quite magical, but string theory is further noted that the Calabi-Yau geometry of space determines the nature of the universe and the laws of physics. Which particles can be exist, what is the mass, how they interaction, even some constants in nature, they are determined by Calabi-Yau space or the book called "inner space" shape. Theoretical physicists use Dirac operator to study the properties of the particle. Through the analysis of the spectrum of an operator, can be estimated the species of particle.

Space-time have 11-dimensions, it is the product of a 4-dimensional space and 7-dimensional Calabi-Yau space. Therefore, when we use the method of separation of variables to solving operator spectrum, it will be affected by Calabi-Yau space. The diameter of Calabi-Yau space is very small then nonzero spectrum becomes extremely large. Such particles should not be observed, because they appear only in extremely high-energy state. On the other hand, particles having a zero spectrum may be observed, they vary depending on the topology of the Calabi-Yau space. This shows that the topology of the small 6-dimensional space in physics is how important. Einstein pointed out that gravity just is a reflection of the space-time geometry.

The chord scientists The shape of this 6-dimensional space is very important. Unfortunately, there are almost infinite kinds of such convoluted 6-dimensional spaces that satisfy the requirements. String scientists have worked tirelessly on this issue, devoting themselves to research and devotion.

Based on the above, the conclusion drawn from the philosophical point of view is that the more questions can be explained, the simpler the theory is, the closer it is to the truth. Comparisons can distinguish good from bad. String theory is too complex, which is definitely not good news. Now, the beautiful mathematical jacket of string theory is far more beautiful than the soul of physics. The important physical properties of basic particles, such as mass, spin, charge and so on, are hidden in the redundant dimension space, just like covering dust with carpet, the problem has not been solved. For the unobservable 6-dimensional curled Karabi-Yao spatial structure, it is similar to the "ether theory" which uses pinion to explain Maxwell's equation. The direction of string theory is wrong.

Vacuum space-time unification theory is 4-dimensional Space-time, vacuum has no dimension, the generation of space-time dimension is the effect of vacuum strain (Fig. 8.3.2). Space-time only has four dimensions, there is no additional dimension, this conclusion makes vacuum unification theory extremely simple. However, this conclusion also negates many imaginations, such as time reversal, wormhole and so on. String theory is to roll the extra dimension into tiny space and finally get 4-dimensional space-time. The dimension of space is innate and has nothing to do with the deformation of space. The extra dimension can be curled, which means that space and time have elasticity and can be deformed, so what constitutes a multi-dimensional space? The basic hypothesis of string theory is more than that of vacuum super-unified theory.

Fig. 8.3.2 Generation of space-time dimension is due to the deformation of the vacuum field effect

2. Electrons and quarks in String theory

The aim of superstring/M theory is to achieve the super-unification of physics. The basic idea is that when observing nature at 10^{15}—10^{19} GeV scales, the final composition of matter is

Chapter 8 Supplementary Notes on Some Basic Issues

not a particle or field, but a string. Strings have mass.

Vacuum fluctuation: In the unified vacuum theory, the vacuum outside the intrinsic space of quantum field is smooth. Close to the observation itself will lead to strong disturbance of vacuum. In order to observe natural phenomena in the scale of 10^{15}—10^{19} GeV, observer is strongly disturbance source to the measured object, resulting in quantum field wavelength shortening to the limit, namely the Planck scale 10^{-33} m, that is. The string is created by the observer short distance observation. In other words, the high-energy observations lead to the generation of the quantum field, we can see mass charge its form is string. More simply, the high-energy input observation objects will be molded into 10^{-33} m string.

The basic entity in string theory (i.e., the mass of the charge of the vacuum) is no longer idealized point particles, but 1-dimensional strings, 2-dimensional film, and extended object of high-dimensional generalized membrane. Superstring with supersymmetry, but different modes of vibration of the strings is representation the entire particle spectrum series all action unified. Like a string in the action of tension will be an infinite number of modes of vibration.

The basic unit of vacuum excitation transition resulting in vacuum defect form a hole, this hole and the surrounding vacuum deformation consistsan electronic, the free state basic unit transition out of the field compression around the vacuum would constitute an anti-electron. For the quark, as electronic, different is the transition is not 1 basic unit, but the 1/3 basic unit with string.

A vibration mode can be described by quantum numbers such as mass, spin… The basic idea of string theory is each vibration modes of string carries a group of quantum number, this set of quantum numbers are corresponding certain distinguishable particles. This is a kind of ultimate unity. This is ultimate unity, that is, all the elementary particles we know are described by a string.

3. Comparison of two theories

Furdamentals of elementary particles (Fig. 8.3.3) exist in the form of waves, similar to string theory.

But string theory avoids many basic problems, such as coupling constants, mass, charge, parity, dark matter, dark energy and so on. These theories all have logical self-consistent explanations and related calculations.

4. The basic structure of string theory

Surface of Mass charge space charge can be arranged on the coordinates of its network to describe. Because it is 2-dimensional curved surface (1-dimensional time, 1-dimensional space), to determine a point by $\sigma(\sigma^1, \sigma^2)$. Here, $\sigma(\sigma^1, \sigma^2)$ is the particle's intrinsic 2-dimensional space-time frame, x_μ is observed space-time frame. In order to indicate arbitrary point of string on the curved surface at any particular moment in the place, you must have a clear rules to set up the link between a point on internal space curved surface of mass charge

The Vacuum Super-unified Theory

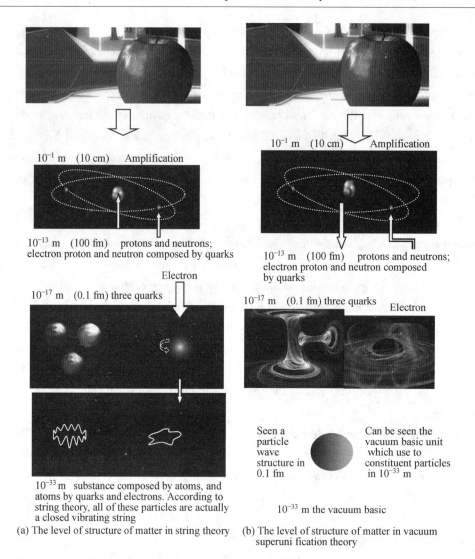

Fig. 8.3.3 Model comparison

and another point of observation space.

$$x_\mu = x_\mu(\sigma^1, \sigma^2) \qquad (8.3.1)$$

In the point particle theory, the elementary particle time-space configuration is described by its 4-dimensional coordinates $x^\mu(\mu=0,1,2,3)$, there is no intrinsic space, this is because the basic particle is considered as a point it no intrinsic structure. Particle is no longer a point in string theory which introduced the intrinsic structure of string, string with 2-dimensional space coordinate σ^1, σ^2, space coordinates $x_\mu(\sigma^1, \sigma^2)$ dependence σ^1, σ^2 space coordinate A and B, caused the character is different between string model and particle theory.

The 2-dimensional curved surface of intrinsic space is changes, in order to describe the surface, we need to introduce two dimensional gauge $g_{\alpha\beta}(\sigma)$, $\alpha, \beta = 1,2$ describe the surface.

Chapter 8 Supplementary Notes on Some Basic Issues

Two adjacent points on the surface σ and $\sigma+d\sigma$, the length between them is

$$ds = \sqrt{g_{\alpha\beta}(\sigma) d\sigma^\alpha d\sigma^\beta} \tag{8.3.2}$$

In order to link the experimental and observations, it is necessary to calculate the interaction probability amplitude (probability amplitude squared can be got the probability of collision course), you have to take a weighted average to all the possible way in collision course.

In string theory, This means that must sum to all possible of historical development process of the 2-dimensional surface. The historical evolution of a 2-dimensional surface by the above two functions $x_\mu(\sigma)$ and $g_{\alpha\beta}(\sigma)$. In order to calculate the probability amplitude, actually you need to calculate each $S[x,g]$ correspond to historical process x, g of the 2-dimensional surface and $e^{-I[x,g]}$ is the sum of all possible surfaces. $I[x,g]$ called the action quantity, it is the functional of $x_\mu(\sigma)$ and $g_{\alpha\beta}(\sigma)$, is given by

$$S[x,g] = \frac{1}{4\pi\alpha'} \int d\sigma d\tau \sqrt{-g(\sigma)} g^{\alpha\beta}(\sigma,\tau) \frac{\partial x^\mu(\sigma)}{\partial \sigma^\alpha} \frac{\partial x^\nu(\sigma)}{\partial \sigma^\beta} \eta_{\mu\nu} \tag{8.3.3}$$

The above formula, α, β are 2-dimensional space-time (σ,τ) vector index, $\mu,\nu = 0,1,2,\cdots,D-1$ is a D-dimensional space-time vector index, $\eta_{\mu\nu}$ is the D-dimensional space-time Min's gauge, namely

$$\eta_{ij} = (+,-,-,\cdots,-) \tag{8.3.4}$$

$\frac{1}{2\pi\alpha'} = T$ is the tension of plastic deformation of string. Why it's in D-dimensional space-time, rather than 4-dimensional space-time? Because when a string moves in space-time, due to the movement of the string is so complex that the 3-dimensional space has been unable to accommodate its trajectory, there must be as high as 11-dimensional space to meet its motion (11-dimensional space is the mathematical equation results). We can observe the actual space is 3-dimensional space and a time dimension, this 6-dimensional space of the excess should by compactification (Cards-Yau space) crimped in the space of 10^{-33} cm. Interior space of point particle is not 3-dimensional, there may be many dimensions. For example, the surface of water pipe is 2-dimensional, but when we look at it from a distance, it's like a 1-dimensional line. The dimension circle around the tube is very short, "curling up" not easily to find. In order to see around the circle that dimension you must be approached the pipes. Therefore, there are two ways of spatial dimension: one is that it can be extended very far and can be displayed directly; second, it is very small and difficult to see when it curls up. In the minutest scale, scientists have proven our universe spatial structure have extended dimension and have curled dimension. these extra curling dimension tightly huddled in a tiny space, using the most sophisticated instruments we also did not detect them (10^{-33} cm). The most exciting of this theory was that it first gives us a no infinite theory of gravitation (this infinite will appear in all previous attempts to describe theory of gravity). Surfaces that contain long and thin tube will be given such result, is the intermediate process between the initial state and final state particles will launch a beam of radiation pulse, it corresponds to a massless and spin 2 particles (Fig. 8.3.4), gravitons have exactly the nature. In a sense it can be said in string theory discovered gravity.

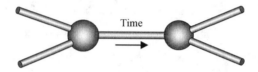

Fig. 8.3.4 Cross section of a string containing the emission and reabsorption of the spin-2 mass particle

String theory in general can only be given the integer spin excitedparticles, which contains only bosons, if you want to introduce fermions, it is necessary to introduce supersymmetry (symmetry between bosons and fermions, supersymmetry, SUSY). String theory which introducing supersymmetry called superstring theory.

5. Degree of freedom of vacuum unified theory

In vacuum unified theory, quantum field has fiber structure, which can be described by intrinsic space-time ξ^μ. x^μ is the observational space. The intrinsic space-time and observational space-time appear in the field function at the same time, so that the macro and micro information of quantum field can be expressed at the same time. Some intrinsic properties can be shown in observation space, and the relationship between this part of intrinsic space and observable space.

$$x_0 = \breve{\gamma}_0 \begin{pmatrix} \xi_0 & 0 & 0 & 0 \\ 0 & 0 & 0 & 0 \\ 0 & 0 & 0 & 0 \\ 0 & 0 & 0 & 0 \end{pmatrix}; x_1 = \breve{\gamma}_1 \begin{pmatrix} 0 & 0 & 0 & 0 \\ 0 & \xi_1 & 0 & 0 \\ 0 & 0 & 0 & 0 \\ 0 & 0 & 0 & 0 \end{pmatrix}; x_2 = \breve{\gamma}_2 \begin{pmatrix} 0 & 0 & 0 & 0 \\ 0 & 0 & 0 & 0 \\ 0 & 0 & \xi_2 & 0 \\ 0 & 0 & 0 & 0 \end{pmatrix}$$

$$x_3 = \breve{\gamma}_3 \begin{pmatrix} 0 & 0 & 0 & 0 \\ 0 & 0 & 0 & 0 \\ 0 & 0 & 0 & 0 \\ 0 & 0 & 0 & \xi_3 \end{pmatrix}; x_4 = \breve{\gamma}_4 \begin{pmatrix} \xi_0 & 0 & 0 & 0 \\ 0 & \xi_1 & 0 & 0 \\ 0 & 0 & \xi_2 & 0 \\ 0 & 0 & 0 & \xi_3 \end{pmatrix} \quad (8.3.5)$$

The elementary particle internal space can be extended to the observation space, which makes the particle there is intrinsic spatial structure, at the same time theory can be simplified. The particle is no longer a point particle of classical field theory, it's similar as chord theory.

Two adjacent points ξ and $\xi+d\xi$ in internal space, between them the length is

$$ds = \sqrt{g_{\alpha\beta}(\xi)\,d\xi^\alpha d\xi^\beta} \quad (\alpha,\beta = 0,1,2,3) \quad (8.3.6)$$

Chapter 8 Supplementary Notes on Some Basic Issues

Due to the intrinsic space frame and space-time degrees of freedom frame are coincide, thus extending to the N-dimension (i. e. N degrees of freedom), between them the length is $ds = \sqrt{g_{ij}(x)dx^i dx^j}$, considering the degrees of freedom space is flat space.

$$ds = \sqrt{\eta_{ij}(x)dx^i dx^j}, i,j = 0,1,2,3,\cdots,N \qquad (8.3.7)$$

Similarly, we have to calculate probability amplitude. Particles from a to b, in this process, from t_a, x_a to t_b, x_b, probability $P(b,a)$ is absolutely square probability amplitude.

$$P(b,a) = |K(b,a)|^2 \qquad (8.3.8)$$

This probability amplitude is the sum of the contributions $\varphi[x(t)]$ of each path.

$$K(b,a) = \sum_{\text{All paths from } a \text{ to } b} \varphi[x(t)] \qquad (8.3.9)$$

The contribution of a single path having an phase angle $\varphi[x(t)] = \text{const} e^{(i/\hbar)S[x(t)]}$ which proportional to action S, and thus

$$S[a,b] = \int_a^b L(\dot{x},x,t)dx \qquad (8.3.10)$$

$$S[x,g] = \frac{1}{2}\int_a^b d\varphi \eta^{ij} D_i\varphi(x) D_i\varphi(x) \qquad (8.3.11)$$

The above formula, i, j is vector index in N DOF space, $i,j = 0,1,2,\cdots, N$; η_{ij} is a Min's gauge in D-dimensional space-time metric, namely

$$\eta_{ij} = (+,-,-,\cdots,-) \qquad (8.3.12)$$

Because fiber structure quantum intrinsic space-time are coincidence with observation space-time, 4-momentum in particle intrinsic space is equal to 4-momentum in observation space-time. Simplified intrinsic wave function equal to particle space wave function, with the field operator in the freedom space $\gamma_n \partial_n (n = 0,1,2,\cdots, N)$, so that the mathematical formulation of vacuum reduces to form of expression of classic quantum field, making theory greatly simplified. Interestingly, in terms relative to the, N-dimensional space of freedom, the extra dimensions, which beyond the 4-dimensional space-time is restricted in intrinsic space, which is somewhat similar, chord theory.

7. Supersymmetry

The introduction of supersymmetry is extremely important for the superstring theory. If you do not introduce supersymmetry, string theory cannot describe fermions. Vacuum does not involve the concept of supersymmetry.

8.3.2 Particle interactions

1. Interaction between elementary particles in String theory

String itself is very simple, just a tiny wire, string can be closed into a circle (closed strings), can be opened like hair (open strings). A string can be decomposed into more tiny strings, also can be composed a longer strings by collide with other string. For example, an open string can be split into two smaller open string; also form an open string and a closed string; a closed string can be split into two small closed strings; two strings collisions can generate two new string.

The interaction between elementary particles are regarded as the ring together, then separated again. To describe such an event can be used by the surface of 2-dimensional space-time, this is entirely because when a string moving in space will sweep out the space-time in a 2-dimensional surface (a tube). Any property between particles in reaction can be regard as the process of various splitting and recombination of a 2-dimensional surface. The reaction exists in this process that the chord ring is absorbed in initial state, while the transmitter the chord ring in the final state.

For example, two particles incident in and three particle emission out, such the scattering process can be described by 2-dimensional curved surface, it has two long tube into (on behalf of initial particle), there are three long tube extending (represents the final state particles). Between these two surfaces, itself may have quite complex topology (Fig. 8.3.5).

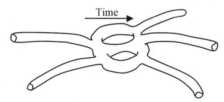

Fig. 8.3.5 Two particles become three particles scattering process in string theory

2. The interaction between elementary particles in Vacuum unified theory

Now let us consider vacuum unified theory interaction between particles.

(1) Static field interaction.

On our observations, the form of this interaction exist in the gravitational field and electromagnetic field. Gravity is the bending space. The gravitational force between two objects is caused by the curvature of space, the curvature tensor

$$R^{\mu}_{\alpha\beta\gamma} = \frac{\partial \Gamma^{\mu}_{\alpha\gamma}}{\partial x^{\beta}} - \frac{\partial \Gamma^{\mu}_{\alpha\beta}}{\partial x^{\gamma}} + \Gamma^{\mu}_{\sigma\beta}\Gamma^{\sigma}_{\alpha\gamma} - \Gamma^{\mu}_{\sigma\gamma}\Gamma^{\sigma}_{\alpha\beta}$$

Showed the presence of gravity.

Electromagnetic space bending performance exist in a macro static electromagnetic field, For relatively stationary or slow moving charged body to achieve interaction by the electric flux line between two body that is described by classical electromagnetic theory. Electromagnetic space curvature is $F^{\mu\nu}$ (Antisymmetric second-order tensor) $F^{\mu\nu} = -F^{\nu\mu} = \partial^{\mu}A^{\nu} - \partial^{\nu}A^{\mu}$. In this case, the Lagrangian density of electromagnetic field is $\mathscr{L} = -\frac{1}{4}F_{\mu\nu}F^{\mu\nu}$.

Chapter 8 Supplementary Notes on Some Basic Issues

(2) Interaction is achieved by exchanging intermediate boson.

From the perspective of microscopic, strong, weak, electric, gravitational interactions can be achieved by exchange intermediate boson. Wherein the strong and weak interactions exist only in the particles intrinsic space are short-range forces, and therefore space of strong and weak interaction space curved in a micro scope. Particle interacts by exchanged boson is entirely equivalent with the space bending generated force field to achieve interaction.

The electromagnetic interactions is achieved by mutual exchange of photons (Fig. 8.3.6), then Lagrangian density of electromagnetic field is $L = -\frac{1}{2}(\partial_\nu A_\mu)(\partial^\nu A^\mu)$. Similar to gravitational interactions between two fermions can be achieved by mutual exchange the graviton.

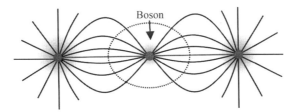

Fig. 8.3.6 Two particles interacting through the electric flux lines to exchange Boson to realize interaction in vacuum unified theory (Boson with fiber structure)

It is necessary to note that Fermions want to launch a Boson, then the vibration speed to superluminal. For the rest mass of the particles, the intrinsic center of the particles reach strain limit, from the perspective of particle intrinsic spatial (see the section of rest mass in this book), any movements will be superluminal, which can be fired boson, of course, also contains the graviton.

The interaction of vacuum unified theory is different string theory, but similar to classical and quantum field theory. The difference lies in the introduction of fiber structure, which allows us to estimate the coupling constant.

8.3.3 Comparison of physical mechanism of unified four force field

There is a direct link between the gravitational effects and particle string of vibration modes. The same correlation stored in between the string vibration mode and other properties of force, a string with electromagnetic, weak and strong interaction are completely determined by its vibration mode. In the vacuum unified theory, gravity, electromagnetic force, weak force and strong particle all from the vacuum deformation, and these differences are caused by different of fiber structure. In the physical image, fiber image of 4-force field is more intuitive than the vibration modes image of string theory.

1. Wave-particle duality

According to string theory, the performance of each elementary particles are derived from its different vibration modes of internal string. Each elementary is composed by a string and all the strings are absolutely the same. Different elementary particles actually are playing different "tone" on same string. Consisting of the universe by countless vibrating strings, like a great symphony.

Volatility of particle is caused by center point vibration at perpendicular to the direction of propagation in particle propagation. Due to the quantum field with fiber structure, can simplify to the particles that to simplify the vibration "string". Different to string theory is the length of simplify vacuum "string" is the intrinsic space particle diameter, meet $\Phi \cdot p = \hbar$, p is the momentum of a particle, $\Phi \gg$ the string length of string theory.

2. Particle mass

String associated with the particle mass is easy to understand. The more intense vibration of string, then greater the energy of particle, the more gentle vibration then smaller the particle energy. This is our familiar phenomenon: when we force pluck at the strings, the vibration will be very intense; gently struck it, the vibration will be very gentle. In addition, according to Einstein's mass-energy principle, energy and mass like a coin has two sides, the different manifestations of the same thing: large energy means great mass, small energy means small mass. Therefore, the particle mass which vibration more intense is large, on the contrary, vibration softer particles are smaller mass. the strings there are a mass , particle mass come from the string, mass is a mystery in string theory.

Particle mass derived from the propagation characteristics of intrinsic spatial in vacuum unified theory, completely abandon the mass parameters, jumped out of the logic circulation of mass explain mass. However, in string theory, static string remains tiny masses, vibration makes the energy increases form larger mass, the mass interpretation of mass is not out of the logic circulation of mass. Vacuum can explain the origin of mass and enable us to understand the essence of mass. Mass to meet: $mc^2 \cdot R = h_f$, R is the intrinsic spatial radius of Fermion. It describes the relationship between center of mass and the intrinsic space, but we are unable to calculate and determine the mass size, static quality itself is the effect of quantum field center points to deformation limit, then the deformation is plastic deformation. In order to determine the quality only by studying the plastic deformation. Want to study plastic deformation, it must enter into the center smaller area of Fermion, the size of a small space is h_f, this is studying scope of superstring/M theory. Physics is perfect only by research the nature of the center of quantum field and its neighborhood.

3. Granular structure of space-time

The spatial structure of superstring theory is discrete rather than continuous. In our daily experiences, space and time are always infinitely divisible, but the reality is very different

Chapter 8 Supplementary Notes on Some Basic Issues

large. Space and time are the minimum value of their own: the minimum scale of space is 10^{-33} cm, the minimum value of time is 10^{-43} s. Because when the space has been small enough to 10^{-33} cm, time and space will be integrated, spatial dimension will be up to ten dimensions, in this case, even if space is also divided, that is we can't understand.

The quantum theory: Energy discontinuity is due to space has the smallest and indivisible unit, will affect the way of energy emission of elementary particles.

In the vacuum unified theory, the vacuum constitutes by the basic unit field, and the size of the base unit is $\hbar_f = 2\hbar$. A basic unit shows in three dimensions after deformation. The dimensions orientation is arbitrary before deformation.

Clearly, there is a huge difference between the vacuum unified theory and string theory.

8.4 Vacuum interpretation of Dirac's "large number hypothesis"

In 1937, Dirac proposed "large number hypothesis" that is perhaps one of the most daring hypothesis in 20th Century. We know that the ratio of static electricity and gravitation in hydrogen atoms is

$$e^2/Gm_p m_e = 2.3 \times 10^{39} = \alpha_1$$

To measure the age of the universe by theunits of the atomic:

$$m_e c^3/e^2 H = 7 \times 10^{39} = \alpha_2$$

Expressed the total mass of Universe by units of the proton mass:

$$8\pi\rho c^3/3m_p H^3 = 1.2 \times 10^{78} = \alpha_3$$

$h = \dfrac{eH}{4}$ fiber unidirectional; $h_f = \dfrac{eH}{2}$ fiber spherical; $h_g = \dfrac{4\pi}{3}\left(\dfrac{H}{2}\right)^3$ geometry $h_f = \dfrac{1}{2}h_g$

(deformation)　　　　　　(deformation)　　　　　　(spherical deformation)

Wherein, G is Newton's gravitational constant, H is Hubble constant. There is an interesting connection between these three numbers:

$$\alpha_1 = \alpha_2 = (\alpha_3)^{1/2} \approx 10^{39}$$

where α_2 is the age of universe, it should be proportional with t, α_1, α_3 are constant.

From the perspective of this book, we put the whole universe analogous to a hydrogen atom, and then analyzed. The ratio of Static electricity and gravitation in hydrogen atoms is the ratio of electrostatic deformation and gravitational deformation. That is the ratio of the strain, the strain corresponding to the intrinsic momentum. Atomic intrinsic time is the time which atomic from scratch, denoted by Φ_{e0}, atomic intrinsic space denoted by Φ_e(diameter); the entire universe as a giant sphere. Analog atomic, age of the universe is the time of universe from scratch, the corresponding atomic intrinsic time, denoted as Φ_0, the universe diameter denoted as Φ, the intrinsic momentum of universe can be written as p_G, the electron intrinsic momentum denoted by p_e, meet

$$p_e \cdot \Phi = E_e \cdot \Phi_0 = h_f = \frac{eH}{2} \qquad (8.4.1)$$

For the gravitational field, to meet

$$P_G \cdot \Phi = E_G \cdot \Phi_0 = h_g = \frac{4\pi}{3}\left(\frac{H}{2}\right)^3 \qquad (8.4.2)$$

As shown in the Figure, that approximate $h_f = h_g/2$, there are following relationship:

$$2p_e \cdot \Phi = 2E_e \cdot \Phi_0 = p_G \cdot \Phi = E_G \cdot \Phi_0 \qquad (8.4.3)$$

To get such a conclusion:

$$\frac{2p_e}{p_G} = \frac{2E_e}{E_G} = \frac{\Phi}{\Phi_e} = \frac{\Phi_0}{\Phi_{e0}} = \alpha \qquad (8.4.4)$$

Atomic ratio of static electricity and gravitation is equivalent to the intrinsic atomic time and age of the universe is α, and get new conclusions: the ratio of electrostatic energy with gravitational energy is α in unit space-time equal to the ratio of universe space size and atomic intrinsic space size.

There is no h_f in the present theory, but only h. $2h = h_f$, so we can get $h = h_g/4$. In the experiment, we can only be measured along a direction, h_f is no observational. Static electricity of hydrogen atom to meet

$$p_{e'} \cdot \Delta x = E_{e'} \cdot \Delta t = h \qquad (8.4.5)$$

Take the same space-time scales was comparable, i.e. $\Phi = \Delta x$, $\Phi_0 = \Delta t$, it can get $p_e = 2p_{e'}$, to get

$$\frac{4p_e}{p_G} = \frac{e^2}{Gm_p m_e} = 9.2 \times 10^{39} = \alpha_1$$

age of the universe (i.e. ratio of cosmic time Φ_0 and atomic time Φ_{0e}).

$$\frac{\Phi_0}{\Phi_{e0}} = \frac{m_e c^3}{e^2 H} = 7 \times 10^{39} = \alpha_2$$

It is not exactly the same. This is because our universe is still in the forming zone, and need time $2.2 \times 10^{39} \Phi_{e0}$ the universe can be formed completely, from this perspective, we are now in the period of the life 2/3 of the universe, the total life of the universe is $9.2 \times 10^{39} \Phi_{e0}$.

Now continue to understand the total mass of the universe question which total mass expressed by proton mass units. Due to there are the following relationship between space diameter of universe and diameter of atomic space:

$$\frac{\Phi}{\Phi_e} = 7 \times 10^{39} = \alpha_2$$

Here, take $\Phi_{e0} = 1$. Approximately equal to the mass of the smaller proton. For a solid sphere,

Chapter 8 Supplementary Notes on Some Basic Issues

if the universe composed by atoms, then the following relationship exists

$$\frac{M_G}{m_p} = \frac{\rho \frac{1}{6}\pi\Phi^3}{\rho \frac{1}{6}\pi\Phi_e^3} = \Phi^3$$

For a spherical shell structure, if the universe composed by atoms, then the following relationship exists.

$$\frac{M_G}{m_p} = \frac{\rho \frac{1}{4}\pi\Phi^3}{\rho \frac{1}{4}\pi\Phi_e^2} = \Phi^2 \approx \alpha_2^2$$

This suggests that the universe and the proton have a shell structure, instead of solid body structure.

From the above discussions, the universe and atoms are isomorphic.

References

[1] DEVANATHAN V. Relativistic quantum mechanics and quantum field theory[M]. New Delhi: Narosa Publishing House, 2011.

[2] COTTINGHAM W N, GREENWOOD D A. An introduction to the standard model of particle physics[M]. 2nd ed. Cambridgeshire: University Press, 2010.

[3] GUY D, COUGHLAN J E, et al. The ideas of particle physics: an introduction for scientists[M]. 3rd ed. Cambridge: Cambridge University Press, 2006.

[4] FOSTER J, NIGHTINGALE J D. A short course in general relativity[M]. 2nd ed. New York: Springer-Verlag, 1998.

[5] WALTER G, REINHARDTJ. Field quantization[M]. Beijing: Beijing World Publishing Corporation, 2003.

[6] XU J J. Quantum field theory[M]. Shanghai: Fudan University Press, 2004.

[7] MUKHANOW V F, WINITZKI S. Introduction to quantum effect in gravity[M]. Beijing: Beijing World Publishing Corporation, 2010.

[8] HOOTF G. 50 years of young-mills theory[M]. Beijing: Science Press, 2007.

[9] ANDREW R, LIDDLE D H. Cosmological inflation and large-scale structure[M]. Cambridge: Cambridge University Press, 2009.

[10] OHANIAN H C, REMO RUFFINI. Gravitation and space-time[M]. XIANG S P, FENG L L, translate. Beijing: Science Press, 2006.

[11] HE H X. An introduction to nuclear chromo dynamics: quantum chromo dynamics and its applications to the system of nucleon and nuclear structure[M]. Hefei: China University of Science and Technology Press, 2009.

[12] CAO C Q. Quantum non-abel gauge field theory[M]. Beijing: Science Press, 2008.

[13] XUE X Z. Guidance of quantum vacuum physics[M]. Beijing: Science Press, 2005.

[14] ZHAO X, ZHAO S S. Strong interaction quantum field principles[M]. Beijing: Science Press, 2012.

[15] ZOU P C. Quantum mechanics[M]. Beijing: Higher Education Press, 1990.

[16] ROZER N. Vacuum dynamics—a new framework of physics[M]. Shanghai: Shanghai Science Popularization Press, 2003.

[17] XU B Y, LIU X S. Applied elastoplastic mechanics[M]. Beijing: Tsinghua University Press, 1995.

[18] LU M W, LUO X F. Elastic theory[M]. Beijing: Tsinghua University Press, 1990.

References

[19] SCHUTZ B F. Geometric methods in mathematical physics[M]. FENG C T, LI S Q, translate. Shanghai: Shanghai Science and Technology Literature Publishing House, 1986.

[20] HU Y G. Gage Theory[M]. Shanghai: East China Normal University Press, 1984.

[21] ZENG J Y. Quantum mechanics (Volume 1)[M]. Beijing: Science Press, 1999.

[22] WANG Z X. Concise quantum field theory[M]. Beijing: Peking University Press, 2008.

[23] BIYOKEN J D, DRELL S D. Relativistic quantum mechanics[M]. JI Z L, SU D C, translate. Beijing: Science Press, 1984.

[24] SU R Q. Quantum mechanics[M]. Shanghai: Fudan University Press, 1997.

[25] ZHANG L. Advances in modern physics[M]. Beijing: Tsinghua University Press, 1997.

[26] HE X T. Observational cosmology[M]. Beijing: Beijing Normal University Press, 2007.

[27] ZHANG N S. Particle physics (Volume 2)[M]. Beijing: Science Press, 1987.

[28] WEBER. General relativity and gravitation waves[M]. CHEN F Z, ZHANG D W, translate. Beijing: Science Press, 1979.

[29] ZHOU J S. Tensor preliminary[M]. Beijing: Higher Education Press, 1985.

[30] WEINBERG S. Principles and applications of gravity and cosmology-general relativity [M]. ZOU Z L, ZHANG L N, translate. Beijing: Science Press, 1984.

[31] QIUZ P. Modern quantum field theory[M]. Shanghai: East China Normal University Press, 1992.

[32] FEYNMANR P, SIBUS A R. Quantum mechanics and path integral[M]. ZHANG B G, WEI X Q, translate. Beijing: Science Press, 1986.

[33] RICHARD F S. From antiparticle to final theorem[M]. LI P K, translate. Hunan: Hunan Science and Technology Publishing House, 2003.

[34] CHEN S Q. Vacuum dynamics images of gravitational field and quantum field[M]. Beijing: Publishing House of Electronic Industry, 2010.

[35] LI H J. Introduction to quantum field theory[M]. Kunming: Yunnan Science and Technology Press, 1989.

[36] ZHAO X, ZHAO S S. Principle of strongly interacting quantum field[M]. Beijing: Science Press, 2012.